NUCLEAR PHYSICS

R. PRASAD

formerly at
Aligarh Muslim University
Aligarh, India

Pearson

ISBN978-93-325-2265-7

FirstImpression, 2014

PublishedbyPearsonIndiaEducationServicesPvt.Ltd,CIN:U72200TN2005PTC057128.

HeadOffice:15thFloor,Tower-B,WorldTradeTower,PlotNo.1,Block-C, Sector16, Noida201 301, UttarPradesh, India.RegisteredOffice:7thFloor,SDB2,ODC 7,8&9,SurveyNo. 01ELCOTIT/ITES-SEZ,Sholinganallur, Chennai–600119,Tamilnadu,India.
Phone: 044-66540100
Website:in.pearson.com,Email:companysecretary.india@pearson.com

Printer Manipal Technologies Limited, Manipal

Contents

Preface

Nuclear Physics has been written for graduates and postgraduates, especially the ones studying 'nuclear physics' or 'nuclear, particle and cosmic ray physics'. Written in a simple language, the book comprises a judicial balance between theory and experiments. It includes latest topics like the heavy ion reactions, radioactive ion beams, pre-equilibrium emission, complete and incomplete fusion, reactions important for nucleon synthesis, Higg's mechanism and God particle, Feynman diagrams etc. An attempt has been made to develop each topic from the very fundamental level to the higher level which may join smoothly with the requirements of advanced studies in the subject. Many solved examples, with a graded level of difficulty, have been included for the practice of the students. A very special feature of the book is the inclusion of MCQs (multiple-choice questions) after every chapter. Some of these questions have more than one correct alternative. A complete answer, in such cases, requires picking up all the correct alternatives that ensures comprehensive understanding of the subject.

The book contains twelve chapters. The first chapter, titled 'The birth of the nucleus', gives a brief review of the scientific temper and important scientific discoveries made till the time when pioneering alpha scattering experiments were carried out by Rutherford and his group. This may help in appreciate the ingenuity of the alpha scattering experiments. Detailed and stepwise derivation of Rutherford scattering formula is then derived using the basic conservation laws of non-relativistic mechanics. The chapter ends with an introduction to the basic properties of the nucleus and the modern quark model of the nucleus.

Having introduced the basic nuclear properties in chapter 1, more detailed description of these properties and methods of their experimental determination have been discussed in chapter 2. Energy and momentum filters are used not only in mass spectrographs but also in other nuclear instruments like accelerators, atomic and molecular beam systems etc. A detailed discussion of these is, therefore, included in this chapter. Any measurement is incomplete without an estimate of the accuracy. It is with this view that the topic of doublet method of mass spectrometry is discussed in details.

Nearly constant and exceptionally large density of nuclear matter points towards the special nature of nucleon-nucleon force. Basic properties of charge—independence, charge symmetry, spin dependence and saturation of two body nucleon-nucleon force as derived from the systematic of stable nuclei—have been discussed in chapter 3.

The properties of microscopic systems are adequately described only by the quantum mechanical treatment. In order to prove this, quantum mechanical analysis of the simplest stable nucleus 'deuteron' has been worked out in chapter 4. Frequently required tool of partial wave analysis has been discussed in sufficient details in this chapter. In nuclear interactions, involving charged particles, the reactants face a Coulomb barrier. In classical physics, the reactants cannot cross through the barrier if their kinetic energy is less than the height of the barrier. However, in case of quantum mechanical treatment it can be shown that there is a finite probability of barrier transmission even when the kinetic energy is less than the barrier height. This very important aspect of quantum mechanics, which explains many nuclear phenomena, has also been developed in chapter 4. Development of quantum mechanical treatment at this stage is essential for a comprehensive understanding of other topics like the alpha decay, nuclear fission and fusion etc that follow.

The characteristics of stable and radioactive nuclei provide a basic set of data that a good nuclear model must be able to reproduce. An inventory of such characteristics is compiled in chapter 5. Independent particle model with an ad hock assumption of spin-orbit coupling led to the nuclear shell model that can explain the magic numbers. Similarly, liquid drop model, based on the similarity between the drops of a liquid and the nucleus has been quite successful in explaining nuclear fission. The Fermi gas model is one of the simplest nuclear models that may correctly predict the dependence of the stability of a nucleus on the difference in the number of neutrons and protons in it. On the other hand collective models, that involve collective motion of several nucleons together, have been able to correctly predict the lower excited states of nuclei. All these models, with details needed for each, have been discussed in chapter 5.

After discussing the characteristics and the models for stable nuclei, Chapter 6 deals with the unstable radioactive nuclei. The chapter details the stability criteria for nuclei, neutron and proton drip lines, laws of natural radioactive decay and its application to the successive series decay leading to the Batman equations. Methods of activation analysis using stacked foil technique for the study of nuclear reactions and carbon-14 dating for determining the archeological age of artifacts have been discussed in this chapter. With the help of tools developed in chapter 4, the elementary theory of alpha, beta and gamma decays are also provided in this chapter. Non conservation of parity in beta decay, its experimental verification and consequences are also outlined. This chapter also includes description of Mossbauer effect and its use as an important experimental tool in different branches of science.

Detecting of nuclear radiations, their identification, measurement of their intensity and energy distribution etc are the basic requirements of experimental nuclear physics. A detailed discussion of energy loss processes that play very important role in devising detectors, for different nuclear radiations is presented in chapter 7. Characteristics of different detectors, their selection criteria etc are also discussed in this chapter. Some special detector set-ups for special needs are also discussed. Neutrinos, emitted in weak nuclear interactions have very little interaction with matter and, therefore, are very difficult to detect. However, neutrinos carry valuable information about their source of origin. Neutrinos created at the time of big bang are expected to be present in the universe and earth, without any interaction/alteration, even today. In view of this, dedicated efforts s worldwide, are being made to develop neutrino detection facilities. India has also planned a large neutrino facility. Details of Indian initiative along with other international efforts are also outlined in chapter 7.

Binary nuclear reactions and reaction mechanisms are discussed in chapter 8. After discussing the conservation laws with the help of some examples, reaction Q-equation is developed, which is then used to discuss cases in which two groups of particles with different kinetic energies are emitted at a given angle. Reaction mechanisms, including compound reaction mechanism, pre-equilibrium emission, direct reactions, optical model and thermodynamic approaches to nuclear reactions are discussed in this chapter in sufficient details. Elementary treatment of Giant resonances produced as a result of the interaction of electromagnetic radiations with the nucleus is also presented in this chapter. Detailed discussions of heavy ion reactions, complete and incomplete fusion, reactions with radioactive ion beams, halo nuclei, and reactions important for nucleosynthesis which has been discussed in chapter 8.

Chapter 9 describes both the direct current and the cyclic accelerators. Information about important international accelerator facilities including the large hadron collider and associated experimental setups etc is also provided in this chapter.

Release of energy in nuclear fission and fusion, fission reactors, role of fast and slow fission neutrons, new concept of accelerator driven energy amplifiers, confinement of plasma, magnetic and inertial confinement, uncontrolled nuclear devices etc have all been discussed in chapter 10 on nuclear energy.

The topics of elementary particles and cosmic rays are included in the nuclear physics syllabus of some universities. Hence, in order to cover the entire course, chapter 11 titled 'Fundamentals of elementary particles' and chapter 12 titled 'Cosmic rays' have been included. Discoveries of some important elementary particles, their classification, eight-fold way, conservation laws, characteristics of fundamental interactions, quark structure of hadrons, electro weak interactions and unification, standard model, Higg's field and God particle, natural units and an introduction to Feynman diagrams etc have all been discussed in a simplified way in chapter 11.

The last chapter of the book gives a detailed account of the discovery, classification, composition, variation with different parameters, origin and impact of cosmic rays on human life. Different experimental set-ups used to study different aspects of cosmic rays have also been described in this chapter. The contribution of two well known Indian scientists, Prof. P.S. Gill and Prof. H.J. Bhabha, in the field of cosmic rays, has also been highlighted in this chapter.

Comments and suggestions for further improvement of the book are most welcome.

About the Author

R. Prasad has more than 40 years experience of teaching physics and nuclear physics to graduate and postgraduate students. He is an ex-professor of nuclear physics at the Aligarh Muslim University, Aligarh, India. Throughout his career, Prof. Prasad supervised half a dozen Ph.D, about two dozen M.Phil, large number of M.Sc projects, eleven research projects funded by various agencies in India and carried out post doctoral research at many international and national institutes/universities including the First Institute of Experimental Physics, University of Hamburg, Germany and Atom Institute, Technical Universities of Austria, Vienna, Austria among many. He has also attended and chaired sessions of a large number of international and national conferences, seminars and symposia and delivered invited talks. He has published more than 80 research papers in various reputed international and national journals and presented six science-based television films under the UGC higher education programme. He is a recipient of prestigious DAAD (German) Fellowship, Post-doc fellowship of the Government of Austria, and Emeritus fellowship of UGC, India. He is a life member of many academic societies of the country.

Acknowledgements

I wish to thank my wife, Sushma Mathur, for her understanding and support since without her it would not have been possible to complete this work. I also wish to thank my amazing students and colleagues who persuaded and encouraged me to write the book. Prof. B.P. Singh deserves special thanks for very useful discussions, reading the manuscript, and suggesting improvements. All members of our research group, including Drs. H.D. Bhardwaj, M. Musthafa, Unnati, M.K. Sharma, D.P. Singh, P.P. Singh, Abhisheik, Vijairaj and others have contributed directly or indirectly in developing the material presented in the book. I, therefore, sincerely thank them all. Thanks are due to the Aligarh Muslim University for extending all possible help and facilities which proved immensely beneficial to complete the book.

R. Prasad

1

The Birth of the Nucleus

1.1 BACKGROUND

The fact that every atom has a nucleus at its centre was established by Ernest Rutherford in 1911. He made this discovery while explaining the result of the alpha scattering experiment carried out by two of his students Hans Geiger and Ernest Marsden. Rutherford is, therefore, called the father of nuclear physics. To fully appreciate the alpha scattering experiment and the great discovery made by Rutherford from the experiment, it will be worthwhile to recall the status of scientific knowledge around the year 1911. A very brief review of the relevant scientific discoveries made till then is as follows:

1. In the year 1803, John Dalton gave the atomic theory according to which elements are made up of atoms, and atoms of the same elements are identical while those of different elements are different. The theory inherently assumed that atoms are the ultimate particles and cannot be broken into smaller parts. Dalton's atomic theory got well established by 1900 as it was supported by many experiments.
2. Dmitri Mendeleev, in the year 1869 arranged different elements in groups called periods, according to their chemical properties. This is called the periodic table. The atoms of different elements that fall in one column of the periodic table have similar chemical behaviour, indicating some similarity in the structure of these atoms.
3. In 1895, Wilhelm Röntgen discovered X-rays, which were later used to study the details of the atom
4. In 1896, Henri Becquerel discovered radioactivity. He found that some atoms undergo spontaneous disintegration giving out different type of radiations. These radiations were identified as positively charged alpha particles, both negative and positively charged beta particles and uncharged electromagnetic radiations called gamma rays. This discovery immediately raised questions about the perception that atoms are unbreakable.
5. In 1898, Marie Curie and Pierre Curie discovered a new radioactive element, which they named as polonium. Polonium emits alpha particles of $\approx$6 MeV energy. Later, in 1900, Friedrich E. Dorn discovered a radioactive gas, radon-222, which also emitted alpha particles of nearly 6 MeV and had a long half life. Both polonium and radon have been used as a source of alpha particles.
6. In 1897, J.J. Thomson discovered electron, a negatively charged particle that was found to be present in every atom. He measured the charge of the electron to be -1 unit, now denoted by -1e, and mass to be around 1/2000 that of a hydrogen atom.

The discovery of electron by Thomson and the fact that electrons are present in every atom together with the discovery of radioactivity by Becquerel clearly indicated that the atom can be broken. Because atom as a whole is neutral and electrons are present in every atom, this required the presence of positive charge in the atom. Later, in 1913, Henry Moseley using X-rays determined the amount of positive charge in different atoms. J.J. Thomson, in 1904, proposed a model for the structure of atom, which was known as 'plum-pudding' or 'watermelon' model. In this model, it was assumed that atom is a solid spherical body of positive charge and the positive charge is uniformly distributed over the volume of the atom. The negatively charged electrons were assumed to be fixed within this positively charged jelly at different locations like the seeds of the watermelon or the currants inside the plum pudding. The total positive charge of the atom was considered equal to the number of electrons inside to guarantee the *over*all electrical neutrality.

1.2 GEIGER AND MARSDEN'S ALPHA SCATTERING EXPERIMENT

With the view to investigate the interior of the atom, Geiger and Marsden devised a very ingenious experiment. They took a source of radioactive gas radon-222, which emits alpha particles of about 6 MeV. The gas was kept in a small lead container with a small window in the front to allow alpha particles to come out. Thus, they could get a collimated beam of alpha particles. The collimated beam of alpha particles was made to hit a thin gold foil placed in front of the beam perpendicular/normal to the beam direction. The experiment was conducted in a vacuum chamber so that the alpha particles of the beam do not interact with the molecules of the air and get deflected. A microscope with its objective lens painted with fluorescent material such as zinc sulphide was used to detect the alpha particles. When alpha particle hits the fluorescent material, a tiny flash of light is emitted. A count of the number of light flashes in a given time gave the number of alpha particles detected during the time. The microscope used in the experiment was capable of rotation in the horizontal plane so that the intensity of alpha particles at any desired angle with respect to the direction of the incident beam could be monitored. A sketch of the experimental lay out is shown in figure 1.1. The experiment was very difficult as it has to be carried out in a dark room so as to identify the tiny flashes. It used to take the eyes of an observer almost half an hour to get adjusted in the dark room to see the tiny flashes and to count them. Moreover, it was so strenuous for the eyes that a person could not take readings for more than a few minutes or so. Later Geiger developed a counter to record the scattered alpha particles that made measurements easy.

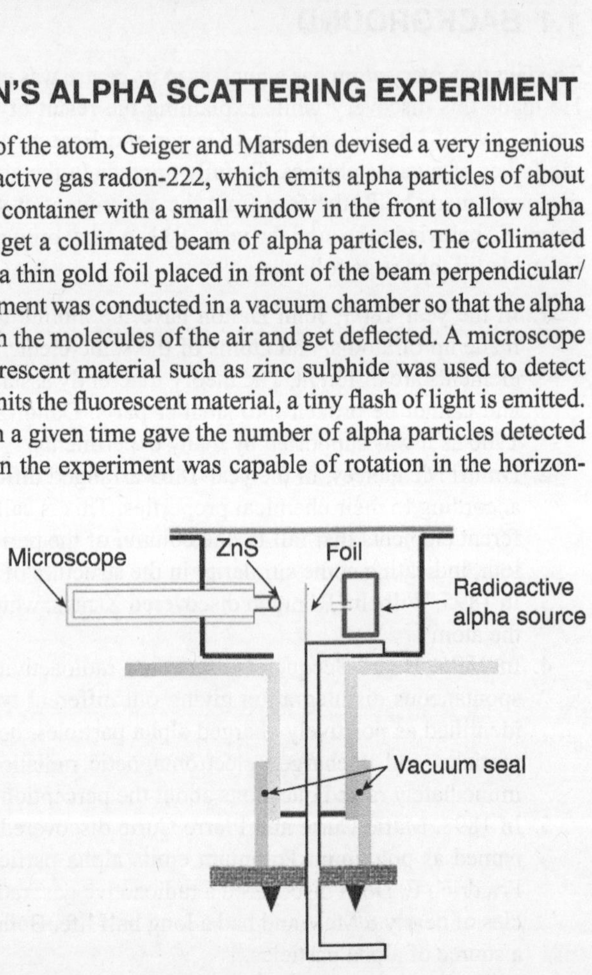

Figure 1.1 *Experimental layout of alpha scattering experiment*

As may be observed from the figure, the only hurdle that the alpha particles in the incident beam coming from the polonium source face is the gold foil or the atoms of gold in the foil. Because alpha particles have two units of positive charge, they may get deviated from their original direction of motion because of the interaction with the atoms of gold in the foil. This deviation of particles from their original direction of motion is called scattering. Geiger and Marsden recorded the number of particles scattered at different angles with respect to the beam direction. Their findings were as follows:

1. Large number of incident alpha particles came out on the other side of the gold foil undeviated, as if they have not encountered any charge during their passage through the foil.
2. Many alpha particles showed small deviations, that is, they got scattered at small scattering angels of the order of 1 or 2 degrees.
3. An extremely small number of alpha particles bounced back or got scattered by more than 90°.
4. Later experiments where foils of materials other than gold were used as target showed that the probability of scattering of alpha particle at a given angle increases with the atomic number Z of the target foil.

The first two observations can be easily explained by Thomson's model of the atom.

In Thomson's model, the scattering by large angle may happen if the same alpha particle is scattered successively by many atoms in different successive layers of the target foil. This is called multiple scattering. Multiple scattering is possible in sufficiently thick foils that contain several layers of atoms. To avoid multiple scattering, Geiger and Marsden used very thin gold foils with only few layers of atoms. In these experiments, the effect of the presence of electrons on the scattering of alpha particles was ignored as electrons have very little size and negligible mass as compared with the alpha particle.

Thomson's model of the atom completely failed to explain the third and the fourth observations of scattering experiment. The results, particularly the observation of scattering of a few particles by large angles (more than 90°) and occasional scattering at 180° completely baffled scientists so much that Rutherford remarked

> *"It was quite the most incredible event that has ever happened to me in my life. It was almost as incredible as if you fired a 15-inch shell at a piece of tissue paper and it came back and hit you. On consideration, I realized that this scattering backward must be the result of a single collision, and when I made calculations I saw that it was impossible to get anything of that order of magnitude unless you took a system in which the greater part of the mass of the atom was concentrated in a minute nucleus. It was then that I had the idea of an atom with a minute massive centre, carrying a charge."*
>
> **—Ernest Rutherford**

This was the moment of the birth of the NUCLEUS.

1.3 RUTHERFORD'S MODEL OF NUCLEAR ATOM

To explain the large angle scattering of alpha particles, Rutherford (1911) assumed that all the positive charge and almost all the mass (excluding the very small mass of all the electrons) of the atom is contained in a very small spherical volume at the centre of the atom, which he called the nucleus. He indicated that the size of the nucleus was almost 1/10,000 the diameter of the atom. The electrons were assumed to be revolving around the nucleus within the volume of the atom. As such, Rutherford's atom was largely empty with a very tiny massive positively charged nucleus at the centre and almost point-like negatively charged electrons moving around the nucleus. The motion of the electrons was essential otherwise they will all be attracted by the nucleus and fall into it through the empty space in the atom. Such a model clearly explains why most of the incident alphas pass through the empty space in the atom without any deviation. Only those alphas that pass very close to the nucleus get scattered, scattering angle depending on their impact parameter. Alpha particles hitting directly head on to the nucleus get reflected back at 180°. Rutherford also derived the expression for the intensity of the particles scattered at a certain scattering angle. The scattering formula of Rutherford has since been tested using both the charged particles of different energies obtained from the radioactive sources as well as the charged particle beams obtained from particle accelerators.

There was, however, one problem with Rutherford's atom. It was its stability. Electrons revolving round the nucleus in Rutherford's atom have acceleration, as motion in a circular path is an accelerated motion, with acceleration always normal to the direction of motion. Because accelerated charge radiate energy (in the form of electromagnetic radiations), the kinetic energy of these electrons must go on reducing. This will result in the shrinking of the radius of the path of the motion and eventually electrons falling into the positively charged nucleus. A Danish scientist, Niels Bohr took the next step 2 years after Rutherford put his theory of nuclear atom. Bohr suggested that electrons in an atom move in fixed orbits or shells. These shells were such that electrons do not radiate while moving in them. Bohr's model is also called the planetary model.

1.4 RUTHERFORD'S SCATTERING FORMULA

The scattering of a charged particle by another charged body is called Rutherford scattering or Coulomb Scattering. Rutherford developed the mathematical formulations for the number of alpha particles scattered at different scattering angles by a massive stationary positive charge, the nucleus of the target atom. It may be remarked that both the nucleus and the incident alpha particle are sub-atomic or microscopic entities and in principle their scattering should be treated quantum mechanically, however, Rutherford in his derivation used classical non-relativistic mechanics (Newton's laws of motion). It was later shown that the results obtained by Rutherford were quite accurate and not much different than those obtained using relativistic quantum mechanical treatment.

1.4.1 Assumptions Made by Rutherford

In deriving the scattering formula, Rutherford made the following assumptions about the scattering process.

1. The nucleus and the alpha particle are considered as point particles.
2. Electric repulsion between the alpha particle and the target nucleus is the only force present in the scattering.

3. The nucleus is so massive in comparison to the alpha particle that it does not move during the scattering. As a result, the energy of the scattered alpha particle does not change.
4. The foil containing the target is so thin that each alpha particle suffers only one scattering.
5. The nuclei in the target foil are separated from each other and do not overlap.

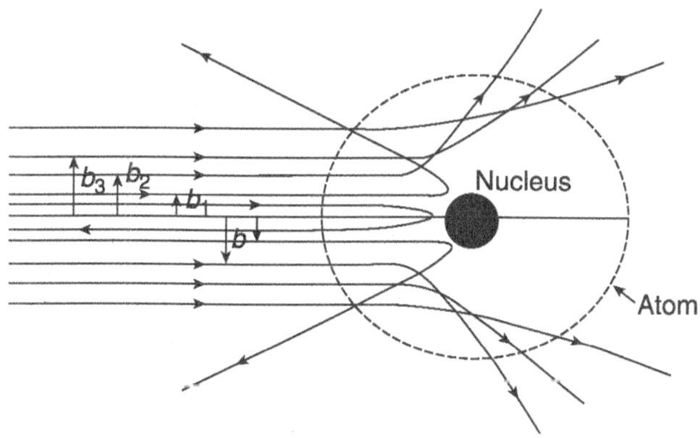

Figure 1.2 *A collimated beam of alpha particles having different impact parameters moving towards the nucleus*

The objective of the following derivation is to obtain an expression for the number of alpha particles scattered by a foil of some material, for example gold, in terms of known parameters such as the kinetic energy of incident particles, their number, and the charge on the nucleus of the target foil. A collimated beam of alpha particles each with mass m, velocity v, charge +2e having kinetic energy $E = \frac{1}{2} mv^2$, moving towards a nucleus in the target foil is shown in figure 1.2. As shown in the figure, different alpha particles of the beam are moving at different distances b_1, b_2, b_3 … from the axis of the figure and are deviated or scattered by different angles by the target nucleus situated on the axis. The distance of the alpha particle from the centre of the nucleus when no force is acting on the particle is called the impact parameter and is denoted by b in the figure. It may be observed that particle with smaller value of b get scattered by a large amount. The particle with $b = 0$ will be scattered back. Let us now look at a particular incident alpha particle, for instance, the one that has the impact parameter b. As the particle approaches the target nucleus (figure 1.3), it experiences a force F, the magnitude and the direction of which changes with the instantaneous position of the alpha particle. The magnitude of the repulsive coulomb force at a distance r from the nucleus is $F = [(2e)(Ze)/4\pi\varepsilon_0)r^2]$, where Ze is the charge on the nucleus. As the particle moves towards the nucleus, the force F initially increases, attains a maximum value at the distance of closest approach, and then decreases till at infinitely large distance it becomes zero. The net result of the repulsive Coulomb's force of variable magnitude is that the particle is scattered at an angle θ with respect to

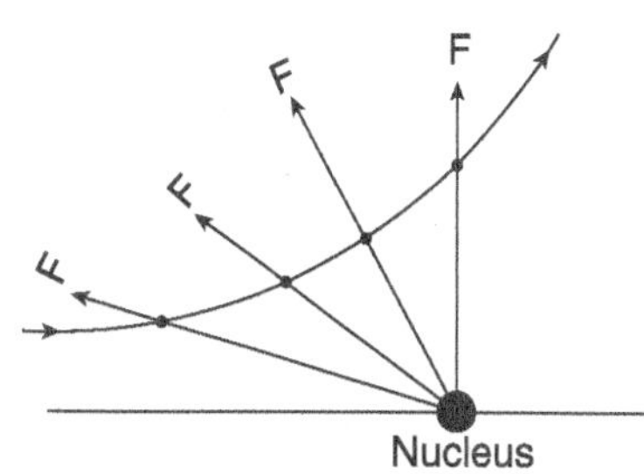

Figure 1.3 *The magnitude and direction of force F at different instants*

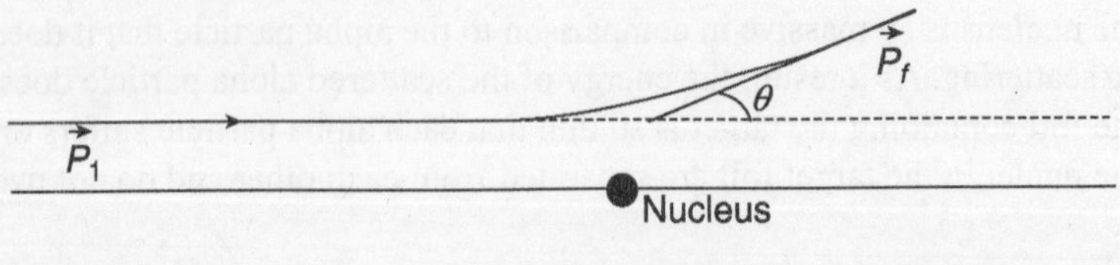

Figure 1.4 *The incident alpha particle is scattered at an angle θ*

its original direction as is shown in figure 1.4. It may be noted that the scattering angle θ is not produced in a single step or in a single push, but has build up gradually in small incremental steps. As such the force F has given the alpha particle an angular velocity ω.

Let us now analyse the effects of the Coulomb's repulsive force on the properties of the system consisting of the alpha particle and the target nucleus.

1.4.1.1 Conservation of linear momentum

As it is assumed that the target nucleus is much heavy and does not move, the kinetic energy, E, of the alpha particle does not change during the scattering. As such the magnitude of the initial total linear momentum of the system P_i is equal to the magnitude of the final total linear momentum P_f. However, the directions of the initial linear momentum and the final linear momentum are different. The force F has produced a change in the linear momentum ΔP [= $(P_f - P_i)$], which may be obtained from the vector diagram shown in figure 1.5.

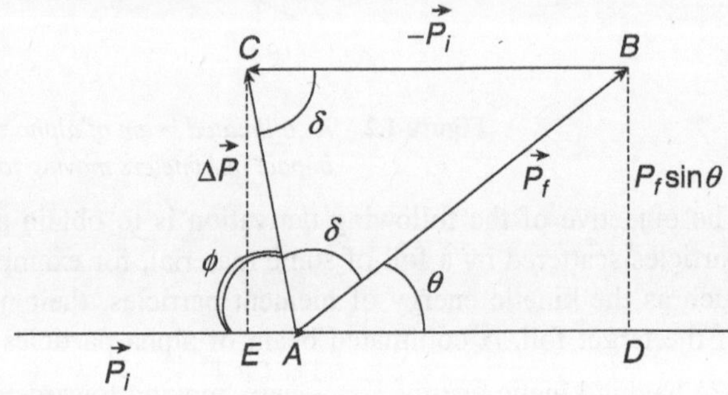

Figure 1.5 *Vector representations of the initial and final linear momentum*

In figure 1.5, AC shows the magnitude and direction of the change in linear momentum ΔP, which makes an angle φ with the initial direction of motion of the alpha particle. Triangle ABC has sides AB and BC of equal lengths (as $|P_f| = |-P_i|$) and, therefore, the angles BCA and BAC, represented by δ, are equal. In addition,

$$\Phi + \delta + \theta = \pi \text{ and } 2\delta + \theta = \pi. \text{ Hence, } \varphi = (\pi/2 - \theta/2) \quad (1.1)$$

Since, $CE = BD$,

$\Delta P \sin\varphi = P_f \sin\theta$ or $\Delta P \sin(\pi/2 - \theta/2) = P_f \sin\theta$, which on expansion gives

$$\Delta P \cos \theta/2 = P\,(2 \sin \theta/2 \cos \theta/2)$$

or

$$\Delta P = 2P \sin \theta/2 = 2\, mv \sin \theta/2 \quad (1.2)$$

1.4.1.2 Conservation of angular momentum

The initial system consisting of the incident alpha particle and the target nucleus has the angular momentum $L_i = r \times P = b \cdot P = bmv$. After scattering, it also has the angular momentum L_f, which is related to the angular velocity ω by the relation $L_f = \mathcal{J}\omega$, where $\mathcal{J}$ is the moment of inertia of the alpha particle with respect to the target nucleus. If the alpha particle is at a distance r from the target nucleus, the value of $\mathcal{J} = mr^2$. Now, during the scattering no torque has been applied on the alpha particle. (Remember that a torque is generated by two parallel and opposite forces and in the present case only one force F has worked.) The angular momentum of the system is, therefore, conserved, which means that $L_i = L_f$; therefore

$$b \cdot mv = mr^2\omega, \text{ hence, } bv = r^2\omega$$

or

$$\omega = (bv/r^2) \tag{1.3}$$

1.4.1.3 Application of the law of impulse

We now apply the law of classical mechanics that states that the change in linear momentum of a system is equal to the impulse of the force responsible for the change of momentum. Accordingly, for the system under consideration, we have

$$|\Delta P| - \int F\,dt \tag{1.4}$$

where the integral has the limits $t = -\infty$ to $+\infty$. As the magnitude of F as well as its direction changes at each instant, it is difficult to find the value of the integral $\int F\,dt$. To simplify the problem, it is necessary to change the variable parameter of the integral. It may be mentioned that ΔP, the total change in momentum during the scattering event has a fixed magnitude and direction.

In figure 1.6, the direction of ΔP is along AB, A is the instantaneous position of alpha particle at instant t when it is at a distance r from the target nucleus O. The instantaneous force F on the alpha particle makes an angle γ with the direction of ΔP as shown in the figure. The component of F along ΔP is $(F \cos\gamma)$ and in perpendicular direction is $(F \sin\gamma)$. The Eq. (1.4) may now be written as

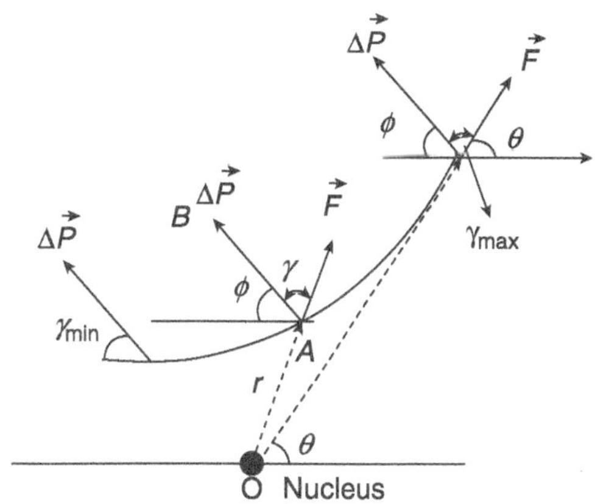

Figure 1.6 *At instant t the force F on alpha particle makes an angle γ with the direction of vector ΔP*

$$|\Delta P| = \int F\,dt = \int F\cos\gamma\,dt + \int F\sin\gamma\,dt \tag{1.5}$$

It may now be seen that the second integral in Eq. (1.5) vanishes in the limit of minus infinity to plus infinity as it gets negative contributions from minus infinity to the point of closest approach and an equal amount of positive contribution from there on to plus infinity. Eq. (1.5) may now be written as

$$|\Delta P| = \int F\,dt = \int F\cos\gamma\,dt = \int F\cos\gamma\,d\gamma\,(dt/d\gamma) \tag{1.6}$$

However, the angular velocity $\omega = (d\gamma/dt)$ and, therefore, $(dt/d\gamma) = (1/\omega)$. Substituting this in Eq. (1.6), we get

$$|\Delta P| = \int_{-\infty}^{\infty} F \cos \gamma \, d\gamma .(1/\omega) \tag{1.7a}$$

It may be noted that the variable inside the integral is now dγ in place of dt. Let us now find the upper and lower limits, i.e. $\gamma_{\max}$ and $\gamma_{\min}$ for the integral. As can be seen from the figure, for $t = +\infty$, $\gamma_{\max} = \pi - (\theta + \varphi) = (\pi/2 + \theta/2)$ and for $t = -\infty$, $\gamma_{\min} = -\varphi = (-\pi/2 + \theta/2)$.

Substituting these limits for the integral and the values of F, and ω in Eq. (1.7), we get

$$|\Delta P| = \int_{\left(\frac{-\pi}{2}+\frac{\theta}{2}\right)}^{\left(\frac{\pi}{2}+\frac{\theta}{2}\right)} \frac{2Ze^2}{4\pi\varepsilon_0 r^2}\left(\frac{r^2}{bv}\right)\cos\gamma \, d\gamma$$

$$|\Delta P| = (2Ze^2/4\pi\varepsilon_0 bv) \int_{-(\pi/2+\theta/2)}^{(\pi/2+\theta/2)} \cos\gamma \, d\gamma$$

$$|\Delta P| = \left(\frac{2Ze^2}{4\pi\varepsilon_0 bv}\right)\cos\frac{\theta}{2} \tag{1.7b}$$

Substituting the value of $|\Delta P|$ from Eq. (1.2), we get

$$2mv \sin\theta/2 = (2Ze^2/4\pi\varepsilon_0 bv)\cos\theta/2$$

or
$$b = \left(\frac{Ze^2}{4\pi\varepsilon_0 E}\right)\cot\frac{\theta}{2} \tag{1.8}$$

An important observation: Eq. (1.7a) tells that for fixed values of the scattering angle θ the impact parameter 'b' and the velocity of the incident particle 'v' (which means the fixed energy E of the incident particle), the value of the change in momentum $|\Delta P|$ is fixed and it is related to the charge Ze of the target nucleus. Thus, the change in momentum $|\Delta P|$, also called 'momentum transfer' in Coulomb's scattering (elastic scattering by Coulomb's force) may be used to determine the charge on the target nucleus, if other parameters, such as the energy of the incident particle, its charge and the angle of scattering are known. This fact is used in determining the nuclear charge distribution or the nuclear radius, as will be discussed in chapter-2.

1.4.2 The Concept of Cross-section

In science, the term cross-section is very frequently used. Cross-section, generally denoted by the symbol σ, for an event is a measure of the probability of occurrence of that event. The cross-section is defined in such a way that its dimensions are that of area (m^2). Cross-section is a quantity that can be calculated theoretically and can also be obtained experimentally. As such cross-section provides a meeting ground between theory and experiment. Because cross-section has the dimension of area, in the case of the interaction between an incident beam of particles and

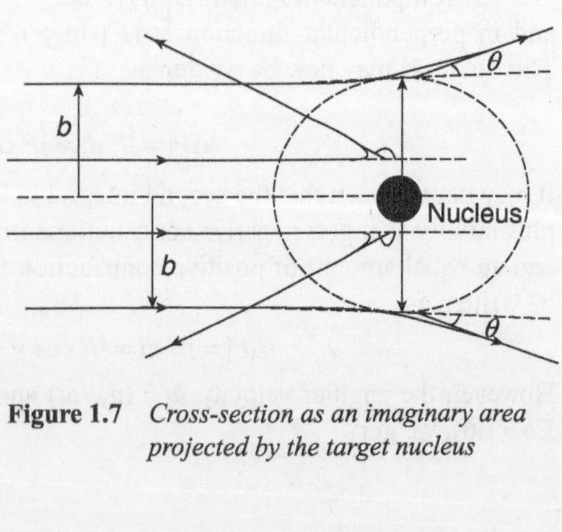

Figure 1.7 *Cross-section as an imaginary area projected by the target nucleus*

the target nucleus, it may be considered equal to the imaginary area that may be associated with the target nucleus such that all incoming particles hitting that area undergo the particular event.

For example, in the present case of alpha scattering, we may associate a cross-section $\sigma_{(\geq\theta)}$ for those scattering events in which the alpha particles are scattered by an angle θ or larger than θ. In figure 1.7, each target nucleus projects an imaginary circular area of radius b to the incident alpha particles such that all particles hitting on this area are definitely scattered by an angle θ or more. This imaginary area of magnitude (πb^2) is, therefore, the desired scattering cross-section $\sigma_{(\geq\theta)}$.

$$\text{or } \sigma_{(\geq\theta)} = (\pi b^2) \tag{1.9}$$

Let us assume that there are N_o target nuclei per unit area in the foil and A is the area of the foil on which the beam of alpha particles is falling. Therefore, the total number of nuclei is N_0A. If each of the nucleus projects an imaginary area $\sigma_{(\geq\theta)}$, than the total area projected to the beam by all the target nuclei is $N_0A\ \sigma_{(\geq\theta)}$. It means that out of all the particles falling on the total area A of the foil, those falling on the area $N_0A\ \sigma_{(\geq\theta)}$ get scattered by an angle θ or more than that. As such the fraction f_θ of the incident particles scattered by θ or more is equal to ratio of the two areas

$$f_\theta = N_0A\ \sigma_{(\geq\theta)}/A = N_0\ \sigma_{(\geq\theta)} = N_0\ (\pi b^2) \tag{1.10}$$

On substituting the value of b from Eq. (1.8), we get

$$f_\theta = \left\{\frac{N_0 \pi Z^2 e^4}{(4\pi\varepsilon_0 E)^2}\right\}\cot^2\frac{\theta}{2} \tag{1.11}$$

On differentiating Eq. (1.11), we may get the fraction of the incident alpha particles that get scattered at angle θ and $d\theta$.

$$df_r = -N_0 Z\left(\frac{Ze^2}{4\pi\varepsilon_0 E}\right)^2 \cot\frac{\theta}{2}\operatorname{cosec}^2\frac{\theta}{2}d\frac{\theta}{2} \tag{1.12}$$

Let N_i be the total number of incident alpha particles so that the number of particles scattered at angle θ and $d\theta$ may be given as $N_i|df_r|$. If the counting of scattered particles is done at a radial distance of R from the target nucleus, the area ΔA of the annular ring where the particles scattered at angle θ and $d\theta$ will fall is given by (see Figure 1.8),

$$\Delta A = (2\pi R\sin\theta)(Rd\theta) = 4\pi R^2 \sin\frac{\theta}{2}\cos\frac{\theta}{2}d\theta \tag{1.13}$$

Thus, the number of alphas $N(\theta)$ scattered per unit area within small angle $d\theta$ at an angle θ with respect to the incident beam direction is

$$N(\theta) = (N_i|df_r|/\Delta A) = N_i\{N_0\ \pi(Ze^2/4\pi\varepsilon_0 E)^2 \cot\theta/2 \operatorname{cosec}^2\theta/2\ d\theta/2\}\ /(4\pi R^2 \sin\theta/2 \cos\theta/2\ d\theta)$$

$$\text{or} \qquad N(\theta) = \frac{N_i N_0 Z^2 e^4}{(8\pi\varepsilon_0)^2 R^2 E^2 (\sin\theta/2)^4} \tag{1.14}$$

The expression represented by Eq. (1.14) is called Rutherford scattering formula. It may be noted that $N(\theta)$ falls rapidly with θ because of the $\sin^4\theta/2$ term, for a fixed value of the kinetic energy of the alpha particles and decreases as $1/E^2$ with energy E at a fix value of θ. It can be

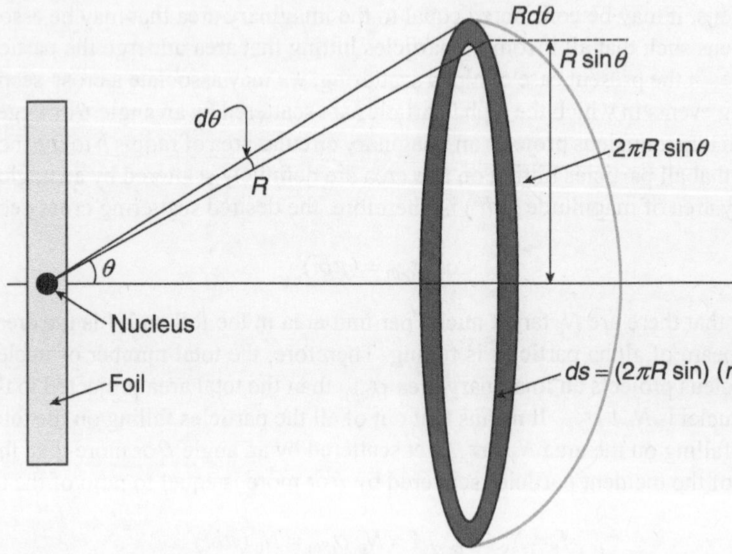

Figure 1.8 *Particles scattered at angle θ and dθ will fall on the annular ring of length 2πRsinθ and of width Rdθ*

shown that the number of scattered particles falls to almost 1% to their value at $\theta = 1°$ with in $\theta = \pm 10°$. However, a very small number (not zero) of particles that are projected head on ($b = 0$) are scattered by 180°. Rutherford formula has been verified using both the alpha particles emitted in natural radioactivity as well as the alpha particles obtained from accelerators. Further, more detailed derivations using relativistic mechanics and allowing the motion of the target nucleus give result that is not much different from the Rutherford's formula.

The following may be observed from Eq. (1.14):

1. $N(\theta)$ is proportional to Z^2 (square of the charge) of the target nucleus, which means that if the foil is replaced by foil of a material of higher atomic number, the number of alpha particles scattered at the same angle will increase.
2. $N(\theta)$ is inversely proportional to the square of the incident energy E of the alpha particles and to the square of the radial distance R of the point of observation from the centre of the target nucleus. As E increases, the magnitude of b (from Eq. (1.8)) decreases, in turn the cross-section for scattering (πb^2) also decreases and hence fewer particles are scattered at the angle θ. On increasing R, the area of the strip of on which the scattered particles fall increases and therefore the number of scattered particles per unit area goes down.

All the experimental observations of Geiger–Marsden's experiment could be explained on the basis of Rutherford's model of the atom having a central nucleus. As such the Rutherford's model of atom got well established. Alpha scattering experiment not only established the nuclear atom but also opened a new method of determining the amount of positive charge (Z) on the nucleus of the atom. The exact value of the atomic number Z for a large number of elements, for almost half of the elements in the periodic table, has been determined using alpha scattering method.

1.5 NUCLEAR ATOM

The results of alpha scattering experiment carried out by Geiger and Marsden were nicely explained by the theory proposed by Rutherford. Particularly, the back scattering of few alphas beyond 90° and up to 180° can be explained only if it is assumed that there is a massive central nucleus which contains all the positive charge of the atom. Rutherford's analysis of alpha scattering data drastically changed the picture of the atom from plum pudding to the centrally located nucleus having all the positive charge and almost 99.9% mass with electrons orbiting in empty space around it. Rutherford's discovery of the nucleus opened a new area of research called nuclear physics. The main objective of research in nuclear physics is to decipher the constituents of the nucleus, the forces that hold them together, their energy states, re-arrangement of nuclear constituents during reactions and so on. It is almost 100 years since the birth of the nucleus but our knowledge of nuclear physics is still incomplete. As a matter of fact even the basic nature of nuclear force that binds the nucleus is not well understood till now, which makes nuclear physics a live, challenging and exciting subject for research even today.

Soon after the nuclear atom was established, scientists started speculating about the structure of the nucleus. Discovery of radioactivity by Becquerel has already indicated that negatively charged beta particles, which were nothing but electrons, are emitted in the disintegration of some nuclei. In 1919, Rutherford announced the discovery of proton, the nucleus of hydrogen atom, and showed that protons are present in every nucleus. As such the proton and electron model of the nucleus was put forward. However, the electrons of ≈ 5–6 MeV energy, which are generally emitted in radioactive beta decay, cannot be accommodated inside the nuclei, which have radii of only a few Fermi. Further, the statistics of many nuclei predicted on the basis of proton + electron hypothesis does not match their experimental value. This led to the rejection of the hypothesis. Around the year 1924, Heisenberg, on the basis of his quantum mechanical approach has already predicted a particle having zero charge and of mass nearly equal to that of a proton. In 1932, Chadwick finally established the presence of neutron, a particle having no charge and mass nearly equal to that of proton. Since neutrons were also emitted during artificial transmutation of elements, it was realized that it is a constituent of the nucleus. The present theory about the constitution of the nucleus is the neutron–proton hypothesis. According to this hypothesis a nucleus, ${}^{A}_{Z}X$ has Z protons and $(A\text{-}Z)$ number of neutrons. The neutron and proton together are called nucleons. A is the total number of nucleons in the nucleus and is called the atomic mass number.

Although it is not possible to directly see the details of a nucleus by any instrument, information about nuclear structure are gathered indirectly, from the study of the systematics of stable nuclei, their relative abundance, nuclear decay, from nuclear reaction studies, and so on. These studies have indicated that neutrons and protons inside the nucleus are in the state of continuous motion, like the electrons in an atom. With the invention of particle accelerators it has become possible to hit a given nucleus by accelerated nuclear particles such as protons, alphas, deuteron and other heavy ions. The experimental results are then analysed assuming a certain structure of the nucleus and the mechanism of the reaction. Thus, the study of nuclear reactions gives insight into both the structure of the nucleus and the mechanism of reaction.

A nucleus being a microscopic system requires quantum mechanical approach. In quantum mechanics, a system is described in terms of a wave function. The nucleus acquires the properties of the wave function that describes it. As a rule, there is no classical analogue for any quantum mechanical property, in spite of which classical analogies for quantum concepts are often given to create some sort of a picture of the concept for better understanding. It is because the human brain is trained in this way from the childhood through the observations of macroscopic objects and large-scale processes.

Solved examples

Example 1.1 The quantity $\frac{e^2}{4\pi\varepsilon_0}$, where e stands for the magnitude of the electron charge is often used in the calculation of Coulomb's barrier and Coulomb's force between charged particles. Calculate the value of the quantity in units that are frequently used in nuclear physics.

Solution. In SI system of units, the unit of charge is Coulomb and electron charge e has a value of 1.6×10^{-19} Coulomb (or C). The quantity $\frac{1}{4\pi\varepsilon_0}$ has the value 9×10^9 Nm^2 C^{-2} = 9×10^9 JmC^{-2}. If these values are substituted, then

$$\frac{e^2}{4\pi\varepsilon_0} = 9\times10^9 \text{ JmC}^{-2} \times (1.6\times10^{-19})^2 \text{ C}^2 = 9\times10^9 \times (1.6\times10^{-19})^2 \text{ Jm} \quad \text{(A)}$$

The SI unit of the above quantity are joule (J)-meter(m). However, in nuclear physics, energy is generally measured in MeV and distance in Fermi (F). Now, 1.6×10^{-13} J = 1MeV; and 10^{-15} m = 1F; Using these conversion factors one may get,

$$\frac{e^2}{4\pi\varepsilon_0} = \frac{9\times10^9\times(1.6\times10^{-19})^2}{1.6\times10^{-13}} \text{ MeV } m = 1.44\times10^{-15} \text{ MeV m} = 1.44 \text{ MeV F}$$

Ans: The quantity $\frac{e^2}{4\pi\varepsilon_0}$ is equal to 1.44 MeV*Fermi.

Example 1.2 A 5 MeV alpha particle is projected on a $^{208}_{80}$Pb nucleus. Calculate the distance of closest approach of the alpha particle to the target nucleus and compare it with the radius of the nucleus.

Solution. Let us assume that the distance of closest approach of the alpha particle from the nucleus is '*d*'. As the alpha particle moves towards the target nucleus its kinetic energy decreases but the potential energy in the electric field of the target nucleus increases. At distance '*d*', the kinetic energy of alpha particle becomes zero and the total energy becomes potential energy. The alpha particle will now start moving backwards. Therefore, at a distance *d*, the total kinetic energy of the alpha particle is converted into potential energy. Therefore,

Potential energy of alpha at distance d from the target $= \frac{1}{4\pi\varepsilon_0}\frac{Ze\cdot 2e}{d}$ = Initial kinetic energy = 5 MeV

We use SI units; therefore, $\frac{1}{4\pi\varepsilon_0} = 9\times10^9 \text{ Nm}^2\text{C}^{-2}$, $Z = 80, e = 1.6\times10^{-19}$ C, 1eV $= 1.6\times10^{-19}$ J

$$\text{Therefore, } d = \frac{9\times10^9\times80\times2\times(1.6\times10^{-19})^2}{5\times10^6\times1.6\times10^{-19}} \text{ m} = 4.608\times10^{-14} \text{ m}$$

Now the size or the radius, R, of the nucleus $^{208}_{80}$Pb is, $R = r_0 A^{\frac{1}{3}} = 1.2\times10^{-15}\times(208)^{\frac{1}{3}}$ m = 7.1×10^{-15} m. It may be observed that the distance of closest approach is about 9 times (or almost an order of magnitude) larger than the nuclear radius.

Ans: The distance of closest approach is 4.608×10^{-14} m and the radius of the nucleus = 7.1×10^{-15}.

Example 1.3 In an alpha scattering experiment 1030 alpha particles of 5 MeV energy were made to hit 1 cm^2 area of a gold foil of mass 1 mg and dimensions 1 cm × 10 cm. Calculate (a) the number of gold nuclei in the foil, (b) the number of gold nuclei hit by alpha particles, (c) the number of alpha particles scattered by 10° or more, (d) number of alpha particles scattered per unit area at 10° at a distance of 10 cm from the foil, (e) if the alpha beam was on for 5 hours, the flux of the beam.

Solution. Note: since all constants and others required in calculations are in SI units, all other quantities must be converted into SI units. Given quantities: (i) Z of gold = 79, (ii) A for gold 197, (iii) electron charge e = 1.6×10^{-19} C, (iv) $(1/4\pi\varepsilon^0) = 9\times10^9$ Nm^2/C^2, (v) sin5° = 0.087, (vi) cot5° = 11.43.

(a) The number of gold nuclei in the full foil N_1 may be calculated using the mass of the foil and the Avogadro number $A_{av} = 6.02\times10^{23}$, which gives the number of atoms of a substance in 1 g mole. Hence, N_1 = {(mass of the foil × A_{av})/molecular weight} = $\{(1\times10^{-3}\times6.02\times10^{23})/179\} = 3.06\times10^{18}$.
(b) Out of the total area of 10 cm^2 only 1 cm^2, 1/10 area was hit. Therefore, N_i, the number of gold nuclei hit = 3.06×10^{17}.
(c) Let us calculate the impact parameter b for which the alpha particle will be scattered by 10^0, using the formula $b = \{Ze^2/4\pi\varepsilon_0 E\}\cot(10/2)$, Now, $E = 5$ MeV $= 5\times10^6\times1.6\times10^{-19}$ J. Substituting the values of the required quantities, we get $b = 26.0\times10^{-14}$ m. Now, each gold nucleus will expose an area πb^2 such that any alpha falling on this area will be scattered by the angle of 10° or more. $A_1 = \pi b^2\times N_i$ gives the total area projected by all the target nuclei such that alphas hitting on this area will undergo scattering of 10° or more. The number of alphas hitting area A_1 is calculated by using the fact that a total of 10^{30} alphas hit an area 1 cm^2 = 10^{-4} m^2. Therefore, the number of alphas hitting the area $A_1 = (\pi b^2\times N_i\times10^{30}/10^{-4}) = N_2$. Substituting the calculated values of b and N_i, the value of N_2 may be obtained as 6.5×10^{26}. Note that out of 10^{30} alphas only 6.5×10^{26}, just 0.065% are scattered by 10° or more. Rest of them got scattered at angles less than 10°.
(d) Desired number of alphas may be calculated using the expression (1.14). All other quantities are either given or have been calculated. The value of $R = 10$ cm $= 1\times10^{-1}$ m needs to be used. Calculations will yield $N(10°) = 44.16/\text{m}^2$.
(e) The flux of a beam of particles is defined as the number of incident particles per unit area (area being normal to the beam) per second. In the present case, 10^{30} alpha particles hit 1×10^{-4} m^2 area in a time of 5 hours. Therefore, the flux = $(10^{30}/10^{-4}\times5\times60\times60) = 5.56\times10^{29}$ alphas per m^2 per second.

Exercises

Exercise p-1.1: Calculate the quantities asked in the above-mentioned problem for a foil of tantalum ($^{181}{}_{73}$Ta, abundance 100%) instead of gold. Explain the difference in the case of gold and tantalum.

Exercise p-1.2: In a scattering experiment, you concentrate on alpha particles with impact parameters $b = 0, 1 \times 10^{-10}$, and 1×10^{0} mm. For which b will the behaviour of the alpha particle be very different in Thomson's and Rutherford's models and for which b will it be nearly similar? Explain your answer.

Exercise p-1.3: While deriving Rutherford's scattering formula, it was assumed that there is no change in the angular momentum. Discuss the validity of this assumption.

Exercise p-1.4: If the probability of alpha particles with energy 5 MeV to be scattered through an angle θ on passing through a foil of some element is 10^{-4}, what is its value for 10 MeV protons passing through the same foil?

Cross Section

Suppose a beam of some particle 'a' is hitting a target X and as the result of the interaction between X and a sometimes particle b is emitted, sometimes particle b_1 is emitted, sometimes particle b_2 is emitted, and so on. Hitting by 'a' and emission of b is one type of event, emission of b_1, b_2... each is a different type of events. These different events may occur with different frequencies. The frequency of occurrence for these events will depend on (i) the number of targets X, (ii) number of incident particles a, and (iii) the time duration for which the particles a hit target X. Events with higher frequency have higher probability. Cross-section σ may now be assigned to each event such that it is proportional to the probability of its occurrence. Cross-section for the event $\sigma_{(a,b)}$ is defined as the interaction probability leading to event (a,b) per unit target per unit time per unit flux of the incident particles. As such

$$\sigma_{(a,b)} = \frac{\text{Number } N_{ab} \text{ of events of type } (a,b) \text{ occured in time } t}{\text{Number } N_X \text{ of target nuclei } X \times \text{ flux of the incident beam} \times \text{ time t}} \tag{1.21}$$

In Eq. (1.21), N_{ab} and N_X are pure numbers and have no unit or dimensions. Flux F = number of particles/area × time. The dimensions of F are $[M^0L^{-2}T^{-1}]$. Dimensions of time t are $[M^0L^0T^{+1}]$. On substituting the dimensions of different quantities, we get $\sigma_{(a,b)} = [L^2]$, the dimensions of cross-section are that of area.

Eq. (1.21) is generally used in experimentally measuring the cross-section for some event. The cross-section for the same event may also be calculated theoretically assuming a certain interaction mechanism. As such cross-section is an important physical quantity that provides a meeting ground for the theory and the experiment.

1.5.1 Fundamental Properties of a Nucleus

Every nucleus has the following properties:

1.5.1.1 Nuclear representation

The nucleus of the atom of an element X is represented by ${}^{A}_{Z}X$, where X is the chemical symbol of the element, Z, the atomic number of the element and $+Ze$, the amount of the positive charge on the nucleus. A is the atomic mass number, which gives the total number of nucleons in the nucleus. Nuclei having same Z but different A are called ISOTOPES. Isotopes have same number of protons but different number of neutrons. Nuclei having same number of nucleons A are called ISOBARS. A family of isobars have nuclei that have same A (= $Z + N$) but different numbers of protons and neutrons. Nuclei with identical number of neutrons are called ISOTONES. A pair of isobar nuclei in which the number of neutrons in one is equal to the number of protons in the other are called MIRROR nuclei. The pair $^{13}_{6}C$ and $^{13}_{7}N$ is a mirror pair. Nuclei with same value of (N–Z) are given the name ISODIAPHERES. $^{12}_{6}C$, $^{16}_{8}O$ and $^{11}_{5}B$, $^{15}_{7}N$ are pairs of isodiapheres. Nuclei that have even number of proton and even number of neutron are called EVEN–EVEN, those with even number of protons and odd number of neutrons EVEN–ODD, odd number of protons and even number of neutrons ODD–EVEN, and with odd numbers of both protons and neutrons as ODD–ODD nuclei. Sometimes nuclei with same values of Z and A are found with different decay modes. They are called ISOMERS. For example $^{137}_{56}Ba$ has two isomers, one that is stable and the other metastable state that decays with a half life of 2.6 minutes.

1.5.1.2 Nuclear mass

Every nucleus has a mass, which is generally denoted by (M or m) and is measured in a unit called unified mass unit or simply mass unit, denoted by 'u'.

$$\mathbf{1}u = \frac{\mathbf{1}}{\mathbf{12}}\ \textbf{of the mass of a carbon} - \mathbf{12}\ \textbf{atom} = \mathbf{1.66043 \times 10^{-27}\ kg} = \mathbf{1.492232 \times 10^{-10}\ J} = \mathbf{931.480\ MeV}$$

(Million electron volt, written as MeV, is a unit of energy, 1 MeV = 1 × 10^6 eV, where 1 eV is the energy gained by a unit positive charge in passing through 1 volt potential difference. Using Einstein mass–energy relation $E = mc^2$, mass can be written in terms of energy.)

In principle, the mass of a nucleus should be equal to the sum of the masses of all the nucleons inside it. However, measurements have shown that the actual mass of the nucleus is less than the sum of the masses of all its nucleons. If the masses of a neutron and a proton are represented by M_n and M_p, respectively, then the mass M_x of the nucleus $^{A}_{Z}X$ may be written as

$$M_x = Z\,M_p + (A - Z)M_n - \Delta M_x; \text{ or } \Delta M_x = [Z\,M_p + (A - Z)M_n] - M_x$$

This disappeared mass ΔM_x, called *Mass excess*, is equivalent to energy ($\Delta M_x c^2$) that has disappeared in binding of the nucleons together in the nucleus. It means that to break the nucleus X into its constituent nucleons an energy equivalent to ($\Delta M_x c^2$) must be supplied from outside. Energy ($\Delta M_x c^2$) is called the *binding energy* of the nucleus X and often denoted by B_x. The binding energies for different nuclei are different from each other. For example, the binding energy of the nucleus deuteron $^2{}_1H$ is about 2.8 MeV while that of an alpha particle, the nucleus of helium atom, $^4{}_2He$ is 28 MeV. If binding energy of the nucleus is divided by the number of nucleons in the nucleus a quantity called binding energy per nucleon, denoted by

$\frac{B}{A}$ may be obtained. Variation of $\frac{B}{A}$ with A, shown in figure 1.9, is of interest. As may be seen in this graph, the value of $\frac{B}{A}$ is maximum (of the order 8.8 MeV) for nuclei around $A \approx 60$. This means that these nuclei are more tightly bound than the other nuclei. On further increasing A, $\frac{B}{A}$ decreases with increasing value of A, having a value of 8.6 MeV at $A \approx 100$ and about 7.6 MeV for $A \approx 200$. Similarly, for light nuclei, with a few sharp peaks in the beginning, $\frac{B}{A}$ increases with A. The basis of energy released both in the case of fusion of light nuclei and in the fission of heavy nuclei can be easily understood from this graph.

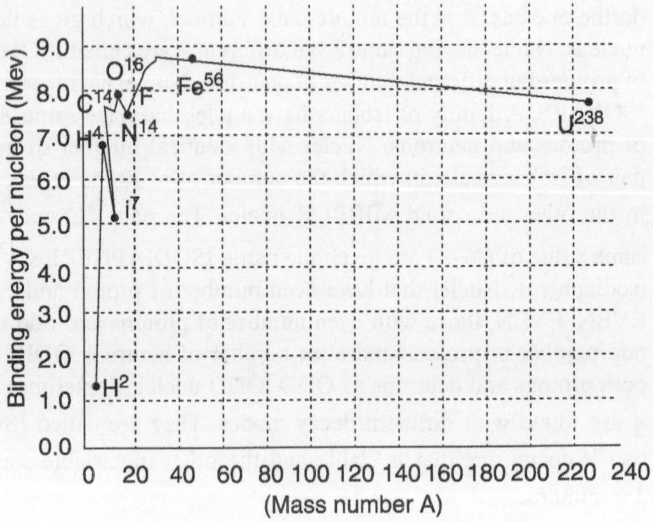

Figure 1.9 *Graph showing the variation of B/A with A*

Suppose we have a nucleus around $A = 200$ where the B/A is of the order of 7.6 MeV. The total binding energy of this nucleus is 200×7.6 MeV. If this nucleus is somehow broken down into two equal or nearly equal halves, each broken half will have A around 100 and the total binding energy of each half will be 100×8.6 MeV. Hence, the sum of the binding energies of the two halves will be $2 \times 100 \times 8.6$ MeV. Thus, in the process of breaking down of a heavy nucleus into two nearly equal parts, the total binding energy has changed by the amount $[2 \times 100 \times 8.6 - 200 \times 7.6] = 200$ MeV. This difference of energy is released in the breaking process as the kinetic energy of the products. In nuclear fission, heavy nuclei undergo either spontaneous breaking (called self fission) such as ^{252}Cf nucleus, or some nuclei such as ^{235}U may be broken by hitting them with very low energy (≈ 0.025 eV) neutrons called thermal neutrons. On an average 200 MeV of energy is released per fission of such nuclei. This is not a big amount of energy but when very large number of nuclei ($\approx 10^{17}$ or more) fission out simultaneously huge amount of energy may be released. Similarly, when two very light nuclei fuse together to make a bigger nucleus, energy is released. There are several other important inferences that may be drawn from Figure 1.10. Some other parameters often used to give the mass of a nucleus $^A{}_ZX$ are mass defect ΔM and packing fraction f defined as follows:

Mass defect ΔM or mass excess = actual mass of the nucleus − atomic mass number A

$$\text{Packing fraction } f \equiv \frac{(\textit{Actual mass of the nucleus} - \textit{Atomic mass number } A)}{\textit{Atomic mass number } A} = \frac{\Delta M}{A}$$

1.5.1.3 Nuclear size

Most of the stable nuclei are spherical, with a distribution of positive charge and of mass in a spherical volume. The radius of the nuclear charge and matter distributions are actually slightly different from each other but generally considered equal. These radii are so small that they are represented in a special unit of distance denoted by '*F*' called Fermi, or femtometer.

$$1\ \mathrm{F} = 10^{-15}\ \mathrm{m}$$

It may, however, be said that there is no sharp boundary within which the charge/mass of the nucleus is uniformly distributed and beyond that it is zero. Experiments have shown that the charge distribution in a nucleus looks like as shown in figure 1.10. It is clear from the figure that there cannot be a unique definition of the radius of the nucleus. The distance from the centre at which the charge density falls to 50% of the central value is generally called the nuclear radius.

It is interesting to note that the central charge density in all nuclei, big or small, is nearly the same. The difference lies in the so-called thickness of the skins of the nuclei. Another important observation is that the radius of a nucleus having A nucleons is related to A by the relation $r_A = r_0 A^{1/3}$, where r_0 is a constant, called *unit nuclear radius*, with values in the range of 1.2–1.5 F. If it is assumed that the mass M_A of the nucleus is proportional to A, the number of nucleons in it and the nucleus is spherical, then $M_A = mA$, where m is the mass of the nucleon and the volume V_A of the nucleus becomes $\frac{4}{3}\pi r_A^3 = \frac{4}{3}\pi r_0^3 A$. As such the density of the nuclear matter becomes $M_A/V_A = \dfrac{m}{\frac{4}{3}\pi r_0^3}$ a constant value for all nuclei. This constant density of nuclear matter leads to the liquid drop model of the nucleus in which it is assumed that all nuclei are like the drops of nuclear matter of different sizes just like the drops of a liquid of different size.

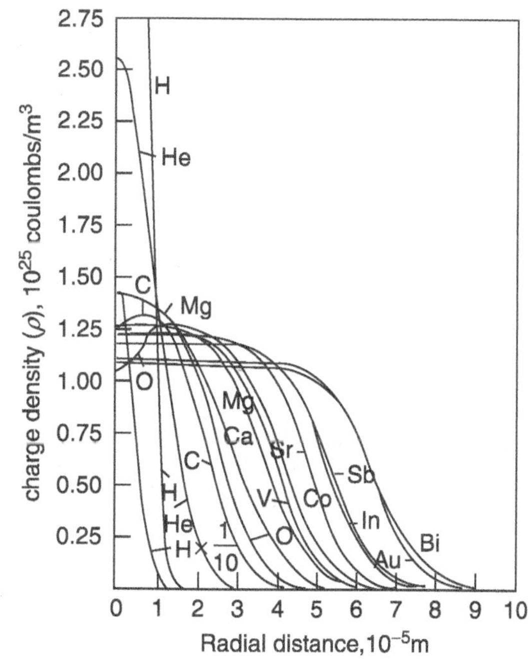

Figure 1.10 *Charge distribution of a nucleus*

1.5.1.4 Nuclear angular moment or spin

Every nucleus has a spin or angular momentum and the associated spin or angular momentum quantum number. The spin quantum number is generally denoted by J or S. The square of the spin angular momentum, denoted by (PJ^2), is given by the quantum mechanical expression

$$PJ^2 = J(J+1)\,\hbar^2, \text{ here } \hbar \text{ is the Planck's constant}/2\pi = 1.05450 \times 10^{-34} \text{ J-sec}$$

The absolute value of the angular momentum j is, therefore, $j = [j(j+1)]^{1/2}\,\hbar$. If the value of J is taken along some other direction, for example in the direction of some externally applied magnetic field, then from quantum mechanics, the observable value of the component of j will be $m_j\hbar$, where m_j is called the magnetic quantum number and can have $(2J+1)$ different values from $-J, -J+1, -J+2 \ldots J-3, J-2, J-1$, J. The maximum possible value of m_j is equal to j. The direction in which the spin J of nucleus points defines the body fixed axis of the nucleus.

It may, however, be noted that there are several nuclei for which $J = 0$. There are also nuclei with both integer and half-integer values of J. Nuclei with integer spin values obey Bose–Einstein statistics and are called bosons, while those with half-integer spins follow Fermi–Dirac statistics and are called fermions. Often, the nuclei for which J is non-zero are compared to a spinning top which also has spin angular moment.

The nucleus $^A{}_ZX$ has Z protons and N $(= A - Z)$ neutrons. Both neutron and proton have intrinsic spin angular moment s_p and s_n with the value of $1/2\hbar$. Besides, these nucleons are also assumed to execute orbital motion inside the nucleus with orbital angular momentum $l\hbar$. The orbital angular momentum quantum number l may have only integer values, 0, 1, 2, ... It may be assumed that the total angular momentum J of the nucleus builds up from the quantum mechanical sum or coupling of the spins s_p of all protons, spins s_n of all neutrons and their orbital angular moments l_p and l_n. Quantum mechanical addition of vectors is quite different from that of the usual vector addition. If two vectors A and B are quantum mechanically added, the result is not a unique single value. All values from $(A + B)$ to $(A \sim B)$ in step of 1 are possible values of the quantum mechanical addition. It has been observed that nuclei with even number of nucleons have either zero or integer spins, while those with odd A have half integer spins.

1.5.1.5 Magnetic dipole moment

Apart from the electric field around them owing to the positive charge, nuclei also have a magnetic field. The magnetic field of the nucleus may be resolved into components due to magnetic dipole and other higher multipoles. However, the physically important component of the magnetic field is due to the dipole term and, therefore, often the moment of the magnetic dipole is simply referred to as magnetic moment of the nucleus, dropping the term dipole. The magnetic dipole moment of the nucleus is denoted by 'μ' and is measured by unit called 'nuclear magneton' which is denoted by (nm).

$$\mathbf{1\ nm} = \left(\frac{e\hbar}{2m_p}\right), \text{ where } m_p \text{ is the mass of a proton}$$

$$= 5.0505 \times 10^{-27}\ (\text{J-m}^2/\text{Wb})$$

In classical terms, the magnetic field of a nucleus may be compared to the field produced by a tiny bar magnet. The magnetic moment of the nucleus is positive if it points in the direction of the spin J of the nucleus and is negative if it is in opposite direction. Like spin, the magnetic moment of a nucleus builds up from the magnetic moments of nucleons inside. A proton has an intrinsic magnetic moment $\mu_P{}^s$, which points in the direction of its spin s_P and is, therefore, positive. Orbital motion of proton inside the nucleus constitutes current element and, therefore, produces additional magnetic moment called orbital magnetic moment. Classically, a current I flowing in a circular loop enclosing an area S has a magnetic moment = IS. Similarly, a proton with +1e charge moving with a velocity v in a circular orbit of radius r will make N revolutions per second, such that $N = v/(2\pi r)$. The number of revolution per second N multiplied by the charge of the proton gives the magnitude of the current I produced by the circular motion of the proton. The area of the current loop generated by the orbital motion of the proton $S = (\pi r^2)$. Thus, the dipole moment produced due to the orbital motion of the proton is equal to $eNS = \frac{evr}{2}$. As such the classical value of the magnetic dipole moment due to the orbital motion of the proton, $\mu_p^l = \frac{evr}{2}$. The angular momentum L_P of this proton is $L_P = m_P r v$. The ratio $\left(\frac{\mu_p^l}{L_p}\right) = \gamma = \frac{e}{2m_p}$, is called the gyromagnetic ratio and is denoted by γ. In practice, however, when an attempt is made to measure the magnetic moment, an external magnetic field for instance in the Z-direction is applied. Now, the measurement will give the Z-component of the magnetic moment $(\mu^l_P)_z$. Assuming that the ratio $\left\{\frac{(\mu_p^l)_z}{(L_p)_z}\right\} = \frac{e}{2m_p} = \gamma$, we get

$$(\mu_p^l) = \left(\frac{e}{2m_p}\right)(I_{,p}) \tag{1.15}$$

In quantum mechanism, the z-component of the angular momentum has the value $m_l\,\hbar$, where m_l may have values from $-L_P$ to $+L_P$ in step of 1; m_l can be a positive or negative integer including zero with the maximum magnitude of L_P. Substituting the value of $(L_P)_Z$ in Eq. (1.15), we get

$$(\mu_p^l) = \left(\frac{e\hbar}{2m_p}\right) m_l = m_l(nm) \tag{1.16}$$

since $\left(\frac{e\hbar}{2m_p}\right) = 1$ nuclear magneton.

It may thus be seen that a proton moving inside the nucleus has the intrinsic magnetic moment μ^S_P and an orbital magnetic moment $(\mu^l_P)_Z$. If it is assumed that classical expression (1.15) also holds for the intrinsic spin motion of the proton, then we get

$$\mu_P^s = s \text{ nuclear magneton} \tag{1.17}$$

which means that a proton with intrinsic spin $\frac{1}{2}$ should have a spin magnetic moment of $\frac{1}{2}$ nuclear magneton magnitude.

On one hand, the measured values of intrinsic magnetic moments of several particles such as neutron, proton, electron, and that of several nuclei do not agree with the corresponding values given by expression (1.17). On the other hand, the experimental values of magnetic moments due to orbital motions of nucleons are correctly given by Eq. (1.16). To overcome this difficulty, dimensionless quantities g_l or g_s are multiplied on the right hand side of Eqs. (1.15) and (1.16), respectively. The modified equations are as follows:

$$\mu_P^l = g_l^P\, m_l \text{ and } \mu_P^s = g_s^P\, s \text{ for proton } \mu_N^s = g^N S \text{ and } \mu_N^l = g_l^s\, m_l \text{ for neutron}$$

From the above equations, the nuclear factor g may be defined as the ratio of the magnetic moment (measured in unit of nuclear magneton) to the corresponding angular moment (measured in unit of $\hbar$). The value of g_l^P for orbital motion is 1 for proton and g_N^l is zero for neutron. For intrinsic spin, g_P^s and g_N^s are not equal to one as may be inferred from the following experimental values of the intrinsic spin magnetic moments for proton and neutron.

$$\mu_P^s = 2.7927 \text{ (nm)};\ \mu_N^s = -1.9135 \text{ (nm)};\ g_P^s = 5.5854,\ g_N^s = 3.8270$$

A neutron also has an intrinsic magnetic moment ($-\mu^{S}_{N}$). The negative sign indicates that the direction of the magnetic moment vector is opposite to the vector spin S. Although there is no definite reason for the magnetic moment of a neutron, it may be explained by assuming that the neutron has a distribution of equal amounts of both positive and negative charges with the negative charge distribution towards the surface. Orbital motion of neutron does not produce any current and hence no magnetic moment.

Like nucleons, an atomic electron also has spin and orbital magnetic moments μ^{s}_{e} and μ^{l}_{e} expressed as

$$\mu^{s}_{e} = -g^{s}_{e}\, s = -1.001145 \text{ Bohr magneton}, \mu^{l}_{e} = -g^{l}_{e}\, l = -m_{l} \text{ Bohr magneton}$$

Bohr magneton (Bm) is the unit in which magnetic moments of atomic particles are measured and is equal to $(e\hbar/2m_{e})$. Bm is nearly 1/1850 of Nm. g^{s}_{e} is nearly 2 and g^{l}_{e} is 1, as that in case of proton.

The magnetic moment of the nucleus may be looked as due to the quantum mechanical addition of the intrinsic spin magnetic moments of all nucleons plus the orbital magnetic moments of all protons.

1.5.1.6 Electric quadrupole moment

The electric field of an arbitrary charge distribution may be decomposed into several components arising due to electric dipole (positive and negative charges of equal amounts separated by a small distance), electric quadrupole (two electric dipoles anti-symmetrically placed along each other), octupole (two quadrupoles placed side by side), and other higher multipoles. The moment of the electric quadrupole assumes importance in nuclear physics as it is related to the shape of the positive charge distribution in the nucleus. Nuclear electric quadrupole moment, in short called quadrupole moment, is represented by the symbol 'Q'. Q is defined in such a way that its units are that of area. In nuclear physics, area is also measured in units called 'barn' denoted by b.

$$\textbf{1 barn (b)} = 1 \times 10^{-24}\ \text{cm}^2 = 1 \times 10^{-28}\ \text{m}^2$$

Q = 0 means that the charge distribution of the nucleus is spherically symmetric

Q = +ve means that the charge distribution is elongated along the direction of the spin J of the nucleus and is cigar shaped. This shape is also called prolate shape.

Q = −ve means the charged distribution is elongated in the direction perpendicular to the direction of the spin J like a knife-edge. This shape of the nucleus is called oblate.

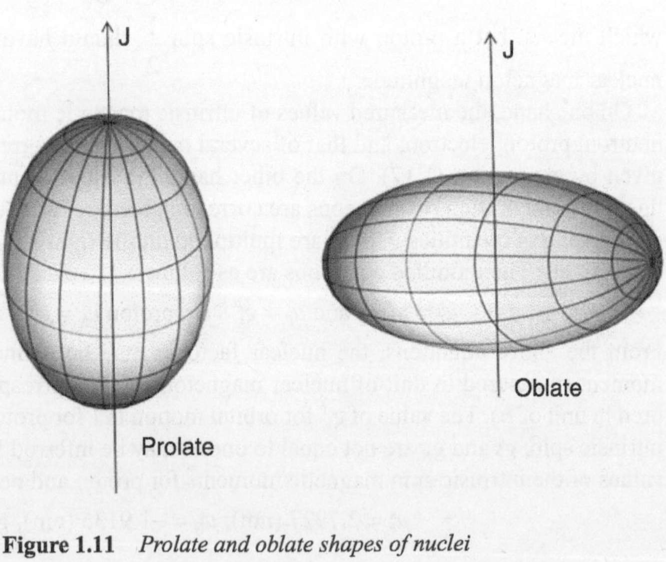

Figure 1.11 *Prolate and oblate shapes of nuclei*

If a nucleus possesses a volume density of positive charge $\rho(r)$ at a radial distance r from the centre, the volume integral $\int r\,\rho(\boldsymbol{r})\mathbf{d}v$ gives the value of the electric dipole moment of the nucleus. Because of the symmetry of charge distribution around the centre the above integral gets contributions of equal but of opposite sign from $\boldsymbol{r}$ and $-\boldsymbol{r}$ and vanishes. As such electric dipole moment is nonexistent for nuclei.

Quantum mechanically the electric quadrupole moment of the nucleus with respect to the body fixed axis is defined by the following relation:

$$Q_0 = \left(\frac{1}{e}\right)\int \rho(r)\,\{3Z^2 - r^2\}\mathrm{d}v \tag{1.18}$$

The quantity $\left(\frac{1}{e}\right)$, where e is the unit charge, is included in the definition to make Q_0 independent of charge. If the charge distribution is spherically symmetric, $r^2 = 3Z^2$ as $X = Y = Z$ and Q_0 is zero. For a charge distribution stretched along the Z axis, Q_0 is +ve, while for charge distributions stretched either along the X or the Y directions, it is negative. The dimensions of Q_0 are that of area. The value of Q along an externally defined direction, defined by an external magnetic field, in which the magnetic quantum number is m_J, is given by

$$Q(m_J) = \left[\frac{3m_j^2 - j(j+1)}{j(2j-1)}\right]Q_0 \tag{1.19}$$

From Eq. (1.19), it may be seen that nuclei that have $j = 0$ or $\frac{1}{2}$ do not show quadrupole moment in the state where $m_J = j$. This does not mean that the nuclei with spin $\frac{1}{2}$ necessarily have spherically symmetric charge distribution. It simply means that the component of Q_0 along the direction of the maximum projection is zero.

The effect of the electrical quadrupole moment is found in the hyperfine structure of atomic and molecular spectra. The effect of higher electrical multipoles has not been detected in hyperfine structure so far.

1.5.1.7 Nuclear parity

Parity is a purely quantum mechanical concept. In quantum mechanics, a single particle is defined by its wave function ψ, which is a function of space coordinates x, y, and z. The wave function or the particle is said to have an even or positive parity if the wave function does not change sign on reflection through the origin. Reflection through the origin essentially means replacing $\boldsymbol{r}$ by $-\boldsymbol{r}$ in the wave function. If it changes sign, the parity is odd or negative.

$$\begin{aligned}\psi(x, y, z) &= \psi(-x, -y, -z) \text{ .... even or +ve parity}\\ &= -\psi(-x, -y, -z) \text{ ....odd or −ve parity}\end{aligned}$$

The parity of a system of particles is obtained by multiplying the parities of individual particles. As an average nucleus has several nucleons, its parity is decided by the parities of other nucleons. The parity of a nucleus is indicated by putting a + or – sign on the J value of the nucleus. Reflection through origin essentially means producing the mirror image of the system and parity, therefore, is related to the mirror symmetry of the system. As nuclei possess symmetry about the centre of mass (or the nuclear centre), they have definite parity. In most of the nuclear reactions, parity is conserved. Conservation of parity in a process is guaranteed if the process and its mirror image are both experimentally possible processes. In some nuclear reactions, such as nuclear beta decay, parity is not conserved. The parity P of a system of particles having individual parities $p_1, p_2, p_3 \ldots p_n$ and moving in states of orbital angular momentum $l_1, l_2, l_3 \ldots l_n$, respectively, is given by

$$P = p_1.p_2.p_3 \ldots p_n.(-1)^{\ell_1}.(-1)^{\ell_2}.(-1)^{\ell_3} \ldots (-1)^{\ell_n}$$

Systems that do not have mirror symmetry do not have a definite parity, instead have helicity. An example is helices. If a system or a particle has total angular momentum J, angular momentum L, linear momentum P, and spin angular momentum S, then the helicity $\mathcal{H}$ is defined as the projection of J on P. Denoting the unit vector in the direction of P by p

$$\mathcal{H} = J{\cdot}p = (L + S){\cdot}p = L{\cdot}p + S{\cdot}p = S{\cdot}p$$

Since L and P are always perpendicular to each other. Helicity is positive when $\boldsymbol{S}$ and $\boldsymbol{P}$ are in the same direction and negative for the opposite direction. Neutrinos emitted in beta decay have helicity and, therefore, parity is not conserved in beta decay.

1.5.1.8 Nuclear statistics

Although we talk about a single nucleus, in practice it is almost impossible to have a single isolated nucleus. It is because even a milli-microgram of an average atomic mass element contains more than 10^{10} atoms and nuclei. Similarly, a nucleus of an average atom contains many nucleons. As such, in nuclear physics, one mostly deals with an assembly of microscopically identical particles. According to quantum mechanics, assemblies of identical microscopic particles follow some rules regarding the distribution of particles in different quantum states. These rules define the statistics of the assembly or of the system. In quantum mechanics, all microscopic particles may be divided in two classes. One class of particles called bosons follow Bose–Einstein statistics and the other class of particles or systems called fermions follow Fermi–Dirac statistics. The statistics of a system cannot change from one type to the other with time and is a constant of motion.

In quantum mechanism, a particle may be represented by a wave function ψ_α, where α stands for all the quantum numbers specifying the particle. The wave function χ_N for the assembly of, for instance N such identical particles may be generated by taking the multiplication of N different wave functions, one for each particle.

$$\chi_N = (\psi_{\alpha 1})_1\,(\psi_{\alpha 2})_2\,(\psi_{\alpha 3})_3 \ldots (\psi_{\alpha i})_i \ldots (\psi_{\alpha i})_j \ldots (\psi_{\alpha N})_N$$

Further, it is known that not only χ_N but other possible combinations $\chi'_N, \chi''_N, \ldots$ obtained by interchanging all the quantum numbers of two particles and their linear combinations are also solutions of the wave equation and are, therefore, valid wave functions for the system of N identical particles. To simplify the problem, let us consider a system of only two particles. Let the wave functions of the two particles be $(\psi_{\alpha 1})_1$ and $(\psi_{\alpha 2})_2$, where α_1 and α_2 define all quantum numbers including that of space and spin, for particles 1 and 2, respectively. The wave function of the system of these two particles may be written as

$$\chi_2 = \frac{1}{\sqrt{2}}\,[(\psi_{\alpha 1})_1\,(\psi_{\alpha 2})_2 \pm (\psi_{\alpha 1})_2\,(\psi_{\alpha 2})_1] \tag{1.20}$$

The factor $\left(\frac{1}{\sqrt{2}}\right)$ is included in expression (1.20) so that the probability density $[(\psi_{\alpha1})_1 (\psi^*_{\alpha1})_1]$ for particle 1 and the similar quantity for particle 2 does not change. In Eq. (1.20), the second term $(\psi_{\alpha1})_2 (\psi_{\alpha2})_1$ is obtained by interchanging all the quantum numbers, including space and spin coordinates of the two particles. In case of bosons $(\psi_{\alpha1})_2 (\psi_{\alpha2})_1 = (\psi_{\alpha1})_1 (\psi_{\alpha2})_2$, while in case of fermions $(\psi_{\alpha1})_2 (\psi_{\alpha2})_1 = -(\psi_{\alpha1})_1 (\psi_{\alpha2})_2$. As such the wave functions for the bosons and fermions are as follows:

$$(\chi_2)_{\text{boson}} = \left(\frac{1}{\sqrt{2}}\right)[(\psi_{\alpha1})_1 (\psi_{\alpha2})_2 + (\psi_{\alpha1})_2 (\psi_{\alpha2})_1]$$

$$(\chi_2)_{\text{fermion}} = \left(\frac{1}{\sqrt{2}}\right)[(\psi_{\alpha1})_1 (\psi_{\alpha2})_2 - (\psi_{\alpha1})_2 (\psi_{\alpha2})_1]$$

The wave function of bosons is symmetric with respect to the exchange of two particles while that of fermions is anti-symmetric. An anti-symmetric wave function for fermions is in confirmation to the Paul's exclusion principle, according to which no two identical fermions in a system can have all their quantum numbers same. In other words, not more than one fermion can be accommodated in a given quantum state. No such restriction applies on bosons. It is possible to find more than one boson in a given quantum state in a system of bosons. As has already been indicated, the parity of a system is related to the reflection of the space coordinates through the centre of mass, while the statistics of a system depends on what happens to the system when the quantum numbers, including space and spin, of two particles are interchanged. It has now been well established that particles such as nucleons, neutrinos, and may other systems such as nuclei that have half-integer spin follow Fermi–Dirac statistics, whereas photons, pi-mesons, and nuclei with integer spins follow Bose–Einstein statistics.

1.5.1.9 Nuclear isobaric (or iso) spin

In view of the fact that neutrons and protons have almost identical properties except the charge, Heisenberg gave the concept that neutrons and protons are the two states of a particle called nucleon. In the absence of electromagnetic field, neutrons and protons behave as the same particle; however, as soon as the electromagnetic field is switched on, the nucleon state gets degenerated into two different states: proton and neutron. This concept is based on the analogy of the splitting of electronic levels subjected to external electrical or magnetic fields. Extending the analogy further, it is assumed that there is a hypothetical isobaric space in which a nucleon has an isobaric spin quantum number, the component T_3 of the isobaric spin is arbitrarily assigned the value $+\frac{1}{2}$ for the proton and $-\frac{1}{2}$ for the neutron. The component T_3 for a nucleus with N neutrons and Z protons is $\frac{1}{2}(Z - N)$. All member nuclei of an isobar family have same A but different values of $(N - Z)$, and hence have different values for the component T_3. For example, $^{14}_{6}C$, $^{14}_{7}N$, and $^{14}_{8}O$ are members of an isobaric family with T_3 values 1, 0, and −1. This clearly shows that the isobaric spin of the system to which the trio belongs has the isotopic spin $T = 1$, with components 1, 0, and −1. The isotopic spin, such as the angular momentum of the nucleus, is also conserved in nuclear reactions. The projection of the isospin T_3 for a nucleus having Z

protons and N is equal to $\frac{1}{2}(Z-N)$. Therefore, the total charge $Z = T_3 + A/2$. Since electric charge is conserved in a nuclear reaction and the charge Z is related to the isospin, it is also conserved in nuclear reaction. It may be remarked that the assignment of +1/2 or −1/2 to the component of the isospin for proton and neutron is quite arbitrary and in some literature −1/2 is used for proton and +1/2 for the neutron.

1.5.1.10 Nuclear ground and excited states

According to quantum mechanics, any quantum system, such as a nucleus, may have many possible discrete energy eigen values, $U_1, U_2, U_3, \ldots U_n \ldots$. One of these energy values, for example, U_0 is minimum. The quantum state of the system in which it has the minimum energy U_0 is called the ground state of the system. A nucleus has nucleons moving inside. The total energy of the nucleus is due to its mass and the energy of the motion of nucleons. One particular pattern of the motion of nucleons corresponds to the lowest energy ground state of the nucleus. If adequate energy is supplied to the nucleus, it may shift to the other higher energy states. All states with energy greater than the ground state are called the excited states of the nucleus. Different states of a nucleus differ in the pattern of motion of the nucleons. First information about nuclear excited states was obtained from the energies of gamma rays emitted in natural radioactive decay. In some cases, it was found that the same nucleus decay by emitting three or more gammas of energies γ_1, γ_2, and γ_3 such that the energy of γ_3 is equal to the sum of the energies of γ_1 and γ_2, i.e. $\gamma_3 = \gamma_2 + \gamma_1$. This indicated that there is an excited state of the nucleus, which is fed by the gamma ray of energy γ_1 (or γ_2), this excited state then decays by emitting γ_2 (or γ_1). Energy, spin or angular momentums, parity, magnetic moment, quadrupole moment, and so on of the excited states are generally different from the ground state. The excitation energy of an excited state is measured with respect to the ground state, taking the energy U_0 of the ground state as the zero of the excitation energy scale. As such the ground states of two different nuclei have different ground state energies in absolute terms, but the excitation energies for each nucleus are measured with respect to its own ground state, not in absolute energies.

For example, the ground state of $^{13}_{7}N$ is about 2.22 MeV higher in absolute energy than the ground state of $^{13}_{6}C$, but the energies of the excited states of the two nuclei are measured with respect to their corresponding ground states as shown in figure 1.12.

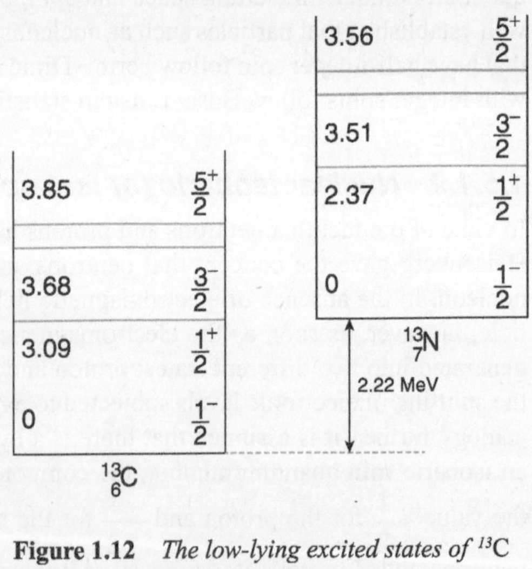

Figure 1.12 *The low-lying excited states of ^{13}C and ^{13}N nuclei*

In this figure, the states of the two nuclei are shown by horizontal lines with their excitation energies on the left side and angular momentum and parity on the right. The characteristic life time of nuclear excited states is of the order of 10^{-12} seconds. However, sometimes due to selection rules and/ or other hindering factors, the decay process becomes slow resulting in excited states of longer half lives. Any nuclear excited state, which has a half-life of more than 10^{-9} s, is called a meta-stable state. Meta-stable states with half-lives ranging from 10^{-9} s to several years have been observed.

Example 1.4 Calculate the binding energy B and the binding energy per nucleon B/A for ^{12}C nucleus, given that the masses of neutron and protons are, respectively, 1.008665 u and 1.007277 u. Also calculate the masses of neutron and proton in MeV.

Solution. It follows from the definition of the unified mass unit that the mass of ^{12}C is 12 u. A ^{12}C nucleus has 6-neutrons and 6-protons. The total mass of 6-neutrons and 6-protons = 6(Mass of a neutron+Mass of a proton) = 6(1.008665 +1.007277)u = 12.095652u . As such a mass = 0. 095652u has disappeared when 6-neutrons and 6-protons joined together to form ^{12}C nucleus. Therefore, the binding energy of ^{12}C nucleus, B = 0. 095652u. Now 1u = 931.49 MeV; Hence,

$B = 0.\ 095652u = 0.\ 095652\,\text{u} \times 931.49 = 89.0950554\,\text{MeV}$

$B/A = 89.0950554/12 = 7.82458795$ MeV.

Mass of a neutron = 1.008665 u = 1.008665 × 931.49 = 939.56 MeV

Mass of a proton = 1.007277 u = 1.007277 × 931.49 = 938.26 MeV

Ans: B = 89.10 MeV; B/A = 7.82 MeV;

Neutron and proton masses = 939.56 MeV and 938.26 MeV, respectively.

1.6 ELECTRON CANNOT BE A CONSTITUENT OF THE NUCLEUS

1.6.1 Nuclear Statistics

Soon after the establishment of the heavy, centrally placed positively charged nucleus in every atom and the discovery of protons from artificial transmutation experiments by Rutherford, it became apparent that protons are present in every nucleus. Emission of negatively charged electrons in natural radioactive decay gave the idea that both protons and electrons constitute the nucleus. A nuclear structure based on proton and electrons was proposed. For example, in case of the nucleus of $^{14}_{7}N$, it was assumed that it contains 14 protons and 7 electrons so that the total charge of the nucleus is $(14e - 7e =) + 7e$ and its mass is mostly due to 14 protons (since the mass of 7 electrons is negligible as compared to the mass of 14 protons). This structure has successfully explained the observed mass and charge of the nucleus but faced problems because of the statistics of the nucleus. According to the proton + electron hypothesis, the nucleus $^{14}{}_{7}N$ that has 21 fermions should follow Fermi–Dirac statistics. However, experimentally it was found that it follows Bose–Einstein statistics. The discrepancy was resolved when neutron was discovered and it was established that the $^{14}{}_{7}N$ nucleus has 7 protons and 7 neutrons in all 14 fermions. A system with even number of fermions behaves like a boson. As a matter of fact, any nucleus with odd number of neutrons will acquire a wrong statistics if derived assuming the proton + electron constitution. Apart from the statistics, the following facts also rule out the possibility of electron being a constituent of the nucleus.

1.6.2 Compton and de-Broglie Wave Lengths of Electron

The Compton wave length λ_c for a particle of mass m is defined as $\lambda_c = \left(\frac{\hbar}{mc}\right)$ where c is the velocity of light. Compton wave length in a way gives the minimum distance in which the particle can be confined. Compton wave length of electron is 2.4262×10^{-10} m, which is of the size of the atom. Nucleus is almost 10^4 times smaller than the atom and therefore electrons cannot live inside a nucleus.

Further, in natural beta decay, electrons with maximum energy of the order of 5 MeV are emitted. A 5 MeV electron has a de-Broglie wave length $\lambda = \dfrac{h}{mv}$ much larger than the size of the nucleus and as such electrons cannot be a constituent of the nucleus. It is, however, possible that electron is created inside the nucleus and is immediately emitted.

1.6.3 Uncertainty in the Momentum of Electron

According to Heisenberg's uncertainty principle, the multiplication of the uncertainties in the momentum Δp and the distance Δx should be of the order of $\hbar$. Supposing that an electron lives inside the nucleus, one can assign a maximum uncertainty in its location Δx a value equal to the size of the nucleus, that is of the order of 10^{-14} m. As such the minimum uncertainty $(\Delta p)_{min}$ in the momentum of the electron should be

$(\Delta p)_{min} = \left(\dfrac{\hbar}{10^{-14}}\right)$, and

$(\Delta p)_{min}\, c = \left(\dfrac{\hbar c}{10^{-14}}\right)$ = the uncertainty in the kinetic energy of relativistic electron = $19.7 \approx 20$ MeV

This shows that if an electron is confined in a space of the order of the size of the nucleus, the minimum uncertainty in the energy of the electron will be as large as 20 MeV or so. Obviously, the energy of the electron if they are inside the nucleus must be much larger than the uncertainty in it. There is, however, no evidence that electrons of very high energies are present in the nucleus. Experimentally, it has been found that electrons of less than 10 MeV are emitted in beta decay.

1.6.4 Magnitude of Magnetic Dipole Moment of the Nucleus

Nuclei also have magnetic (dipole) moments that are measured in the unit of nuclear magneton. In addition, electrons possess magnetic moments that are much larger and are measured in Bohr magneton. Bohr magneton is almost 1850 times larger than nuclear magneton. If electrons were present in the nucleus then the magnetic moment of the nucleus should also be of the order of Bohr magneton. However, the measured values of nuclear magnetic moments are of the order of nuclear magneton only.

1.6.5 Angular Momentum of the Odd A Nuclei

Proton–electron model of the nucleus gives wrong values for the angular momentum J of odd neutron nuclei. For example, the angular momentum of ^{14}N was very accurately determined to be $1\,\hbar$, but the $P + e$ model predicts a half integer value.

1.6.6 Nuclei that May Emit both β^+ and β^- Particles

The proton–electron model for the nucleus was put forward on the basis that electrons (β^-) are emitted by many radioactive nuclei and therefore they must be present in the nucleus. Some radioactive nuclei such as ^{64}Cu may emit β^+ and also β^-. If both types of beta particles are present in the nucleus of the copper isotope, then these betas will combine together (annihilate) emitting energy.

1.6.7 Beta-Neutrino Pair

In beta decay, the beta particle is always accompanied by neutrino. The magnetic moment of a neutrino is measured to be less than 10^{-7} Bm. As such if both electron and neutrino are present in the nucleus then the magnetic moment of the nucleus should be at least of the order of 1 Bohr magneton. This is contrary to the measured values of the nuclear magnetic moments, which are always of the order of nuclear magneton.

All the above facts indicate that electron cannot be a constituent of the nucleus.

Example 1.5 The mass excess for $^{166}_{66}$Dy nucleus is –67193 μu. Considering the mass excesses for the neutron and the proton as 8650 μu and 7277 μu, respectively, calculate the binding energy per particle of $^{166}_{66}$Dy.

Solution: We know that $1\ \mu u = 1\times10^{-6}$ u; the mass M = mass number A + Mass excess; Hence the mass of $^{166}_{66}\text{Dy} = 166 - 0.067193 = 165.932807\,\text{u}$; and the masses of neutron and proton, respectively, as $M_n = 1 + 0.008650 = 1.008650\,\text{u}; M_p = 1 + 0.007277 = 1.007277\,\text{u}$.

The binding energy of $^{166}_{66}$Dy, denoted by $B = (100\times M_n + 66\times M_p) - \text{Actual mass of } ^{166}_{66}\text{Dy}$

Therefore, B = (167.345282)–(165.932807) u = 1.412475 u = 1.412475 × 931.49 MeV = 1315.70634 MeV

Binding energy per particle $= \dfrac{B}{A}$ = 1315.70634/166 = 7.93 MeV

Ans: The binding energy per particle of $^{166}_{66}$Dy is 7.93 MeV per nucleon.

Exercise p-1.5: List the various quantum properties of a nucleus and discuss two of them in detail.

Exercise p-1.6: Give two examples each of isobars, isotopes, isotones, mirror, even–odd, odd–odd nuclei.

Exercise p-1.7: What can be a possible test to verify if parity is conserved in a process?

Exercise p-1.8: What is isobaric spin and in what respect it is comparable to the nuclear angular momentum J.

Exercise p-1.9: State Pauli's exclusion principal and indicate which property (ies) of the nucleus is (are) influenced by it.

Exercise p-1.10: Using the $\dfrac{B}{A}$ versus A curve, calculate the approximate value of the total binding energy of ^{235}U nucleus.

Exercise p-1.11: The $\dfrac{B}{A}$ decreases with the A for heavy nuclei. What may be the possible reason for the decrease?

Exercise p-1.12: Calculate the value of 1 micro unified mass unit (1 μu) in MeV and kilogram.

Exercise p-1.13: Using the data given in solved example -1.5, calculate the value of the packing fraction for $^{166}_{66}Dy$ in MeV.

Exercise p-1.14: If M_1 and M_2 represent the masses and B_1, B_2 the binding energies of two mirror isobars of mass number A than show that the sum of the mass of a neutron and a proton is equal to

$$\frac{1}{A}[(M_1 + M_2) + (B_1 + B_2)]$$

Exercise p-1.15: A nucleus has a spin $I = \frac{5}{2}$. What is the absolute value of the spin and what are the values of the possible magnetic quantum numbers?

1.7 THE NUCLEUS TODAY

Rutherford's scattering experiments carried out by Geiger and Marsden not only proved the existence of the positively charged heavy nucleus at the centre of each atom but also opened a new method of studying microscopic systems by bombarding them with some high energy particles. Accelerators, machines that may impart large kinetic energies to particles such as electrons, protons, pi, and other mesons were made and more detailed investigations about the structure of the nucleus were done. These studies have revealed that protons and neutrons are themselves made of another type of particles called quark. Quarks are very interesting particles for many reasons. First, they cannot be found in free state. Electrons, protons, and neutrons for example can be produced as a beam in a laboratory but it is not possible to produce a beam of quarks. The reason is that inside a nucleon quarks are very strongly bound together. An idea about the strong binding of quarks in a proton may be obtained from the mass of proton (938 MeV/c^2) and the mass of the three quarks present inside the proton is just (20 MeV/c^2). The difference in the mass is a measure of the binding energy of quarks in proton. Another rather amazing property of quarks is that they have fractional electrical charge. Till recently, it was believed that unit charge e, charge on electron or proton cannot be broken into fractional charges. However, now it is proved that quarks have $\left(-\frac{1}{3}e\right)$ and $\left(\frac{2}{3}e\right)$ charges. Two different types (called flavours) of quarks have been identified. Scientists, however, like to talk about them in pairs of: up quark and down quark. The first member of the pair has an electric charge $\left(+\frac{2}{3}e\right)$ while the second member has an electric charge $\left(-\frac{1}{3}e\right)$. There is another property associated with quarks that is named 'colour charge'. Not only nucleons but all other elementary particles are made up of combinations of different quarks. Quarks are bound together with extremely strong force inside the nucleon; much larger than the force that binds nucleons inside a nucleus. The very strong quark–quark binding force decreases very rapidly with the distance and a relatively weaker component of this force leaks out of the nucleon to provide residual two body force for the binding of nucleons in a nucleus.

In spite of the great advances made in understanding the structure of nucleus during the last 100 years, since the discovery of atomic nucleus by Rutherford, however, the fact remains: even the most fundamental force between nucleons is still not fully understood. This is the beauty of nature—it never reveals itself.

Multiple choice questions

Note: In some of the following questions more than one alternative may be true. Tick all the correct alternatives in such cases for the complete answer

Exercise M-1.1: An odd Z nucleus behaves like a boson, the neutron number of the nucleus must be

(a) even (b) odd (c) either even or odd (d) half integer.

Exercise M-1.2: The quadrupole moment of a nucleus of spin $\frac{3}{2}\hbar$ in state m_j is one fourth of its value with respect to the body fixed axis. The value of m_j is

(a) $\frac{5}{2}$ (b) $\frac{3}{2}$ (c) $\frac{1}{2}$ (d) $-\frac{3}{2}$

Exercise M-1.3: Imagine an excited nucleus which has one neutron and one proton in relative orbital angular momentum state $\ell = 3$. The possible spin quantum number of the excited state may be

(a) 0, 1, 2 (b) 1, 2, 3 (c) 2, 3, 4 (d) 3, 4, 5

Exercise M-1.4: The radius of a nucleus is 6.0 F. Given that the unit nuclear radius is 1.2 F and the nucleus has 73 neutrons, the atomic number of the nucleus is

(a) 35 (b) 45 (c) 52 (d) 63

Exercise M-1.5: Natural uranium ore contain 0.72 % ^{235}U isotope and 99.28% ^{238}U isotope. The approximate number of ^{235U} nuclei in 1 g of ore is

(a) 18.2×10^{16} (b) 18.2×10^{18} (c) 18.2×10^{20} (d) 18.2×10^{22}

2 Basic Properties of the Nucleus and Their Determination

As the nucleus of the atom is very small in size, it is not possible to directly see it even with the microscope of highest amplification. The properties of the nucleus are, therefore, determined by indirect means. Another problem is that one cannot isolate a single nucleus for study. There are always a large number of nuclei in a sample. Further, one mostly deals with atoms rather than the nuclei as it is almost impossible to remove all the electrons of all the atoms of a sample, more so in the case of heavy atoms that contain many electrons. Samples used for the determination of the properties of nuclei generally contain a large number of neutral atoms of the material. In some studies, ionized gaseous samples may also be used. In view of the above-mentioned limitations, designing, carrying out, and interpreting the results of experiments in nuclear physics require extreme care, experimental acumen, and sound understanding of the subject.

2.1 DETERMINATION OF NUCLEAR CHARGE

Today, it is known that the atomic number of an element in the periodic table gives the number of protons and the units of positive charge on the nucleus of the atom. It was, however, Van den Broek who in 1913 suggested that the charge on the nucleus is equal to the atomic number. To maintain the neutrality of the atom, it was necessary to take the number of electrons in an atom equal to the units of positive charge on the nucleus. Adapting the suggestion of Broek, Bohr developed the quantum theory of the structure of atom and successfully explained the origin of the atomic spectra.

The following methods may be used to determine the charge on the nucleus of an atom.

1. X-ray attenuation
2. Moseley's law for the frequency of K-series X-ray lines
3. Scattering of α particles

2.1.1 X-ray Attenuation

In this method, the number of electrons in an atom is determined by measuring the attenuation of X-rays when they pass through the foil of the material. As the number of electrons in an atom is equal to the units of positive charge in the nucleus, the nuclear charge is determined indirectly.

When X-rays fall on a foil of a material, a part of the intensity of the incident X-rays get scattered in all directions. J.J. Thomson studied the scattering of X-rays in detail and explained that X-rays falling on the electron of an atom makes the electron oscillate. Oscillating electrons re-emit X-rays of same frequency as the original X-ray, but in all directions. As such a part of the incident X-ray energy is lost from the incident beam. Thomson showed that the rate at which the energy is scattered by each electron is given by

$$\text{Rate of energy scattered per electron} = \sigma_e I$$

where I is the intensity of the incident X-ray beam and $\sigma_e = (8\pi/3)(e^2/m_e c^2)$ is called the Thomson electronic cross-section and has the dimensions of area. The σ_e may be considered as the effective area projected by each electron to the incident X-rays such that X-rays falling on this area get definitely scattered. Here, m_e is the rest mass of electron. It may be noted that σ_e has a constant value (= 0.666 barn = 2/3 b) independent of the frequency of the incident X-ray. This is true only when the energy of the incident X-ray is more than the binding energy of the electron and less than its rest mass energy (0.511 MeV). Let us consider a foil of thickness Δx of some material having N atoms per unit volume (N is called the number density of the material), each atom having Z electrons. The number of electrons per unit area of the foil becomes $N_{\text{area}} = NZ\Delta x$, which is the number of scattering centres per unit area of the foil. If a collimated beam of X-ray of intensity I falls on the foil, N_{area} scattering centres will scatter out for instance ΔI amount of energy from the incident beam. ΔI energy will not pass on to the other side of the foil and will be lost to the surroundings due to the scattering by electrons. As ΔI is a loss, it is written with a negative sign.

$$\Delta I = -N_{\text{area}}\sigma_e I = -NZ\Delta x \sigma_e I$$

or

$$\Delta I/I = -(NZ\sigma_e)\Delta x = -\mu\Delta x \tag{2.1}$$

where $\mu(=NZ\sigma_e)$, called the linear attenuation (or absorption) coefficient, has the dimension of length inverse (L^{-1}).

In the case of limit $\Delta I \rightarrow 0$ and limit $\Delta x \rightarrow 0$, they may be replaced by their derivatives dl and dx and Eq. (2.1) may be written as

$$\frac{dI}{I} = -\mu dx \tag{2.2}$$

On integration, Eq. (2.2) gives the intensity of the X-ray beam after traversing a thickness x of the foil as

$$I = I_0 e^{-\mu x} \tag{2.3}$$

where I_0 is the intensity of the beam when $x = 0$, i.e. the incident intensity. Eq. (2.3) states that the intensity of the beam decreases exponentially with the thickness of the foil. If ρ is the density of the material, then Eq. (2.3) may be written as

$$I = I_0 e^{-\left\{\frac{\mu}{\rho}(\rho x)\right\}} \tag{2.4}$$

The quantity μ/ρ with dimensions $M^{-1} L^2$ is called mass attenuation coefficient and is denoted by μ_m and (ρx) is a measure of the thickness of the foil in units of mass per unit area and may be denoted by x_m. Rewriting Eq. (2.4), we get

$$I = I_0 e^{-\{\mu_m x_m\}} \tag{2.5}$$

On taking log of both sides of Eqs (2.3) and (2.5), we get

$$\ln I = \ln I_0 - \mu x \quad \text{or} \quad \ln I = \ln I_0 - \mu_m x_m \tag{2.6}$$

Eq. (2.6) shows that a graph between $\ln I$ and x will be a straight line with slope μ or (μ_m).

In actual experiment, a collimated beam of X-rays from a source is made to fall normally on a foil of known thickness of the material (see figure 2.1). An X-ray detector is placed on the other side of the foil in line with the X-ray beam. The detector readings I_1, I_2, I_3 … for the intensity of the transmitted X-rays are taken for different thicknesses d_1, d_2, d_3 … of the foil. A graph is then plotted between log I and thickness d of the foil (figure 2.2). The data points on the graph are fitted with a straight line. The slope of the line gives the value of the attenuation coefficient. If the thickness of foil is taken in units of length, the slope of the straight line gives the linear absorption coefficient or the mass attenuation coefficient if thickness is in unit of mass per unit area. As the attenuation coefficient is related to Z, (see Eq. (2.1)) the number of electrons per atom and hence the charge on the nucleus of the atoms of the foil can be determined.

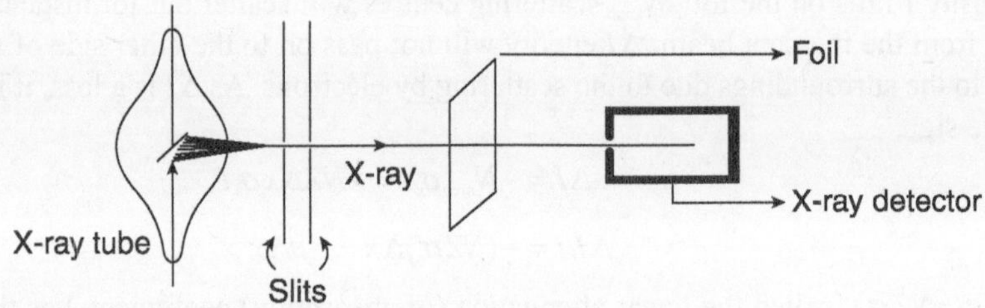

Figure 2.1 *Setup for measuring the attenuation of X-rays*

Exercise p-2.1: The linear attenuation coefficient of a lead foil (density 11.35 g cm^{-3}) for gamma rays of 662 keV is 1.15 cm^{-1}. What is the value of the mass attenuation coefficient of lead for the same gamma rays in SI unit?

Exercise p-2.2: Why is it better to take several readings for the X-ray intensities for different foil thicknesses and to draw a graph, instead of taking two readings, one without a foil to get I_0 and another for a known thickness x to get I and use Eq. (2.3)? In an experiment, the linear attenuation coefficient for a metal foil for X-rays of a given frequency was found to be 0.216 cm^{-1}. The mass attenuation coefficient for the same X-rays and for the same foil was 0.80 cm^2/g. Calculate the density of the material of the foil in SI units. Can you guess the material of the foil?

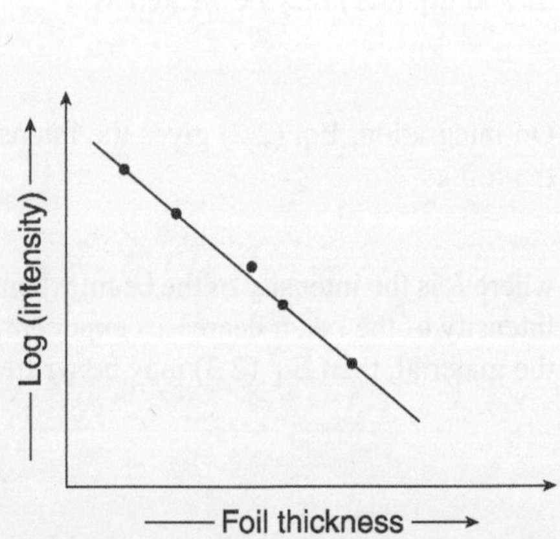

Figure 2.2 *Log (intensity) versus foil thickness graph*

2.1.2 Moseley's Law for the Frequency of the Characteristic X-ray K-lines

X-rays are emitted when high-energy electrons produced in the electron gun of an X-ray tube fall on the anode of the tube. A typical X-ray energy spectrum is shown in figure 2.3. The broad gross part of the spectrum is called the continuous spectrum and depends essentially on the kinetic energy and the intensity of the accelerated electrons. Continuous spectrum is produced due to the deceleration of electrons on hitting the anode. Superimposed on the continuous spectra are sharp spectral lines. The frequency (or the energy) of these lines depends on the material of the anode on which the electrons hit. These sharp lines constitute the characteristic X-ray spectrum, specific to the material of the anode. The characteristic lines are produced due to electronic transitions in the atoms of the anode material. High-energy incident electrons create vacancies at various electron levels and electrons from higher levels fall back emitting characteristic X-ray lines. The K X-ray lines are emitted when electrons from higher levels jump to the vacancies in the *K*th level.

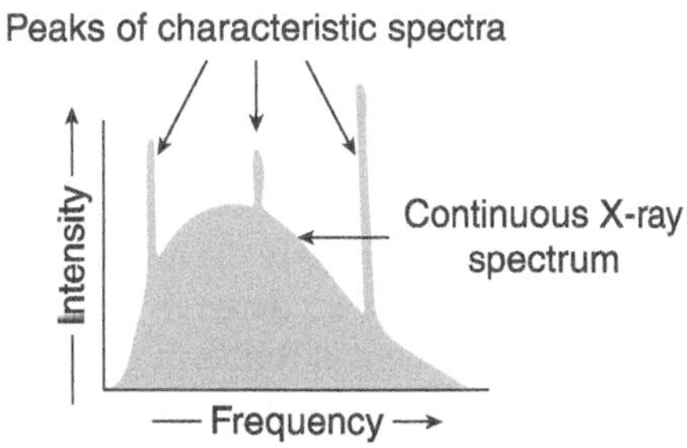

Figure 2.3 *Typical X-ray spectrum*

Moseley in 1913 measured the frequency ν of X-ray K lines for several elements and found the following empirical relationship between the frequency and the atomic number Z of the element.

$$\nu = K(Z-1)^2, \text{ where } K \text{ is a constant.}$$

He extended his studies further and found a general relation

$$\nu = K_1(Z-K_2)^2$$

The constants K_1 and K_2 have different values for different spectral lines. Later, it was argued that the constant K_2, which has a value of 1 for K-lines, arises because of the screening of the nuclear charge by electrons.

Moseley's determination of this relationship provided a means to arrange the elements in the periodic table according to their atomic number – earlier elements were arranged in the periodic table according to their masses.

Thus, using the empirical relation of Moseley, the atomic number Z or the charge on the nucleus of the atom can be found.

Exercise p-2.3: In your opinion, the value of K_2 will be larger or smaller than 1 for L X-rays? Give reason(s) in support of your answer.

2.1.3 Scattering of α Particles

As has already been pointed out, scattering of α particles by foils of different materials not only proved the presence of the positively charged and massive nucleus at the centre of each atom but has also opened a new and very accurate method of finding out the atomic number Z of the nucleus.

Rutherford, on the basis of his nuclear atom, has shown that the differential scattering cross-section $(d\sigma/d\Omega)$ (θ), which gives the cross-section for the scattering of particles within a solid angle $d\Omega$ at scattering angle θ, may be given as

$$\frac{d\sigma}{d\Omega}(\theta) = \frac{Z^2z^2e^4}{16E^2}\frac{1}{(\sin\theta/2)^4} \tag{2.7a}$$

where Ze is the charge on the nucleus of the atoms of the foil, ze the charge on the particles of the incident beam (=$2e$ in case of α particles), E the kinetic energy of incident beam particles, and θ the scattering angle. From the experimental point of view, the differential scattering cross-section may be written as

$$\frac{d\sigma}{d\Omega}(\theta) = \frac{\text{Number of particles scattered at angle } \theta \text{ with in a solid angle } d\Omega}{\text{Number of scattering centres} \times \text{incident flux}} \tag{2.7b}$$

In the experimental set up used by Geiger and Marsden, a beam of α particles hit a foil normally; it was difficult to measure the differential cross-section. It is because very small number of incident α particles are scattered in a small solid angle at a given value of the scattering angle θ. Chadwick, a student of Rutherford, designed an experimental setup that increased the intensity of scattered α particle in a given solid angle at a given scattering angle by many folds. Chadwick's experimental arrangement is shown in figure 2.4.

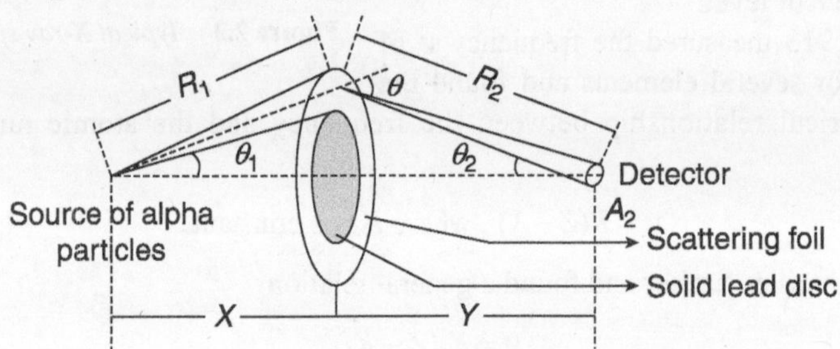

Figure 2.4 *Experimental setup of Chadwick for measuring differential scattering cross-section*

In this setup, an annular ring shaped foil of the scattering material was used. The direct beam from the α particle source was totally blocked by a thick disc of lead at the centre of the annular ring, and thus the detector was shielded from the direct beam. In this geometry of the experiment, only those α particles that are scattered in a solid angle dΩ at a mean angle θ from all parts of the annular ring reach the detector. The number of scattering centres can be easily calculated from the known mass of the annular foil, the incident flux from the intensity of the α source, and the number of particles scattered per unit time can be obtained from the detector readings. As such the differential cross-section can be directly measured. Using the experimentally measured value of the scattering cross-section in Eq. (2.6), the charge Ze or the atomic number of the foil material can be determined. Chadwick very accurately measured the absolute charges (atomic numbers) of three elements platinum, silver, and copper using α scattering.

Solved example S-2.1

In an experiment carried out using Chadwick setup, a beam of α particles falls normally on annular silver ($^{108}_{47}$Ag) foil. A counter placed behind the foil and shielded from the direct α particles records the scattered α particles. What will happen to the count rate of the counter if the silver foil is replaced by a gold ($^{179}_{79}$Au) foil having same mass and area as the silver foil?

Solution. From Eqs. (2.7a) and (2.7b), it follows that
Detector count rate α number of particles scattered per unit time in direction θ within solid angle $d\Omega$

$$\propto \frac{d\sigma}{d\Omega}(\theta) \times \text{Number of scattering centres} \times \text{ incident flux}$$

$$\propto \frac{Z^2 z^2 e^4}{16E^2} \frac{1}{(\sin\theta/2)^4} \times \text{Number of scattering centres} \times \text{ incident flux} \qquad \text{(A)}$$

The number of scattering centre $= \dfrac{\text{mass of the foil} \times \text{Avogadro's number}}{\text{mass number } A}$, substituting this in Eq. (A), we get

$$\text{Count rate} \propto \frac{Z^2 z^2 e^4}{16E^2} \frac{1}{(\sin\theta/2)^4} \times \frac{\text{mass of the foil} \times \text{Avogadro's number}}{\text{mass number } A} \times \text{incident flux} \qquad \text{(B)}$$

In case of silver and the gold foils, all other factors except their charge Z and mass number A are identical; hence,

$$\frac{\text{Count rate for gold}}{\text{Count rate for silver}} = \frac{\dfrac{Z^2_{\text{gold}}}{A_{\text{gold}}}}{\dfrac{Z^2_{\text{sil}}}{A_{\text{sil}}}} = \frac{Z^2_{\text{gold}} A_{\text{sil}}}{Z^2_{\text{sil}} x A_{\text{gold}}} = \frac{79^2 \times 108}{47^2 \times 179} \qquad \text{(C)}$$

From Eq. (C) we get,

$$\frac{\text{Count rate for gold}}{\text{Count rate for silver}} = 1.70$$

Ans: On replacing the silver foil by the gold foil of same mass and area, the count rate will increase by a factor of 1.70.

Exercise p-2.4: Determine the average solid angle subtended by the detector of area A_2 on the annular ring in terms of R_2 and θ. (see fig. 2.4)

Exercise p-2.5: If the mass of the annular scattering ring of a material of atomic weight X is m microgram in Chadwick's experiment, calculate the number of scattering centres in the ring. Take Avogadro's number as 6.0×10^{26} atoms/kilo mole.

Exercise p-2.6: What is the advantage of using an annular shaped ring in Chadwick's setup?

2.2 DETERMINATION OF NUCLEAR MASS

For most of the atoms, except for the very light ones, it is not possible to remove all the atomic electrons; therefore, it is the mass of the atomic or molecular ion instead of the mass of the nucleus that is generally determined by experiments. The mass of the atom of atomic number Z and atomic mass number A, denoted by $M_A(Z, A)$, is related to the mass of the nucleus $M_N(Z, A)$ by the relation

$$M_N(Z, A) = M_A(Z, A) - (ZM_e + B_{ez})$$

where M_e is the mass of an electron ($= 9.1 \times 10^{-31}$ kg, or 511 keV) and B_{ez} the binding energy of Z electrons in the atom. Although every electron in the atom has a different value for the binding energy, but an estimate of the total binding energy of Z electrons in an atom may be made from the formula $B_{ez} = 15.73\ Z^{7/3}$ based on Thomas–Fermi model.

Atomic and nuclear masses are generally measured in a unit called 'unified mass unit', which is denoted by 'u'. It is defined in terms of the mass of the atom of ^{12}C.

$$1\ \text{u} = 1/12\ (\text{the mass of } ^{12}\text{C atom}) = 1.66043 \times 10^{-27}\ \text{kg} = 1.492232 \times 10^{-10}\ \text{J}$$

$$= 931.480\ \text{MeV}$$

Earlier another unit called 'atomic mass unit' and denoted by amu was used, which is based on the mass of ^{16}O.

Accurate determination of nuclear masses is very important from the point of view of identifying different isotopes of elements, isolating medically or agriculturally important isotopes and also for determining the amount of rare isotopes in archaeological samples for the purpose of finding their age.

Out of the several possible techniques, the following techniques are frequently used for determining the nuclear masses.

1. Mass spectroscopy
2. Accelerator mass spectrometry (AMS)
3. Nuclear mass from reaction data
4. Weizsacker's semi-empirical mass formula
5. Penning trap

We shall, however, discuss in detail only the mass spectroscopic method and the method of mass determination from reaction data, which are most widely used for the determination of masses. Weizsacker's semi-empirical formula will be discussed in detail elsewhere in the book.

2.2.1 Mass Spectroscopy

A block diagram of a mass spectroscopy set up is shown in figure 2.5. There are three basic components of the system: the ion source, the filter system, and the detector. All the three components of the system are suitably coupled and are kept under high vacuum.

2.2.1.1 The ion source

An ion source produces positive ions of the atoms or molecules of the mass of interest. Several types of ion sources are available depending on the type of the material to be studied. In some

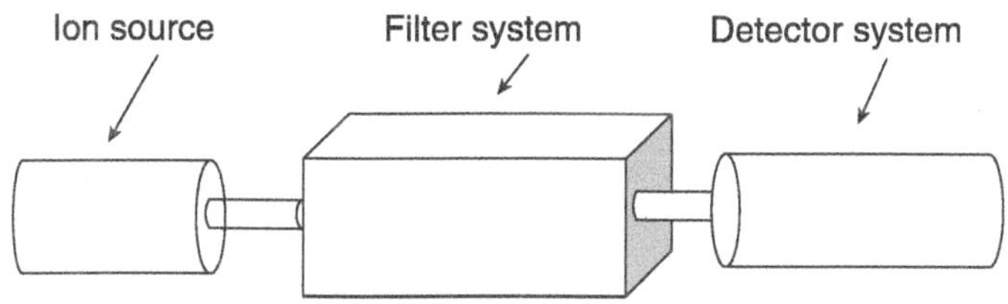

Figure 2.5 *Block diagram of a mass spectroscopy setup*

cases, just heating the substance produces its ions. In case of some gases, applying high frequency radio waves produces ions of the gas atoms/molecules. Small electrical discharge tubes with high-energy electron gun for bombarding the small quantities of the material have also been used as ion source. High-energy electrons impinging on the material vaporize it and produce ions. Most of the ion sources also use a hollow metallic capillary, which is kept at a negative potential to attract positive ions for extraction from the ion source. An ion source provides a divergent beam of species of ions with different velocities V, masses M, and positive charges q. It may, however, be noted that the charge on each ion $q = ne$, where n is an integer and e is the charge on a proton.

2.2.1.2 The filter system

The second component of the mass spectrometry is the filter system. A filter system consists generally of two filters that may separate the ions according to their energy, velocity, momentum, or angular velocity and so on as required. Combinations of filters are generally used to disperse ions of different charge to mass ratios at different positions for collection. It is also possible to collect ions of different charge to mass ratio at a fixed point by changing the settings of the filter components. If ions of different charge to mass ratio are simultaneously collected at different positions, the instrument is called a mass spectrograph. If ions of different charge to mass ratio are all collected at the same position, one after the other by changing the settings of the filter parameters, the instrument is called a mass spectrometer. Each component of mass spectroscopy is coupled to each other and the total setup is kept under high vacuum of the order of 10^{-5} mm of mercury to avoid the loss of ions by scattering with air molecules. Filters generally employ electric fields and/or magnetic fields of special characteristics. Most of the time magnetic fields are produced from electromagnets. However, permanent magnets of special shapes have also been used.

Energy filter: Two types of energy filters are commonly used: curved cylindrical condenser and high voltage energy filter.

***Curved cylindrical condenser*:** As shown in figure 2.6, a uniform electric field of intensity ξ between the curved plates of the condenser keeps an ion of mass M and charge q, moving with the velocity V in a circular path of radius r_E such that the electric force F_E balances the centripetal force F_{CP}, where

$$F_E = q\xi = ne\xi = F_{cp} = \frac{MV^2}{r_E}$$

The kinetic energy E_{kin} of the ion is then given by

$$E_{kin} = \frac{1}{2}MV^2 = \frac{1}{2}(ne)\xi r_E \quad (2.8)$$

It is clear that only those ions for which the velocity, mass, and charge satisfy Eq. (2.8) will come out of the narrow slit at the end of the condenser. This arrangement selects the energy of the ions coming out of it.

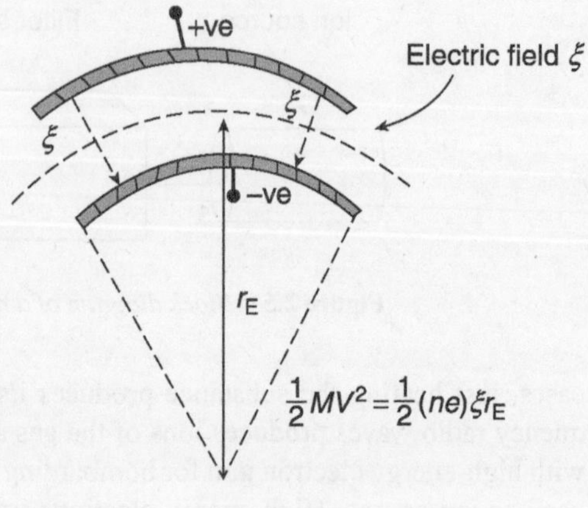

Figure 2.6 *Curved cylindrical condenser as energy filter*

High-voltage energy filter: Ions coming out of the ion source have different energies, but the energy difference is not very large, at most only of few keV. If they are passed through a large potential difference of, for instance U volts, each ion will gain an energy qU. If qU is much larger than the original energy spread of the ions, it can be neglected and all ions may be considered to be of energy qU (figure 2.7).

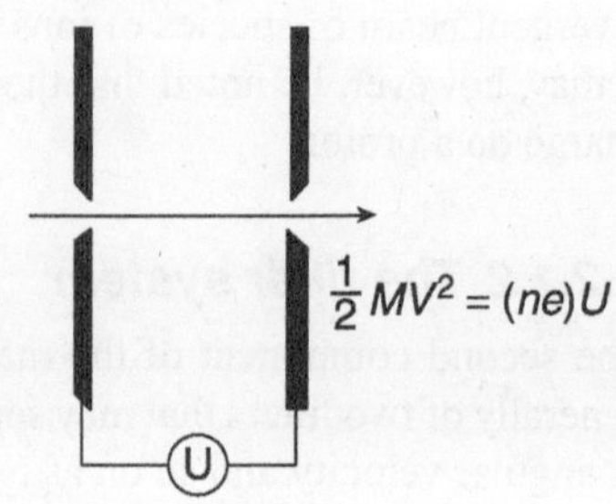

Figure 2.7 *High-voltage energy filter*

Momentum filter: A magnetic sector with uniform magnetic field of strength B works as a momentum filter (see figure 2.8). The magnetic field exerts a force F_M on an ion of charge q, mass M, and velocity V moving perpendicular to the field and constrains it in a circular path of radius r_M such that

$$F_M = BqV = MV^2/r_M$$

or

$$MV = Bqr_M = (ne)Br_M \quad (2.9)$$

Thus, ions with a fixed value of linear momentum, defined by the radius of curvature of the sector and the strength of the magnetic field can pass through the filter.

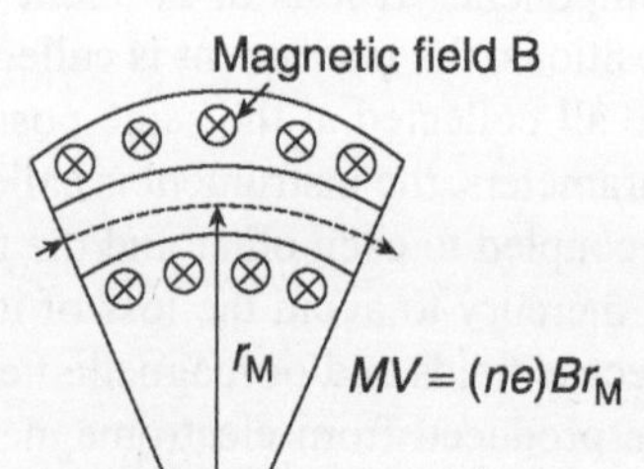

Figure 2.8 *Magnetic momentum filter*

Velocity filter: When charged ions of velocity V pass through mutually perpendicular uniform electric field of strength ξ and magnetic field of strength B, applied simultaneously such that the magnetic force F_M is just counter balanced by the electric force F_E, then

$$F_E = q\xi = F_M = BqV$$

or

$$V = (\xi/B) \quad (2.10)$$

The system of crossed electric and magnetic fields allows all ions with same velocity to pass through it irrespective of their charge. The velocity of the ions coming out of the filter is decided by the ratio of the electric to magnetic field strengths. A velocity filter is shown in figure 2.9.

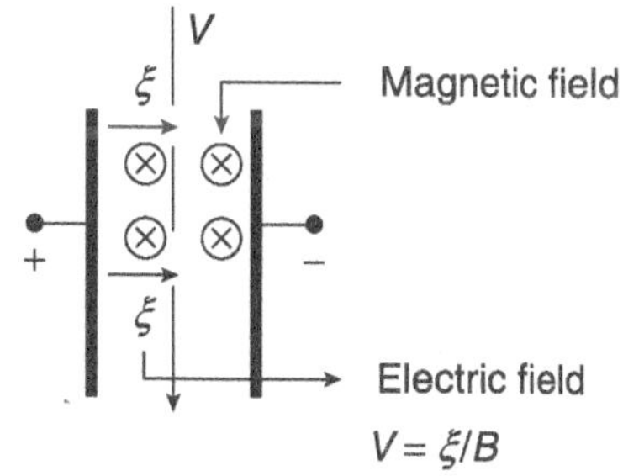

Figure 2.9 *Velocity filter*

Trochoidal filter: In a velocity filter that uses crossed electric and magnetic fields all ions having velocity V defined by Eq. (2.10) ($V = \xi/B$) pass through the filter, but ions with velocities other than V are deflected and follow a path that is trochoidal in a plane normal to the magnetic field. The important point, however, is that all ions with same value of charge to mass ratio irrespective of their velocities come to a focus at the same point. This property may be used to collect and analyse the ion of a given mass.

Angular velocity filter: When charged ions of velocity V enter a region of uniform magnetic field B, normal to the direction of the field, they are constrained to move in circular orbit of radius R such that

$$MV^2/R = BqV;\ \omega(\text{the cyclic frequency}) = V/R = Bq/M \quad (2.11)$$

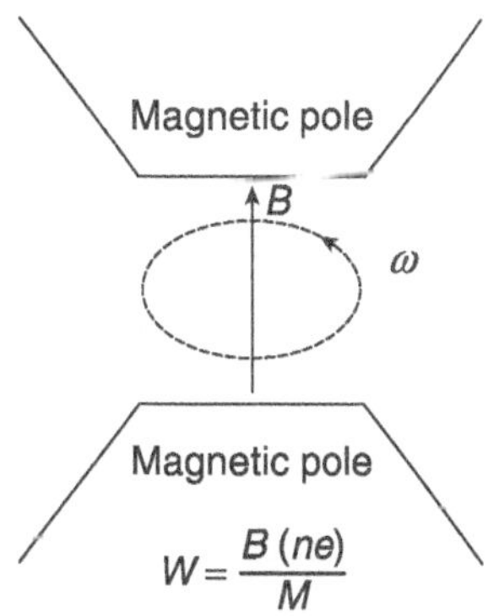

Figure 2.10 *Angular velocity filter*

Eq. (2.11) states that ions with same value of specific charge to mass ratio (q/M) will move with same cyclic frequency (figure 2.10) and will have same time period. This is the condition of resonance cyclotron frequency.

In most of the mass spectrometers/spectroscopes, two of the above-mentioned filters are used one after another to select the desired ions.

2.2.1.3 The detector system

Detector system is an important component of the set up. In most of the earlier mass spectroscopes, photographic plates were used to record the ions separated by the filter system. Different isotopes of a given element produce separate dark lines on the photographic plates. The thickness or darkness of these lines is proportional to the relative abundances of the isotopes. In mass spectrometers, a current-measuring sensitive ammeter, an ionization chamber, or some other electronic detector such as scintillation counter, gas-filled counter, or solid-state detector is used. The pulse height/count rate of the detector gives the relative abundance of different isotopes brought to the detector one after another by changing the settings of the filter elements.

2.2.1.4 Focusing Properties of Electric and Magnetic Sectors

While working with the combination of filters, three types of focusing should be considered: (1) directional focusing, (2) velocity focusing, and (3) mass focusing.

Cylindrical electrostatic energy filter allows ions of a fixed kinetic energy (decided by the strength of the electric field) and moving along the median path to pass through. However, ions that enter at small angles with respect to the median path are also focused on coming out from the electrostatic sector as shown in figure 2.11. In a similar way, a magnetic momentum filter also focuses ions that enter at small divergent angles as shown in figure 2.12. It can be shown that both the electrical and magnetic filters behave as a lens of focal length $f = \dfrac{1}{K\sin(KR\theta)}$ and the first and second principal planes of the lens lie at a distance $D = \dfrac{\tan\left(KR\theta/2\right)}{K}$ from the edges inside the sector.

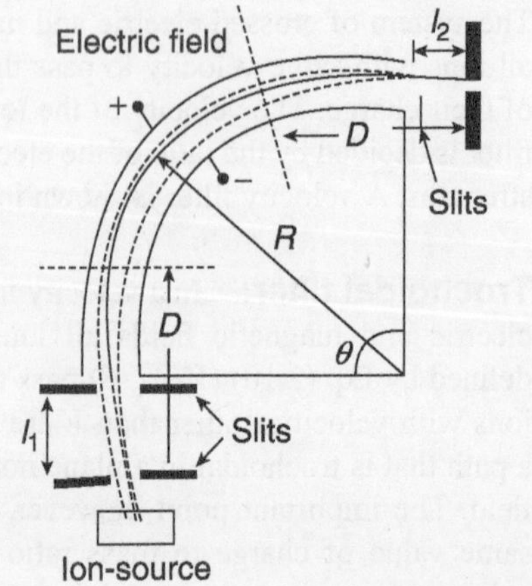

Figure 2.11 *Focusing in electrostatic filter*

Here, $K = \sqrt{2}/R$ for the electrostatic energy filter and $K = 1/R$ for the magnetic momentum filter, R being the radius of curvature of the sector. As both type of filters behave as lens, it is, therefore, possible to arrange the combination of an energy filter and a momentum filter in such a way that the dispersion of ions produced by the one is annulled by the other and the ions are brought to a focus at one point as they exit from the system to reach the detector. This is called the directional focusing. Directional focusing may be achieved for a given pair of filters by suitably adjusting their relative separation.

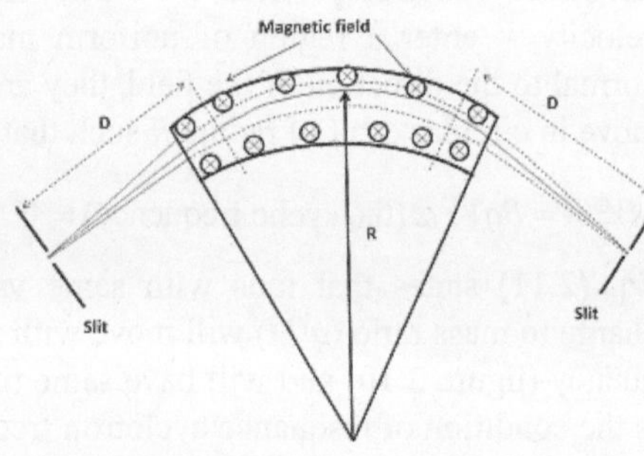

Figure 2.12 *Focusing in magnetic filter*

Apart from directional focusing, there may be two other types of focusing: the velocity focusing and the mass focusing. The question of velocity and mass focusing arises because ions with slightly different velocities and masses emerge from any filter. An energy filter disperses the ions according to the value of mv^2 and a momentum filter disperses them according to their value of mv. The dimensions of electrical and magnetic sectors and distances between slits may be adjusted in such a way that there is no net directional dispersion and no net velocity dispersion at the final slit. Mass dispersion remains. Thus, ions of different masses reach the exit slit at different lateral positions. Instruments in which both directional and velocity focusing are employed are called *double focusing* instruments.

A 127° electrostatic energy filter with a 60° momentum filter completely compensates for the velocity dispersion and provided a linear mass scale in Bainbridge's mass spectrograph shown in figure 2.13. In this instrument, ions having a mass difference of the order of 1% are collected at about 5 mm apart. The mass resolution ($\Delta M/M$) is $\approx 10^{-4}$.

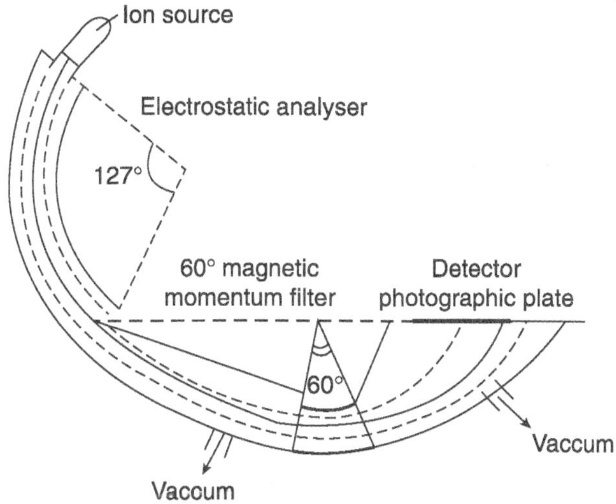

Figure 2.13 *Bainbridge and Jordon mass spectrograph*

Nier and Roberts used a 90° energy filter along with a 60° momentum filter in their mass spectrometer. In this spectrometer, ions coming out from the ion source are accelerated by a potential of about 4 kV before entering the energy filter (figure 2.14). Ions of different charge to mass ratio may be brought to the exit slit either by changing the strength of the magnetic field or the strength of the electrical field. Changing of the electric field strength is easier and also free from problems such as the residual magnetic field because of the hysteresis, faced if magnetic field is changed. They used two ion collectors in their instrument to detect and record the ions reaching the exit slit. By comparing the ion currents in the two ion collectors, the relative abundance of the two nearby isotopes could be determined. Figure 2.15 shows the folded paths of ions with different masses and velocities. The ions passing through the median plane are shown by the central straight line.

Aston's mass spectrograph, the first of the double focusing type built in 1919, was capable of separating ions of 1% mass difference with the accuracy of about 0.1 Dumpster, Nier and others have also developed single focusing mass spectrometers, which employ only a momentum filter. Single focusing instruments are for routine use and do not have high resolution.

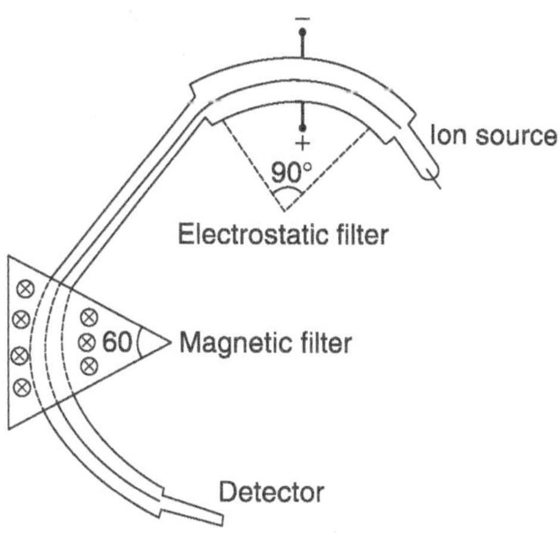

Figure 2.14 *Nier and Roberts mass spectrometer*

2.2.1.5 The mass synchrometer

The angular frequency filter may also be used as a mass spectrometer. It is because the cyclotron frequency $\omega = Bq/M$ for a given B depends on the mass of the ion. If the cyclic frequency of an ion in a known magnetic field is measured, the mass of the ion can be found. An instrument based on this principle has been built at the Brookhaven National Laboratory and is called the mass synchrometer. In the original design, the frequency of the ion was determined from the time of flight of the ion. However, in the new setup it is more accurately determined by applying a radio frequency electric impulse to the ions and studying their phase relations.

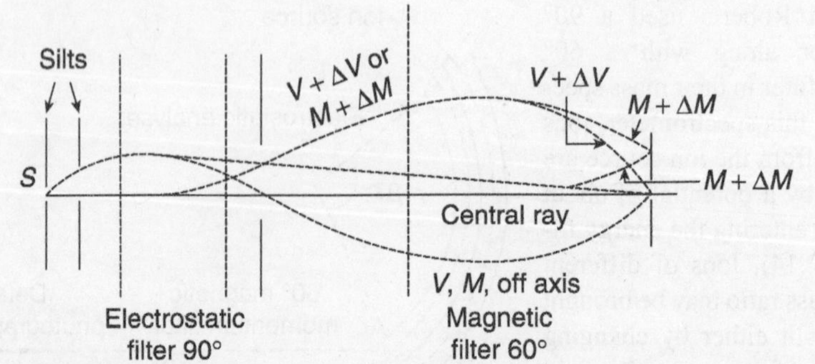

Figure 2.15 *Folded paths of ions with different velocities and masses*

Solved example S-2.2

A narrow beam of singly charged ^{10}B and ^{11}B ions of energy 10.0 keV pass through a slit in to a uniform magnetic field of 2000 Gauss and after a deviation of 180°, the ions are recorded on a photographic plate. Calculate (i) the spatial separation of the images and (ii) the mass resolution of the system. Take the mass of the ion equal to A u, where A is the mass number of the ion.

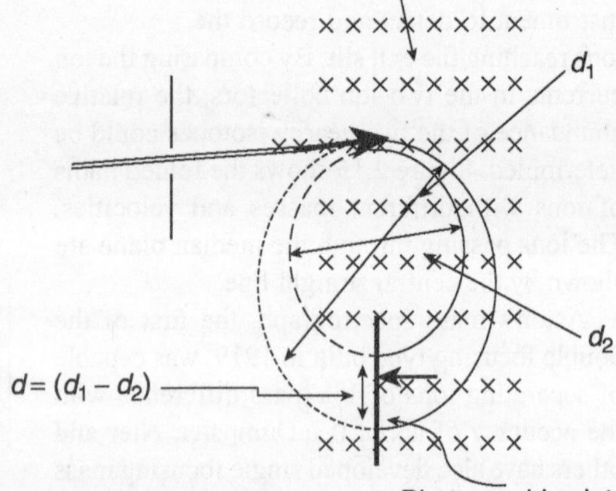

Solution. The experimental arrangement is shown in the figure. It may be noted that the direction of motion of the ions has deviated by 180° from the original direction of motion when they hit the photographic plate. The magnetic induction B (= 2000 Gauss = 0.20 T (Tesla)) that is normal to the velocities of the ions bent in circular paths of radii r_1 and r_2. As is clear from the figure, the spatial separation $d = (d_1 - d_2) = 2(r_1 - r_2) = 2\Delta r$. Let M_1, M_2, V_1, and V_2 denote, respectively, the masses and the velocities of ion-1 (^{10}B) and ion-2 (^{11}B). The kinetic energies of the two ions are equal E = 10 keV = 10 × $10^3 \times 1.6 \times 10^{-19}$ J = 1.6×10^{-15} J. When ions enter the magnetic field they are subjected to the magnetic force $F_m = qBV$, which is balanced by the centripetal force$F_c = MV^2/r$, that is

$$\frac{MV^2}{r} = qBV;\ \text{so}\ r = \frac{MV}{qB} \qquad \text{(A)}$$

Also, $E = \frac{1}{2}\ MV^2 E = E = \frac{1}{2} M V^2$

Therefore, $V = \left(\frac{2E}{M}\right)^{1/2}$ substituting this value of V in Eq. (A), we get

$$r = \frac{M}{qB}\left(\frac{2E}{M}\right)^{1/2} = \frac{(2EM)^{1/2}}{qB} \quad \text{(B)}$$

We now substitute the values of M for the two ions and calculate r_1 and r_2

$(r_1 - r_2) = \frac{(2E)^{1/2}}{qB}\left[\sqrt{M_1} - \sqrt{M_2}\right] = \Delta r$. We put $E = 1.6 \times 10^{-15}$ J; $M_1 = 11, u = 11 \times 1.6 \times 10^{-27}$ kg; $M_2 = 10, u = 10 \times 1.6 \times 10^{-27}$ kg; $q = 1e = 1.6 \times 10^{-19}$ C; $B = 0.2$ T. Note that all quantities are now in SI units. The value of Δr will be in metres.

$$\Delta r = \frac{(2 \times 1.6 \times 10^{-15})^{1/2}}{(1.6 \times 10^{-19} \times 0.2)}\left(\sqrt{1.1 \times 1.6 \times 10^{-26}} - \sqrt{1.6 \times 10^{-26}}\right) = 0.109 \times 10^{-1}\ \text{m} = 1.09\,\text{cm}$$

The spatial separation $d = 2\Delta r = 2.18$ cm

From Eq. (B), we have

$$r^2 = \frac{2E}{(qB)^2}M \quad \text{(C)}$$

Differentiating Eq. (C), we get

$$2r\,\delta r = \frac{2E}{(qB)^2}\delta M$$

Dividing Eq. (D) by Eq. (C), we get

$$2\frac{\delta r}{r} = \frac{\delta M}{M} \quad \text{(D)}$$

Hence, the mass resolution

$$\frac{\delta M}{M} = \frac{2\Delta r}{\text{mean value of } r_1 \text{ and } r_2} = \frac{2.18\ \text{cm}}{22.91\,\text{cm}} = 0.95$$

$$\textit{Percent mass resolution} = \frac{\delta M}{M} \times 100 = 95\%$$

Ans: (i) Spatial separation = 2.18 cm (ii) percent mass resolution = 95%.

2.2.1.6 Doublet method of mass spectroscopy

In a mass spectroscope, the mass-separated ions generally fall on a photographic plate and make a permanent record. Distance on the photographic plate is calibrated in terms of mass using some standard, preferably mono isotopic ions. Similarly, in a mass spectrometer the electric field is calibrated in terms of mass with the help of standard ions. However, to determine the mass of some ion of unknown mass accurately the doublet method is used. In this method, ions of two or more different molecules (or atoms) that have same atomic mass number and one of the molecule containing the desired atom of unknown mass, are simultaneously produced in the ion source and their mass spectrum is measured. Because the ions have same mass number but slightly different masses, their mass spectrum consists of closely spaced as many lines (or peaks) as the number of molecular ions. For example, let us consider the three pairs (doublet) of ions

$(CH_4)^+$ and O^+, both have atomic mass number $(12 + 4 = 16)$, (16)

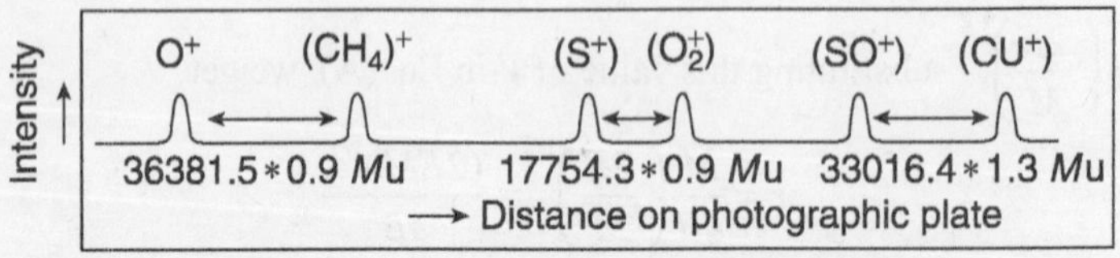

Figure 2.16 *Doublets corresponding to (O, CH_4), (S, O_2), and (SO, C_4)*

$(O_2)^+$ and $(S)^+$, both have atomic mass number ($2 \times 16 = 32$), (32)

$(C_4)^+$ and $(SO)^+$, both have atomic mass number ($4 \times 12 = 48$), ($32 + 16 = 48$)

If all the six ions are simultaneously passed through the mass spectroscope or spectrometer, the spectrum will show three pairs of doublet peaks as shown in figure 2.16.

Since the distance/electric field strength on the *X*-axis is already calibrated in terms of mass, the separations between these peaks, *M*, *N*, and *Q* are known in terms of unified mass unit u, we get

$$(CH_4)^+ - (O)^+ = M\,\text{u} \tag{2.12}$$

$$(O_2)^+ - (S)^+ = N\,\text{u} \tag{2.13}$$

$$(C_4)^+ - (SO)^+ = Q\,\text{u} \tag{2.14}$$

To accurately determine the mass of hydrogen atom, the following simple algebraic operations may be carried out.

Multiply Eq. (2.12) by 3 and add to Eq. (2.13) to get

$$3(CH_4) - 3(O) + 2(O) - (S) = 3M + N$$

or

$$3C + 12H - (O) - (S) = 3M + N \tag{2.15}$$

On subtracting Eq. (2.14) from Eq. (2.15), we obtain

$$12H - C = 3M + N - Q \tag{2.16}$$

Therefore, the mass of one hydrogen atom

$$= 1/12\,[\text{mass of one carbon atom in } u + 3M + N - Q]\,\text{u}$$

$$= 1\,\text{u} + 1/12\,[3M + N - Q]\,\text{u} \tag{2.17}$$

In Eq. (2.17), all quantities on the right hand side are known and therefore the value of the mass of hydrogen atom can be determined.

Solved example S-2.3

The main advantage of doublet method is the much less absolute error in the value of mass determined from this method. This may be clear from the following example in which mass differences for the following doublet pairs were determined as follows:

$$(H_2)^+ - (D_1)^+ = (1.54825 \pm 0.001) \times 10^{-3}\,\text{u} \tag{A1}$$

$$(D_3)^+ - \frac{1}{2}(C_1)^{++} = (42.3065 \pm 0.004) \times 10^{-3}\,\text{u} \tag{A2}$$

In the above-mentioned equations, the subscript after the chemical symbol of the atom denotes the number of atoms in the molecule and the superscripts of positive sign the charge on the ion. In case of carbon, a factor of ½ is included as the ion is doubly charged. If Eq. (A1) is multiplied by three and added to Eq. (A2), we get

$$6\text{H} - \frac{1}{2}\text{C} = 45.9512 \times 10^{-3}\ \text{u or } H = 1.0\ \text{u} + \frac{1}{6}\ (45.9512 \times 10^{-3})\ \text{u}$$

The mass of one hydrogen atom = 1.00782 ± Δ

where Δ is the error in the calculated value of the mass of hydrogen atom and depends on the errors involved in Eqs. (A1) and (A2). From the theory of the propagation of errors $\Delta = 1/6[(3 \times 0.001)^2 + (0.004)^2]^{1/2} \times 10^{-3}u$

$$\approx 0.0008 \times 10^{-3}u$$

and the mass of hydrogen atom = 1.00782 ± 0.0000008 u. It may be seen that the accuracy of the result has increased at least by fourfold.

2.2.2 Accelerator Mass Spectrometry (AMS)

This is a relatively new method of identifying and separating ions of different masses. As the name suggests, in this method, an ion accelerator is used. The accelerator has an ion source that may produce negative ions. It is now possible to have an ion source that may produce negative ions of any element starting from lithium up to uranium. The accelerator is generally a Van de Graff generator with a high voltage of 0.2 MV to as high as up to 15 MV. The negative ions from the ion source are first accelerated to the high-voltage terminal potential for instance V volts. Then, these high-energy negative ions are passed through a 'stripper', which remove the negative charge from the ions and make them positively charged. The strippers are either thin metallic sheets or gases at low pressure. When negative ions pass through the stripper they lose their electrons and become positively charged. The positive ions at the high-voltage terminal of the accelerator can be accelerated once again by the same potential V to the ground potential. In this way, the same potential is used twice to accelerate first the negative ions and then the positive ions. This is called the principle of tandem acceleration. If an ion of charge q is accelerated twice by the potential V it gains an energy = $(q + 1)V$ eV. When the potential V is in million volts the energy gained by the charged ion is in MeV. This pre-acceleration of ions to a high kinetic energy by a tandem accelerator works like an energy filter. The accelerated positive ions are then analysed with the combination of electrostatic energy filter and magnetic momentum filters, though of much bigger dimensions. By suitably adjusting the parameters of the filters, ion of any desired charge to mass ratio may be brought at the exit slit. High-resolution solid-state detectors that may also sense the charge of the ion are used to detect them.

The main advantages of AMS are as follows:

1. The AMS system has the great power of separating rare isotope from an abundant neighbouring one, for example ^{14}C can be distinctly separated from the more abundant ^{12}C (98.89%). This property is called the abundance sensitivity.
2. The molecular isobars are completely eliminated.
3. It is possible to detect long-lived radio-isotopes of very low abundances of the order of 10^{-12} to 10^{-18} such as ^{36}Cl, ^{26}Al, and ^{14}C. The AMS out performs the other competing techniques of decay counting for all radio-isotopes with long half-lives.

2.2.2.1 Applications of AMS

There are several applications of AMS, the most important being its use for carbon dating of archaeological samples to determine the age of the sample by measuring the amount of ^{14}C atoms. ^{3}H, ^{14}C, ^{36}Cl, and ^{129}I isotopes have been used as hydrological tracers. In biomedical applications, ^{41}Ca has been used to measure bone re-absorption in human beings. All these rare isotopes are easily identified and separated by the AMS.

2.2.3 Nuclear Mass from Reaction and Decay Data

2.2.3.1 Nuclear mass from reaction data

Masses of short-lived isotopes that are produced in nuclear reactions are often determined from the experimental data recorded in the reactions. Masses of some nuclear particles that undergo self-decay are determined from the measured kinetic energies and masses of the final decay products.

In a typical nuclear reaction, a nuclear particle or a nucleus '$^{A_a}_{Z_a}a$' hits a target nucleus $^{A_X}_{Z_X}X$. As soon as the incident particle comes within the range of the nuclear field of the target nucleus interactions between the two takes place in a time interval of the order of 10^{-16} s or less. What exactly happens during the interaction time is not exactly known. However, at the end of the reaction a final residual nucleus $^{A_Y}_{Z_Y}Y$ and a lighter nuclear particle $^{A_b}_{Z_b}b$ are emitted. The reaction shown in Eq. (2.18) is in short written as *X(a,b)Y*.

$$^{A_a}_{Z_a}a + ^{A_X}_{Z_X}X = ^{A_Y}_{Z_Y}Y + ^{A_b}_{Z_b}b \tag{2.18}$$

Here, A_a, A_X, A_Y, A_b, Z_a, Z_X, Z_Y, and Z_b are, respectively, the atomic mass numbers and atomic numbers of a, X, Y, and b. The left hand side in Eq. (2.18) represent the entrance channel and the right hand side the exit channel. It is assumed that a, X, Y, and b are all in their ground states. Let us represent the masses and the kinetic energies of the particles, respectively, by M_a, E_a; M_x, E_x; M_y, E_y; and M_b, E_b. In nuclear reactions, the sum of mass energy and kinetic energy is conserved, therefore,

$$[M_x + M_a]c^2 + E_a + E_x = [M_y + M_b]c^2 + E_y + E_b \tag{2.19}$$

Or

$$[(M_x + M_a) - (M_y + M_b)]c^2 = [(E_y + E_b) - (E_a + E_x)] \equiv Q \tag{2.20}$$

Or

$$\begin{aligned} Q &= \text{(initial total mass energy − final total mass energy)} \\ &= \text{(final total kinetic energy − initial total kinetic energy)} \end{aligned} \tag{2.21}$$

Eqs. (2.20) and/or (2.21) define the Q value of a reaction. If the difference in the initial and final mass energies is positive, the reaction has a positive Q value and in principle can initiate with zero kinetic energy in the entrance channel. Further, for positive Q value reactions the total kinetic energy in the exit channel is more than the total kinetic energy in the entrance channel. In such reactions, a part of the entrance channel mass is converted into kinetic energy. It may, however, be kept in mind that in positive Q value reactions some kinetic energy may have to be supplied so that the target and the incident particles overcome the Coulomb barrier and reach close enough to be in the range of nuclear forces to initiate nuclear reaction. Positive Q value reactions are called 'exoergic reactions' as kinetic energy is released in these reactions.

In case the total mass in the exit channel is more than the total mass in the incident channel, the Q value of the reaction is negative. The reaction cannot get initiated unless excess kinetic energy is supplied in the entrance channel to create mass. Negative Q value reactions are called 'endoergic' or threshold reactions, as a minimum kinetic energy called the threshold energy, slightly more than the magnitude $|Q|$ of the Q value has to be supplied to start the reaction. The threshold energy $>|Q|$, because a part of this energy ($=|Q|$) goes in creating the excess mass and the remaining part provides kinetic energy to the target to recoil that is necessary for the conservation of linear momentum.

Q value for a reaction can be determined experimentally from the kinetic energies of the outgoing reaction products in case of positive Q value reactions and from the measurement of the threshold energy in negative Q value reactions. If the value and the masses of three nuclear particles/nuclei out of the four (a, X, Y, and b) are known, the mass of the fourth particle/nucleus can be determined. It may once again be emphasized that whenever one talks about the masses of the nuclei in Eq. (2.19) generally it refers to the masses of the neutral atoms, as the masses of electrons cancel out on the two sides of the equation and the difference in the binding energies of electrons is neglected being too small.

Solved example S-2.4

The Q value of the exoergic reaction shown below was determined from the measured values of the kinetic energies of the outgoing neutron and the α particle and is found to be 17.3 MeV.

$$^{2}_{1}\text{H} + {}^{3}_{1}\text{H} = {}^{4}_{2}\text{He} + {}^{1}_{0}\text{n} + 17.3 \text{ MeV}$$

The masses of $^{2}_{1}$H, $^{3}_{1}$H, and $^{4}_{2}$He are, respectively, 2.014102 u, 3.016050 u, and 4.002603 u. Using the above data, determine the mass of neutron.

Solution. The mass of the neutron may be calculated as

$$M_n = [M_{2H} + M_{3H} - M_{4He} - Q] = 1.008869 \text{ u}$$

It may be noted that before substituting the value of Q in the above-mentioned equation it should be converted from MeV to the unified mass unit u.

2.2.3.2 Nuclear mass from decay data

Many nuclei decay spontaneously. The spontaneous decay is also a type of nuclear reaction with the only difference that in the entrance channel there is only one nucleus, generally at rest and in the exit channel there may be two or three particles including gamma quanta. Instead of the Q value one talks of the decay energy in the case of spontaneous decay. The mass of neutron cannot be determined using spectrographs or spectrometers as it is electrically neutral. However, spontaneous decay of neutron into proton, β^- and antineutrino may be used to determine its mass. As a matter of fact the mass difference ($^{1}_{0}$n – $^{1}_{1}$H) is an important constant often required for calculations of masses of other nuclei. From the neutron decay, we get

$$M_n - M_H = ({}^{1}_{0}\text{n} - {}^{1}_{1}\text{H}) = Q_{\beta-} \text{ or decay energy for } \beta^- \text{ decay}$$

$$= \text{Maximum kinetic energy of } \beta^- \text{ (antineutrino has zero kinetic energy)}$$

The end point energy or the maximum kinetic energy of β^- particles may be accurately determined from the measured β particle energy spectrum and the value of the mass difference of neutron and hydrogen atom has been determined as 0.782 ± 0.013 MeV.

In nuclear experiments, the kinetic energies can be measured to the accuracy (ΔE_{kin}) of the order of 5 keV. This corresponds to a mass uncertainty of

$$\Delta M = 5 \times 10^{-3} \text{ MeV} = 0.005/931 \text{ u} = 5 \times 10^{-6} \text{ u}$$

Solved example S-2.5

A π^+ meson at rest decays into a μ^+ muon and a muonic neutrino ν_μ according to the following equation;

$$\pi^+ = \mu^+ + \nu_\mu$$

Using the following data, calculate (a) the rest mass of π^+ meson both in MeV and u units, (b) the kinetic energies of emitted μ^+ and ν_μ.

Data: (i) rest mass of $\mu^+ = 105.66$ MeV;
(ii) Q-value of the reaction 33.91 MeV;
(iii) rest mass of neutrino $\nu_\mu = 0$

Solution. Before carrying out the solution, it is necessary to see which type of mechanics should be used. In the present problem, one of the emitted particle is neutrino that has negligible or zero rest mass. No matter whatever kinetic energy it takes, it will always be relativistic. Hence, in this problem, we have to use relativistic mechanics. In general, when very light particles such as electron, muon, and neutrinos are involved in a reaction, relativistic calculations are required. Further, we denote the rest mass of particles by M, their total energy by E^t, their kinetic energy by E^{kin}, and their linear momentum by P, with the appropriate subscripts. Also, we know that the relativistic expression for the total energy of a particle is given,

$$(E^t)^2 = (Pc)^2 + M^2c^4 \tag{A}$$

And,

$$(E^t)^2 = (E^{kin} + Mc^2)^2 \tag{B}$$

Hence

$$(Pc)^2 = (E^{kin} + Mc^2)^2 - M^2c^4 = E^{kin^2} + 2Mc^2E^{kin} \tag{C}$$

(a) In this problem, it is given that the Q value of the reaction is 33.91 MeV.
Therefore, 33.91 MeV $= [M_\pi - (M_\mu + M_\nu)]c^2$

$$= [M_\pi - M_\mu] \text{ MeV} = [M_\pi - 105.66 \text{ MeV}]$$

The rest mass of $\pi^+, M_\pi = 33.91 + 105.66 = 139.57$ MeV

As 931.49 MeV = 1 u = 1.66×10^{-27} kg, $M_\pi = 0.1498\ u = 2.48 \times 10^{-28}$ kg

(b) At rest π^+ decays into μ^+ and ν_μ; therefore, the magnitudes of their linear momentums will be equal, that is

$$|P_\mu| = |P_\nu|$$

Or

$$(P_\mu c)^2 = (P_\nu c)^2 \tag{D}$$

We know that the Q value is equal to the total energy released in the decay, which is sheared between the muon and the neutrino as their kinetic energies. Therefore, the kinetic energy of muon may be obtained by subtracting the energy of the neutrino from the Q value. In case of neutrino, the total energy is equal to the kinetic energy as its rest mass is zero. As such,

$$E_\mu^{kin} = Q - E_\nu^{kin} = Q - P_\nu c \tag{E}$$

Hence,

$$(P_\nu c)^2 = (Q - E_\mu^{kin})^2 = Q^2 + E_\mu^{kin^2} - 2QE_\mu^{kin} \quad \text{(F)}$$

Now, from Eq. (C)

$$(P_\mu c)^2 = E_\mu^{kin^2} + 2M_\mu c^2 E_\mu^{kin} \quad \text{(G)}$$

Equating Eqs. (F) and (G), we get

$$Q^2 + E_\mu^{kin^2} - 2QE_\mu^{kin} = E_\mu^{kin^2} + 2M_\mu c^2 E_\mu^{kin}$$

Or

$$(2M_\mu c^2 + 2Q)E_\mu^{kin} = Q^2$$

Or

$$E_\mu^{kin} = \frac{Q^2}{(2M_\mu c^2 + 2Q)} = \frac{33.91^2}{2(105.66 + 33.91)} = \frac{1149.8881}{279.14} = 4.12 \text{ MeV}$$

Ans: Thus, the kinetic energy of the muon (μ^+) is 4.12 MeV and the kinetic energy of the neutrino is 33.91 − 4.12 = 29.79 MeV.

2.2.3.3 Reaction cycles and chains

A little thinking will tell that the mass of any unknown nucleus can be found in terms of the masses and the Q values of several independent nuclear reactions linked to each other. Masses particularly of light nuclides have been determined in this way by evaluating the linked chains of nuclear reactions. There is good agreement between the masses determined from the mass spectroscopy and from the reaction data giving a direct confirmation of Einstein's mass–energy equivalence and also proving that naturally occurring elements are exactly identical to the one synthesized through nuclear transformations.

Solved example S-2.6

Alpha particles of kinetic energy 4.18 MeV and β^- particles with maximum kinetic energy of 1.25 MeV are emitted in the two-step decay chain as follows. The atomic masses of nuclides $^{234}_{91}$Pa and $^{4}_{2}$He are respectively 234.0423u and 4.0026u. Calculate (i) the disintegration energies for the β and α decays and (ii) the atomic masses of the nuclei $^{234}_{90}$Th and $^{238}_{92}$U.

$$^{238}_{92}\text{U} \rightarrow {}^{234}_{90}\text{Th} + {}^{4}_{2}\text{He} \text{ Kinetic energy} = 4.18 \text{ MeV}$$

$$\downarrow$$

$$^{234}_{91}\text{Pa} + \beta^- \text{ Maximum Kinetic energy} = 1.25 \text{ MeV}$$

Solution. Let us denote the nuclear masses of ^{238}U, ^{234}Th, ^{234}Pa, and ^{4}He, respectively, by M_U, M_{Th}, M_{Pa}, and M_{He} and their linear velocities, respectively, by V_U, V_{Th}, V_{Pa}, and V_{He}.

(i) Q value or the disintegration energy for α decay: When the ^{238}U nucleus decays at rest into an α particle and the thorium-234 nucleus, the two decay products fly in opposite directions with equal linear momentum to conserve the total linear momentum. Therefore,

$$M_{\text{He}}V_{\text{He}} = M_{\text{Th}}V_{\text{Th}} \text{ and hence } V_{Th} = \frac{M_{\text{He}}}{M_{\text{Th}}}V_{\text{He}}$$

Also, the kinetic energy of nucleus ^{234}Th,

$$E_{\text{Th}} = \frac{1}{2}M_{\text{Th}}V_{\text{Th}}^{\;2} = \frac{1}{2}M_{\text{Th}}\left(\frac{M_{\text{He}}}{M_{\text{Th}}}V_{\text{He}}\right)^2 \tag{I}$$

Similarly, the kinetic energy of ^{4}He,

$$E_{\text{He}} = \frac{1}{2}M_{\text{He}}V_{\text{He}}^{\;2} \tag{II}$$

Therefore, from Eqs. (I) and (II), the ratio

$$\frac{E_{\text{Th}}}{E_{\text{He}}} = \frac{M_{\text{He}}}{M_{\text{Th}}} \tag{III}$$

Hence, the kinetic energy of ^{234}Th

$$\text{E}_{\text{Th}} = \frac{M_{\text{He}}}{M_{\text{Th}}}\text{E}_{\text{He}} = \frac{4}{234}\text{E}_{\text{He}} = 0.017 \times 4.18\ \text{MeV} = 0.07\ \text{MeV}$$

It may be noted that the kinetic energies of the emitted nuclei are inversely proportional to their masses.

The disintegration energy or the Q value for α decay $= Q_{He} = E_{He} + E_{Th} = 4.18 + 0.07 = 4.25$ MeV

Disintegration energy or Q value of β^- decay, Q_{β^-}: In case of β decay, the neutrino which has zero rest mass is also emitted with a β particle. However, when β particle is emitted with maximum kinetic energy, the neutrino has zero kinetic energy, and again the β particle and the residual nucleus ^{234}Pa are emitted in opposite directions with equal magnitude of linear momentum. The kinetic energy of ^{234}Pa can be calculated in terms of the maximum kinetic energy of the β particle as has been done in the case of α decay. However, the mass of the β particle is negligible as compared to the mass of ^{234}Pa; hence, the kinetic energy of ^{234}Pa is negligibly small and can be taken to be zero. As such,

$$Q_{\beta^-} = \text{Maximum kinetic energy of } \beta^- = 1.25\ \text{MeV}.$$

(ii) Atomic masses of ^{234}Th and ^{238}U: The disintegration energy or the Q value for β^- decay is equal to the difference of the atomic masses of the parent and daughter nuclei. Therefore,

$$M_{\text{Th}} = M_{\text{Pa}} + Q_{\beta^-} = 234.0423\ \text{u} + 1.25\ \text{MeV} = 234.0404\ \text{u} + \frac{1.25}{931.49}\text{u} = 234.0436\ u$$

Similarly, $$Q_{\text{He}} = M_U c^2 - [M_{\text{Th}} + M_{\text{He}}]c^2,$$

Therefore, $$M_{\text{U}} = \frac{Q_{\text{He}}}{c^2} + [M_{\text{Th}} + M_{\text{He}}]$$
$$= 0.00456 + 234.0436 + 4.0026 = 238.0507\ \text{u}$$

Ans: Atomic mass of ^{238}U = 238.0507 u; atomic mass of ^{234}Th = 234.0436 u; Q_{He} = 4.25 MeV; Q_β = 1.25 MeV.

Exercise p-2.7: What do symbols 'amu' and 'u' stand for? Which of the two is presently used and why?

Exercise p-2.8: What is meant by double focusing in mass spectroscopy?

Exercise p-2.9: In what respect are electric and magnetic sector filters like a lens?

Exercise p-2.10: What is the advantage of doublet method?

Exercise p-2.11: What are the advantages and uses of AMS?

Exercise p-2.12: Why is the threshold energy more than the magnitude of the Q value for negative Q value reactions?

Exercise p-2.13: What is really measured: the mass of the nucleus, the mass of the ion, or the mass of the neutral atom?

Exercise p-2.14: Determine the Q value for the following reaction both in unified mass unit u and MeV.

$$^{14}{}_{7}\mathrm{N} + {}^{2}{}_{1}\mathrm{H} = {}^{15}{}_{7}\mathrm{N} + {}^{1}{}_{1}\mathrm{H} + \mathrm{Q}$$

Using the doublet mass data,

$$(^{14}\mathrm{N}^{1}\mathrm{H})^{+} - (^{15}\mathrm{N})^{+} = 0.01074 \pm 0.0002 \text{ u and}$$

$$2(^{1}\mathrm{H})^{+} - (^{2}\mathrm{H})^{+} = 0.00153 \pm 0.00004 \text{ u}$$

2.3 DETERMINATION OF NUCLEAR RADIUS

There is enough evidence, which suggests that

(a) Most of the nuclei are spherical in shape. Nucleons inside the nucleus are held together by strong nuclear forces and spherical shape having minimum surface area for a given volume is most suitable to provide strong binding. There are some nuclei that are not spherical, which are called deformed nuclei. Their shape is like an ellipsoid, but even in that case the difference between the semi-major and semi-minor axis is not more than 20%.

(b) The distribution of protons or of the positive charge inside the nucleus is almost uniform throughout. Protons are not concentrated more at the surface. This shows that nuclear force between proton and proton is much stronger than the Coulomb repulsion between them.

(c) The density of nuclear matter is nearly constant. This means that the nuclear matter is almost un-compressible. If there is any compressibility it is not more than 20%.

(d) There is no sharp boundary such that nuclear charge and/or nuclear matter are confined within it and just outside the boundary they are zero. As a matter of fact, experiments have revealed that the nuclear charge (or matter) density ρ_0 is highest at the centre, remains constant up to a certain distance from the centre and then falls off with the increasing distance from centre. The distance from the centre up to which the density remains constant is called the 'central volume' and the region where the density undergo change the 'skin' of the nucleus. Big and small nuclei differ in the thickness of their skin but the central volume density is nearly same for all nuclei.

(e) As there is no sharp nuclear boundary, the term radius of the nucleus is rather vague. The distance from the centre of the nucleus where the charge density (or the matter density) falls to half of its central value is generally called the radius of the nucleus. The distance in which the charge density falls from 90% to 10% is called the skin thickness of the nucleus.

(f) In the nucleus, matter density appears to follow the charge density distribution but extends a little further. As such the nuclear charge radius is slightly smaller than the radius for matter distribution (also called the nuclear force radius).

(g) The radius R of a nucleus having A nucleons may be given as $R = r_0 A^{1/3}$, where r_0 is a constant and is called the unit nuclear radius. The unit nuclear radius for charge distribution is $\approx 1.2 \times 10^{-15}$ m or 1.2 F, while the corresponding quantity for nuclear matter or force distribution is ≈ 1.4 F.

In principle, there are nine methods by which the radius of the nucleus can be measured. They are as follows:

1. Elastic scattering of fast electrons by nuclei
2. Coulomb energy difference between mirror isobars
3. Anomalous scattering of α particles
4. Life times of α emitters
5. Isotopic shift in line spectra
6. Decay of mu-mesonic atom
7. Elastic scattering of fast neutrons by nuclei
8. Cross-sections for charged particle induced nuclear reactions
9. Fine structure splitting of X-ray levels in heavy atoms.

However, only the first four important methods are discussed in detail here.

2.3.1 Elastic Scattering of Fast Electrons By Nuclei

We see objects through the light scattered by them. Further, for a detailed study of an object the wavelength of the light should be of the order of the dimensions of the object. In a similar way, in order to study the details of charge distribution in a nucleus the matter waves produced by the fast moving charged particle such as electron may be used. As the size of the nucleus is around 10 F, high energy electrons are frequently used to study the charge distribution of nuclei. Another advantage of using electron is that it does not interact with the nuclear field of the

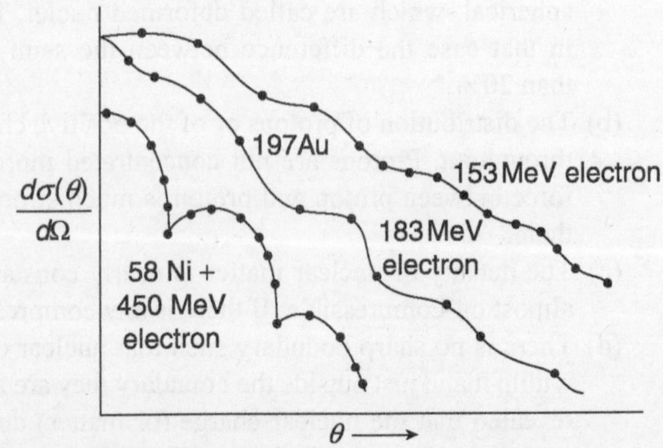

Figure 2.17 *Differential elastic scattering cross section versus scattering angle*

nucleus and is scattered only by the Coulomb field of the nucleus. Very precise experiments on the elastic scattering of high-energy electrons have been done by Hofstadter and co-workers. In a typical experiment, 183 MeV and 153 MeV electrons from a linear accelerator were made to fall on a thin gold foil. The electrons scattered by the foil were analysed for the energy to identify the elastically scattered electrons. The intensity of only elastically scattered electrons at different scattering angles was recorded. A typical graph of differential scattering cross-section against the angle of scattering is shown in figure 2.17. Results of some other electron scattering experiments with higher electron energies are also shown in the figure. Assuming that the target gold foil in the experiment contains a single 2D layer of nuclei of gold and that each nucleus is a hard disc of diameter 'd', the foil behaves as a 2D grating of discs. If the light of wavelength λ is incident on this 2D grating, it will produce a diffraction pattern shown in figure 2.18 with the first minima at 1.22 λ/d. The diffraction pattern actually produced by the elastically scattered electrons is shown in figure 2.17, which resembles the diffraction pattern of figure 2.18, but the minima's are not so pronounced. It is because the nuclei are not like hard disc, they are not opaque to the incident electrons and also the boundary of the nucleus is not sharp as that of a disc.

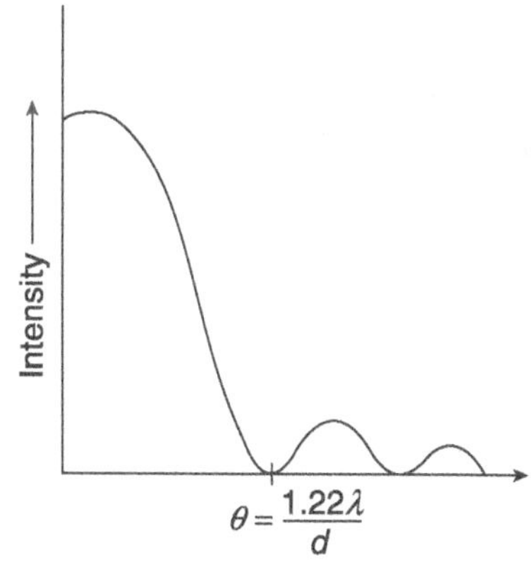

Figure 2.18 *Diffraction pattern of a 2D grating made of opaque discs*

Elastic scattering of electrons by the target nuclei of gold is comparable to the elastic scattering of α particles in Rutherford's experiment. The main difference is that in the case of electron scattering the Coulomb force is of attraction while in α scattering the Coulomb force was of repulsion. As such the basic concepts of Coulomb scattering developed in the study of Rutherford's scattering may also be applied in the present case of electron scattering. In the case of Rutherford's scattering, it was observed that the Coulomb scattering produces only a change ΔP (in the direction) in the linear momentum of the incident particle. This change in momentum, also called the 'linear momentum transfer' is often represented by q in literature. It is also pointed out in Chapter 1 that the magnitude of the linear momentum transfer ΔP (or q) depends on the angle of scattering as well as on the charge seen by the incident particle. As such, it is possible to deduce information about the charge distribution in the nucleus from the study of the linear momentum transfer distribution of elastically scattered electrons. Figure 2.19 shows a typical distribution (differential cross-section) of linear momentum transfer 'q' obtained in fast electron scattering experiment. As the magnitude of linear momentum transfer is related to the magnitude of the charge sensed by the incident particle (electron), it is possible to deduce the nuclear charge distribution from the measured distribution of linear momentum transfer.

We now give a broad outline of the quantum mechanical treatment of the elastic scattering of electrons by nuclei. For simplicity, let us not consider the spin of the electron and the target nucleus and consider only the elastic scattering of electron by the Coulomb field of the target nucleus. In quantum mechanics, every system is represented by a wave function. Let ψ_i and ψ_f denote the initial and the final wave functions of the electron and V the Coulomb potential of the nucleus. An important parameter that describes the interaction between the incident electron and the Coulomb field is the transition matrix element M_{if} given by the volume integral:

$$M_{if} = \int \psi_i V \psi_f \, d\zeta \tag{2.22}$$

Here, $d\zeta$ represents the volume element. In quantum mechanical description, linear momentum is quantized. Let us assume that the electron has initial linear momentum $k_i\hbar$ and final momentum $k_f\hbar$. It is further assumed that an electron can be represented by a plane wave, which has the wave function $\psi = e^{-ik\cdot r}$. Under these assumptions, the initial and the final electron wave functions may be written as $\psi_i = e^{-ik_i \cdot r}$; $\psi_f = e^{-ik_f \cdot r}$. Substituting these values in Eq. (2.21)

$$M_{if} = \int e^{-ik_i \cdot r} V e^{-ik_f \cdot r} \, d\zeta = \int V e^{iqr} d\zeta \tag{2.23}$$

where $q = k_f - k_i$ is the linear momentum transfer. It may, however, be kept in mind that the scattering is elastic; therefore $|k_i| = |k_f| = k$ but the direction of the final momentum has changed by the scattering angle θ. However,

$$q^2 = (k_i - k_f)^2 = [2k^2 - 2k^2 \cos\theta] = 2k^2[1 - \cos\theta] = 2k^2[1 - \{1 - 2\sin^2\theta/2\}]$$

Or $$q = 2k\sin\theta/2 \tag{2.24}$$

Coming back to Eq. (2.23), the differential elastic scattering cross-section at angle θ, denoted by $d\sigma(\theta)/d\Omega$ is proportional to the square of the matrix element $|M_{if}|^2$ and therefore

$$\frac{d\sigma(\theta)}{d\Omega} \propto |M_{if}|^2 = \left|\int V e^{iqr} d\zeta\right|^2 \tag{2.25}$$

The integral on the right hand side in Eq. (2.25) is nothing but three dimensional Fourier transform of potential V. The Fourier analysis and Fourier transform are standard mathematical operation that one studies in mathematical physics.

Next, let us consider the special case when the nucleus is a point charge, in that case $V = (1/4\pi\varepsilon_0)(Ze^2/r)$ where Ze is the charge on the target nucleus and $(-e)$ is the charge of the electron. Substituting this value of V in Eq. (2.25), we get

$$\left[\frac{d\sigma(\theta)}{d\Omega}\right]_{\text{point nucl.}} \propto |M_{if}|^2_{\text{point nucl.}} = \left(\frac{Ze^2}{4\pi\varepsilon_0}\right)^2 \left|\int \frac{e^{iqr}}{r} d\zeta\right|^2$$

Further simplification gives

$$[d\sigma(\theta)/d\Omega]_{\text{point nucleus}} \propto (Ze^2/\varepsilon_0)^2 (1/q^2) = \left(\frac{Ze^2}{2k\varepsilon_0}\right)^2 \left(\frac{1}{\sin^2\theta/2}\right) \tag{2.26}$$

Next, let us consider the real nucleus, not a point charge but an extended charge distribution with charge density at a distance r' from the centre being represented by $\rho_c(r')$. Now, the potential V' felt by the electron is given by

$$V' = (Ze^2/4\pi\varepsilon_0) \int \rho_c(r')/(|r - r'|) \, d\zeta$$

When the above value of V' is substituted in Eq. (2.23) and after some simplification, we get

$$|M_{if}|^2 = (Ze^2/\varepsilon_0)^2 (1/q^2) \, |\int \rho_c(r') e^{iqr} d\zeta|^2 \tag{2.27}$$

The integral in Eq. (2.27) is the 3D Fourier transform of the charge distribution and is called the *charge form factor* of the nucleus. It is represented by $F(q)$.

$$[d\sigma(\theta)/d\Omega]_{\text{real nucleus}} \propto |M_{if}|^2_{\text{real nucleus}} = (Ze^2/\varepsilon_0)^2(1/q^2) \, |\int (e^{iq\cdot r}/r) \, d\zeta|^2$$

$$= |M_{if}|^2_{\text{point nucleus}} \, |F(q)|^2 \tag{2.28}$$

It may also be observed in Eq. (2.28) that the square of the matrix element for real nucleus is related to the square of the matrix element for point nucleus and the square of the form factor. As such

$$[d\sigma(\theta)/d\Omega]_{\text{real nucleus}} = [d\sigma(\theta)/d\Omega]_{\text{point nucleus}}\,(F(q))^2 \tag{2.29}$$

Using the measured values of the differential cross-section and the calculated value of the differential cross-section for point nucleus, the form factor $F(q)$ can be determined. It may be noted that $F(q)$ depends on the value of momentum transfer q, and, therefore, data from different experiments and at different energies may be reduced to obtain the form factor for same q. Once the form factor is known, the inverse Fourier transform of $F(q)$ gives the charge distribution.

$$\rho_c(r')d\zeta = \text{Inverse Fourier transform } F(q)$$

The nuclear charge distribution obtained from the analysis of the momentum transfer q using the inverse Fourier transform $F(q)$ is shown in figure 2.19.

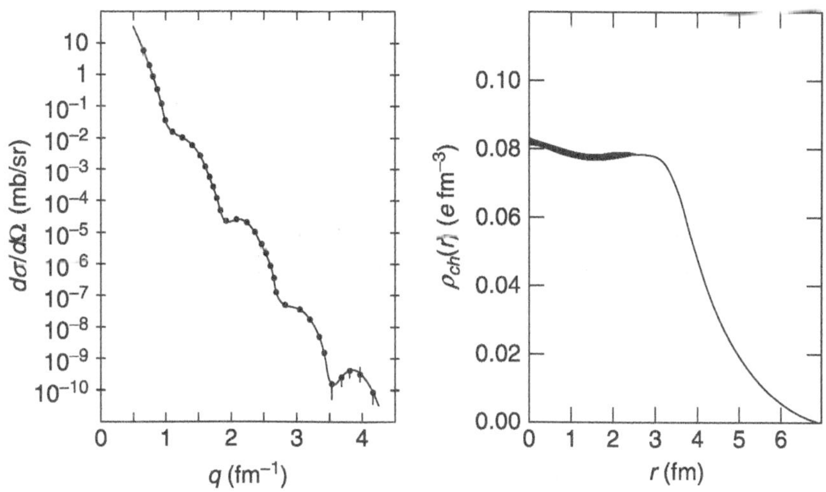

Figure 2.19 *Nuclear charge distribution obtained from experiments*

The measured differential scattering cross-section data for most of the energies of incident electrons could be best reproduced by a nuclear charge distribution of the shape.

$$\rho(r) = \frac{\rho(0)}{\left\{1+e^{\left(\frac{r-R_e}{z_1}\right)}\right\}} \tag{2.30}$$

where $\rho(0)$ is the central charge density at very small values of r. At $r = R_e$, the charge density falls to $\frac{1}{2}\rho(0)$, half of its central value. The skin thickness t (distance between the points where charge density falls from 90% to 10%) is given by $t = 4z_1\text{In}3$. A rough sketch of the charge density distribution given by Eq. (2.30) is shown in figure 2.19. A plot of the measured differential scattering cross-section $\left(\frac{d\sigma}{d\Omega}\right)$ with the linear momentum transfer (q) is also shown in this figure.

The charge density and the nucleon density determined from high energy electron experiments for different nuclei are shown in figures 2.20 and 2.21. These figures show that the central parts of all nuclei, big or small, are almost identical. The difference lies in their skins.

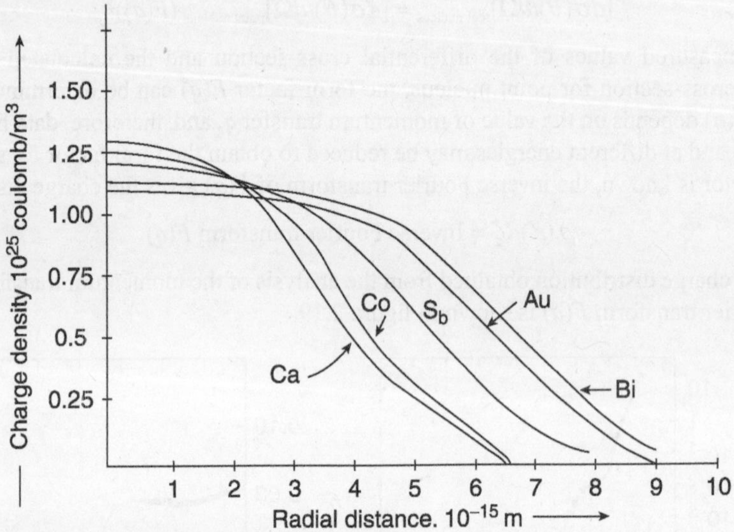

Figure 2.20 *Charge density of different nuclei determined from high energy electron scattering*

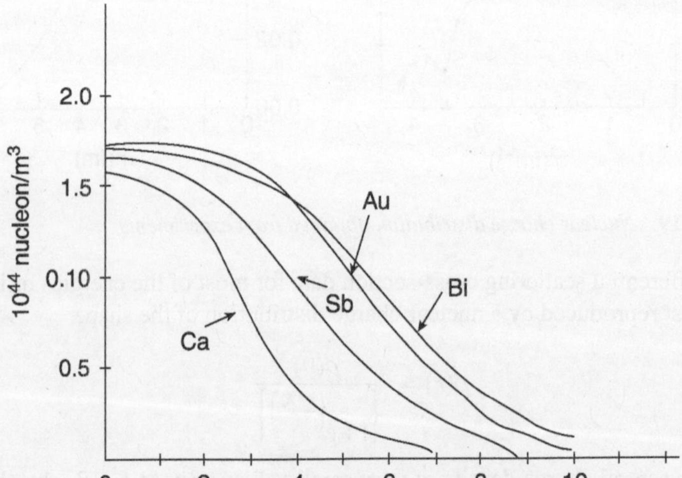

Figure 2.21 *Nucleon density for different nuclei obtained from high-energy electron scattering data assuming that the ratio of the proton to neutron density is constant*

2.3.2 Coulomb Energy Difference between Mirror Isobars

The electrostatic energy of a uniformly charged sphere of radius R with a total charge Q is given by the expression $\frac{3Q^2}{5R}$. The derivation of this formula can be found in any book on electricity. The electrostatic energy E_c of a nucleus with Z protons uniformly distributed within a spherical volume of radius R may, therefore, be written as

$$E_c = \frac{3Z^2e^2}{5R} \tag{2.31a}$$

This is based on the assumption that protons have all smeared as a positively charged jelly of uniformly charged density. If it is assumed that the protons are still separate entities, positively charged point particles uniformly distributed over the volume of the sphere and interacting with each other except with itself, then E_c is given by

$$E_c = \frac{3Z(Z+1)e^?}{5R} \tag{2.31b}$$

As may be observed from Eqs. (2.31a) and (2.31b), for large values of Z there is not much difference between the values of Coulomb energy.

Many isobaric mirror nuclei pairs differ by only one unit of charge and the higher charge isobar decays into the lower charge isobar by β^+ emission. Further, for each pair the total number of nucleons A is same and $A = (2Z - 1)$, where Ze is the charge on the nucleus of the higher charge isobar. Some examples are ${}^{13}_{7}\text{N} \rightarrow {}^{13}_{6}\text{C} + \beta^+ + \nu$; ${}^{15}_{8}\text{O} \rightarrow {}^{15}_{7}\text{N} + \beta^+ + \nu$; ${}^{23}_{12}\text{Mg} \rightarrow {}^{23}_{11}\text{Na} + \beta^+ + \nu$; and so on.

In each of these cases, a positron and a neutrino are emitted as the result of the decay. The emitted positron has a continuous energy spectrum with the maximum kinetic energy denoted by E^{β}_{max}, which E^{β}_{max} has a different value for each pair of mirror nuclei and can be measured quite accurately. In the case when positron is emitted with maximum kinetic energy E^{β}_{max}, the neutrino is emitted with zero kinetic energy. The rest mass of neutrino is also zero. Hence, when the β particle is emitted with maximum energy, it takes all the available energy and no energy is carried by the neutrino. Further, it is reasonable to assume that the final nucleus (the lower charge isobar) remains stationary and does not recoil as its mass is much larger than the mass of the emitted β particle. Hence, the total energy available from the decay is carried only by the positron in the form of its kinetic energy E^{β}_{max}.

In positron decay, a proton in the nucleus is converted into a neutron, a β^+ particle, and a neutrino. The neutron remains inside the nucleus and the positron and the neutrino escape. This may be represented by the following equation.

$$\text{P} = \text{N} + \beta^+ + \nu$$

In each of the above-mentioned decays, the parent isobar with higher charge undergoes decay. Let us assume that the energy available for the decay is ΔE_{decay}. A contribution ΔE_c to the decay energy ΔE_{decay} will come from the Coulomb energy difference of the parent and the daughter isobars. A contribution ΔE_{nu} to the decay energy ΔE_{decay} may also come from the binding energy difference of a proton and a neutron in the nucleus. It may be remembered that in each of the above-mentioned decay, a proton in the parent mirror nucleus is converted into a neutron in the daughter. Therefore, if the binding energy of a proton in the parent nucleus is different from the binding energy of a neutron in the daughter, then only ΔE_{nuc} will be non-zero. Next, let us calculate how this decay energy ΔE_{decay} will be consumed.

It may be remembered that a neutron is heavier than a proton and, therefore, an amount of energy ΔE_{NP}, equal to the mass difference of neutron and proton $(M_N - M_P)c^2$ (=1.29 MeV) is consumed in the conversion of a proton into a neutron. Further, a positron (β^+) and a neutrino are also created in the decay process. The amount of energy $\Delta E_\beta = m_o c^2 = 0.511$ MeV is required for creating the positron and the rest mass energy of a neutrino is zero. The maximum kinetic energy $E^\beta{}_{max}$ is also supplied by the decay energy. Thus, the energy balance may be represented as

$$\Delta E_{decay} = \Delta E_c + \Delta E_{nuc} = \Delta E_{NP} + \Delta E_\beta + E^\beta{}_{max}$$

Or
$$E^\beta{}_{max} = \Delta E_c + \Delta E_{nuc} - (\Delta E_{NP} + \Delta E_\beta) \tag{2.32}$$

However, ΔE_c = electrostatic energy difference of the parent nucleus with Z protons and the daughter nucleus with $(Z-1)$ protons

$$= \frac{3e^2}{5R}[Z^2 - (Z-1)^2] = \frac{3e^2}{5R}[2Z-1]$$

However, in light nuclei, the number of neutrons is nearly equal to the number of protons, that is $N = Z$, and, therefore, $2Z = A$. Further, neglecting 1 in comparison to $2Z$ and substituting $R = r_0 A^{1/3}$ in the above-mentioned equation, we get

$$\Delta E_c = \frac{3e^2}{5r_0} A^{2/3}$$

And $(\Delta E_{NP} + \Delta E_\beta) = 1.29 + 0.511 = 1.81$ MeV

Substituting the above values in Eq. (2.32), we get

$$E^\beta{}_{max} = \frac{3e^2}{5r_0} A^{2/3} + \Delta E_{nuc} - 1.81 \tag{2.33}$$

Experimental values of $E^\beta{}_{max}$ for different systems of mirror isobars when plotted against $A^{2/3}$ fell on a straight line that made an intercept of −1.81 MeV on the energy axis. Three important conclusions can be made from this observation.

1. $\Delta E_{nuc} = 0$. This means that the nuclear binding energies in the mirror nuclei for proton and neutron are identical, which in turn means that the nucleon–nucleon force is charge independent.
2. The assumption that nuclear radius $R = r_0 A^{1/3}$ is correct.
3. The nuclear charge is uniformly distributed over the volume of the nucleus.

The value of r_0 determined from the slope of the straight line gives the value 1.45 F. The radius determined from this value is called the classic Coulomb radius. Instead of considering the uniform charge distribution, if the protons are considered as point particles then

$$E^\beta{}_{max} = (3e^2/5r_0)\,[A^{2/3} - A^{-1/3}] - 1.81 + \Delta E_{nuc}$$

A graph between E^{β}_{max} and $(A^{2/3} - A^{-1/3})$ still remains a straight line making an intercept of −1.81 MeV on the energy axis and thus the conclusions drawn earlier remain valid.

If the electrostatic energy of the nucleus is calculated quantum mechanically for a nucleus for which $A = 2Z$, we get

$$E_c = (3e^2/5R)[Z(Z-1) - 0.77Z^{1/2}]$$

The value of r_0 obtained using the above-mentioned expression for Coulomb's energy is about 10% smaller than the classical value. Further, correction for the Pauli exclusion principle, applicable to protons in the nucleus, reduces the value of r_0 by another 10%. Thus, the quantum mechanically obtained value of r_0 from the decay data of mirror isobars is about 20% smaller than the classical value.

Solved example S-2.7

${}^{27}_{14}Si$ and ${}^{27}_{13}Al$ are mirror nuclei, the former is a positron emitter of maximum energy 3.48 MeV. Calculate r_0.

Solution. From Eq. (2.33), we have

$$E^{\beta}_{max} = (3e^2/5r_0)A^{2/3} - 1.81 + \Delta E_{nuc}$$

And experiments have shown that ΔE_{nuc} is zero, hence

$$(3e^2/5r_0)A^{2/3} = E^{\beta}_{max} + 1.81 = 3.48 + 1.81 = 5.29 \text{ MeV}$$

Or

$$\frac{3 \times 27^{2/3}}{5r_0}\left(\frac{e^2}{4\pi\varepsilon_0}\right) = 5.29 \text{ MeV}$$

The factor of $\left(\frac{1}{4\pi\varepsilon_0}\right)$ comes up when quantities are measured in SI units.

However,

$$\left(\frac{e^2}{4\pi\varepsilon_0}\right) = 1.44$$

Substituting the above values, we get

$$r_0 = \frac{27 \times 1.44}{5 \times 5.29} = 1.47 \text{ F.}$$

Ans: The value of r_0 is = 1.47 Fermi = 1.47×10^{-15} m

2.3.3 Anomalous Scattering of α Particles

Soon after proposing the theory of nuclear atom, Rutherford himself observed deviations in the measured α scattering cross-sections from the theoretical values predicted by his scattering formula. These deviations were more pronounced for low mass targets and higher α energies. He used the term *anomalous scattering* for these deviations. Several workers carried out detailed study of these deviations, particularly for even–even targets. Even–even nuclei are important because they have zero spin and since the spin of α particle is also zero, there is no interference coming from the spins of the interacting nuclei in the scattering process. Figure 2.22 shows a rough sketch of the anomalous α scattering results.

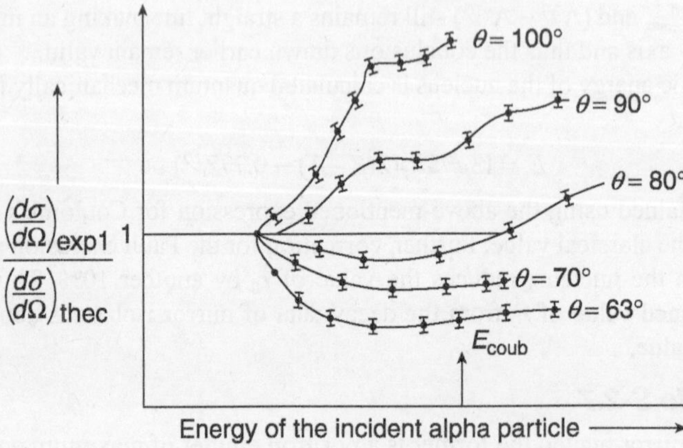

Figure 2.22 *Rough sketch of anomalous α scattering results*

In this figure, the ratio of experimental scattering cross-section to the cross-section predicted by Rutherford's formula is plotted as a function of α particle energy. The Coulomb barrier height for the system is indicated by E_{coub} in the figure. The horizontal line is drawn for the value of the cross-section ratio = 1. Any deviation of experimental data from the horizontal line indicates anomalous scattering. The following observations may be made from the inspection of figure 2.22.

1. Anomalous scattering starts at energies less than the Coulomb barrier.
2. It takes off at all angles almost simultaneously.
3. Anomalous scattering shows peak structures at different energies for different angles.

The anomalous scattering cannot be understood in terms of classical physics. According to classical physics, an incoming α particle experiences a Coulomb barrier as it approaches the positively charged nucleus. In most of the cases studied earlier, the kinetic energies of incident α particles were much less than the barrier heights and classically it was not possible for α particles to overcome the potential barrier to reach the nucleus. As a result, only Coulomb scattering, described by Rutherford's formula was observed. With the availability of higher energy α particles, it became possible to observe anomalous scattering. The explanation for anomalous scattering can be given in terms of quantum mechanical tunnelling of the Coulomb barrier by those α particles that have energies smaller than the Coulomb barrier. A rough sketch of the potential between an α particle and a target nucleus is shown in figure 2.23.

Repulsive Coulomb barrier increases in height as the α particle moves nearer to the nucleus. However, at a distance very close to the nucleus, the strong attractive nuclear force between the α particle and the nucleus starts building up rapidly and within a very short distance the strong nuclear attractive force overcomes the repulsive Coulomb force. Beyond that point (B and B′ in figure 2.23), the α particle experiences only an attractive nuclear force. An attractive force may be represented by a potential well. For simplicity, the attractive nuclear potential is represented by a square well in the figure. The Coulomb barrier for this case is the maximum height of Coulomb potential at point B. In classical physics, no α particle with energy less than the Coulomb barrier height may reach the nuclear potential, the boundary of the square well.

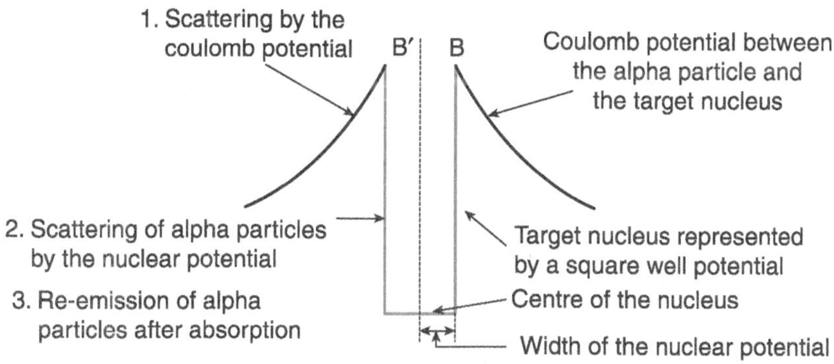

Figure 2.23 *Rough sketch of the potential between an α particle and the target nucleus*

In quantum mechanics, it is, however, possible that an α with energy less than the Coulomb barrier may tunnel through the barrier and may reach the nuclear potential well. Alpha particles that tunnel through the Coulomb barrier show anomalous scattering at all angles first. As the energy of incident α particle increases, more and more particles tunnel through the barrier, increasing the magnitude of anomalous scattering. Three components contribute to the scattering of α particles. The first is the scattering from the Coulomb potential, the second is the scattering from the nuclear potential, and the third is the re-emission of α particles of the same energy as the incident αs after their absorption by the target nucleus. For some energy of α particles, at a given angle, the contributions from the above-mentioned three processes add up in phase called coherent scattering and show peaks.

In this quantum mechanical description of the α scattering, the radius of the target nucleus enters through the width of the nuclear potential well and the minimum incident energy at which the anomalous scattering takes off. The finite size of the α particle does affect the nuclear radius. Analysis of the anomalous scattering data with correction for the fine size of the α particle has shown that the radius of a nucleus with A nucleons may be given as $R = r_o A^{1/3}$ with $r_0 = 1.3 \pm 0.1$ F.

It is to be noted that this method gives the nuclear force radius.

2.3.4 Nuclear Radius from the Life Time of α Emitters

Most of the nuclei with atomic mass number $A > 140$ decay by α emission. However, the energies of the emitted α particles and their mean lives are different for each. Let us look into the reverse process of α decay, that is an α particle being projected on the daughter nucleus ${}^{A-4}{}_{Z-2}X$. The impinging α particle will experience a repulsive Coulomb barrier $V_{cou} = \{(Z-2)e \cdot 2e/r\}$, which will increase with the decrease of the relative separation r between the nucleus and the α particle. However, at very close distance $d \approx R_{nuc} + R_{\alpha}$, the sum of the radii of the nucleus and the α particle, the attractive potential of nuclear strong force will overtake the repulsive potential and the net potential will become an attractive deep well. The shape of the potential experienced by the α particle is shown in figure 2.24.

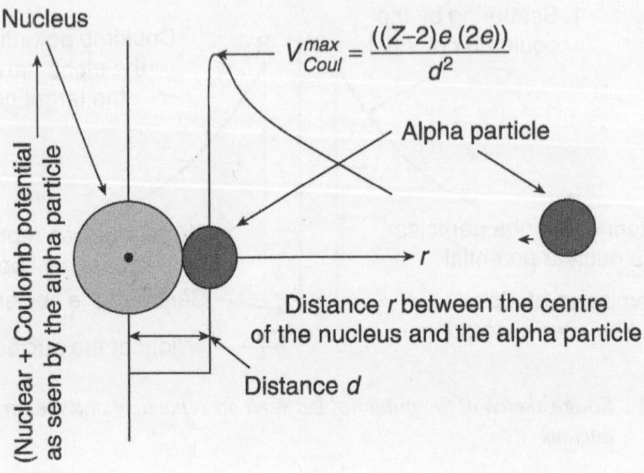

Figure 2.24 *Sketch of the potential seen by an a particle moving towards the target nucleus of charge (Z – 2)e*

The approximate maximum height of the repulsive Coulomb barrier V^{max}_{cou} can be calculated if the values of the radii of the target nucleus and the α particle are known. It is clear from this figure that an incident α particle will be absorbed by the target nucleus if it could reach a distance smaller than d. However, classically only an α particle with kinetic energy larger than V^{max}_{cou} will be able to overcome the potential barrier and reach the attractive part of the potential. Let us now look at the α particle, which is inside the nucleus. It will also experience the same potential if it tries to come out. Classically, it will not be possible for it to come out of the nucleus unless it has a kinetic energy inside the nucleus either equal to or greater than the maximum barrier height. Experimentally, it has been found that α particles with kinetic energies much less than the Coulomb barrier are emitted by nuclei that undergo α decay. It was also observed that the half-life or the decay constant depends very strongly on the energy of the emitted α. These observations could not be explained on the basis of classical physics. It was Gamow who for the first time applied quantum mechanics to the problem of α decay. Gamow argued that inside the nucleus sometimes two neutrons and two protons come close enough and form α particles. He associated a probability P for the formation of α particles inside the nucleus, which he called preformed α particles. These preformed α particles move around in the volume of the nucleus reaching the nuclear surface repeatedly after a certain time interval. The frequency 'f' of their reaching the nuclear surface may be given by $f = (2R/v)$, where R is the radius of the residual nucleus (once α particle has been formed, the remaining nucleus is the residual nucleus) and v the velocity of the preformed α particle. When an α particle reaches the surface of the nucleus it encounters the potential barrier. Since the kinetic energy of the α particle inside the nucleus is less than the barrier height, it makes several thousand attempts to break through the barrier and in one of them it succeeds. This is called tunnelling. The probability of tunnelling is represented by a factor T called the barrier transmission coefficient. T depends very strongly on the kinetic energy of the α particle. Accordingly, the decay constant λ is build up of two multiplicative factors λ_0 and T. Here, λ_0 is the decay constant had there been no potential barrier for α particle at the nuclear surface and is called the reduced width. T is the transmission coefficient or the probability of tunnelling through the barrier.

Or $$\lambda = \lambda_0 T = (P \cdot f)T \tag{2.34}$$

Rough estimates suggest a value of $P \approx 25(4/A)^3$ and for heavy nuclei the magnitude of P is of the order of 10^{-3}. The frequency f can be calculated if the energy of the α particle and the radius of the nucleus are known. It is generally of the order of 10^{21} s^{-1}. As such the factor $(P \cdot f)$ is $\approx 10^{18}$ s^{-1}.

The most important factor is the transmission coefficient T. Quantum mechanical calculations carried out by Gamow show that T may be given by the relation

$$T = e^{-\gamma}$$

Here, γ is called 'Gamow factor'.

In figure 2.25, the Coulomb potential between the nucleus of charge $(Z-2)e$ and the α particle of charge $2e$ is shown by the curve ab. If the radius of the nucleus is R, then the maximum height of the Coulomb potential for the α particle is V_{Cou}^{max}, as shown in the figure. The horizontal line gh in the figure shows the kinetic energy E_{kin} of the preformed α particle inside the nucleus. There is an area marked as B in the figure. According to Gamow theory of α decay, it can be shown that the Gamow factor 'γ' is proportional to the area B, which is bounded by the kinetic energy of the preformed α particle at the bottom, the line defining the radius of the nucleus on the left and the curve showing the Coulomb potential on the right. The transmission coefficient T is equal to the exponential raised to the negative power of 'γ'. A little decrease in 'γ' increases the magnitude of T by a large factor because of the exponential dependence. As shown in the figure, the magnitude of the area B decreases if the radius R of the decaying nucleus is increased and/or when the kinetic energy E_{kin} of the emitted α particle is large. In both these cases, the transmission coefficient increases by a large factor. Thus, a slight change in the kinetic energy of the emitted α particle changes T by a large amount which in turn changes the decay constant λ or the half-life. The value of the decay constant or the half-life of an α emitter is, therefore, very sensitive to the radius of the decaying nucleus.

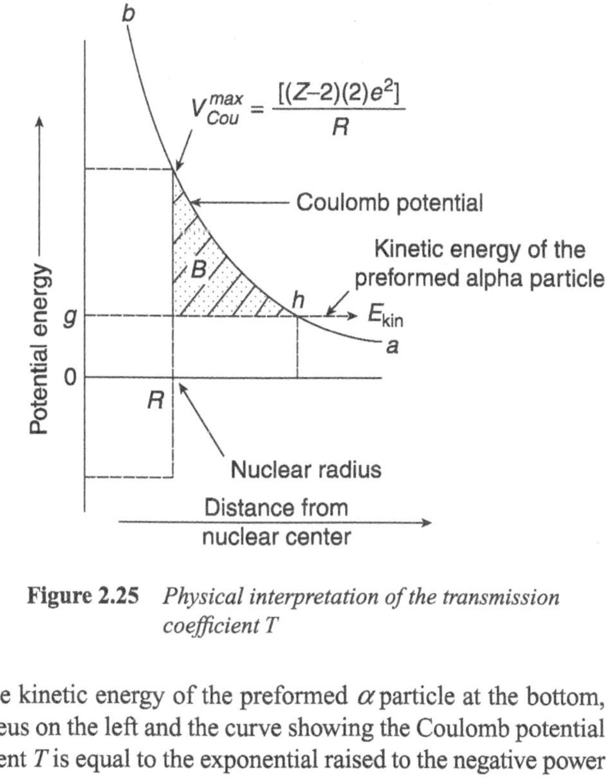

Figure 2.25 *Physical interpretation of the transmission coefficient T*

The experimental data on the decay constants (or the half lives) for a large number of α emitters have been analysed to obtain the radii, R, of the decaying nuclei. It has been found that the nuclear radius R may be written as $R = r_0 A^{1/3}$. The value r_0 in the range of 1.2–1.4 F reproduces the experimental data on the half lives of α emitters well, which once again indicates that the nuclear density is constant. The unit radius r_0 determined from the decay constant is often called the nuclear force unit radius.

Exercise p-2.15: List three methods to determine the electromagnetic radius of the nucleus and give details of one of them.

Exercise p-2.16: Which method of determining the nuclear electromagnetic radius is best in your opinion? Give reasons for your choice.

Exercise p-2.17: What will happen if instead of high-energy electrons high-energy π^- mesons are used for determining the radius of the nucleus?

2.4 NUCLEAR ANGULAR MOMENTUM, MAGNETIC DIPOLE MOMENT, AND ELECTRIC QUADRUPOLE MOMENT

Before we discuss methods of determining the nuclear angular momentum, magnetic dipole moment and electric quadrupole moment, we briefly review the classical and quantum mechanical approach to these properties.

2.4.1 Angular Momentum

Most of the subatomic particles may have two types of angular momentums: one due to the orbital motion and the other intrinsic angular momentum, which is also called the intrinsic spin of the particle. The classical orbital angular momentum can be quantized using the correspondence principle. The quanta of orbital angular momentum turns out to be an integer multiple of $\hbar$. The intrinsic angular momentums may have values that are both integer and half-integer multiples of $\hbar$. The half-integer values for angular momentum cannot be generated by any orbital motion of a reasonably sized subatomic particle. It is because to generate half-integer values for the angular momentum the speed of surface charge on the particle exceeds the velocity of light, which is not possible. The classical orbital angular moment L is defined as:

$$L = r \times p$$

It may be translated into quantum mechanical operator using the correspondence principle

$$p \rightarrow -i\hbar \cdot \nabla;\ L_z = -i\hbar(x\partial/\partial y - y\partial/\partial x) = -i\hbar\partial/\partial\varphi$$

where φ is the Azimuthal angle in polar coordinates. The state of a particle with a given orbital angular momentum is characterized by the quantum numbers l and m belonging to operators.

$$L^2\psi_{lm} = l(l+1)\hbar^2\psi_{lm} \quad l \text{ can have integer values}$$

$$L_z\psi_{lm} = m\hbar\,\psi_{lm} \quad -l \le m \le +l$$

Orbital angular momentum is quantized and can have absolute value $L_{abs} = [l(l+1)\hbar^2]^{1/2}$. The magnetic quantum number m can have $(2l + 1)$ values from $-l$ to $+l$ in steps of unity.

The doublets (or fine structure) observed in alkali atom spectra cannot be explained on the basis of integer values of angular momentum. Pauli, Uhlenbeck, and Goudsmit assigned an intrinsic angular momentum or spin $\left(\frac{1}{2}\right)\hbar$ to the electron and argued that in an external magnetic field the intrinsic angular momentum can have two orientations giving rise to the observed doublets.

The components of the orbital angular momentum L fulfil the commutation rules

$$[L_i, L_j] = i\hbar\varepsilon_{ijk}L_k$$

Taking these algebraic identities as the *definition* of the intrinsic angular momentum (and not quantising the classical orbital angular momentum) half-integer values for the quantum mechanical angular momentum can be obtained. If J denotes the intrinsic or quantum mechanical angular momentum then

$$[J_i, J_j] = i\hbar\varepsilon_{ijk}J_k \qquad \text{(defining commutation rule)}$$

$$J^2\psi_{Jm_J} = J(J+1)\hbar^2\psi_{Jm_J} \qquad (J = 0, \tfrac{1}{2}, 1, 3/2, 2,\ldots)$$

$$J_z\psi_{Jm_J} = m_J\hbar\psi_{Jm_J} \qquad (-J \le m_J \le +J)$$

2.4.2 Magnetic Dipole Moment (μ)

Magnetic dipole moment of a particle is proportional to its angular momentum, orbital or inherent, or the combination of both. Charged particles and those neutral particles that have charged constituents exhibit dipole moments, for example neutrons and paramagnetic atoms. The unpaired valence electron in paramagnetic atoms has both orbital and inherent spin angular momentums and corresponding to those magnetic moments. Particles that do not have angular moment do not have magnetic moments as well.

Classically, a current of strength i passing through a loop of area A produces a magnetic moment μ_{cla}.

$$\mu_{\text{cla}} = i \cdot A$$

The direction of μ_{cal} is the same as that of the vector area A. In case the current is constituted by the revolution of a single particle of charge q, and mass m moving in a circular orbit of radius r with the frequency $f = v/(2\pi r)$, the current i is

$$i = qv/(2\pi r)$$

And the magnetic moment $\mu_{\text{cla}} = \{qv/(2\pi r)\}\ (\pi r^2) = qvr/2 = (q/2m)\ (mvr) = (q/2m)\ L$ as shown in figure 2.26.

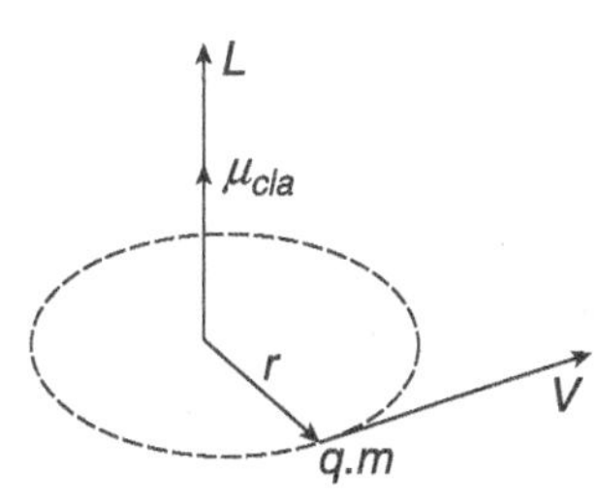

Figure 2.26 *Classical magnetic moment*

In quantum mechanical description, a direction is defined by the angular momentum of the particle. The projections of the angular momentum and the magnetic moment on this direction are the measurable variables. As such the quantum mechanical magnetic dipole moment μ is proportional to I, the maximum value of the angular moment of the particle. Therefore,

$\mu = \gamma I$, where proportionality constant γ is called the gyromagnetic ratio. For a particle having a unit charge e, one may also write

$$\mu = g(\mathcal{M}/\hbar)I$$

where $\mathcal{M} = e\hbar/2m$ is called a magneton. The value of magneton for electron is much different from that of a proton because of the mass m. In case of electron $e\hbar/2m_e$, where m_e is the mass of the electron, is called the Bohr magneton denoted by Bm and has the numerical value 1 Bm = $5.788381749 \times 10^{-5}$ eV T^{-1}. In case of proton, $e\hbar/2m_p$, where m_p is the mass of the proton, is called nuclear magneton denoted by nm. 1 nm = $3.152451233 \times 10^{-8}$ eV T^{-1}.

The 'g-factor' measures the deviation of the real magnetic moment from its value expressed in magneton. It is a number of the order of 1 and has different values for the orbital and intrinsic angular momentum. For the orbital motion g_l is equal to 1 for electron and proton, but zero for neutron. For intrinsic angular momentum of $\frac{1}{2}\hbar$, Dirac's theory predicts a value of 2 for the g-factor g_s. However, widely different values for g^s_p and g^s_n the g-factors respectively for proton and neutron, have been found, probably due to the difference in the charge distributions inside these particles.

$$g^s_p = 5.5856946 \text{ and } g^s_n = -3.8260854$$

2.4.3 Electric Quadrupole Moment Q

Electric quadrupole moment of the nucleus is related to the distribution of the positive charge inside the volume of the nucleus as seen by an external observer. In principle, there are two ways in which information about the nuclear charge distribution may be obtained. One is to find out the magnitude of the electric potential produced by the nuclear charge distribution at the location of the observer and then to see if the potential tells something about the nuclear charge distribution. The other possibility may be to put the nucleus in some other external electric field and determine the electrostatic energy of the nucleus in this electric field. Brief out lines of both these methods are as follows. A typical nucleus with uniform charge density ρ per unit volume is shown in figure 2.27.

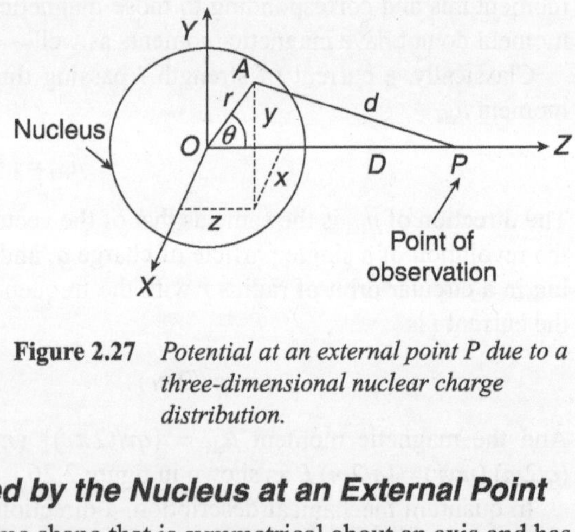

Figure 2.27 *Potential at an external point P due to a three-dimensional nuclear charge distribution.*

2.4.3.1 Electric Potential Produced by the Nucleus at an External Point

Let us consider a nucleus as a 3D body of some shape that is symmetrical about an axis and has a uniform distribution of positive charge. In the case of a nucleus, at least one axis of symmetry is required since nuclei have definite parity (i.e. $\psi(r) = \mp\psi(-r)$ *or a sort of* mirror symmetry). As such there are three possible shapes for the nuclear charge distribution: spherical, prolate, and oblate. Let us consider a small charge $(\rho d\tau)$ at point A, which is at a distance r from the centre 'O' of the nucleus, ρ is the density of nuclear charge at point A and $d\tau$ the element of volume. Let P be the point of observation. We shall first calculate the electric potential $d\phi$ at point P due to the charge $(\rho d\tau)$ at point A. Next, we will integrate $d\phi$ over the whole volume of the nucleus to get the potential at P due to the complete nucleus.

$$d\phi = \frac{\rho\, d\tau}{d} = \frac{\rho\, d\tau}{(D^2 - 2rD + r^2)^{\frac{1}{2}}} = \rho\, d\tau \left(\frac{1}{D}\right)\left[1 - \left(\frac{2r}{D} - \left(\frac{r}{D}\right)^2\right)\right]^{-\frac{1}{2}} \tag{2.34a}$$

The big bracket at the extreme right in the above equation can be expanded using binomial theorem using the formula $(a+b)^n = a^n + na^{n-1}b + \frac{n(n-1)}{2}a^{n-2}b^2 \cdots$ and substituting $a = 1$, $b = -\left\{\frac{2r}{D} - \left(\frac{r}{D}\right)^2\right\}$, and $n = -\frac{1}{2}$. The quantity $\left(\frac{r}{D}\right)$ is small and the magnitude of the higher powers of the quantity decrease rapidly. It is, therefore, sufficient to retain terms up to $\left(\frac{r}{D}\right)^2$ in the expansion. This gives

$$d\phi = \frac{\rho\, d\tau}{d} = \rho\, d\tau\left(\frac{1}{D}\right)\left[1 + \frac{r}{D}\cos\theta + \left(\frac{r}{D}\right)^2\left(\frac{3}{2}\cos^2\theta - \frac{1}{2}\right) + \cdots\right.$$

Or

$$d\phi = \frac{\rho\, d\tau}{D} + \frac{\rho r\cos\theta\, d\tau}{D^2} + \frac{\rho r^2\left(\frac{3}{2}\cos^2\theta - \frac{1}{2}\right)d\tau}{D^3} + \cdots \tag{2.34b}$$

Integrating both the sides of Eq. (2.34b) over the volume of the nucleus gives,

$$\Phi_P = \frac{1}{D}\iiint \rho\, d\tau + \frac{1}{D^2}\iiint \rho r\cos\theta\, d\tau + \frac{1}{D^3}\iiint \rho r^2\left(\frac{3}{2}\cos^2\theta - \frac{1}{2}\right)d\tau + \ldots \tag{2.34c}$$

In Eq. (2.34c) $\iiint$ represents volume integration. It follows from Eq. (2.34c) that the potential Φ_P at point P due to whole of the nucleus may be represented as the sum of several terms. The first term on the RHS of Eq. (2.34c) decreases as $(1/D)$, second term as $(1/D^2)$, third term as $(1/D^3)$, and so on. However, it is known that the potential due to a charge placed at a point decreases as $(1/D)$, where D is the distance of the point of observation from the charge. Hence, the first term in Eq. (2.33c) gives the potential at point P that will be produced if the total charge of the nucleus ($\iiint \rho\, d\tau$) is placed at the centre O of the nucleus. The first term is sometime called the monopole term. Further, it is known that the potential due to a dipole at a distance D from the centre of the dipole decreases as $1/D^2$. It may be recalled that a dipole consists of two equal and opposite charges separated by a very small distance and, therefore, the net charge of a dipole is zero. The second term on the RHS of Eq. (2.34c), therefore, gives the contribution to the potential at P, as if an imaginary dipole is placed at the center of the nucleus. To calculate the magnitude of the dipole term, we need to evaluate the volume integral. The function ρr in the volume integral is an odd function of r, which means that contributions of equal magnitude but of opposite sign come to the integral from r and $-r$, and therefore, the volume integral vanishes. The same fact may also be stated by saying that since all nuclei have a definite parity, the electric dipole contribution to the potential at an external point is zero. One may also say that the electric dipole moment for nuclei does not exist. The third term of Eq. (2.34c), that decreases as $(1/D^3)$ gives the contribution of the quadrupole term. A quadrupole is formed when two dipoles are placed side by side in opposite directions. The net charge of a quadrupole is also zero. In general, it may, therefore, be said that a distribution of charges in a region of space produces a potential at an external point, which is the sum of the potentials produced by a monopole, a dipole, a quadrupole, and so on, as shown in figure 2.28.

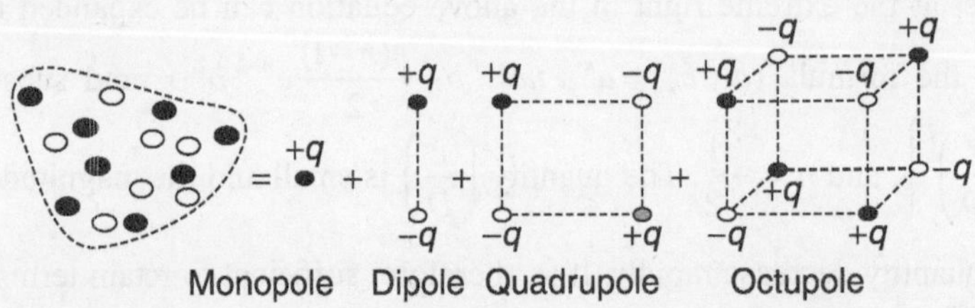

Figure 2.28 *Multipole expansion of charge distribution*

The component of the classical electric quadrupole moment of a nucleus in the *z*-direction is defined as

$$Q_{clas} = \iiint r^2 \rho(r)\left[\frac{3}{2}\cos^2\theta - \frac{1}{2}\right]d\tau = \frac{1}{2}\iiint \rho(r)d\tau \iiint (3\cos^2\theta - r^2)d\tau$$

However, from figure 2.27, $r\cos\theta = z$.
Hence,

$$Q_{clas} = \frac{1}{2}\iiint \rho(r)(3z^2 - r^2)d\tau \tag{2.34d}$$

It may be remembered that Eq. (2.34d) gives the *z*-component of the classical electric quadrupole moment. The classical quadrupole moment of the nucleus is taken along the direction of the angular momentum or spin *I* of the nucleus. In the body fixed direction of *I*, the absolute magnitude of the angular momentum of the nucleus $\left(\sqrt{I(I+1)}\hbar\right)$ The unit of the classical quadrupole moment is 'e-barn', where e is the charge on the electron and 1 barn = 10^{-24} cm^2 = 10^{-28} m^2.

It is clear from Eq. (2.34d) that if more charge is contained along the *z*-direction, ($3z^2$) dominates over the (r^2) term and, hence, Q_{clas} is positive. Such a charge distribution elongated along the *z*-direction is called prolate. In the other case, if the charge distribution is spread more in the *x* and *y* directions, the (r^2) term over rides the ($3z^2$) term, and Q_{clas} is negative. This makes an oblate charge distribution.

2.4.3.2 Electrostatic Energy of a Nucleus Placed in an External Electric Field

As has already been stated, the quadrupole moment of a nucleus may also be defined in terms of the electrostatic energy that a nucleus acquires when it is placed in a static external electric field. As a matter of fact such electric fields, at the location of the nucleus, are always present in any piece of matter. These fields are produced by the electrons of the atom.

Let us assume that an electric potential $\Phi(x, y, z)$ may be defined in a region of space around the origin of the coordinate system (*x*, *y*, *z*). A nucleus with charge density $\rho(x, y, z)$ is now placed in this region of space with its centre at the origin. The total electrostatic energy W produced by the interaction of potential Φ and charge ρ may be given as,

$$W = \iiint \Phi(x, y, z)\, \rho(x, y, z) d\tau \tag{2.35a}$$

The integral in Eq. (2.35a) is an integral over the volume of the charge distribution and $d\tau$ is the volume element. To evaluate the above integral, it is convenient to expand the potential Φ about the origin using Taylor series expansion. If $\Phi(0)$, $\left(\frac{\partial^n \Phi}{\partial x^n}\right)_0$, and so on represent, respectively, the value of the potential at origin and the values of the nth derivatives of Φ evaluated at the origin, then

$$W = \Phi(0)\iiint \rho\, d\tau + \left(\frac{\partial \Phi}{\partial x}\right)_0 \iiint x\rho\, d\tau + \left(\frac{\partial \Phi}{\partial y}\right)_0 \iiint y\rho\, d\tau + + \left(\frac{\partial \Phi}{\partial z}\right)_0 \iiint z\rho\, d\tau \left(\frac{\partial \Phi}{\partial z}\right)_0 +$$
$$\frac{1}{2}\left(\frac{\partial^2 \Phi}{\partial x^2}\right)_0 \iiint x^2 \rho\, d\tau + \frac{1}{2}\left(\frac{\partial^2 \Phi}{\partial y^2}\right)_0 \iiint y^2 \rho\, d\tau + \frac{1}{2}\left(\frac{\partial^2 \Phi}{\partial z^2}\right)_0 \iiint z^2 \rho\, d\tau +$$
$$\left(\frac{\partial^2 \Phi}{\partial x \partial y}\right)_0 \iiint xy\rho\, d\tau + \left(\frac{\partial^2 \Phi}{\partial x \partial z}\right)_0 \iiint xz\rho\, d\tau + \left(\frac{\partial^2 \Phi}{\partial y \partial z}\right)_0 \iiint yz\rho\, d\tau + \tag{2.35b}$$

The first term in expression (2.35b) gives the interaction energy of a monopole, the next three terms the interaction energy of a dipole, and the next six terms the interaction energy due to the quadrupole. The dipole terms vanish because of the symmetry of the charge distribution about one of the axis, for instance the z-axis. Out of the six quadrupole energy terms the last three terms with products such as $\int xy\rho\, d\tau$, $\int xz\rho\, d\tau$, $\int yz\rho\, d\tau$ also vanish as they contain odd functions. W_{qua}, the electrostatic energy due to the quadrupole may be written as,

$$W_{qua} = \frac{1}{2}\left(\frac{\partial^2 \Phi}{\partial x^2}\right)_0 \iiint x^2 \rho d\tau + \frac{1}{2}\left(\frac{\partial^2 \Phi}{\partial y^2}\right)_0 \iiint y^2 \rho d\tau + \frac{1}{2}\left(\frac{\partial^2 \Phi}{\partial z^2}\right)_0 \iiint z^2 \rho d\tau \tag{2.35c}$$

As it has been assumed that the charge distribution is symmetrical about z-axis, the contributions from the x and y directions in Eq. (2.35c) will be equal. Eq. (2.35c) then reduces to

$$W_{qua} = \frac{1}{2}\left(\frac{\partial^2 \Phi}{\partial z^2}\right)_0 \iiint z^2 \rho\, d\tau + \frac{1}{2}\left(\frac{\partial^2 \Phi}{\partial x^2} + \frac{\partial^2 \Phi}{\partial y^2}\right)_0 \iiint \left(\frac{x^2 + y^2}{2}\right) \rho \partial\tau \tag{2.35d}$$

We now use the relations

$$r^2 = x^2 + y^2 + z^2 \text{ and, therefore } \left(\frac{x^2 + y^2}{2}\right) = \left(\frac{r^2 - z^2}{2}\right)$$

Also, from Laplace's relation,

$$\left(\frac{\partial^2 \Phi}{\partial x^2} + \frac{\partial^2 \Phi}{\partial y^2}\right)_0 = -\left(\frac{\partial^2 \Phi}{\partial z^2}\right)_0$$

Substituting the values in Eq. (2.35d), we get

$$W_{\text{qua}} = \frac{1}{2}\left(\frac{\partial^2 \Phi}{\partial z^2}\right)_0 \iiint z^2 \rho \, d\tau + \left[-\frac{1}{2}\left(\frac{\partial^2 \Phi}{\partial z^2}\right)_0 \iiint \left(\frac{r^2 - z^2}{2}\right) \rho \, d\tau\right]$$

$$= \frac{1}{4}\left(\frac{\partial^2 \Phi}{\partial z^2}\right)_0 \iiint (3z^2 - r^2) \rho \, d\tau$$

$$= \frac{1}{2}\left(\frac{\partial^2 \Phi}{\partial z^2}\right)_0 Q_{\text{clas}}$$

The derivation shows that the quadruple moment of a nucleus may also be related to the energy acquired by the nucleus when it is placed in an external electric field.

2.4.3.3 Quantum Mechanical Representation of the Quadrupole Moment

In quantum mechanical representation of the quadrupole moment, the factor of $\left(\frac{1}{2}\right)$ in Q_{clas}, the classical value is dropped and it is divided by the electronic charge 'e'.

We represent $\left(\frac{2}{e}\right) Q_{\text{clas}}$ by Q_{body} as the quadrupole moment attached with the body of the nucleus. As has already been stated, the classical quadrupole moment Q_{clas} is taken about the z-axis that is fixed with the body of the nucleus. The body-fixed z-axis points in the direction of angular momentum I having the absolute value $\sqrt{I(I+1)}\ \hbar$. The quantum mechanical quadrupole moment Q_I is taken along the space-fixed axis in the direction of the maximum projection of the nuclear spin I. The magnitude of the nuclear magnetic moment in this space-fixed direction is $(I\hbar)$. Let us denote the direction of the space-fixed axis by z' and the angle between z' and the body-fixed axis z by 'β'. It is clear from figure 2.29 that,

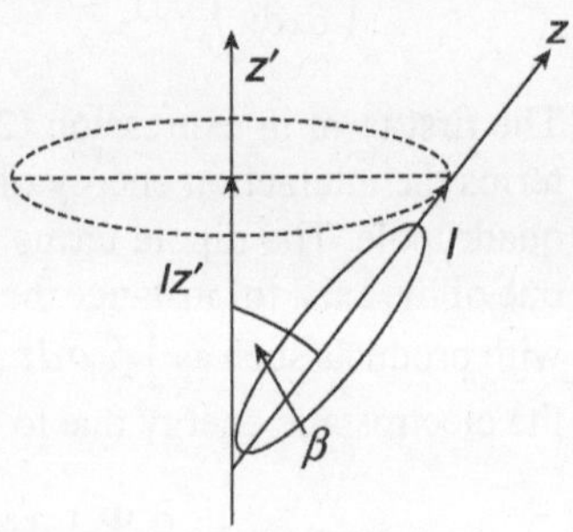

Figure 2.29 *Resolving the body-fixed quadrupole moment along the direction of maximum projection of spin I.*

$$\cos\beta = \frac{I_{z'}}{I} = \frac{I\hbar}{\sqrt{I(I+1)}\hbar} = \frac{I}{\sqrt{I(I+1)}} \qquad (2.35e)$$

The quantum mechanical quadrupole moment in the direction along Q is then given by

$$Q_I = \text{projection of } Q_{\text{body}} \text{ on the } z' \text{ axis}$$

$$= \text{Quantum mechanical operator for the projection} \times Q_{\text{body}}$$

$$= \frac{1}{2}(3\cos^2\beta - 1) \times Q_{\text{body}} = \frac{1}{2}\left\{3\left(\frac{I^2}{(I(I+1)}\right) - 1\right\} Q_{\text{body}}$$

$$= \frac{1}{2}\left(\frac{2I-1}{I+1}\right)Q_{\text{body}} = \left(\frac{I-\frac{1}{2}}{I+1}\right)Q_{\text{body}} \quad (2.35f)$$

Eq. (2.35f) states that the quantum mechanical quadrupole moment of a nucleus, which has $I = \frac{1}{2}$, is always zero, although it may have a finite value of the body-fixed quadrupole moment Q_{body}. The nuclei with $I = 0$ have no spin and, therefore, no preferred direction attached to the body. The time average of the charge distribution of such nuclei is zero. Thus, both type of nuclei with $I = 0$ and $I = 1/2$ do not have quadrupole moments.

The quantum mechanical quadrupole moment is the expectation value of

$$Q(I) \equiv <(3z^2 - r^2)> = <\Psi|3z^2 - r^2|\ \Psi^*> m_I = I$$

where the wave function belonging to magnetic quantum number $m_I = I$ has to be chosen.

One may also consider the expectation values for m_I that are less than I, giving the various projections of the intrinsic quadrupole moment on the axis of the nuclear angular momentum I. This leads to the general expression

$$Q(m_\ell) = \frac{\{3m_\ell^2 - I(I+1)\}}{(I(2I-1))} Q(I) \quad (2.36)$$

It may be noted that both for $+m_\ell$ and $-m_\ell$ values, $Q(m_l)$ has the same value. As a result, W_Q has the same sign and magnitude for $+m_\ell$ and $-m_\ell$

Solved example S-2.8

Derive an expression for the quadrupole moment for a charge distribution of an ellipsoidal shape with semi-major and semi-minor axes of lengths 'a' and 'b'. Assume a uniform distribution of charge in the ellipsoid.

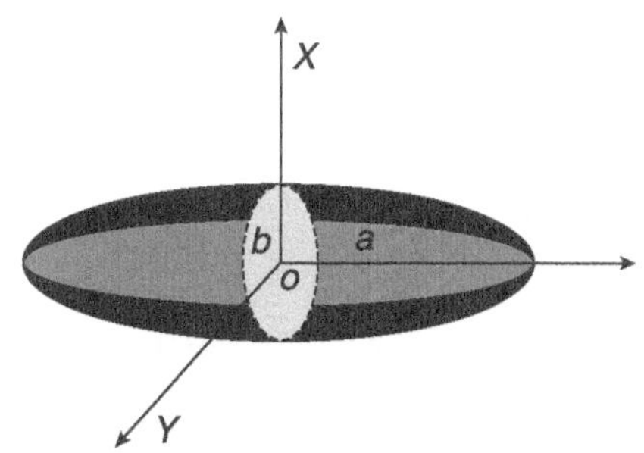

Solution. Assume that the total charge in the ellipsoid is Ze. The volume of the ellipsoid is $\left(\frac{4\pi}{3}ab^2\right)$. The uniform charge density $= \frac{Ze}{\left(\frac{4\pi}{3}ab^2\right)}$. The equation of an ellipsoid in Cartesian coordinates is

$$\frac{x^2}{b^2} + \frac{y^2}{b^2} + \frac{z^2}{a^2} = 1,$$

However, in this case, it is more convenient to work in the cylindrical coordinate system defined by the coordinates (ℓ, z, φ), where

$\ell^2 = x^2 + y^2$, and the equation of ellipsoid becomes $\frac{\ell^2}{b^2} + \frac{z^2}{a^2} = 1$. Further, the volume element $d\tau = d\ell\, dz\, d\varphi$. Also, $r^2 = \ell^2 + z^2$

The quadrupole moment $Q = \frac{1}{e}\iiint \rho\,(3z^2 - r^2)d\tau = \frac{1}{e}\rho \iiint (2z^2 - \ell^2)\ \ell d\ell dz\, d\varphi$ (A)

The variable z has the limit $-a$ to a; ℓ from 0 to $\left[b\sqrt{1-\frac{z^2}{a^2}}\right]$ and φ from 0 to 2π

The integral given by Eq. (A) can now be evaluated within the above-mentioned limit to get

$$Q = \frac{1}{e}\rho \iiint (2z^2 - \ell^2)\ \ell d\ell\ dz\ d\varphi = \frac{Z}{\left(\frac{4\pi}{3}ab^2\right)}\left[\int_0^{2\pi} d\varphi\right]\left[\int_{-a}^{a}\int_0^{b\sqrt{1-\frac{z^2}{a^2}}} \left((2z^2\ell - \ell^3)\, d\ell\, dz\right)\right]$$

$$= \frac{3Z}{2\,ab^2}\int_{-a}^{a}\left[b^2 z^2\left(1-\frac{z^2}{b^2}\right) - \frac{b^4}{4}\left(1-\frac{z^2}{b^2}\right)^2\right]dz = \frac{2}{5}Z\,(a^2 - b^2)$$

$Q = \frac{2}{5}Z\,(a^2 - b^2)$ barn if a and b are in cm; m^2 if a and b are in m.

Ans: The quadrupole moment of an ellipsoid is $\frac{2}{5}Z(a^2 - b^2)$, where Ze is the total charge in the ellipsoid.

Solved example S-2.9

The electric quadrupole moment of $^{177}_{72}$Hf nucleus is +3.0 barn. Calculate the lengths of the semi-major and semi-minor axes of the nucleus, considering the value of the unit nuclear radius as 1.2 F.

Solution. $Q = 2/5[Z(a^2 - b^2)]$, putting $Q = 3$ and $Z = 72$, we get

$$(a^2 - b^2) = \frac{3\times10^{-28}\times5}{2\times72} = 1.042\times10^{-29}\ \text{m}^2 \quad \text{(A)}$$

Also, the average radius R of the nucleus is related to a and b $R^2 = ½\{a^2 + b^2\}$; therefore, $(a^2 + b^2) = 2R^2 = 2(r_0A^{1/3})^2 = 2\{1.2 \times 10^{-15} \times (177)^{1/3}\}^2$

Or $$(a^2 + b^2) = 2\{1.2 \times 10^{-15} \times 5.61\}^2 = 9.079 \times 10^{-29}\ \text{m}^2 \quad \text{(B)}$$

From Eqs. (A) and (B), we get

$$2a^2 = 1.012 \times 10^{-28}\ \text{m}^2, \text{ and } a = 7.11 \times 10^{-15}\ \text{m} = 7.11\ \text{F}$$

Substituting this value of a in Eq. (A), the value of b may be calculated as 6.34 F

Ans: $a = 7.11$ F; $b = 6.34$ F

Solved example S-2.10

An ellipsoidal nucleus has a charge Ze, uniformly distributed over its volume. If R is the average radius of the nucleus and Δ the difference between the semi-major and semi-minor axes of the nucleus, calculate the quadrupole moment of the nucleus. Also calculate the fractional deviation from spherical symmetry.

Solution. $Q = \frac{2}{5}Z(a^2 - b^2)$, the average radius $R = \frac{1}{2}(a + b)$ and $\Delta = (a - b)$

$$Q = \frac{2}{5}Z(a+b)(a-b) = \frac{4}{5}ZR\Delta = \frac{4}{5}ZR^2\left(\frac{\Delta}{R}\right)$$

Ans: Fractional deviation from spherical symmetry $= \left(\frac{\Delta}{R}\right) = \frac{5Q}{4R^2}$

Exercise p-2.18: Estimate the ratios of the major to minor axes for ${}^{181}_{73}Ta$ & ${}^{123}_{51}Sb$ given that their quadrupole moments are respectively +6 barn & − 1.2 barn. Consider $r_0 = 1.5 \times 10^{-15}$ m.

Also it can be shown that $\frac{a}{b} \approx 1 + \Delta$. Using these expressions, it is easy to show that a/b for Ta is 1.143 and for Sb 0.947.

Exercise p-2.19: A charge 'q' is distributed uniformly along a straight wire of length 'a'. This linear charge distribution is held along the z-axis with its centre at the origin 'O' as shown in the figure A. Calculate the quadrupole moment of the charge distribution along (i) x-axis and (ii) z-axis.

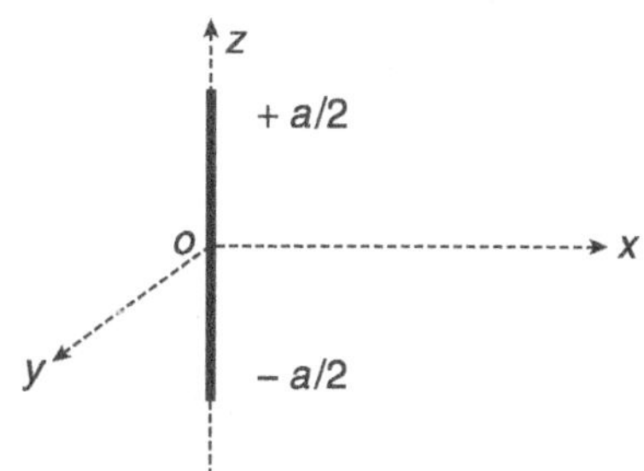

Figure A *Change distribution of Exercise p-2.19*

2.5 DETERMINATION OF NUCLEAR ANGULAR MOMENTUM I

As the methods of determining the nuclear angular momentum (also called nuclear spin), magnetic dipole moment (often called dipole moment), and the electric quadrupole moment (called simply quadrupole moment) are all inter related, it is better to consider them simultaneously. It has already been stated that it is generally not possible to have an isolated bear nucleus, and actually it is either atoms or molecules of different elements that are used to study the properties of the nucleus. Now, once there is an atom or a molecule, along with the nucleus there are atomic electrons as well. Like the angular momentum of the nucleus, the atomic electrons also possess angular momentum the magnitude of which is comparable to the magnitude of the nuclear angular momentum and the effects of it could not be neglected. The negatively charged electrons of the atom also produce an electric field at the location of the nucleus, the electric field being non-uniform over the volume of the nucleus. Besides these electric fields, both the nucleus and the atomic electrons may have magnetic moments. Although the magnetic moment of the nucleus is much smaller (measured in nuclear magneton which is about 2000 times smaller than the Bohr magneton) than the magnetic moment of the atomic electrons but in special cases the interaction of nuclear and atomic magnetic moments may become significant. Another complication comes from the external magnetic and/or electric fields that are usually applied for measuring the desired nuclear property. In view of this, let us consider how each nuclear property behaves under the complex situation where other fields due to atomic electrons are present. It is, however, interesting to note that electron magnetic moments cancel out in pairs and, therefore, only the unpaired valence electron in the atom has a magnetic moment.

Let us consider the angular momentum first. To differentiate nuclear angular momentum from the atomic angular momentum, the former is denoted by I and the later by J. It may be remembered that when it is said that the nucleus has a spin I it means that the absolute value of the angular momentum $I_{abs} = \{I(I + 1)\}^{1/2}\hbar$. Now, suppose an external magnetic or electric field is applied in the direction of the Z-axis. According to quantum mechanics, the observable values of the projection of the angular momentum in the Z-direction are given by m_I, called the magnetic quantum number, which can have following $(2I + 1)$ values:

$$I, (I-1), (I-2), (I-3), \ldots -(I-3), -(I-2), -(I-1), -I$$

The maximum measurable value of the spin is $I\hbar$. The relation between m_I and I_{abs} for a system of $I = 3$ is represented in figure 2.30. From the figure, it follows that

$$m_I\hbar = (I_{abs}\cos\Phi)\hbar$$

Exactly, same is true for the spin J of the electrons.

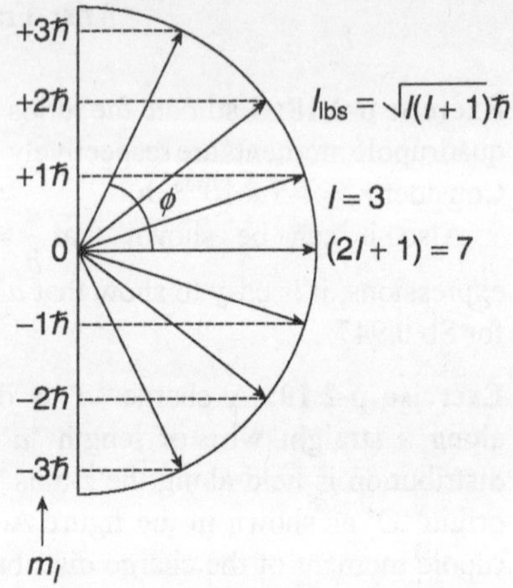

Figure 2.30 *Relationship between m_I and I_{abs}*

2.5.1 Nuclear Angular Momentum from the Multiplicity of the Hyper Multiplet

Let us now consider an atom in which there is an unpaired valence electron in state s, as for example the atoms of alkali metals. As both the nucleus and the unpaired valence electron have magnetic moments, the nuclear angular momentum I and the angular momentum J of the electron couple together to form F, the resultant total angular moment of the atom. Both I and J undergo Larmor precession around F, as is shown in figure 2.31. The total angular moment F can have values from $|I + J|$ to $|I - J|$ in steps of unity. The total angular momentum F will have quantized projections m_F along a direction specified by some external field. The absolute or numerical value of $F = \{F(F + 1)\}^{1/2}$. The possible number of different values of F, called the multiplicity of F, will be $(2I + 1)$ if $I < J$ or $(2J + 1)$ if $J \leq I$.

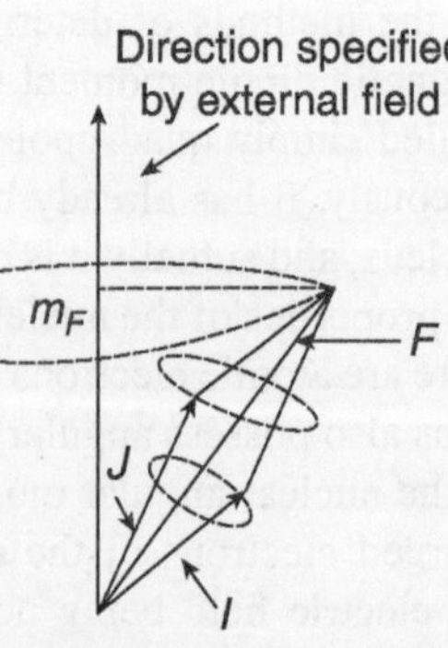

Figure 2.31 *Coupling of I and J*

Fine and hyperfine structures in atomic spectra: Atomic spectra consist of a large number of spectral lines corresponding to various electronic transitions. The energies of these transitions are well reproduced by the difference of electron energies in different quantum states in the central Coulomb field of the nucleus. High-resolution spectrograph showed that some spectral lines are made up of finer lines. This is called the fine structure of spectral lines and is due to the spin of the electron. Use of instruments with better resolution revealed groups of closely spaced spectral lines, called hyperfine multiplets. These hyperfine multiplets originate from the interaction of nuclear magnetic moment with the magnetic moment of the unpaired valence electron. The magnetic interaction between the nucleus and the electron through their magnetic moments create electronic states that differ slightly in energy from each other. Transitions corresponding to these states produce the hyperfine multiplets. The number of lines in a multiplet is equal to the multiplicity of the total angular momentum F of the atom. Hence, in those cases where the nuclear spin I is smaller than J, the direct counting of the number of the members of the multiplet gives the angular momentum of the nucleus.

2.5.2 Interval Rule for Hyperfine Multiplets

The magnetic interaction energy W_{mag} between two magnetic moments μ_I of the nucleus and μ_J of the unpaired valence electron is proportional to $\mu_I \cdot \mu_J \cos(I,J)$, but $\mu_I = g_I I$ and $\mu_J = g_J J$. As such

$$W_{mag} = aIJ\cos(I, J) \tag{2.37}$$

Here, a is a constant and $\cos(I, J)$ is the cosine of the angle between I and J and is given by

$$\cos(I, J) = [(F^2 - I^2 - J^2)/(2IJ)] \tag{2.38}$$

On substituting the value in Eq. (2.37), we get

$$W_{mag} = (a/2)[F(F+1) - I(I+1) - J(J+1)] \tag{2.39a}$$

It is now possible to calculate the magnetic energy for different possible value of F, starting from the highest value of $F = (I + J)$, $(I + J - 1)$, … to the minimum value $|I - J|$. The difference in W_{mag} for two successive values of F is tabulated in table 2.1.

Table 2.1 *Calculated values of the difference of magnetic energy between two successive states*

Larger value of F	Smaller value of F	ΔW_{mag}
$I+J$	$I+J-1$	$a(I+J)$
$I+J-1$	$I+J-2$	$a(I+J-1)$
$I+J-2$	$I+J-3$	$a(I+J-2)$
$I+J-3$	$I+J-4$	$a(I+J-3)$
$I+J-4$	$I+J-5$	$a(I+J-4)$
$I+J-5$	$I+J-6$	$a(I+J-5)$

It may be observed from the table that the energy difference is proportional to the larger of the two values of F. Thus, the separation between the successive hyperfine multiplets goes on decreasing in a regular fashion. In those cases where there is no interference from other fine or hyperfine splitting, the value of I can be determined from ΔW_{mag}, if J is known.

In principle, it is possible to determine the nuclear angular momentum from either the number of members of the hyperfine multiplet group and/or by measuring the relative energy separation of successive members, but in practice it is a very difficult task. It is because of two reasons. First, the magnetic energy differences between members of the hyperfine multiplets are very small of the order of 10^{-6} eV. Second, the electric field of the atomic electrons at the location of the nucleus interacts with the quadrupole moment of the nucleus, if the nucleus is deformed. The energy of this quadrupole interaction is of the order of 10^{-8} eV, much smaller than the energy for magnetic interaction.

2.5.3 Hyperfine Structure and External Magnetic Field

Coupling of the nuclear and the atomic angular momentum gives rise to the hyperfine structure, which is due to the $(2F + 1)$ sub-states corresponding to the possible values of the total angular momentum F. If a magnetic field B is applied now along the Z-axis, a torque $\zeta = \mu_F \times B$ is applied on the magnetic moment μ_F (vector sum of μ_I and μ_J) corresponding to the total angular momentum F. As a result, the total angular momentum F starts to precess around the direction defined by the external magnetic field B, that is along Z direction. Now, there are two magnetic fields present at the location of the nucleus: (i) the internal magnetic field due to the magnetic moment of the unpaired electron that is quite large of the order of 10^5–10^7 gauss. However, the direction and the magnitude of this internal field keep changing with the motion of the electron. As opposed to that, the magnetic field at the location of the electron due to the magnetic moment of the nucleus is much smaller, about 2000 times smaller because of the small value of the nuclear magnetic moment. On applying an external magnetic field B (if the magnitude of B is larger than the average nuclear magnetic field), the electron starts to precess around the direction of the external field B, the average internal magnetic field at the location of the nucleus becomes almost constant and points along the Z-direction, the direction of the external field B. (ii) The external field B and the average internal field both act along the Z-direction and add up. This sufficiently large field aligns the nuclei. Depending on the magnitude of B, three different situations may occur.

(a) **If B is very weak:** An external magnetic field of less than 100 gauss is treated as a very weak field as it is much smaller than the internal magnetic field. In this case, the angular momentum I and J remain coupled to each other and F precesses along the direction Z (of B), each F level will split into $(2F + 1)$ sublevels. This is called the Zeeman effect.

(b) **If B is weak:** A magnetic field larger than about 100 gauss is generally treated as a weak field. The field is strong enough to break the coupling between I and J. As a result, both I and J independently precess about the direction of the applied field. Now, the energy levels are designated in terms of m_J and m_I magnetic quantum numbers, respectively, for the electron angular momentum and the nuclear angular momentum. It is to be noted that for each value of m_J there are $(2I + 1)$ sub-states corresponding to the different values of m_I. Thus, $(2I + 1)$ closely packed m_I multiplets for each value of m_J are separated from each other by a relatively larger energy gap. The number of lines or sub-states for different m_J values remains $(2I + 1)$. This is called the Paschen-Back effect. The spin I of the nucleus can be easily found just by counting the number of lines in multiplets corresponding to different m_J values. Back and Goudsmit determined the angular momentum of bismuth by counting the number of sublevels in each magnetic sub-state of the electron spin J. Figure 2.32 shows the hyperfine structure for a hypothetical system having unpaired $s_{1/2}$ electron and the nucleus with angular momentum $I = 1\hbar$.

The angular momentum of the electron is $J = (l + s) = ½$. In the absence of external field, the system in ground state has two components for $F = I + J = 3/2$ and $I - J = 1/2$. Under very weak field, the $F = 3/2$ splits into four components corresponding to the $(2F + 1)$ multiplicity and the $F = 1/2$ into two components. If the external field is weak, the coupling between J and I breaks and the levels are designated according to the magnetic quantum numbers m_J and m_I. In the present case, there are two possible values of $m_J = +1/2$ and $-1/2$. Possible values of m_I are $(2I + 1)$, that is 3 (1, 0, −1). It is interesting to note that for the weak field (Paschen-Back effect) the ordering of different m_I sub-states for the different m_J branches are in reverse order. Another experimental fact is that the relative separation between sub-states does not depend on the magnitude of the applied field B, proving that the internal magnetic field is much stronger than the external field.

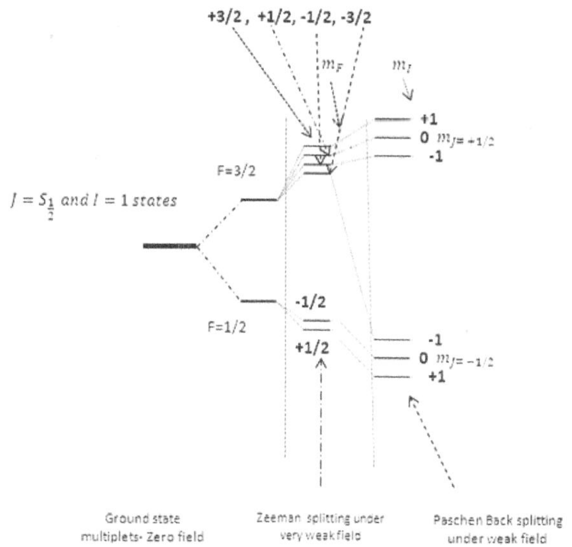

Figure 2.32 *Hyperfine structure under no-field, very weak field, and weak field*

The transitions between the magnetic sub-states follow the following selection rules

In very weak field (Zeeman hyperfine structure):

$$\Delta F = 0, \pm 1 \quad \text{and} \quad \Delta m_F = 0, \pm 1$$

In weak field (Pachen-Back hyperfine structure):

$$\Delta m_J \quad \text{or} \quad \Delta m_I = 0, \pm 1$$

On further increasing the magnitude of B beyond 1000 gauss, the coupling between electron spin angular momentum s and the orbital angular momentum l is also broken.

Exercise p-2.20: What are the main differences between the Zeeman and Pachen-Back effects?

Exercise p-2.21: What is the order of magnitude of the magnetic field produced by the atomic electrons?

2.5.4 Rotational Band Spectra of Homonuclear Diatomic Molecule

Molecules consisting of two identical nuclei (or atoms) are called homonuclear diatomic molecules such as $^{16}{}_{8}O_2$ and $^{14}{}_{7}N_2$. On excitation, these molecules produce three types of distinct emission spectra. The first and the most energetic part of the spectrum is the line spectrum due to electronic transitions. The next in energy is the band spectra due to the vibration of the atoms of the molecule around their common centre of mass. As the vibrations of the molecules can be well represented by a pure or near-pure simple harmonic oscillator, the vibrational spectrum is characteristic of the harmonic oscillator spectra. The third and lowest in energy is the rotational spectra. It is produced by the rotation of the molecule about an axis perpendicular to the symmetry axis. The axis joining the centres of the two nuclei is called the symmetry axis. We concentrate only on the rotational motion of the molecule around the common centre of mass.

The angular momentum I_1 and I_2 (=I, same value) of the two nuclei may be parallel to each other or anti-parallel. In case they are parallel, the molecule is called ortho-molecule and if they are anti-parallel a para-molecule. The coupling of angular momentum I with the angular momentum J of the electron is weak and, therefore, during the rotational motion of the molecule a para-molecule remains para and an ortho. The rotational motion does not affect the symmetry of the molecule.

In the general case, the electric field at the location of the electrons is not central as the molecules are not spherically symmetric. As a result, the electron angular momentum J is not conserved. However, the component of J on the symmetry axis, denoted by K, is conserved (see figure 2.33). The transverse component R_t of J changes with the rotation of the molecule. The rotational excited states of the molecule are generated by keeping K constant and increasing J in steps of unity. It may, however, be noted that the rotational state of lowest energy corresponds to $K = J$, which does not mean that R_t is zero. It is because the absolute value of $J_{abs} = [J(J+1)]^{1/2}$. The rotational energy of the molecule is given by

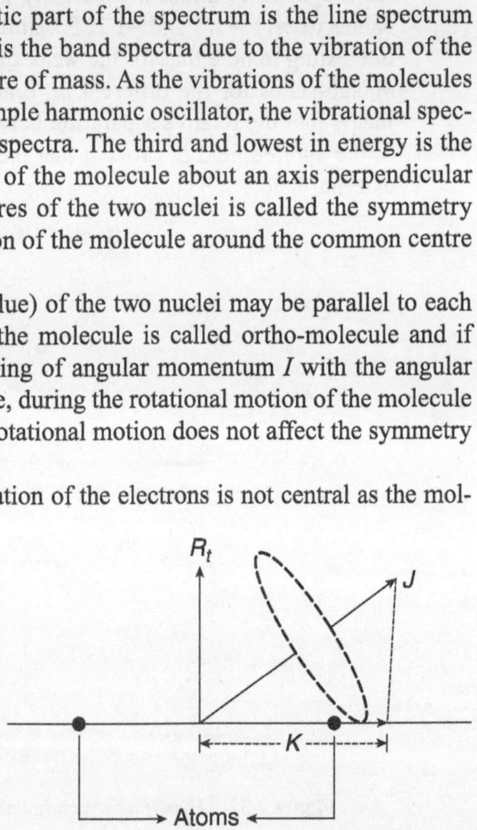

Figure 2.33 *Nutation of a homonuclear diatomic nucleus around J*

$$E_{rot} = \left(\frac{\hbar^2}{2\mathfrak{J}}\right)[J(J+1) - K^2] \tag{2.39b}$$

Here, $\mathfrak{J}$ is the moment of inertia of the molecule about the axis of rotation.

Because there is no torque in the system, the J does not change and the molecule undergoes nutation around J vector. Further, in the ground state of the homonuclear diatomic molecule, $K = 0$.

The space part of the total wave function ψ_{total} of the molecule can be written as the multiplication of four wave functions: (i) ψ_e, the molecular electronic wave function that contains the position coordinates of the nuclei. (ii) ψ_{rot}, the wave function that describes the rotation of the molecule. (iii) ψ_{vib}, that describes the vibration of the molecule and (iv) ψ_{spin}, the wave function that contains the nuclear angular momentum I_1, I_2, and so on. Therefore,

$$\Psi_{total} = \psi_e \psi_{rot} \psi_{vib} \psi_{spin} \tag{2.40}$$

Symmetry of the system plays a crucial role in transitions as it is conserved. Let us, therefore, consider the symmetry of each wave function of Eq. (2.40).

The symmetry of ψ_{total} is decided by the value of the angular momentum I of the nuclei. If I is an integer, ψ_{total} is symmetrical for the interchange of the nuclei as the molecule is a system of two bosons. In the case of I being half-integer, each nucleus is a fermion, and ψ_{total} is anti-symmetric for the exchange of nuclei.

In the electronic ground state of the molecule, ψ_e is always symmetric. However, in excited state the electronic wave function may be symmetric or anti-symmetric.

Similarly, ψ_{vib} is also symmetric in the case of homonuclear diatomic molecule.

The symmetry of ψ_{rot} depends on the value of J. It can have values starting from $J = K$, $J = K + 1, J = K + 2,\ldots$ In the ground state of the rotational motion $K = 0$; therefore, the possible values of J are 0, 1, 2,...Therefore, the rotational wave function ψ_{rot} is symmetrical if J is 0 or even and anti-symmetric for odd values of J.

The spin wave function, ψ_{spin}, is symmetric for ortho-molecule and anti-symmetric for the para-molecule.

Now, in a transition the symmetry is conserved. Out of the four wave functions, the symmetry of ψ_{spin} is fixed (as ortho-molecule remains ortho and a para-molecule remains para), hence in a transition the symmetry of the product $\psi_e \psi_{vib} \psi_{rot}$ should be conserved.

The selection rule for dipole transitions between the sub-states requires $\Delta J = \pm 1$.

In the ground state, ψ_{vib} is symmetric. However, the symmetry of ψ_{rot} depends on the value of J, and when J changes by ± 1 in electric dipole transitions its symmetry changes. If it is assumed that the symmetry of ψ_e does not change, the transition will change the symmetry of the product $\psi_e \psi_{vib} \psi_{rot}$, which is not allowed. It means that pure rotational, pure vibrational, and pure rotational-vibrational transitions (transitions in which the symmetry of ψ_e does not change) are not allowed in case of homonuclear diatomic molecules. Rotational and/or vibrational transitions will always be accompanied by electronic transitions in which ψ_e changes its symmetry to accommodate the change in the symmetry of ψ_{rot} and keeps the symmetry of the product $\psi_e \psi_{vib} \psi_{rot}$ intact.

In case of electronic-rotational transitions, the alternate lines in the band spectrum originate from the even and odd values of J, like J going from $J = 10$ to $J = 9$, J going from 9 to $J = 8$, and so on. As such the alternative lines of the band spectra are associated with the ortho and the para molecules. Thus, the ratio of the intensities of the alternate lines in the band spectrum are equal to the ratio of the number of ortho and para molecules in the system. The number of ortho and para molecule in a given sample depends on their statistical weights.

In statistics, the statistical weight of a state is proportional to the number of sub-states of the state. It is because each sub-state is assigned an equal probability. To determine the statistical weights of ortho and para molecules, we should first find out the number of sub-states for each of them. In the para-molecule the angular momentum I of the two nuclei point in opposite directions. The nuclear spin wave function for para-molecule is anti-symmetric of the form

$$\psi_{\text{spin}} = (1/\sqrt{2})[\psi_i(1)\psi_j(2) - \psi_j(1)\psi_i(2)] \tag{2.41}$$

With each $\psi_i(1)$ and $\psi_j(2)$ describe a state having the absolute angular momentum value of $\{I(I+1)\}^{1/2}\hbar$ and the magnetic quantum number $m_I\hbar$. There are $(2I+1)$ sub-states for particle 1. For particle 2, the number of sub-states remains $2I$, because for anti-symmetric state sub-states of the type $\psi_i(1)\psi_i(2)$ where the particles are in the same state are not permitted. Therefore, the number of total sub-states of the para-molecule becomes $2I(2I+1)$. However, it may be noted that all the sub-states have been counted twice (combinations of the type $\psi_i(1)\ \psi_j(2)$ and $\psi_j(2)\ \psi_i(1)$ represent the same state). Hence, the total number of sub-states for the para-molecule is $\frac{1}{2}[2I(2I+1)] = [I(2I+1)]$.

The number of sub-states for the ortho-molecule may be obtained by using a spin wave function having the same terms as for the para-molecule but with a positive sign instead of the minus in Eq. (2.41). Therefore, for an ortho-molecule in addition to the sub-states of the para molecule there will be another $(2I+1)$ sub-states of the type $\psi_i(1)\psi_i(2)$ where both the particles are in the same state. The number of sub-states for the para molecule are $[I(2I+1)] + (2I+1) = (2I+1)(I+1)$

The ratio of the statistical weights of ortho-molecule to para-molecule

$$= [(2I+1)\,(I+1)]/[I(2I+1)] = (I+1)/I$$

It follows from the above that the intensity of the alternative lines in the rotational band spectrum will be in the ratio of $(I+1)/I$. The value of the nuclear angular momentum I can now be determined from the measured value of the intensity ratio of alternative lines in the rotational band spectra.

Solved example S-2.11

A $D_{5/2}$ term in the optical spectrum of $^{39}_{19}K$ has a hyperfine structure with four components. Determine (a) the spin of the nucleus (b) the total angular momentum of the atom (c) the ratio of the energy interval between the successive members of the multiplet.

Solution. (a) It is given that the angular momentum of the electronic state J is 5/2 since the hyperfine structure belongs to $D_{5/2}$ state. Now, it is known that the number of members of a hyperfine multiplet is equal to either $(2J+1)$ or $(2I+1)$, whichever is smaller. Here, I is the spin of the nucleus. Now, $(2J+1) = (2\times 5/2+1) = 6$. However, the number of members of the multiplet is only four. It means that I is smaller than 5/2 and $(2I+1) = 4$. This gives $I = 3/2$.

(b) The total spin F of the atom is given by the quantum mechanical sum $F = J + I = 5/2+3/2$, The spin F can have different values from $|J+I|$ to $|J-I|$ in step of unity. Therefore, the possible values of F are 4, 3, 2, and 1.

(c) The energy of the successive states is given by Eq. (2.39) as

$$(a/2)\ [F(F+1) - I(I+1) - J(J+1)]$$

$$\text{Therefore, for } F = 4,\ E_4 = a/2\left[4(4+1) - \frac{3}{2}\left(\frac{3}{2}+1\right) - \frac{5}{2}\left(\frac{5}{2}+1\right)\right]$$

$$= a/2[20 - 12.5]$$

Similarly, for $F = 3$, $E_3 = a/2[12–12.5]$, for $F = 2$, $E_2 = [6–12.5]$ and for $F = 1$, $E_1 = a/2[2–12.5]$

The energy differences $\Delta E_1 = E_4 - E_3 = a/2[8]$; $\Delta E_2 = E_3 - E_2 = a/2[6]$ and $\Delta E_3 = E_2 - E_1 = a/2[4]$

Therefore, $\Delta E_1{:}\Delta E_2{:}\Delta E_3 = 8{:}6{:}4 = 4{:}3{:}2$

Ans: The ratio of energy intervals are: 4:3:2.

Exercise p-2.22: Explain why it is not possible to have pure rotational transitions in a homonucleus diatomic molecule?

Exercise p-2.23: Why is the intensity of a line in rotational band proportional to the number of sub-states feeding the transition?

Exercise p-2.24: Which molecule—ortho or para—has larger number of sub-states and why?

Exercise p-2.25: How is the angular momentum defined in quantum mechanics?

2.6 DETERMINATION OF NUCLEAR MAGNETIC MOMENT

2.6.1 Molecular Beam Resonance

Before we go into the details of the molecular beam resonance method, let us first study the effect of a magnetic field B on an object having a magnetic moment μ. The magnetic field produces a torque τ given by

$$\tau = \mu \times B.$$

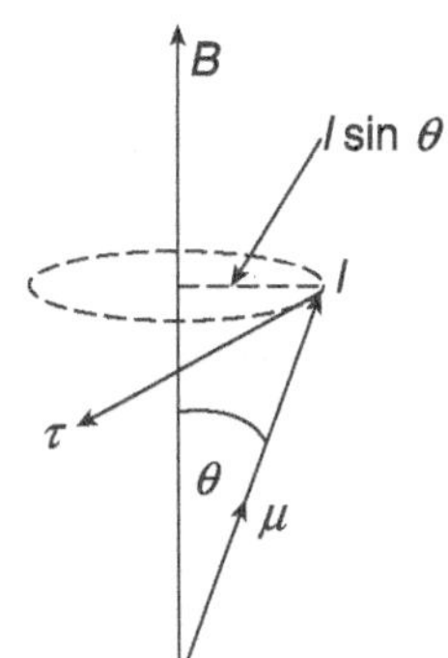

Figure 2.34 *The Larmor precession*

As a result of the torque τ, the magnetic moment starts rotating around the direction of the applied magnetic field B with angular momentum I. As a result of the torque τ, the magnetic moment starts rotating around the direction of the applied magnetic field B with angular momentum ℓ. This rotator motion is called Larmor precession. The torque is numerically equal to the rate of change of angular momentum l. Now, as shown in the figure 2.34, the change of angular momentum is equal to 'l sinθ', and is always perpendicular to the angular momentum l. If ω_{lar} denotes the frequency of The Larmor frequency corresponding to nuclear magnetic moments (which are of the order of few nuclear magnetons) and magnetic fields of the order of few hundred to about a thousand gauss lie in the region of mega cycles per second. This frequency range corresponds to radio waves (RF) of electromagnetic spectrum. If energy is supplied to a magnetic dipole, which is undergoing Larmor precession around a magnetic field, through the radio frequency electromagnetic waves of frequency ν, such that $\hbar\nu = \Delta W_{mag}$, where ΔW_{mag} is the energy difference between the two magnetic sub-states of the nucleus and $\nu = \omega_{lar}$, then the nuclear dipole may absorb energy from the RF electromagnetic field and may shift to the higher energy sub-state. This is called resonance absorption. Resonance absorption is frequently used in nuclear physics to determine the magnetic moment of the nucleus.

Neutral atomic and/or molecular beams can be produced by heating the salt or the element in an oven with a slit. The temperature of the oven should be such that the mean free path of the atoms/molecules is larger than the slit. A beam of neutral atoms or molecules can be taken out through the slit in a vacuum chamber to carry out the experiment. Three different types of experimental arrangements have been used to determine the nuclear magnetic moment. They are: (a) magnetic resonance (b) nuclear para-magnetic resonance, and (c) nuclear resonance induction.

The experimental arrangement of the apparatus used in magnetic resonance method is shown in figure 2.35.

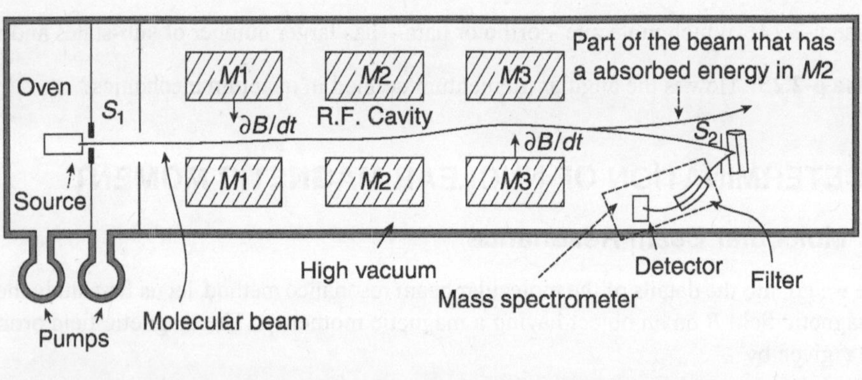

Figure 2.35 *Molecular beam magnetic resonance apparatus*

The essential parts of the apparatus are as follows:

1. An oven where the element or the salt of the material to be studied is heated and a beam of neutral atoms or molecules with thermal energies comes out of the beam defining slit S_1.
2. The beam then passes through three set of magnets which forms the most important part of the apparatus. The first and the third magnets M1 and M3 produce non-uniform magnetic fields in Z-direction. The magnitude of the magnetic gradient in M1 and M3 is same but its direction in M3 is opposite to that in M1. The second magnet M2 produces a uniform magnetic field in Z-direction. However, there is a cavity in the central magnet M2 where a RF coil may be placed. An alternating RF signal of desired frequency may be applied to the coil. The orientation of the coil is such that the direction of the alternating magnetic field produced by the RF is in horizontal direction, that is, perpendicular to the magnetic fields of the three magnets.

3. The detector system is placed behind a slit identical in size to the beam defining slits. It consists of a narrow hot tungsten wire. The neutral molecules of the beam get ionized on hitting the wire. The ions so produced are then attracted by a hollow electrode at an appropriate potential that attracts the ions and direct them to a single focusing mass spectrograph for mass analysis. The mass analysis of the ions is done to confirm that they are ions of the beam. The mass analysed ions then hit a photomultiplier tube with appropriate electronics. The ions are counted as electronic pulses by a counter. The special feature of the detecting system is its very high sensitivity.

Working of the apparatus: The apparatus is used for molecules that have no resulting atomic angular momentum. A beam of neutral atoms or the molecules comes out from the beam defining slit S_1 and passes through the magnet M1. As the magnetic field of M1 is non-uniform in the Z-direction, the molecules are subjected to a force $(\mu_z \partial B/\partial z)$ due to their magnetic moments. It may be remembered that in a uniform magnetic field the magnetic moment produces Larmar precession around the direction of the magnetic field and in a non-uniform magnetic field a net force acts on it. As a result of the force exerted by the non-uniform magnetic field, the beam deviates slightly from its direction of motion. Next, the beam passes through the uniform magnetic field B of magnet M2. If for a minute it is assumed that $B = 0$, the molecules will continue to move on and will pass through the magnetic field of M3. This magnetic field is also non-uniform exactly like the field of M1, but the gradient of the field $(-\partial B/\partial z)$ is just opposite to the gradient of M1. The result is that now the molecules get a push in the opposite direction. The net result of the magnetic fields of M1 and M3 is to give deflections of equal amounts but in opposite directions so that they cancel out and the molecules reach the detector slit as if they have not been deflected from their original path. In this condition, there will be sufficient number of molecules reaching the detector giving a certain high count rate N (say) in the detector counter. If the uniform magnetic field B (in direction Z) in M2 is now switched on, the molecules after getting a slight deflection in M1, will now be trapped in Larmor precession in uniform magnetic field B along the direction Z. The degeneracy in the energy of magnetic sub-states of the molecule due to the different projections of m_I on the Z-axis will be removed. A radio wave of frequency ν is now applied through RF coil in the cavity of magnet M2. If the frequency ν of the applied RF is equal to the Larmar frequency ω_{lar} of the molecule, and $\hbar\nu = \Delta W_{\text{mag}}$, the difference in the energies of the magnetic sub-states, a resonance will take place and the molecule will absorb energy from the RF field to shift to the next higher energy magnetic sub-state. As a result of this energy absorption, the orientation as well as the magnitude of the projection of the magnetic moment μ_z on the Z-axis will change. Now, these molecules will pass through the non-uniform magnetic field of M3. A force will again be applied by the non-uniform magnetic field on the molecules but the magnitude of the force will be different than the force the molecules have experienced while passing through M1. The net result will be that the molecules that have absorbed energy will not be able to reach the detector and the detector count rate will fall. As such, fall in the detector count rate is an indication of the energy absorption by the molecules in the cavity of M2 or of resonance. The conditions of resonance are

$$\Delta W_{\text{mag}} = \hbar\nu \text{ and } \nu = \omega_{\text{lar}} = (\mu_z B)/I\hbar$$

For a given value of the RF frequency ν, the strength of the magnetic field B is changed to get the minimum detector count rate. From the known values of B and ν for minimum detector count rate, the value of the magnetic moment μ_z can be determined if the angular momentum I is known.

The first beautiful demonstration of the application of molecular beam resonance method was made by Stern and Gerlach in 1922. Through this experiment they proved the space quantization. Rabi and his co-workers used LiCl molecules and determined the value of the magnetic moment of ^{7}Li as 3.24 nm (nuclear magneton). The typical curve showing the fall in the beam intensity as a function of magnetic field and the resonance RF frequency is shown in figure 2.36.

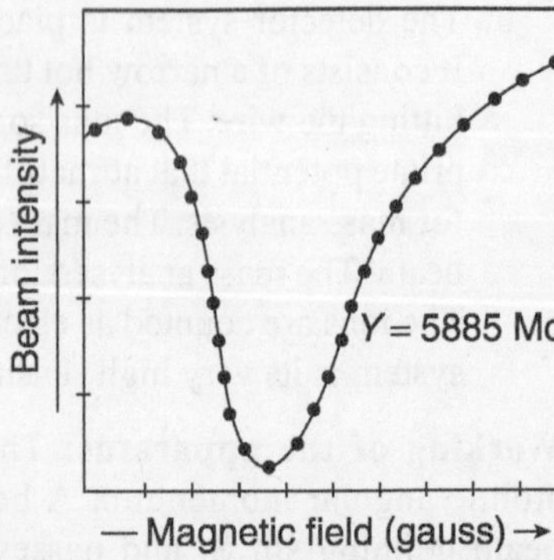

Figure 2.36 *Typical curve showing the fall in beam intensity at resonance*

2.6.2 Nuclear Magnetic Resonance Induction Method

As far back as 1936, C.J. Gorter suggested that if a large number of nuclei are simultaneously forced to change their magnetic sub-state by absorbing energy from the RF field under resonance conditions, it will be possible to measure the Larmor frequency and the magnetic moment of the nucleus. F. Bloch and his co-workers in 1946 carried out an experiment based on the idea of Gorter to determine the magnetic moment of proton. In this method, a very small amount of the sample (about 100 mg) is taken and is placed in an external homogeneous magnetic field B. If the sample material is not ferromagnetic the magnitude of the magnetic field at the location of the sample is almost the same as B or slightly less than B because of the inherent diamagnetism of substances. Internal magnetic fields, if present because of the un-paired electron changes rapidly because of the thermal motion of electrons and cancels out. An RF driver coil produces a magnetic field perpendicular to B. When the RF field is tuned to the Larmor frequency, resonance occurs and proton flips from the magnetic sub-state 1/2 to −1/2. It may, however, be remembered that the resonance condition induces both absorption of energy by the nuclei in lower energy sub-state (proton in sub-state +1/2) and also emission of energy by protons that are initially in higher energy sub-state −1/2. Thus, both absorption of energy from the RF field and emission of energy back to the RF field take place simultaneously at resonance. However, the important point is that in the sample at a given temperature the number of nuclei in higher energy state are always less in number than the number of nuclei in lower energy state. The equilibrium distribution of protons in higher and lower energy states is given by Boltzmann's law according to which the ratio N_{low}/N_{high} is given by

$$N_{low}/N_{high} = e^{(-E/kT)}$$

where E is the energy difference between the higher and the lower energy states and k and T are, respectively, Boltzmann constant and the absolute temperature. For protons at the room temperature ($T = 300$ K) and for a magnetic field of few thousand gauss, the fractional difference in the population of the protons in lower and higher energy sub-states is of the order of 7×10^{-6}.

Although it is a very small number, it is sufficient to produce the energy absorption signal. In the induction method of detecting the absorption signal, another RF coil called the detector or receiving induction coil is used. When energy is absorbed from the field the amplitude of the driver RF signal in the driver coil decreases and this decrease in the amplitude induces signal in the detector/receiver coil placed perpendicular to both the main magnetic field B and the magnetic field of the RF signal. The electronic circuitry further amplifies the signal. It may, however, be noted that the electronic pulse picked up by the detector coil, corresponding to the decrease in the amplitude of the RF signal due to energy absorption, is a transient pulse. To synchronize the pulse on the oscilloscope the pulse should repeat with a fixed frequency. To achieve this, the sample is swept a large number of times through the resonance by modulating the coil and synchronizing the signal with the modulating frequency. Magnetic moments for a large number of nuclei have been measured by this method.

2.6.3 Paramagnetic Resonance Method

Crystals of paramagnetic substances have one unpaired electron per atom in angular momentum state $l = 0$, referred to as S-state. Since $l = 0$, the electron possesses no magnetic moment due to orbital motion and have intrinsic magnetic dipole moment μ_e which is of the order of few Bohr magneton (Bn). When such atoms are placed in an external magnetic field of strength B, they align their spins either along the applied magnetic field or opposite to it. The state in which the spin is parallel to the magnetic field has lower energy as compared to the state with spin in opposite direction. At room temperature and under normal conditions, the states of higher energy lose energy and become low energy states and the low energy states absorb energy and flip to the higher energy states continuously. However, the number of atoms in lower energy state is always more than the atoms in the higher energy state. The energy difference between the higher energy and lower energy states $\Delta W_{mag} = \{\mu_e B - (-\mu_e B)\} = 2\mu_e B$. The frequency ν of the photon corresponding to this energy difference may be given by

$$\nu = \Delta W_{mag}/\hbar = 2\mu_e B/\hbar$$

When realistic values of the quantities like $\mu_e \approx$ few Bohr magneton and $B \approx$ few Weber per square meter are substituted in the above-mentioned equation, the frequency ν comes out to be of the order of 10^{10} cycles/s that lies in microwave range. As such if the crystal is subjected to microwave flux of the appropriate frequency, the atoms of the crystal with spin parallel to the field will absorb energy from the microwave field and those which have their spin opposite to the field may lose energy. This is called induced absorption and emission. However, more transitions from the lower energy states to higher energy states will occur due to the larger number of atoms in lower energy states. If no other magnetic field is present, a single absorption line will appear in the microwave absorption spectrum of the crystal. However, if the nucleus of the atom has the angular momentum I, there will be $(2I + 1)$ possible values of the magnetic quantum number m_I and corresponding to that the single absorption line will spilt in to $(2I + 1)$ components. This splitting is due to the interaction between the magnetic moment of the electron and the nuclear magnetic moment.

It is to be noted that for a fixed frequency microwave flux, the absorption lines for different transitions will appear at different strengths of the magnetic field B. As such in actual experiments, the magnetic field is varied from a lower value to a higher value and absorption lines appear one after another. If the angular moment of the nucleus is $I = 5/2$, in all six absorption lines corresponding to the following transitions will be observed on increasing the magnetic field.

$$m_I = +5/2;\ s = +1/2 \rightarrow m_I = +5/2;\ s = -1/2$$

$$m_I = +3/2;\ s = +1/2 \rightarrow m_I = +3/2;\ s = -1/2$$

$$m_I = +1/2;\ s = +1/2 \rightarrow m_I = +1/2;\ s = -1/2$$

$$m_I = -1/2;\ s = +1/2 \rightarrow m_I = -1/2;\ s = -1/2$$

$$m_I = -3/2;\ s = +1/2 \rightarrow m_I = -3/2;\ s = -1/2$$

$$m_I = -5/2;\ s = +1/2 \rightarrow m_I = -5/2;\ s = -1/2$$

For the fixed frequency of the microwave field, the absorption lines will appear on increasing the strength of the field B as shown in figure 2.37. The magnetic moment of the nucleus may be determined from the measured value of the energy separation between the absorption lines using the known values of the microwave frequency and the strength of the magnetic field B. The number of absorption lines tells about the angular momentum of the nucleus.

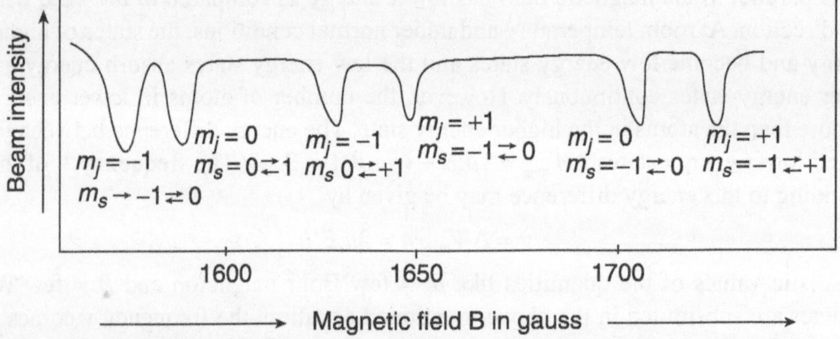

Figure 2.37 *Absorption spectra of a paramagnetic crystal having a nucleus of angular momentum $I = 5/2$*

Exercise p-2.26: Differentiate between ortho and para hydrogen and explain which of the two types of molecules will show hyperfine structure and why?

Exercise p-2.27: In molecular beam resonance method explain the fall in the intensity of the beam at resonance.

Exercise p-2.28: What are the conditions for the resonance in molecular beam resonance method of determining the magnetic moment?

Exercise p-2.29: What is the main advantage of resonance induction method for determining the magnetic moment over the molecular beam resonance method?

Exercise p-2.30: Suppose a particle of charge +1e and rest mass 1000 MeV/c^2 is rotating with angular velocity of 2.6×10^{23} rad/s about its axis. If the angular momentum of the particle is equal to $\hbar$ $(= 6.625\times10^{-34}\ Js)$, calculate the rotational energy of the particle in MeV.

[**Hint:** Let the moment of inertia of the particle about its axis of rotation be $\mathfrak{J}$. Then, the rotational energy $E_{rot} = ½\mathfrak{J}\omega^2$ and the angular momentum $J = \mathfrak{J}\omega$. From these two relations, it is easy to get $E_{rot} = ½J\omega = 1/2 \times 6.625 \times 10^{-34} \times 2.6 \times 10^{23} = 8.6125 \times 10^{-11}$ J. However, 1.6×10^{-13} J = 1 MeV, therefore, 8.6125×10^{-11} J = 538.4 MeV. It may be noted that this classical non-relativistic treatment is valid as the rest mass of the particle is much larger than its kinetic energy.]

2.7 DETERMINATION OF THE NUCLEAR QUADRUPOLE MOMENT *Q*

2.7.1 The Shift in the Energy of Hyperfine Structure

The magnetic interaction energy of the ordinary hyperfine lines is given by the expression (2.39a)

$$W_{mag} = (a/2)\ [F(F + 1) - I(I + 1) - J(J + 1)] \tag{2.39a}$$

Expression (2.39a) holds for the case where the nuclear charge distribution is spherically symmetrical. However, if the nucleus is deformed, the interaction energy W_{mag} get changed by an amount that is proportional to the multiplication of the nuclear quadrupole moment and the gradient of the electrical potential produced by the electrons of the atom at the site of the nucleus. The revised energy W_Q is given by

$$W_Q = B/4\ [\{3/2(C(C + 1) - 2I(I + 1)\ J(J + 1)\}/\{I(2I - 1) - J(2J - 1)\}] \tag{2.36a}$$

where B, called the quadrupole factor, is proportional to the quadrupole moment Q and C is defined as $C = F(F + 1) - I(I + 1) - J(J + 1)$.

In principle, the value of the quadrupole moment of the nucleus can be obtained from this shift in the energy of ordinary hyperfine structure. In actual practice, the gradient of the electrical potential at the site of the nucleus is estimated theoretically and, therefore, the value of the quadrupole moment determined from the energy shift is subject to the errors in the estimation of the electric field gradient. However, sometime different isotopes of the same element show shifting of the hyperfine structure. In such a case, the electric field gradient at the site of the nucleus in different isotopes is almost identical and, therefore, the ratio of the energy shifts gives a more reliable value of the ratio of their quadruple moments. A very good example is in the case of the isotopes of the element osmium. The even-even isotopes 192, 190, 188, and 186 of osmium have $I = 0$ and, therefore, show only a single line in the spectrum. However, each of the line is found to be shifted by a nearly constant amount of 0.03 cm^{-1} per mass unit. This shifting is due to what is called the isotopic shift. The isotope ^{189}Os shows four hyperfine components. Setting the multiplicity $(2I + 1) = 4$, one gets the spin of this isotope to be $I = 3/2$. From the absolute separations between the four hyperfine components, the magnetic moment μ of ^{189}Os has been found to be 0.7 + 0.1 nm. The four components of the hyperfine structure were found to be shifted from the energies predicted by Eq. (2.39a). The quadrupole moment calculated from the energy shift of hyperfine structure gave the value $Q = +(2.0 \pm 0.8)b$.

2.7.2 The Hyperfine Structure of Molecular Rotational Spectra

The absolute difference in the energies of the successive members of the hyperfine multiplet of rotational band spectra of diatomic molecules depends on four factors: (i) the angular momenta of the nuclei, (ii) the angular momentum of the molecular electron distribution, (iii) the quadrupole moment of the nuclei, and (iv) the gradient $(\partial^2\phi/z^2)$ of the electric potential ϕ at the site of the nucleus due to all electrons of the molecule. The energy separation does not depend on the magnetic moment. The quadrupole moment of the deuteron was determined from the energy separation of hyperfine structure of D_2. In a similar way, the angular momentum I and the quadrupole moment Q for ^{14}N, ^{18}O, ^{33,34}S, ^{79,81}Br, and so on have also been determined from the rotational spectral shifts and absolute difference in the energies of rotational hyperfine lines. The rotational bands for some heavy molecules lie in microwave region. As microwaves have been used for radar and other defence related equipment since World War II, the microwave techniques have made great advances in precision and accuracy. As a result, accurate determination of Q and I for diatomic molecules using microwaves have been done.

Multiple choice questions

Note: In some of the following questions more than one alternative may be true. Tick all the correct alternatives in such cases for the complete answer

Exercise p-2.31: Discuss the classical and the quantum mechanical definitions of the quadrupole moment. What are the main differences between the two?

Exercise M-2.1 A thin metal foil of thickness d cm contains N atoms of the metal per cubic centimetre. Each atom has a nuclear cross section σ cm^2 for the scattering of α particles by an angle θ or more. The fraction of normally incident α particles scattered by more than θ is

(a) $N\sigma d$ (b) $N\frac{\sigma}{d}$ (c) $d/N\sigma$ (d) Nd/σ

Exercise M-2.2 $^{235}_{92}$U emits 7 α particles and 4 β^- rays before reaching the final stable product. The final stable nucleus is

(a) $^{207}_{84}$Po (b) $^{207}_{82}$Pb (c) $^{203}_{84}$Po (d) $^{203}_{82}$Pb

Exercise M-2.3 Assuming constant density model of the nucleus with unit radius 1.45 F, the maximum kinetic energy of positron emitted in the decay of $^{23}_{12}$Mg to $^{23}_{11}$Na is approximately

(a) 1 MeV (b) 3 MeV (c) 5 MeV (d) 7 MeV

Exercise M-2.4 The Coulomb barrier height in MeV between a particle of charge ze and the target nucleus of charge Ze, with the unit nuclear radius of 1.5 F, is approximately

(a) $0.76\, z\, Z^{2/3}$ (b) $0.76\, z^{2/3} Z$ (c) $0.76\, z^{2/3}\, Z^{2/3}$ (d) $0.76\, z\, Z^{-2/3}$

Exercise M-2.5 A singly ionized oxygen ion of kinetic energy 20 keV is passed through a radial electrostatic filter of field strength 16 kV/cm. The radius of curvature (in cm) of the path of the ion is approximately

(a) 5 (b) 10 (c) 15 (d) 25

Exercise M-2.6 The quadrupole moment and the magnetic dipole moment of nuclei are measured, respectively, in units

(a) cm, nuclear magneton (b) nuclear magneton, barn
(c) barn, nuclear magneton (d) cm, Bohr magneton

Exercise M-2.7 In the expression $\mu = gJ$, if the magnetic moment is measured in nuclear magneton then the angular momentum should be measured in units of

(a) ergs × m (b) $\hbar$ (c) $\frac{\lambda}{2\pi}$ (d) MeV

Exercise M-2.8 The doublet separation of 0.03637 u between two singly charged ions corresponds to an energy difference (in MeV) of approximately

(a) 14 (b) 24 (c) 34 (d) 44

Exercise M-2.9 In a particular nucleus the outermost proton (mass 1.7×10^{-27} kg) at a distance of 4 F from the centre moves in a circular orbit with orbital angular momentum quantum number $\ell = 1$. The speed of the proton is approximately

(a) 2.2×10^{11} m/s (b) 0.1×10^{10} m/s (c) 0.22×10^{9} m/s (d) 0.22×10^{8} m/s

Exercise M-2.10 If ϑ is the angle between the direction defined by the body axis and the direction defined by the magnetic substrate $m_I = I$ for a nucleus of spin ½. The value of $\cos\vartheta$ is

(a) 1/3 (b) $\sqrt{3}/2$ (c) $\frac{\sqrt{3}}{2}$ (d) $\frac{1}{\sqrt{3}}$

3

Force between Nucleons

3.1 INTRODUCTION

Considering the mass of a nucleus proportional to the number of nucleons in it and the radius equal to $r_0A^{1/3}$, it is easy to calculate the density for the nuclear matter of a spherical nucleus to be of the order of 10^{17} kg/m³. In addition, the density of the densest metal, mercury, is of the order of 10^{10} kg/m³. The very large difference, of the order of 10^7, between the density of nuclear matter and the ordinary matter indicates two facts: (1) the ordinary matter contains large volumes of empty spaces, (2) the force responsible for the binding nucleons inside a nucleus is very much stronger than and may be very different from the forces responsible for the binding of ordinary matter.

3.2 INADEQUACY OF CLASSICAL FORCES

On the first point, it may be mentioned that the ordinary matter is made of molecules and atoms. There is large vacuum in an atom around the nucleus in which electrons revolve and the intra-atomic spaces in a molecule are also empty. With regard to the second point, the forces that bind atoms and molecules in ordinary matter are largely electrical forces and to a much lesser extent gravitational and magnetic forces. It may be remembered that the range, the distance up to which the effect of the force can be felt, of all the three forces (also called classic forces) extends up to very large distances essentially up to infinity. Scattering and other experiments have indicated that the range of the force responsible for the binding of nucleons in nucleus is very small of the order of 2×10^{-15} m or 2 F. Very short-range nuclear force is a major point of difference between the classical and nuclear forces. Another difference is the manner in which atomic electrons are bound in an atom and the binding of nucleons in a nucleus. In an atom, the central positively charged nucleus produces an electric field spreading up to a large distance and electrons, relatively far away from the centre of force, are held in stationary orbits. Electrons are held at relatively weaker part of the central electric field, which is apparent from the fact that the binding energies of outer (valence) electrons are only of the order of few electron volts. The inner electrons in an atom are also not very strongly bound, their binding energies being of the order of few keV. By contrast, in a nucleus there is no source that may produce a central field; nucleons exert force on each other to bind themselves strongly, B/A of the order of 8 MeV. Such a large binding force between two nucleons cannot be provided by classical forces, as the gravitational attraction between two nucleons (a proton and a neutron) at a separation of 2 F may give a potential energy which is at least 10^{36} orders of magnitude smaller than the binding energy of nucleons. There is no question of electrostatic force of attraction between nucleons as neutron has no electric charge and protons will repel each other. A force of magnetic nature may come into play between two nucleons because of their magnetic dipole moments. However, this force will have opposite nature when the two nucleons have parallel and anti-parallel spins. Also, critical analysis of the data on the separation energies of the last nucleon from various nuclei has indicated that the nuclear force between nucleons is attractive for both the anti-parallel and parallel spin orientations. In view of this, classical forces are totally inadequate to provide the strong binding to nucleons in a nucleus.

One way to understand nuclear structure is to start with the force between two nucleons, find a potential for the nucleon–nucleon interaction, develop a structure for the nucleon consistent with the potential, and extend it to construct the nuclear structure. The first step in this approach is to find the properties of nucleon–nucleon force.

Exercise p-3.1: Calculate the gravitational potential energy between two protons at 2 F apart.

Exercise p-3.2: Calculate the electrical potential energy between two protons held 2 F apart.

3.3 TWO-BODY NUCLEON–NUCLEON FORCE

If it is assumed that nucleons exert force on each other in pairs, then the following nucleon pairs are possible: (n–n), (p–p), and (n–p). As both neutron and the proton have intrinsic spins S_n and S_p ($=\frac{1}{2}\hbar$), there are two possibilities for the spin orientations of the two nucleons of the pair. The first possibility is that the spins are parallel making the total angular momentum (j) of the pair 1 $\hbar$ with three possible values of $m_j = 1, 0, -1$. The parallel spin state is, therefore, called the *triplet state* and is denoted by superscript 3. The second possibility is that the spins of the nucleons in the pair are anti-parallel, making the j of the pair zero, and only single possible value for $m_j = 0$. The system with anti-parallel spins is called *singlet state* and is denoted by superscript 1. In view of this, there are six possible states of a two-nucleon system: ^{1}(n–n), ^{1}(p–p), ^{1}(n–p), ^{3}(n–n), ^{3}(p–p), ^{3}(n–p). States ^{3}(n–n) and ^{3}(p–p) are disallowed by Pauli exclusion principle, which states that no two identical fermions can have identical quantum numbers. It may, however, be noted that ^{3}(n–p) state is allowed. (Can you tell which quantum number is different for neutron and proton?) As such only four pairs, three singlet, and one triplet states are left for consideration. The forces corresponding to these states may be denoted by $F_{1(n-n)}$, $F_{1(p-p)}$, $F_{1(n-p)}$, and $F_{3(n-p)}$. A word of caution about $F_{1(p-p)}$ is that it is the specific nuclear force between two protons and does not contain the Coulomb repulsive force, which is in addition to the nuclear force but of very small magnitude as compared with it.

Information about the nucleon–nucleon (n–n) force may be obtained from (1) the systematic of the separation energies of last nucleon in light nuclei, (2) level scheme of mirror nuclei, (3) study of the properties of deuteron, (4) nucleon–nucleon scattering data, and the (5) constant density of nuclear matter.

3.3.1 Separation Energy of the Last Nucleon in Light Nuclei

Energy required to take out the outermost neutron (or proton) from the nucleus is called the separation energy of last neutron (or proton) and is denoted by S_n (or S_p).

$$S_n(Z, N) = [M(Z, N-1) + M_n - M(Z, N)]c^2 \quad \text{and} \quad S_p(Z, N) = [M(Z-1, N) + M_p - M(Z, N)]c^2$$

Here, M represents the mass of the nucleus having neutrons and protons as specified in the bracket, M_n and M_p, respectively, the neutron and the proton masses. The multiplication with c^2 is done to convert mass into energy. The c^2 term can be dropped if masses are given in the unit of energy (MeV) and the nuclear mass may be replaced by the corresponding atomic mass without much error as electron mass ultimately cancels out. Replacement of nuclear masses by atomic masses only brings in some negligible error due to the difference in the electron binding energies in the two atoms. Alternatively, S_n and S_p can also be given in terms of the binding energies of the two nuclei.

$$S_n(Z, N) = B(Z, N) - B(Z, N-1) \text{ and}$$
$$S_p(Z, N) = B(Z, N) - B(Z, N-1)$$

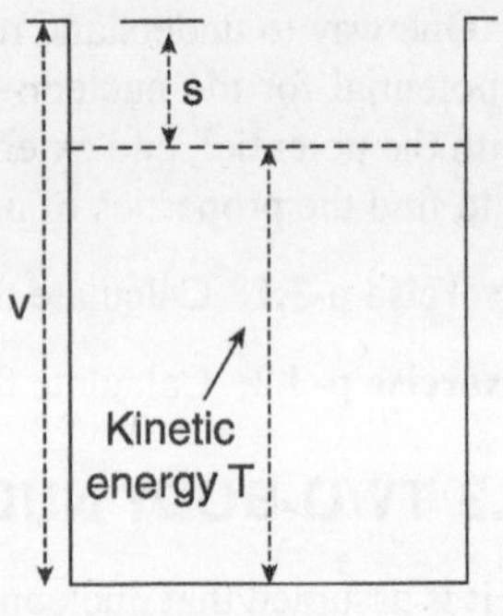

Figure 3.1 *Nuclear potential well and the nucleon separation energy*

To confine a particle within a given space, the de Broglie wavelength of the particle must be of the order of the size of the space. This condition puts a limit on the minimum energy of the particle for confinement in the limited space. For example, to confine a nucleon inside a nucleus of size of the order of 10^{-14} m, the minimum kinetic energy T of the nucleon must be of the order of 20 MeV. This implies that inside the nucleus a nucleon is in a potential well of depth V larger than 20 MeV as shown in figure 3.1. The separation energy $S = (V - T)$. The depth of the potential V must not be less than ≈30 MeV, considering the average binding energy of nucleon to be around 8 MeV.

An analogous quantity in atomic case is the binding energy of the last valance electron. The binding energy of valance electron in an atom shows cyclic variation. This is attributed to the electronic shell closure. Once a given shell is completely filled, the binding of the next electron becomes very weak. Similar cyclic variation is exhibited by S_n and S_p. This and some other systematic trends observed in stable nuclei, such as the extra stability associated with nuclei having the number of neutrons or/and protons equal to the magic numbers 2, 8, 20, 28, 50, 82, 126, and so on, indicates that nucleons inside the nucleus move in some sort of shell structure. Shell structure in atomic case is the outcome of confining particle in a central field. In case of the nucleus, there is no central field as such. To explain the apparent shell structure, it is assumed that all the nucleons inside a nucleus produce a mean central field in which each nucleon moves independently. This is the basis of the independent or single-particle model of the nucleus. The motion of nucleons in single particle approach may be grouped into states of different angular moment l. The experimental data on S_n and S_p states the following:

1. Only nucleons in the same l state (have no relative angular momentum relative to each other with even parity) exert force on each other. Interaction between nucleons in states of different orbital angular momentum is insignificant.
2. In the case of lightest nuclei, $^2_1$H, $^3_1$H, and $^4_2$He, the S_n increases from 2.2 to 5.4 to ≈20 MeV. Similarly, for $^6_3$Li, $^7_3$Li, and $^8_4$Be, S_n increases from 5.4 to 7.2 to ≈18.1 MeV. Similar trends are also observed for S_p. For example, for $^2_1$H, $^3_2$He, and $^4_2$He, S_p increases from 2.2 to 5.2 to 20 MeV. Systematic observations of the above-mentioned type shows that in light

nuclei neutron is less bound in case N is odd, neutron binding increases if the number of neutrons become even and it becomes maximum when the both number of neutron and proton are even. Similarly, S_p is less for odd Z, nucleus is larger in even–odd nucleus and attains maximum value for even–even nuclei. The tight binding of the last neutron or proton when the number of neutrons or protons in the nucleus is even provided the concept of some additional energy associated with the pair of the same kind. This is called the pairing energy and is denoted by δ_n or δ_p, respectively, for the neutron and the protons. Neutron and proton pairing energies are defined as

$$\delta_n = \frac{1}{2}[S_n^{(\text{even-}N)} - S_n^{(\text{odd-}N)}] \quad \text{and} \quad \delta_p = \frac{1}{2}[S_p^{(\text{even-}P)} - S_p^{(\text{odd-}P)}]$$

The magnitudes of S_n and S_p in odd-N and even-Z, and in even-N and odd-Z light nuclei are almost equal. In addition, the magnitudes of δ_n and δ_p are also nearly equal in light nuclei. In the case of heavy nuclei, similar trend in S_n and S_p values have been observed, particularly for nuclei near magic numbers, but they are less pronounced.

These close similarities in separation energies and pairing energies for neutrons and protons suggest that fundamentally the force between any pair of nucleons is independent of the charge.

Solved example S-3.1

Calculate the minimum kinetic energy of a proton that is confined in a nucleus of $^{27}{}_{13}$Al. Consider the value of r_0 as 1.2 F.

Solution. The radius of the aluminium nucleus

$$r = r_0(27)^{\frac{1}{3}} = 3.6 \times 10^{-15} \text{ m}.$$

Considering the de Broglie wavelength λ of the proton to be equal

$$\lambda = 2r = 2 \times 3.6 \times 10^{-15} \text{ m}.$$

Let M_p denote the mass of the proton and E be its kinetic energy. The momentum p of the proton is related to its energy E by the expression $p = \sqrt{2M_pE}$. The de Broglie wavelength λ of a proton of energy E is $\lambda = \frac{h}{p} = \frac{h}{\sqrt{2M_pE}}$. If the proton is confined in a nucleus of radius 3.6 F, then λ should either be smaller or at the most equal to $r = 3.6$ F. Therefore,

$$\left(2\times3.6\times10^{-15}\right)^2 = \frac{h^2}{2M_pE} \quad \text{or} \quad E = \frac{h^2}{2M_p}\frac{1}{\left(2\times3.6\times10^{-15}\right)^2}$$

Substituting the value of $M_pM_p = 1.67 \times 10^{-27}$ kg and $h = 6.62\times10^{-34}$ Js, we get

$$E = \frac{\left(6.62\times10^{-34}\right)^2}{2\times1.67\times10^{-27}}\frac{1}{51.84\times10^{-30}} = 0.25 \times 10^{-11} \text{ J}$$

$$= \frac{0.25\times10^{-11}}{1.6\times10^{-16}} \text{MeV} = 15.78 \text{ MeV}$$

It may be noted that the minimum energy required for confining a particle in given distance *d* is inversely proportional to the square of the distance. Therefore, if the nucleon is confined within a distance *r* (=3.6 F), the minimum energy will be around 64 MeV. Also, the average kinetic energy of nucleons inside a nucleus of average size is considered as 20 or 25 MeV.

Ans: The minimum kinetic energy of proton should be 15.78 MeV.

Solved example S-3.2

The mass defect ($M_{(Z,A)} - A$) for $^{16}_{8}O$, $^{15}_{8}O$, $^{1}_{0}N$, and, $^{1}_{1}P$ in μu are, respectively, −5085, +3070, +8665. and. +7825.

Calculate the binding energies of ^{16}O and ^{15}O nuclei and the separation energies of last neutron from ^{16}O nucleus.

Solution. Mass defects in the problem are given in units of micro-unified mass unit. The masses of the respective nuclides in the unit of u are

$$M_{16O} = 16 - 0.005085 = 15.994915 \text{ u}; M_{15O} = 15 + 0.003070 = 15.003070 \text{ u}$$

$$M_N = 1 + 0.008665 = 1.008665 \text{ u}; M_P = 1 + 0.007825 = 1.007825 \text{ u}.$$

(i) Binding energy of ^{16}O, $B_{16O} = 8\,(M_N + M_P) - M_{16O} = 8(1.008665 + 1.007825) - 15.994915 = 16.13192 - 15.994915 = 0.137005 \text{ u} = 0.137005 \times 931.48 = 127.62$ MeV.

$$\text{Binding energy per nucleon in } ^{16}O = 127.62/16 = 7.97 \text{ MeV}$$

(ii) Binding energy of ^{15}O, $B^{15}{}_O = (7 \times M_N + 8 \times M_P) - M^{15}{}_O$

$$= (7.060655 + 8.0626) - 15.003070 = 0.120185 \text{ u} = 111.95 \text{ MeV}$$

$$\text{Binding energy per nucleon in } ^{15}O = 111.95/15 = 7.43 \text{ MeV}$$

(iii) Separation energy of last neutron from ^{16}O, $S_n = (M^{15}{}_O + M_N) - M^{16}{}_O$

$$= (15.003070 + 1.008665) - 15.994915 = 0.01682 \text{ u} = 15.67 \text{ MeV}$$

The binding energy of the last neutron may also be calculated from the difference of the binding energies of ^{16}O and ^{15}O nuclei

$$= 127.62 - 111.95 = 15.67 \text{ MeV}.$$

It may be observed from the calculations that the average binding energy per nucleon in ^{16}O is larger than for ^{15}O nucleus. Further the separation energy of the last neutron from the nucleus ^{16}O is much larger (15.67 MeV) than the average binding energy per nucleon (7.97 MeV). This is because ^{16}O is a doubly magic nucleus and is very strongly bound.

Exercise p-3.3: A target nucleus $^{A}_{Z}X$ is bombarded by energetic deuterons to produce the following (d, p) reaction:

$$^{A}_{Z}X + ^{2}_{1}D \rightarrow ^{(A+1)}_{Z}X + ^{1}_{1}H + Q_{(d,p)}$$

Show that the separation energy S_n of the last neutron from ${}^{(A+1)}_{Z}X$ is given by

$S_n = Q_{(d,p)}$ + Binding energy of deuteron.

Exercise p-3.4: The mass excess $(M - A)$ for the four stable isotopes ${}^{110}_{48}Cd$, ${}^{111}_{48}Cd$, ${}^{112}_{48}Cd$, ${}^{113}_{48}Cd$ and for neutron and proton are, respectively, −95,072, −95,811, −97,237, −95,591, +8665, and +7825 μu. Calculate the increase in the total nuclear binding energy when one neutron is added each to ^{110}Cd, ^{111}Cd, and ^{112}Cd nuclei.

3.3.2 Energy Levels of Mirror Nuclei

In several pairs of mirror nuclei, such as ${}^7_3Li$, ${}^7_4Be$; ${}^{11}_5B$, ${}^{11}_6C$; and ${}^{13}_6C$, ${}^{13}_7N$, the higher charge isobar decays into the lower charge isobar by emitting β^+ and neutrino. As the atomic mass number A for the two members of a mirror pair is same, they have same radius and same volumes. In the higher charge isobar, the charge Ze and in the lower charge isobar $(Z-1)e$ charge is distributed in the same volume.

The schemes of low-lying level for the mirror pairs ${}^7_3Li$, ${}^7_4Be$ and ${}^{13}_6C$, ${}^{13}_7N$ are shown in figure 3.2a,b. It may be clearly seen in these diagrams that: (1) the angular moment and parity of the levels in the members of a pair are same. (2) The sequence of levels for the two members of a pair is identical, but (3) each level of the higher charge isobar of the pair is shifted in energy by the amount ΔU with respect to the corresponding level of lower charge isobar. The two members of each isobar-pair have two differences; first, their electric charge differs by one unit and second the higher charge isobar has one more proton and one less neutron compared with the lower charge isobar. Except these differences, the two members of a pair have all other properties such as radius, volume,

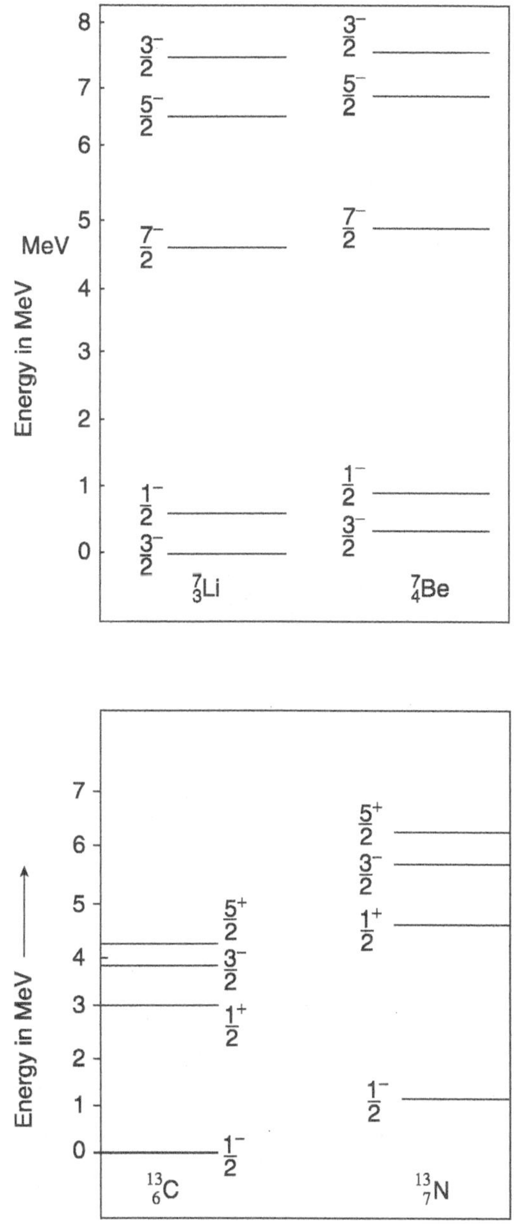

Figure 3.2 *Energy levels of (a) ${}^7_3Li$ and ${}^7_4Be$ and (b) ${}^{13}_6C$ and ${}^{13}_7N$ mirror pairs*

and total number of nucleons same. It is obvious that above-mentioned differences are responsible for the energy difference ΔU. ΔU will contain contributions from three effects:

1. Difference in the electrostatic energies of the two members of the pair ΔU_{Cou}
2. Neutron–proton mass difference ΔU_{mass}
3. Difference in the (nuclear) binding energies of the two members of the pair ΔU_{nuc}

Now, the electrostatic Coulomb energy U_{Cou} of a charge q, which is uniformly distributed in the volume of a sphere of radius r, is given by

$$U_{Cou} = \frac{3}{5}\left(\frac{q^2}{4\pi\varepsilon_0 r}\right)$$

Therefore,

$$\Delta U_{Cou} = \frac{3}{5}\left(\frac{e^2}{4\pi\varepsilon_0 r}\right)\{Z^2 - (Z-1)^2\} = \frac{3}{5}\left(\frac{e^2}{4\pi\varepsilon_0 r}\right)\{2Z-1\}$$

However, $r = r_0 A^{1/3}$, $r_0 = 1.5 \times 10^{-15}$ m, and $(2Z - 1) = A$, which gives

$$\Delta U_{Cou} = \frac{3}{5}\left(\frac{e^2}{4\pi\varepsilon_0 r_0}\right)\{A^{2/3}\}$$

The value of ΔU_{Cou} for any pair can be easily calculated from the above-mentioned equation.

Apart from the difference in the electrostatic Coulomb energy, the mass difference between a neutron and a proton in the parent and daughter nucleus also contributes to the total energy difference between the two isobars. The $\Delta U_{mass} = (M_n - M_p)c^2 = 1.29$ MeV. The nuclear-binding energies of the two members of a pair may be different if the nuclear-binding force for a neutron is different from the binding force for a proton. The term ΔU_{nuc} will arise from that difference.

Putting all these terms together, we get

$$\Delta U = \Delta U_{Cou} + \Delta U_{mass} + \Delta U_{nuc} = \frac{3}{5}\,(e^2/4\pi\varepsilon_0 r_0)\,\{A^{2/3}\} - 1.29 + \Delta U_{nuc} \tag{3.1}$$

Experimentally, the magnitude of ΔU can be obtained from the decay products. The higher charge isobar of the pair decays into other members by β^+ (positron) emission. Neutrino is also emitted along with the positron. When the positron is emitted with maximum energy $E\beta_{max}$ the neutrino has zero kinetic energy. Since neutrino has zero rest mass the sum of $E\beta_{max}$ and the rest mass energy ($m_e c^2$) of the positron must be equal to ΔU.

$$\Delta U = E\beta_{max} + m_e c^2 \tag{3.2}$$

where m_e is the rest mass of the positron (or the electron).

A comparison of ΔU value calculated using Eq. (3.1) with the experimental values obtained from Eq. (3.2) or from other reaction data for each mirror pair, gave ΔU_{nuc} to be zero.

The fact that the shift in the energy of corresponding levels of mirror nuclei can be fully explained by the sum of the electrostatic Coulomb energy difference and the difference in the masses of neutron and proton clearly shows that the replacement of a neutron by a proton

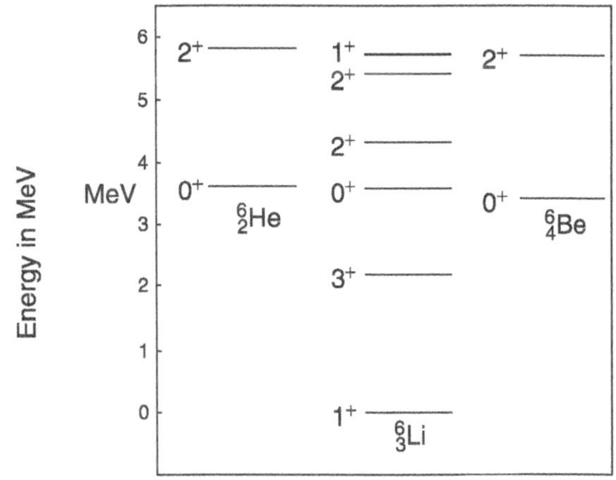

Figure 3.3 *Energy levels of $^6_2$He, $^6_4$Be, and $^6_3$Li*

(or vice versa) does not change the nuclear binding energy. This simply means that nuclear forces do not differentiate between neutron and proton. In nuclear physics, the equivalence of the neutron–neutron force and the proton–proton force ($F_{1(n-n)} = F_{1(p-p)}$) is called charge symmetry and the more general equivalence of neutron–neutron, proton–proton, and the neutron–proton force $[F_{1(n-n)} = F_{1(p-p)} = F_{1(n-p)}]$ is referred to as charge independence.

Next, let us consider mirror nuclei that differ by two unit of charge, such as $^6_2$He and $^6_4$Be. In these nuclei, the total number of nucleons is same and hence their radii are equal. As may be seen in figure 3.3, the levels of the two nuclei are almost identical except a little energy shift. It is, however, important to note that in this pair, $^6_2$He has two (p–n) bonds and one (n–n) bond while in case of $^6_4$Be there are two (p–n) bonds and one (p–p) bond. Clearly, if there is any difference in the nuclear binding energies of the two nuclei, it must be due to the difference in the binding energies of (n–n) and (p–p) bonds. Calculations have shown that the slight energy difference in corresponding levels of the mirror pair can be fully accounted for by the differences in the Coulomb electrostatic energy and the energy corresponding to the difference in the masses of neutron and proton. Once again this shows that the nuclear force between two neutrons is equal to the force between two protons, proving the charge symmetry of nuclear force.

It is interesting to look at the energy levels of the nucleus $^6_3$Li (figure 3.3), which also have six nucleons and same radius as the other two nuclei. Some levels in figure 3.3 correspond to the levels in ^{6}He and ^{6}Be as expected; however, there are some additional levels as well. These additional levels in ^{6}Li appear because in the case of ^{6}He and ^{6}Be nuclei, some (p–p) or (n–n) states are restricted and are not populated due to the Pauli exclusion principle. In addition, additional (p–n) states in which exclusion principle does not apply are populated in ^{6}Li creating additional levels.

In short, the similarity of the low-lying levels of mirror nuclei and the fact that the energy shift between the corresponding levels is fully accounted by the electrostatic energy difference and the difference in the neutron proton masses states that fundamentally nucleon–nucleon force is charge independent.

3.3.3 Difference in the Binding Energies of $^3_1$H and $^3_2$He

The mirror nucleus $^3_1$H has one proton and two neutrons and, therefore, the possible nucleon–nucleon interaction states are ^{3}(n–p) + ^{1}(n–p) + ^{1}(n–n), ^{3}(n–n) state is forbidden by Pauli exclusion principle. In $^3_2$He, the possible states are ^{3}(n–p) + ^{1}(n–p) + ^{1}(p–p); ^{3}(p–p) state is excluded because of exclusion principle. Any difference in the binding energies of the pair will come from the Coulomb energy difference and the binding energy difference of ^{1}(n–n) and ^{1}(p–p) states. The difference in the binding energies ≈0.76 MeV could be fully accounted by the difference of the Coulomb energies of the two nuclei. This again indicates that *nuclear force is charge symmetric*, F^1(n–n) = F^1(p–p).

Exercise p-3.5: Determine the Coulomb energy difference between the mirror isobars $^{13}_{7}$N and $^{13}_{6}$C and hence the maximum energy of positrons emitted by ^{13}N. Consider r_0 = 1.4 F.

3.3.4 Study of Deuteron Properties

Deuteron, $^2_1$H, is the smallest stable nucleus and has one neutron and one proton. The nucleons in deuteron revolve round the centre of mass and their relative separation vary with time. The square root of the mean of the squares of the distance between the centre of mass and the proton is called the r.m.s. electromagnetic radius and is denoted by r_d. Experimental values of some properties of deuteron are: (1) binding energy B_d = 2.22 MeV. (2) r_d = 2.1 F. (3) spin or the angular momentum $J = 1\hbar$. (4) Parity = even or positive. (5) Magnetic dipole moment μ_d = 0.857393 nm. (6) Electric quadrupole moment Q_d = +0.00282 barn.

As the spin of deuteron is 1, the neutron and proton spins are parallel and, therefore, deuteron is an example of triplet neutron–proton state represented as a ^{3}(n–p) state. The deuteron has been analysed under quantum mechanical approach, in which a potential well is assumed to represent the system. In principle potential well of any shape may be used, but the simplest one is a square or rectangular well with two parameters, the depth V and the width b. High-energy scattering experiments have already suggested that there is a hard core of infinite height at a distance c from the centre of a nucleus such that nucleons cannot come closer than the distance c. This is also required for nuclear stability; otherwise nucleons will all collapse into each other under the strong attractive nuclear force between them. The radius of the core c has been estimated to be of the order of 0.4 F. A pictorial representation of the square-well approach is shown in figure 3.4. The appropriate well parameters should reproduce the observed properties of the deuteron, particularly, the binding energy, and the r.m.s. radius. The analysis shows that no unique values for the

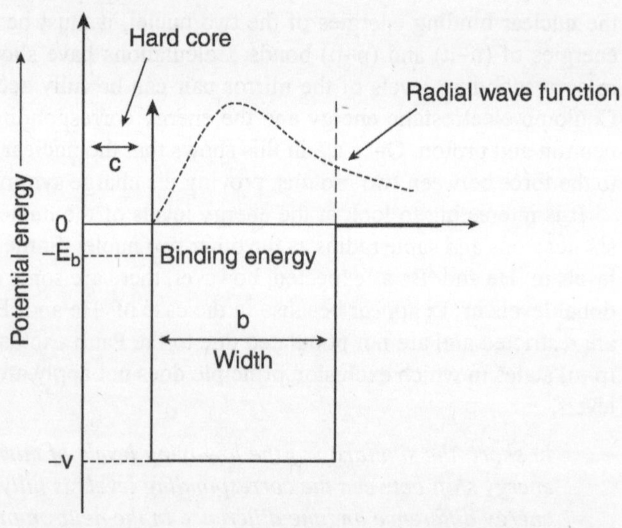

Figure 3.4 *Pictorial representation of square-well approach to deuteron and the shape of the radial wave function*

depth V and the width b could be obtained, rather several combinations of V and b could reproduce the observed binding energy. For example, a square-well potential of depth $V \approx 35$ MeV and width $b \approx 2$ F or $V = 73$ MeV and $b = 1.33$ F can give the correct binding energy of the deuteron. Detailed quantum mechanical analysis of deuteron has given the following information about the two body nucleon–nucleon forces.

1. Triplet (n–p) force $F_{3(n-p)}$ is just strong enough to bind deuteron, while the singlet force $F_{1(n-p)}$ is not enough strong to provide a bound singlet state. It means $F_{3(n-p)} > F_{1(n-p)}$.
2. In deuteron, the two nucleons remain outside the range of the nuclear force for a considerable period of time, resulting in small binding energy of deuteron. This is possible only in quantum mechanical systems where particles may undergo tunnelling, which is not possible in classical systems.
3. To reproduce the quadrupole moment of deuteron, an admixture of S and D states is necessary. The $F_{3(n-p)}$, thus, mixes states.
4. Further, the prolate shape of the deuteron suggests that the force between the neutron and the proton in deuteron is attractive when they are one above the other, and less attractive or even repulsive when they are side by side as shown in figure 3.5.

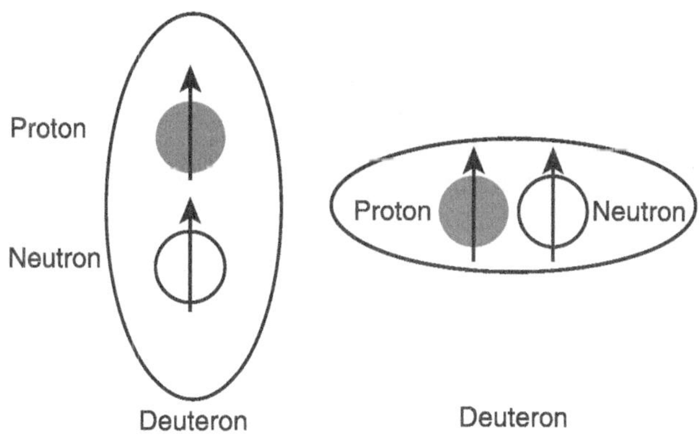

Figure 3.5 *The force between neutron and proton is attractive when they are one above the other and less attractive or even repulsive when they are side by side. That is why the shape of deuteron is prolate. Tensor force depends on the angle between the spins and the line joining the centres of nucleons*

This shows that the force also depends on the angles between the spin directions and the line joining the centres of nucleons. Such force is called a tensor force. The tensor dependence in radial nucleon–nucleon potential is generally represented by the term

$$S_{12} = (1/r^2)\,[3(\sigma_1 r)\,(\sigma_2 r) - (\sigma_1 \sigma_2) r^2]$$

Here, σ_1 and σ_2 are the spin vectors for the two nucleons. The term has different values for the cases when σs are parallel and perpendicular to the radius vector r, even for the same value of $\sigma_{total} = \sigma_1 + \sigma_2$.

Exercise p-3.6: Calculate the value of the tensor term S_{12} for the two configurations shown in figure 3.5 and for the same configurations in case of anti-parallel spins.

3.3.5 Nucleon–Nucleon Scattering

3.3.5.1 Neutron–proton scattering

As there is no bound singlet state of n–p system (system such as deuteron but with spin zero), it is not possible to study the properties of $F_{1(n-p)}$ directly. These un-bound states are, however, formed in scattering of neutrons with protons and their effect can be seen in scattering cross-sections. It may be noted that in scattering both triplet ^{3}(n–p) and singlet ^{1}(n, p) states are simultaneously populated as there is no restriction on any state from Pauli exclusion principle. The triplet state has three sub-states and the singlet state has one. As such the total number of sub-states is four. If it is assumed that each sub-state is populated with same probability, the total scattering cross-section will be equal to the weighted average of the sub-states, which will be given as

$$\sigma^{\text{total}} = \frac{3}{4}\sigma^{\text{triplet}} + \frac{1}{4}\sigma^{\text{singlet}} \tag{3.3}$$

Large number of precise experiments using neutrons from nuclear reactors and from secondary neutron-producing reactions initiated by accelerated charged particles has been conducted to measure the n–p scattering cross-sections. Polyethylene, organic material with carbon and hydrogen, is often used as target. In some experiments at very low energy, hydrogen gas has also been used as target. The measured total (n, p) cross-section between 0.01 eV and 15 MeV is shown in figure 3.6. As may be noticed in this figure, the curve can be divided into three distinct parts: (1) from 0.01 to 10 eV neutron energy, (2) between 10 and 10^4 eV, and (3) from 10^4 eV to 15 MeV. In the large energy range of 10–10^4 eV, the cross-section is nearly constant at a value of about

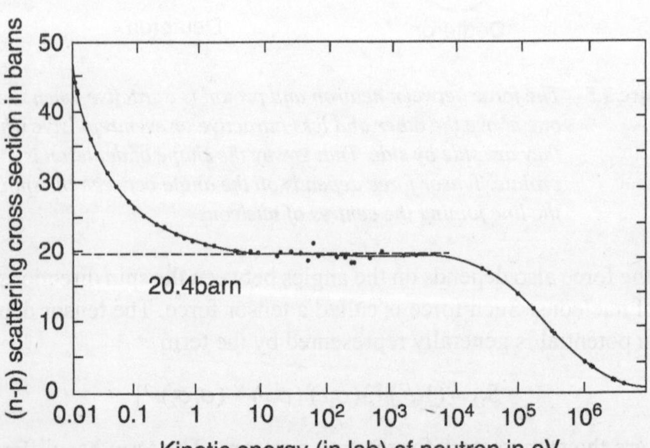

Figure 3.6 *(n, p) cross-section between 0.01 eV and 15 MeV*

20 barn. At lower energies, the scattering cross-section increases rapidly while at higher energies $>10^4$ eV, it decreases slowly. Analysis of the experimental data has been carried out assuming that the protons are fixed at their location and are free. At energies >10 eV, this assumption is true; however, at lower energies where mostly gaseous targets have been used, the protons were neither free nor fixed. At lower energies, the de Broglie wavelength of neutron becomes large and, therefore, the neutron interacts with the hydrogen molecule rather than the proton. The thermal motion of gas molecules also increases the apparent target size. Moreover, the molecular binding energy becomes comparable to the energy of incident neutron. All these effects enhance the measured scattering cross-section at lower energies. On applying corrections for these effects in the measured cross-section values, it turns out that in the lower energy range also the actual scattering cross-section is of the order of 20 barn.

In the quantum mechanical framework, the scattering problem is treated by decomposing the incoming neutron wave into partial waves of different orbital angular momentum l with respect to a fixed target. In elastic scattering, the target potential changes the phases of the incident partial waves without affecting the flux. The net effect of scattering potential is a shift in the phase of the l-th partial wave by an angle d_l. The scattering cross-section for each partial wave then becomes proportional to $\sin^2 d_l$. At low energies below 15 MeV, only $l = 0$ or S-wave contributes to scattering because the higher partial waves lie beyond the range of the nuclear field of the target proton. As such for low energy, the total (n–p) scattering cross-section $s\,^{s}_{(n,p)} = \{4p \sin^2 d_0/k^2\}$, where $k = \frac{\sqrt{2mE}}{\hbar}$ and m is the reduced mass $= M_n/2$ and E the centre of mass energy of incident neutron. In the zero energy limit $k \to 0$, the scattering cross-section can be given in terms of the parameters a_t and a_s, which are the scattering lengths for the triplet and singlet scattering events. Scattering lengths is the intercept of the tangent to the radial wave function at the outer boundary of the potential well. Scattering length a_t is positive for the bound ^{3}(n, p) state and a_s is negative for an unbound state like ^{1}(n, p). Precise values for a_t and a_s determined from coherent and incoherent scattering of neutrons by protons give $a_t = 5.38$ F and $a_s = -23.7$ F. The experimental value of scattering length for triplet state is consistent with the deuteron potential well parameters that give correct binding energy and r.m.s. radius of deuteron.

The negative value of the scattering length for singlet potential is consistent with the fact that the singlet state is not bound. If same well width $b = 1.33$ F that gives correct binding energy for deuteron is used with the singlet scattering length = −23.7 F, the singlet potent depth for zero energy approximation comes out to be 54.7 MeV, while the minimum depth of the well to produce a bound system is about 57 MeV. As such it may be stated that the singlet state narrowly misses to become a bound state by only about 3 MeV or so.

In summary, it may be said that a rectangular well with appropriate width and depth parameters can satisfactorily reproduce the binding energy, r.m.s. radius, scattering lengths for triplet and singlet states obtained from measured scattering cross-sections at low energy and for zero-energy approximation. Also, for higher energies, the cross-sections for both triplet and singlet state scatterings can be represented in terms of two additional parameters, the effective range r_{ot} and r_{os}, respectively, for the triplet and the singlet states. With the parameters $a_t = 5.28$ F, $a_s = -23.7$ F, $r_{ot} = 1.7$ F and $r_{os} = 2.4$ F, the measured total scattering data between 10 eV and 15 MeV can be satisfactorily explained. This shows the great success of quantum mechanical treatment of both the deuteron and the scattering process.

3.3.5.2 Proton–proton scattering

Proton–proton scattering has also been extensively studied. The major points of difference between n–p and p–p scattering are: (1) Triplet ^{3}(p–p) state in not possible due to exclusion principle. (2) Coulomb force is always present at all angles and for all partial waves. To overcome the interference due to Coulomb scattering differential scattering cross-section, $\left(\frac{d\sigma}{d\Omega}\right)$ is measured and the experimental value is corrected for the contribution from Coulomb scattering. (3) As the target and the incident particles are same, it becomes difficult to differentiate the incident particle and the scattered target. As a result, it becomes impossible to differentiate between small-angle and large-angle scattering events. (4) According to Dirac's theory, the vacuum is filled with negative energy electrons. In the presence of strong electrostatic field around proton, the wave functions of negative energy electrons get distorted resulting in the polarization of space and at very small distance from the proton the $1/r^2$ dependence of Coulomb field does not hold. However, at larger distances there is not much effect. Proton–proton scattering has been successfully treated quantum mechanically and vacuum polarization and interference between Coulomb and nuclear scattering terms have been clearly observed. Although the p–p scattering phase shift and n–p scattering phase shifts at the same energy cannot be compared but it is possible to calculate back the potential well parameters for p–p scattering. These calculations have shown that the potential well parameters for singlet p–p scattering are nearly equal to the parameters for singlet n–p scattering, once again confirming that nuclear forces are charge independent.

Nucleon scattering at higher energies confirmed the presence of a hard core. Further, the concept of phase shift has been successfully applied at high energies and consistent set of phase shift parameters have been obtained that can explain all the quantities that are observed in nucleon–nucleon scattering.

The study of deuteron and of the nucleon–nucleon scattering has revealed the following:

1. Nuclear forces are charge independent. The same inference was also drawn from the study of separation energies of last nucleon and the level schemes of mirror nuclei.
2. Scattering at high energies show that there is hard or repulsive core at the centre of the nuclear potential.
3. Nuclear force is spin dependent and of tensor nature.

Exercise p-3.7: In n–p scattering cross-section, triplet state contributes three times more than the singlet state. Explain the reason for this larger contribution of the triplet state.

Exercise p-3.8: The experimentally measured (n, p) scattering cross-section is found to increase with the decrease of energy, in the very low energy region. How can this be explained?

Exercise p-3.9: Clearly bring out the differences in (n, p) and (p, p) scattering.

3.3.6 Constant Density of Nuclear Matter: Exchange Forces

The density of the nuclear matter can be readily calculated by dividing the mass of a nucleus by its volume. Mass of a nucleus with A nucleons = $M_n A$, where M_n is the nucleon mass and the volume = $4/3\pi r^3$, where r is the radius. The radius $r = r_0 A^{1/3}$, where the unit radius r_0 (=1.5 F) is constant. The density of nuclear matter, therefore, is $\frac{M_n}{\frac{3}{4}\pi r_0^3}$ (where M_n is the nucleon mass) a constant value for all nuclei. As all nuclei big or small have same density, they may be compared with the big and small drops of a liquid. The similarity between a liquid drop and a nucleus may be looked in another way as well. In case of a liquid drop, each molecule of the liquid is bound with only few molecules around it and not to all the molecules present in the liquid. As a result, the total binding energy of a liquid drop becomes proportional to the number of molecules in the drop. In a similar way, each nucleon inside a nucleus is bound only to few other nucleons around it and not to all other nucleons present in the nucleus. This results in the total binding energy of the nucleus being proportional to A, the number of nucleons in it. [If each nucleon in the nucleus was interacting and was bound to all other $(A - 1)$ nucleons, then the binding energy of the nucleus should have been proportional to the number of all the bonds which are $A\ (A - 1)$. For heavy nuclei, $A\ (A - 1) \approx A^2$ and, therefore, the total binding energy of heavy nuclei would have been proportional to A^2 instead of A].

The property of the nucleon–nucleon force that restricts the interaction of a nucleon to only a few around it is called the saturation property of the nuclear force in analogy to the saturation of the force binding molecules in liquids. The analogy between the nucleus and the liquid drop may be further explored to understand the binding between the nucleons in a nucleus.

Let us consider a drop of liquid hydrogen. In this drop, two atoms of hydrogen are chemically bound through homopolar binding to make an H_2 molecule. A third atom of hydrogen is not attracted now to join with a molecule of H_2. This is called the saturation of chemical binding force. In homopolar binding, this saturation is achieved by assuming that the electrons continuously exchange their position from one atom to the other in the molecule. Mathematically, this exchange of electrons between the atoms of a molecule can be achieved by using exchange operators.

Extending the concept of exchange operator to the case of nucleon, there may be three types of operators: (1) operator that may exchange the space coordinates of two nucleons, (2) operator that may exchange the spin coordinates of two nucleons, and (3) the operator that may exchange both the spin and space coordinates. Apart from these exchange operators, there may also be a non-exchange operator that does not exchange any property. In simple terms, operators of a force are generally multiplicative factors. For example, if an attractive force is operated by the operator −1, the force becomes repulsive. Operators of exchange forces change the nature of the force between nucleon from attractive to repulsive depending on the value of the relative angular momentum between the nucleons.

1. **Majorana force:** The Majorana force exchanges the position coordinates of the two nucleons. It simply means that the Majorana force operator only interchanges the positions of the two interacting nucleons, while their spins and other properties remain same. Majorana force is attractive for the two nucleons if their relative angular momentum l is zero or even repulsive for odd l states. For example, in S $(l = 0)$ and D $(l = 2)$ states it will give attraction and for the P state $(l = 1)$ it will provide repulsion between the two nucleons.
2. **Bartlett force:** The operator of Bartlett force interchanges the spins of the two interacting nucleons, leaving all other parameters including their positions unchanged. For example, if the first nucleon has its spin up and the other spin down, then as a result of Bartlett interaction the spin of the first nucleon will turn down and that of the other up. (*What will be the effect of Bartlett interaction on two parallel spin nucleons?*). For zero or even angular momentum states, Bartlett force is attractive in triplet and repulsive in singlet states. For odd l states, Bartlett force is repulsive for triplet and attractive for singlet states.

3. **Heisenberg force:** This force exchanges both the spin and the space coordinates of the two interacting nucleons. Naturally, the operator produces the same effect that is produced by the successive operations of Majorana and Bartlett force operators. The Heisenberg force is attractive for even-*l* triplet and odd-*l* singlet states. It is repulsive for the even-*l* singlet and odd-*l* triplet states.
4. **Wigner force:** Wigner force is not a force that provides saturation. This force provides only binding to the two interacting nucleons. Its operator is +1 for states of all *l* values.

Table 3.1 *Exchange operators for different type of forces*

Type of Force	Even-*l* Singlet	Odd-*l* Singlet	Even-*l* Triplet	Odd-*l* Triplet
Majorana	+1	–1	+1	–1
Bartlett	–1	–1	+1	+1
Heisenberg	–1	+1	+1	–1
Wigner	+1	+1	+1	+1

The actual force between two nucleons cannot be totally of Wigner type as it will not be able to provide the saturation. Also, it cannot be totally of Heisenberg type because Heisenberg force is attractive for triplet state if nucleons are in the state of $l = 0$ (S-state). The deuteron fulfils these conditions, and should have been the most tightly bound system if nucleon–nucleon force was totally of Heisenberg type. We know that the binding energy of deuteron is only 2.23 MeV. Alpha particle, the nucleus of helium, is the most tightly bound nucleus with binding energy of 28 MeV. Hence, Heisenberg force alone cannot be the nucleon–nucleon force. As the exchange of π meson between two nucleons corresponds to the exchange of their positions, Majorana force must be a significant component of the total nucleon–nucleon force. The two nucleon system can be represented as a mixture of Majorana and Wigner forces.

3.4 MEDIATION OF NUCLEAR FIELD

How does two nucleons stick together to make a nucleus? The answer to the question is obtained by drawing an analogy from the interaction of two charges. In quantum electrodynamics, a well-understood and developed theory of Coulomb field, two charged bodies interact through Coulomb field. Let us, therefore, first review the mechanism of interaction between two charges.

Quantization of electric field generates photon, which communicates the electric field between charged bodies through their exchange. This mechanism of interaction brings in the concept of the speed of the field and the time delay in action at a distance. In case of electric charges, the speed of the field is the speed of light, as photon, the quanta of the field travels with the speed of light. The mechanism also brings in the concept of virtual photons in case of electric field and virtual field particles in general.

Virtual photons are quite different from the real photons. These are called virtual because they exist within the constraints imposed by the uncertainty principle. According to this principle, an uncertainty ΔE in energy can stay for a time duration Δt such that $\Delta E \times \Delta t \sim \hbar$. As such, if a virtual photon of energy E is created, it will live for a time $\Delta t \sim \hbar/E$. A virtual photon travelling with the speed of light c will cover a distance $r = c\Delta t$ in time Δt. Hence, $r = c(\hbar/E)$. A virtual photon of energy E can be created for time duration $\Delta t = \hbar/E$ and can make a trip of distance $r = \hbar c/E$ during the time interval Δt. After this time interval, the photon must disappear to conserve the energy.

This fluctuation in energy is allowed only within the limits of time Δt, allowed by the uncertainty principle. Hence, a virtual photon of energy E is emitted by an electron, it carries a momentum $p = E/c$. The virtual photon is absorbed by the other charged particle changing its momentum by the amount of p. As such the total change of momentum during the process of exchange (emission and absorption) of virtual photon between two charged particles is $2p$. Now, according to the law of mechanics the force F responsible for the change of momentum (for the exchange of virtual photons in this case) is equal to the rate of change of momentum = $2p/\Delta t$. As such

$$F = 2p/\Delta t = (2E/c)/(\hbar/E) = 2E^2/\hbar c = [2(\hbar c/r)^2)/\hbar c] = (2\hbar c/e^2)\,(e^2/r^2)$$

This is how the electrostatic force between two charges may be derived from the concept of the virtual photon exchange. The electromagnetic force is an infinite range force and this is due to the exchange of mass-less particles, that is photons. However, these photons are not real but virtual. These virtual photons exist only in association with charges and currents and are not to be confused with real photons. However, the 'virtual' photons have real affects and, in that respect, are real. For example, a real photon moving in free space can be represented as in figure 3.7a where the photon dissociates in to a virtual $e^{\pm}$ pair which heals up (or recombines) after a time $\Delta t = \hbar/mc^2$. Similarly, (see figure 3.7b) an electron passing in free space may be represented as a particle that emits a virtual photon, which dissociates into a virtual $e^{\pm}$ pair that eventually heals and emits a virtual photon, which is reabsorbed. The free e^- is thus surrounded by charges (virtual) in the vacuum and the e^- of e pair is repelled while e^+ of $e^{\pm}$ pair is attracted by the real e^-. This is called vacuum polarization and affects the inverse square law in the immediate vicinity of charges. The effect of vacuum polarization is felt in p–p scattering and in other processes.

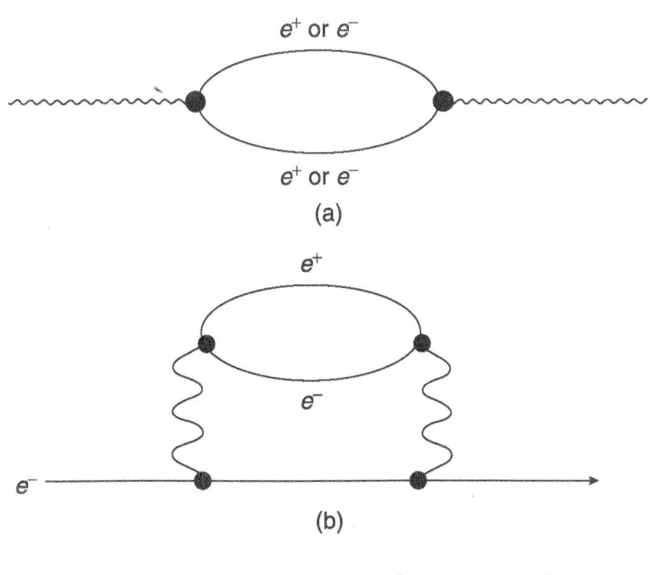

Figure 3.7 *(a) A real photon moving in free space may dissociates into a virtual e^+ and e^- pair. (b) An electron may emit a virtual photon that may create a virtual electron–positron pair which recombine to produce a virtual photon that is absorbed by the electron*

(Virtual photon emitted by a charged particle, an electron in the figure, creates a pair of virtual electron and positron which live for a short time and recombine (heal up) to give back virtual photon which in turn is reabsorbed by the electron)

This analogy to the electrodynamics was used by Yukawa to predict the emission and absorption of a virtual particle for mediation of nuclear force, now known as the p -meson or, pion. The nuclear force that has finite range, obviously, has to be mediated by the exchange of finite mass particle. Its range r = ħc/E = ħ/mc is controlled by the mass of mesons such as pion.

Yukawa considered the following scalar potential equation of electrodynamics

$$\left(\nabla^2 - \frac{1}{c^2}\frac{\partial^2}{\partial t^2}\right)U = 0 \tag{3.4}$$

which reduces to $\nabla^2 U = 0$, for the static field and on integration gives the static solution $U = q/(4\pi\varepsilon_0 r)$. It may be observed that Eq. (3.4) can be obtained from the equation giving energy of a mass less particle by substituting the quantum mechanical operator $(-\hbar^2\nabla^2)$ for the square of the momentum (p^2) and $(-i\hbar\ \partial/\partial t)$ for the energy.

For a mass-less particle:

$$E^2 = p^2c^2 \text{ or } -p^2 + E^2/c^2 = 0 \tag{3.5}$$

On substituting the values of p and E in terms of the operators in Eq. (3.5), one obtains Eq. (3.4). The equation for the energy of a particle of rest mass m is

$$E^2 = p^2c^2 + m^2c^4 \quad \text{or} \quad -p^2 - m^2c^2 + E^2/c^2 = 0 \tag{3.6}$$

Yukawa argued that if quantum mechanical operators for momentum and energy are substituted in Eq. (3.6), it should give the equation for the scalar potential u for the mass m particle that may be exchanged between nucleons. On substitution of operators, we get

$$(\nabla^2 - m^2c^2/\hbar^2 - 1/c^2\ \partial^2/\partial t^2)\,u = 0.$$

which in case of static field (time independent case) reduces to

$$(\nabla^2 - \frac{m^2c^2}{\hbar^2})\,u = 0 \text{ or } (\nabla^2 - \mu^2)\,u \tag{3.7}$$

$$\text{where } \mu = mc/\hbar. \text{ A solution of Eq. (3.7) is } u = g\left(\frac{e^{-\mu r}}{r}\right) \tag{3.8}$$

Here, g is a constant, which plays the same role as is played by charge in electrostatic. g depends on the properties of the source, the nucleon in the present case. In electrostatic case, the mechanical potential resulting from the scalar potential U is $V = q^2/4\pi\varepsilon_0 r$. Extending the analogy, the mechanical potential in the case of two nucleons each of strength g is

$$V = g^2\,\frac{e^{-\mu r}}{r} \tag{3.9}$$

3.4.1 Virtual Mesons

Virtual particles (mesons) are continuously exchanged between nucleons giving binding and saturation. These particles of mass m and energy $E = mc^2$ are virtual because they are created out of the uncertainty principle and live for a time $\Delta t = \hbar/mc^2$. In the interval of time Δt, the virtual particle if travelling with the maximum speed, the speed of light may travel a maximum distance $d = \Delta tc = \hbar/mc$. Thus, d gives the maximum range of the field. Substituting the range of nuclear field $d \approx 2$ F, the mass m of the virtual particle may be calculated to be around 200 m_e, m_e being the mass of electron. It may be remarked that the range of the virtual particle that communicate the field is inversely proportional to the mass of the virtual particle.

Soon after the prediction by Yukawa, a particle named μ-meson (presently called muon) was discovered in cosmic rays. Initially, it was thought that μ-meson is Yukawa's particle. However, it was soon recognized that μ-meson could not be the right candidate for two reasons. First, it weakly interacts with matter and second, it has a spin ½$\hbar$, therefore, cannot be absorbed or emitted from a nucleon without changing its spin. As a matter of fact, all field quantas that mediate fields, such as photon, π-meson, gluon, W and Z bosons, and so on are bosons with zero or integer intrinsic spins.

3.4.2 Yukawa Particle: π-meson

Soon the right Yukawa particle, the π-meson (also called pion) with a mass about 280 m_e was found, again in cosmic rays. There are three types of pions π^+, π^0, and π^-. Charged virtual pions are exchanged between proton and neutron, while neutral pions between neutron–neutron and proton–proton. Substituting the mass of π-meson as 300 m_e the range of the force comes out to be about 0.7 F. There is one problem with π-meson as well. It is that the parity of π-meson is negative. If the virtual π-meson is emitted or absorbed by a nucleon in $l = 0$ (S-state), then the parity of the nucleon will change. Therefore, the π-mesons are emitted or absorbed in $l = 1$ state or in other higher odd l states. This leaves the parity of nucleon unchanged but their spins are flipped, that is changes from 1/2 pointing up to 1/2 pointing down. In case of deuterons, the exchange of charged virtual pions between proton and neutron in $l = 1$ state automatically makes the wave function for the system non-spherical. Thus, tensor force term, which makes deuteron shape prolate and gives rise to additional attraction when the nucleons are one over the other as compared to when they are side by side, evolves naturally. It is also possible that not one but two virtual pions are emitted and absorbed simultaneously. The two-pion potential is of shorter range and varies as $e^{-2\mu r}/r$. Other heavy pions may also be exchanged but only when the nucleons are much closer. Figure 3.8 shows the emission and absorption of virtual pions. A very crude sketch of nucleon–nucleon potential showing relative distances where one-, two-, and three-pion exchanges become important is shown in figure 3.9.

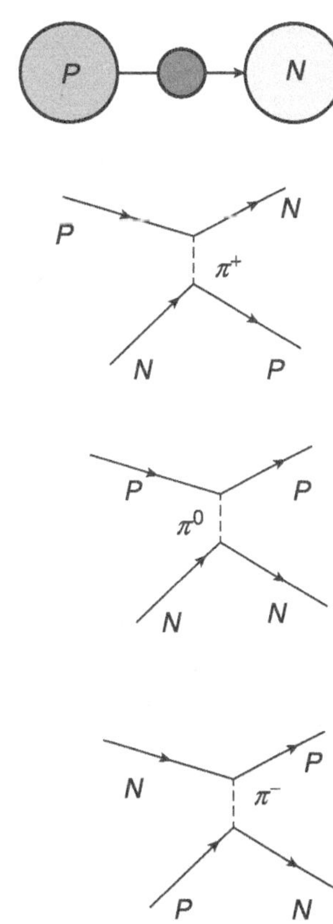

Figure 3.8 *Exchange of virtual pions between nucleons*

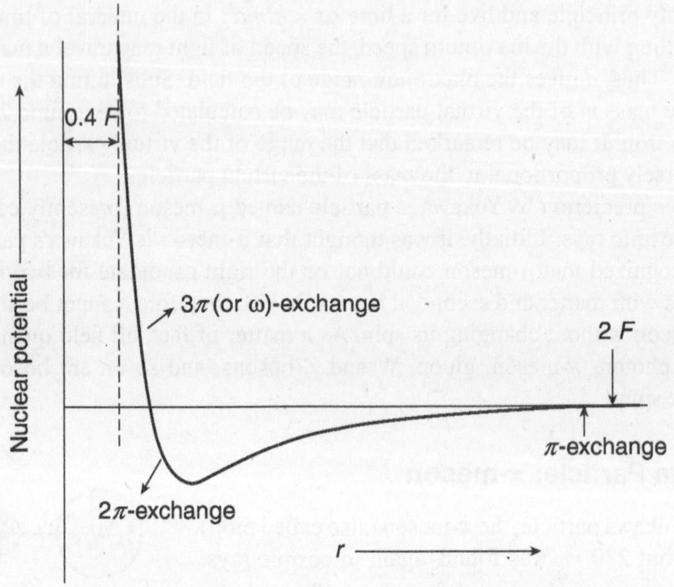

Figure 3.9 *A crude sketch of nucleon–nucleon potential showing distances where one-, two-, and three- pion exchanges become important*

3.5 SPIN–ORBIT DEPENDENCE OF NUCLEAR FORCE

The occurrence of magic numbers suggested that nucleons inside a nucleus move in some sort of shells just like the electrons in an atom. The cyclic behaviour of separation energy for the last neutron and proton further confirm the shell structure. Nuclear shell closures at lower magic numbers Z and/or $N = 2, 8, 20$ could be easily explained on the basis of shell model. However, to achieve the closure of shells at higher neutron or proton numbers 28, 50, 82, and 126 it was necessary to invoke strong coupling between the orbital and the spin motions of each individual nucleon. Detailed study of shell model of the nucleus revealed that the nuclear force also depends on the relative directions of the orbital angular momentum l and the spin angular momentum s of the nucleon. The force is attractive if l and s are parallel and repulsive if anti-parallel. This gives rise to what is called the spin orbit coupling term in nuclear potential. As the orbital angular momentum of a nucleon depends on its velocity of motion, the spin-orbit dependent force is also called the velocity-dependent force.

3.6 NUCLEON–NUCLEON POTENTIAL

Before proceeding further let us once again recapitulate the basic information about nucleon–nucleon force obtained from different sources.

1. They are of short range, charge symmetric, and charge independent.
2. Force is spin dependent. Triplet (n, p) force is strong enough to form a bound state but the singlet (n, p) force is just not strong enough to form a bound state. Singlet (n, p), (n, n), and (p, p) forces are of equal or nearly equal magnitude.

3. Nucleon–nucleon force is of tensor nature that means the force between two nucleons not only depends on the spins of the two nucleons but also on the angles which their spins make with the line joining their centres.
4. Nucleon–nucleon force is saturated, which means that a nucleon interacts with only a few nucleons around it. Saturation of force is brought about by exchange of pions between nucleons. Exchange force can explain saturation as it may also become repulsive depending on the property that is exchanged.
5. The nucleon–nucleon force also depends on the relative orientation of their angular momentum and spins. This is called the velocity dependence of the force.
6. High-energy scattering experiments have demonstrated that there is a hard repulsive core (≈0.4 F) such that two nucleons cannot came closer than this distance.
7. Since total electric charge is conserved in nuclear processes and the isotopic spin is related to charge, it is also conserved like the total angular momentum. All properties shown by angular momentum are expected to be shown by the isotopic spin as well.

As has been indicated earlier, a nucleon–nucleon potential now needs to be developed that may incorporate the above-mentioned properties of nucleon–nucleon force. A very crude shape of the potential is shown in figure 3.9 in which the distance where one, two, and three pion transfer becomes important is also shown. The potential may then be used in the Schrödinger equation of the given system to get its important parameters. The simplest potential is the square-well or rectangular-well potential. Other potential forms such as Woods–Saxon (figure 3.10), Yukawa (figure 3.11), Harmonic oscillator (figure 3.12), and many other have been tried each with limited success.

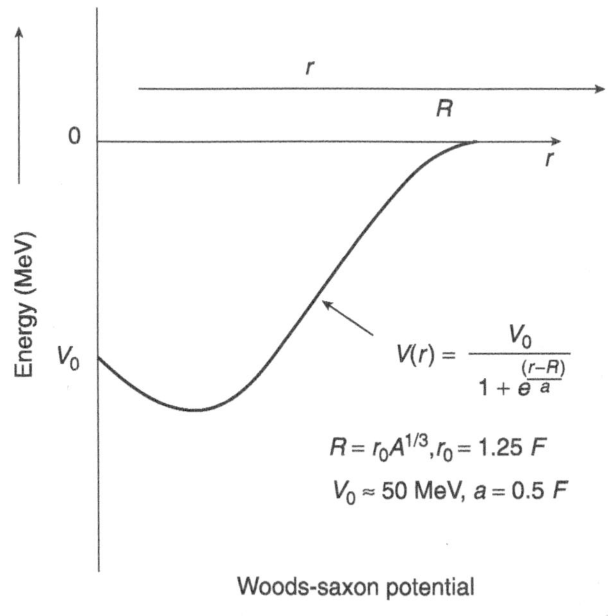

Figure 3.10 *Woods–Saxon potential*

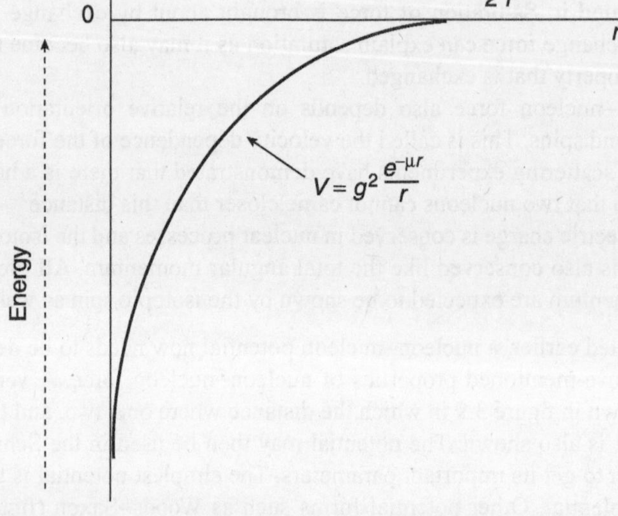

Figure 3.11 *Yukawa potential*

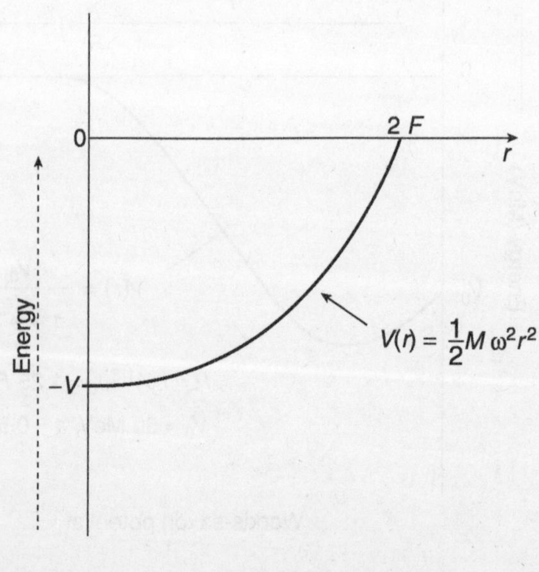

Figure 3.12 *Harmonic oscillator potential*

One form of total potential between two nucleons, in modern parlance called AV8 potential is as follows.

$$V_{12} = V(r) + V_{\sigma}(r)\,(\sigma_1\sigma_2) + V_{\zeta}(r)\,(\zeta_1\zeta_2) + V_T(r)S_{12} + V_{T\zeta}(r)S_{12}(\zeta_1\zeta_2) + V_{T\sigma}(r)\,(\sigma_1\sigma_2)\,(\zeta_1\zeta_2) + V_{ls}(r)\,(ls) + V1_{\zeta}(r)\,(ls)\,(\zeta_1\zeta_2)$$

Here, $V(r)$s are radial potentials which may have square well or Woods–Saxon or any other form for each component of the total potential. Terms I, II, III …VIII of the potential represent, respectively, pure radial, spin–spin, isospin–isospin, tensor, tensor–isospin, tensor spin–spin, spin–orbit, and isospin–orbit interactions. As already indicated, the tensor term S_{12} is generally given as:

$$S_{12} = 1/r^2[3\,(\sigma_1 r)\,(\sigma_2 r) - (\sigma_1\sigma_2)\,r^2]$$

where σ_1 and σ_2 are the spins of the two nucleons and ζ_1 and ζ_2 their isospins. It may be observed that the first term in the expression for S_{12} depends on the angles between the spins and the radius vector while the second term on the relative orientation of the spins of the two nucleons.

3.6.1 Local and Global Nuclear Potentials

Instead of using a two-nucleon potential, attempts have been made to write the potential for the whole nucleus. Optical potential – which contains two terms, the real and the imaginary – has been quite successful in explaining the scattering and nuclear reactions. Nuclear potentials are generally classified as local – which is applicable to a small range of energy and/or the nuclear mass. Global nuclear potentials can be applied for a broader range of energy/nuclear mass. None of the potentials developed so far is applicable to all the stable nuclei. It has also been observed that not only two-body force but three-body forces are also important in case of the nucleus.

3.7 QUARK AND GLUON

It is now well established that all matter in the universe is made of two types of entities: quarks and leptons. Quarks and their anti-particles, anti-quarks, may group together to make particles that may stay free, such as nucleons and π-mesons. Particles formed by the grouping of quarks are called HADROS. Leptons may interact with quarks and with their groupings but they do not group or pair among themselves. Electrons, muons, and their neutrinos are leptons. There are two types (called flavours) of quarks; the up-quark denoted by u has an electric charge of $+2/3e$ and the down-quark denoted by d has an electric charge of $-1/3e$, e being the charge of proton. Their antiparticles denoted by $\bar{u}$ and $\bar{d}$ have $-3/2e$ and $+1/3e$ charges. The quark structure of some of the particles is as follows:

Proton = (uud); neutron = (udd); π^+ meson = (u$\bar{d}$); π^- meson = ($\bar{u}$d)

While Yukawa showed that the strong interaction between nucleons is communicated by the exchange of π-mesons, it has now become clear that the interaction between quarks is more fundamental. The strong force between quarks is communicated through the exchange of a particle called gluon and not by the exchange of π-mesons. The force that is communicated by the exchange of π-meson is now treated as a residual force of gluons, such as the van der Waals force in chemical binding is the residual electrostatic force between the positively charged atomic nucleus and the negatively charged electrons. It may, however, be pointed out that the Yukawa theory of meson exchange still remains valid for the nuclear two-body residual interaction, although the more fundamental strong nuclear force is communicated by gluons between quarks.

The quark structure of particles though quite successful; however, there was no apparent reason why three quarks group up to form a nucleon or a quark and an anti-quark join to make a π-meson. There was another problem: two very short lived particles Δ^{++} and Δ^{-} with quark structures (uuu) and (ddd) were identified. This was in violation with Pauli exclusion principle. To overcome this difficulty, Moo Young Han of Duke University, Yoichiro Nambu (University of Chicago), and Oscar Greenberg (University of Maryland) suggested that a new quantum number called 'colour' may be assigned to each quark. The quantum number colour was assumed to have three components, 'red', 'green', and 'blue' denoted, respectively, by subscripts 'r', 'g', and 'b'. As such there can be six types of quarks: u_r, u_g, u_b, d_r, d_g, and d_b along with their six anti-quarks with inverse colours. These colours were assigned with a specific purpose, red, green, and blue colours on mixing produce white colour – neutrality of colour, in a similar way it was demanded that particles that may stay freely must have a combination of quarks such that the total system is colourless or white. As such Δ^{++} with the structure $(u_r u_g u_d)$ does not violate Pauli principle as each quark is in a different quantum state because of its colour quantum number. Further, the total structure is colour neutral or achromatic. Pions also may stay freely because they make a colourless combination of quarks. In Yukawa theory, the pions carry charge from one nucleon to the other. In case of the quark, the quanta of strong nuclear field gluon exchange colour between quarks. Colour is regarded as the equivalent of electrical charge in quark–gluon physics and the atomic physics where stability demands electrical charge neutrality, the particle physics demands colour neutrality or chromatic neutrality.

Exercise p-3.10: What are virtual particles? Clearly distinguish between a real and a virtual pion and explain how virtual pions mediate between nucleons.

Exercise p-3.11: Briefly describe how Yukawa developed the concept of the mediation of nuclear interaction between nucleons through the exchange of π-meson.

Exercise p-3.12: Using the range of the nuclear field as 2 F, show that the mass of the virtual particle that mediate interaction between nucleons should be around 168×10^{-27} kg.

Multiple choice questions

Note: In some of the following questions more than one alternative may be true. Tick all the correct alternatives in such cases for the complete answer.

Exercise M-3.1: A system of two half-integer spin particles has repulsive singlet-state for all values of relative angular momentum. The most likely exchange force is

(a) Majorana (b) Bartlett (c) Heisenberg (d) Wigner

Exercise M-3.2: The range of a virtual particle of mass m is of the order of

(a) $\dfrac{\hbar}{mc^2}$ (b) $\dfrac{\hbar}{mc}$ (c) $\dfrac{\hbar c}{m}$ (d) $\dfrac{mc}{\hbar}$

Exercise M-3.3: The mass excess $(M - A)$, defined as the difference between the atomic mass M and the atomic mass number A of the nucleus, for four isobars X, Y, M, and N is, respectively, –5432–3248, +3269, and +5671 μm (micro mass unit). The most and the least bound nuclei are, respectively,

(a) N and X (b) M and Y (c) X and N (d) Y and M

Exercise M-3.4: The maximum Coulomb potential energy (in MeV) between $^{27}_{13}$Al and $^{64}_{29}$Cu nuclei is about ($r_0 = 1.2$ F)

(a) 15 (b) 25 (c) 35 (d) 45

Exercise M-3.5: The life time of a virtual particle of mass m cannot be larger than

(a) $\frac{\hbar}{mc^2}$ (b) $\frac{\hbar}{mc}$ (c) $\frac{\hbar c}{m}$ (d) $\frac{mc}{\hbar}$

Exercise M-3.6: Which type of nuclear force is always attractive?

(a) Majorana (b) Bartlett (c) Heisenberg (d) Wigner

Exercise M-3.7: Arrange the nuclides $^{14}_{7}N$, $^{16}_{8}O$, and $^{17}_{9}F$ in the decreasing order of neutron separation energies.

(a) $^{14}_{7}N$, $^{16}_{8}O$, and $^{17}_{9}F$ (b) $^{17}_{9}F$, $^{14}_{7}N$, and $^{16}_{8}O$

(c) $^{17}_{9}F$, $^{16}_{8}O$, and $^{14}_{7}N$ (d) $^{16}_{8}O$, $^{17}_{9}F$, and $^{14}_{7}N$

Exercise M-3.8: Deuteron is considered to be an admixture of S and D states. Deuteron could not be an admixture of S and P states because such an admixture will not reproduce the

(a) mass of deuteron (b) spin of deuteron
(c) quadrupole moment of deuteron (d) parity of deuteron

Exercise M-3.9: Deuteron could not be an admixture of S and F($\ell = 4$) states because such an admixture will not reproduce the

(a) mass of deuteron (b) spin of deuteron
(c) quadrupole moment of deuteron (d) parity of deuteron

Exercise M-3.10: Strong spin orbit coupling is required to explain

(a) stability of nuclei (b) higher magic numbers
(c) n-p scattering data (d) deuteron quadrupole moment

4 Quantum Mechanical Analysis of Some Nuclear Systems

Information about the nuclear force that binds nucleons in the nucleus is obtained essentially from four types of studies: (1) the systematic of stable nuclei, (2) properties of the ground state and excited states of stable and long-lived radioactive nuclei, (3) nucleon–nucleon scattering, and (4) nuclear reactions. The other possible approach is to assume a certain nature for the nuclear force and theoretically generate a nucleus with given numbers of neutrons and protons and then to compare the properties of the theoretically generated nucleus with the one actually existing in nature. Refinements in the assumptions made about the nuclear force may then be carried out to reproduce the experimental data. Both these approaches have been followed in nuclear physics. However, from the perspective of developing a theoretical model of the nucleus, light stable nuclei with only a few nucleons are best suited. The lightest nucleus is the nucleus of hydrogen atom that contains a single proton. There is no excited state of hydrogen nucleus and, therefore, not much information about the nature of nuclear force can be obtained from it. Electron scattering experiments have indicated that the positive charge distribution in proton has a radius of about 0.8 F. Deuteron is the next stable nucleus that has one neutron and one proton. Deuteron does not have any excited state as well. However, the measured values of the spin, magnetic dipole moment, electric quadrupole moment, and the electromagnetic radius of deuteron have given valuable information about the nature of the nuclear force.

Successful application of quantum mechanical approach to the hydrogen atom by Bohr and subsequent development of detailed quantum mechanical frame work by Heisenberg, Schrödinger, and others established beyond doubt its power of describing subatomic systems. In this chapter, elementary quantum mechanical treatment of deuteron, nucleon–nucleon scattering, and barrier transmission is developed.

4.1 DEUTERON AND ITS PROPERTIES

Deuteron is the only stable system with two nucleons, a proton, and a neutron that is found in a free state in nature. Study of the deuteron is, therefore, of considerable value in getting information about the nature and properties of nucleon–nucleon force. Both di-neutron (a system with two neutrons) and di-proton are unstable and not found in nature. The presently accepted values of deuteron parameters along with brief notes on the method of their measurement and significance are as follows:

1. Deuteron represented as $^{2}_{1}D$ or $^{2}_{1}H$ is found only in ground state and its excited states are unstable.
2. **Mass:** $3.34358325 \times 10^{-27}$ kg, or 2.01355532 u; or 1875.61282 MeV.
   It is determined by penning trap technique.
3. **Binding energy:** 2.225 MeV.
   It is determined from the energy of gamma rays emitted in the radiative capture reaction $^{1}_{0}n + ^{1}_{1}p = ^{2}_{1}D + \gamma$ (also written in short as n (p, γ) D). It may be noted that the binding energy per nucleon (B/A) in deuteron is only 1.11 MeV, which is much smaller than the average binding energy per nucleon of ≈8 MeV in the case of heavy nuclei. This indicates that the deuteron is a loosely bound system.
4. **Spin (or angular momentum) $J = 1$.**
   The spin of the deuteron has been determined by counting the number of components in hyperfine multiplet.

   Deuteron contains one neutron and one proton, each with spin ½. A spin value of 1 for deuteron is possible when the spins of the proton and the neutron are parallel and point in the same direction. As the spin $J = 1$ can have three possible values of the magnetic quantum number m_J, the deuteron is a triplet state. The singlet state of a neutron and a proton system, where the spins of the two particles are opposite to each other giving a resultant $J = 0$, does not exist. It means that ^{3}(n, p) state (deuteron) is a bound stable state and the ^{1}(n, p)-singlet state is unbound.
5. **Magnetic dipole moment μ_d:** 0.857393 nm (nuclear magneton).
   It is determined using magnetic resonance method.
   It may be noted that the magnetic dipole moment of deuteron is nearly, *but not exactly equal to* the sum of the magnetic moments of neutron and proton.

   $$\mu_n = -1.913 \text{ nm}; \mu_p = 2.792 \text{ nm. Hence, } \mu_n + \mu_p = 0.879 \text{ nm} \neq \mu_d.$$

   The fact that the magnetic moment of deuteron is nearly equal to the sum of the inherent magnetic moments of neutron and proton suggests that there is almost no contribution to the magnetic moment μ_d by the orbital motion of the proton. That means in deuteron, neutron, and proton are essentially in a state of a relative motion for which the relative orbital angular momentum ℓ is zero. The state of motion for which $\ell = 0$, is called the S-state. In deuteron, the neutron and the proton are in $^{3}S_{1}$-state, the superscript 3 giving the multiplicity of the state and subscript 1 indicating the J value. The $^{3}S_{1}$-state is spherically symmetric.
6. **Electric quadrupole moment Q_d:** 2.86×10^{-3} b (b = barn; 1b = 10^{-24} cm^{2}).
   The electric quadrupole moment of deuteron has been determined from the shift in the energy of lines in the rotational band spectrum. As Q_d is positive, the deuteron has a positive charge distribution of prolate shape.

   Now, there is a contradiction. As mentioned earlier, from the magnitude of the magnetic moment it appears that deuteron is a $^{3}S_{1}$-state. However, S-state is spherically symmetric and, therefore, the quadrupole moment of deuteron should have been zero. As will be discussed later, the deuteron is not a pure S-state. It is an admixture of $^{3}S_{1}$ and $^{3}D_{1}$ states.

If it is assumed that for about 96% time of the motion the neutron and proton remain in $^3S_1$-state but for a very small time, about 4% of the time period, they are in the $^3D_1$-state for which the relative angular momentum $\ell = 2$, but $J = 1$, then the magnitudes of both Q_d and μ_d can be satisfactorily reproduced. When in $^3D_1$-state, the magnetic moment of proton due to orbital motion also contributes to the magnetic moment of the deuteron along with the inherent magnetic moments of neutron and the proton. This explains why $\mu_d \neq \mu_p + \mu_n$. Also, when deuteron changes from $^3S_1$ to $^3D_1$, the spin of the deuteron flips to keep the value of $J = (\ell + s) = 1$. This means that the directions of ℓ and s are parallel but opposite. As such, the magnetic moment due to the orbital motion of proton is negative. This explains why $\mu_d < (\mu_n + \mu_p)$.

7. **Radius of the deuteron:** The root mean square electromagnetic radius, which is defined by the relation $[<r^2>]^{1/2} = [\int r^2 \rho(r) d^3r / \rho(r) d^3r]^{1/2}$, where $\rho(r)$ is the charge density at a distance r from the nuclear centre, has been determined from electron scattering and is about 2.1 F for deuteron.
8. **Parity of deuteron:** Parity of deuteron, deduced from deuteron reactions and theoretical consideration, is even.
9. **Statistics:** Deuteron with two nucleons (in $J = 1$ state) follows Bose–Einstein statistics.

4.2 SCHRÖDINGER EQUATIONS FOR A SYSTEM OF TWO SPINLESS PARTICLES IN SPHERICAL POLAR COORDINATES

We consider the general case of a system of two *spinless* particles of mass M_1 and M_2 with total energy W and potential energy U. The two-body problem can be reduced to a one-body problem by using the centre of mass (CM) coordinates. In CM system, one considers the motion of only one particle of reduced mass M, where $M = \dfrac{M_1 M_2}{M_1 + M_2}$ moving with the velocity V with respect to the centre of mass. Here, V is the relative velocity of M_1 and M_2. The Schrödinger equation for the particle of reduced mass M in Cartesian coordinates is

$$\nabla^2 \psi + \left(\frac{2M}{\hbar^2}\right)[W - U]\psi = 0 \tag{4.1}$$

Here, ψ is the space part of the wave function. Next step is to write Eq. (4.1) in spherical polar coordinates. Spherical polar coordinates are better suited for problems related to the nucleus because of the spherical symmetry. The three spherical polar coordinates r, θ, and φ are related to the Cartesian coordinates x, y, and z through the relations $x = r \sin\theta \cos\varphi$; $y = r \sin\theta \sin\varphi$; and $z = r \cos\theta$, as shown in figure 4.1, the polar axis is taken in z-direction.

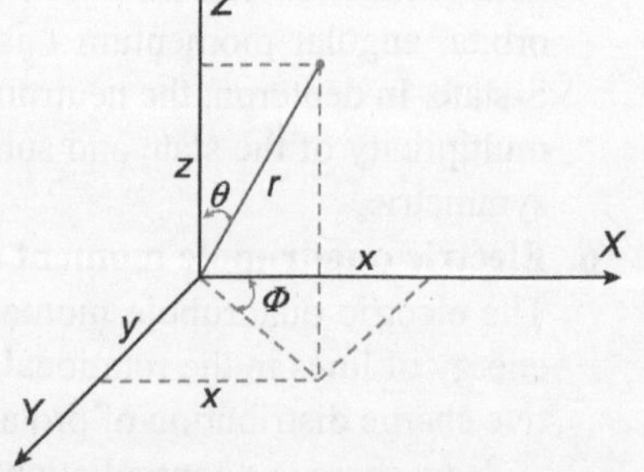

Figure 4.1 *Relation between the spherical polar and Cartesian coordinates*

Schrödinger Eq. (4.1) in spherical polar coordinates is

$$\frac{1}{r^2}\frac{\partial}{\partial r}\left(r^2\frac{\partial\psi}{\partial r}\right)+\frac{1}{r^2\sin\theta}\frac{\partial}{\partial\theta}\left(\sin\theta\frac{\partial\psi}{\partial\theta}\right)+\frac{1}{r^2 sin^2\theta}\frac{\partial^2\psi}{\partial\varphi^2}+\frac{2M}{\hbar^2}[W-U]\psi=0 \quad (4.2)$$

We now make another assumption that the potential U is a function of r alone and does not depend on θ and φ, that is $U = U(r)$, then the wave function $\psi(r, \theta, \varphi)$ may be written in two parts: $\chi(r)$, a function of r only and $Y(\theta, \varphi)$ a function of (θ, φ) only. Therefore, substituting $\psi(r, \theta, \varphi) = \chi(r)Y(\theta, \varphi)$, and dividing by ψ, we get

$$\frac{1}{\chi(r)}\frac{d}{dr}\left(r^2\frac{d}{dr}\chi(r)\right)+\frac{2Mr^2}{\hbar^2}[W-U(r)]=-\frac{1}{Y(\theta,\varphi)}\left[\frac{1}{\sin\theta}\frac{\partial}{\partial\theta}\{Y(\theta,\varphi)\}+\frac{1}{\sin^2\theta}\frac{\partial^2}{\partial\varphi^2}\{Y(\theta,\varphi)\}\right] \quad (4.3)$$

In Eq. (4.3), LHS is a function of r alone and RHS is a function only of θ and φ. As r, θ, and φ are all independent variables, Eq. (4.3) will hold only if the two sides are equal to some constant. Let the constant be $\ell(\ell + 1)$. Thus, the radial part of Eq. (4.3) may be written as

$$\frac{1}{\chi(r)}\frac{d}{dr}\left(r^2\frac{d}{dr}\chi(r)\right)+\frac{2Mr^2}{\hbar^2}[W-U(r)]=\ell(\ell+1)$$

or

$$\frac{1}{\chi(r)}\frac{d}{dr}\left(r^2\frac{d}{dr}\chi(r)\right)+\frac{2Mr^2}{\hbar^2}[W-U(r)-\{(\ell(\ell+1)\hbar^2/2Mr^2\}]=0 \quad (4.4a)$$

And the angular part as

$$-\frac{1}{Y(\theta,\varphi)}\left[\frac{1}{\sin\theta}\frac{\partial}{\partial\theta}\{Y(\theta,\varphi)\}+\frac{1}{\sin^2\theta}\frac{\partial^2}{\partial\varphi^2}\{Y(\theta,\varphi)\}\right]=\ell(\ell+1) \quad (4.4b)$$

The constant of separation $\ell(\ell + 1)$ has been chosen with a purpose. Let us recall that Eq. (4.4) in classical framework represents the motion of a particle of mass M, total energy W, and potential energy $U(r)$ moving with a velocity V in the centre of mass. V is the relative velocity of the particles. The kinetic energy of the particle is $\frac{1}{2}MV^2$. The total energy W is the sum of the kinetic and potential energy. Therefore,

$$W=\left(\frac{1}{2}\right)MV^2+U(r) \quad (4.5)$$

As $U(r)$ is positive, it is derived from a force of repulsion between the two particles. A rough sketch of the classical motion of the particle of mass M around the centre of mass C is shown in figure 4.2.

Let the distance of closest approach of the particle to the centre of mass be x when the particle is at point P. Now, at point P the velocity V of the particle is normal to x and hence the angular momentum J of the particle is given by

$$J = MVx \quad \text{or} \quad (4.6)$$

$$V = J/Mx \quad (4.7)$$

Substituting the value of V in Eq. (4.5), for the total energy of the particle, we get

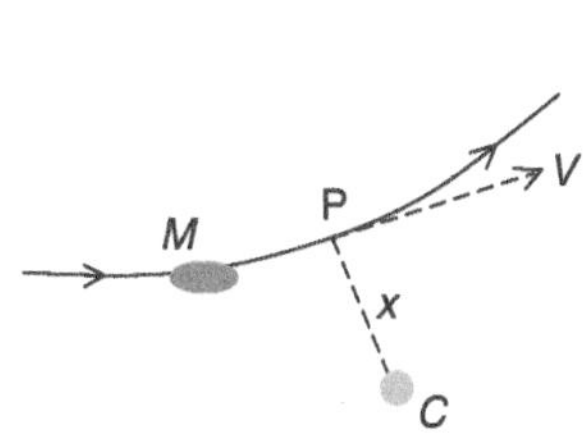

Figure 4.2 *Classical motion of the particle of reduced mass M about the centre of mass C*

$$W\,(\text{at P}) = \frac{1}{2}(MJ^2/M^2x^2) + U(r = x)$$
$$= (J^2/2Mx^2) + U(r = x) = (J^2/2I) + U(r = x) \tag{4.8}$$

Here, I represents the moment of inertia of the particle of mass M about an axis passing through the centre of mass. In Eq. (4.8), the kinetic energy of the linear motion of the particle is replaced by the energy of rotational motion with angular momentum J along an axis perpendicular to the polar axis and passing through the centre of mass. As no torque is present in the system, the angular momentum of the system is conserved. It means that the value of J remains same irrespective of the instantaneous position of the particle. Hence, Eq. (4.8) will remain valid for all values of r. In addition, in going from classical to quantum mechanical frame work $J^2 = \ell(\ell + 1)$ and therefore at any instance, the total energy of the particle may be written as

$$W = [\{(\ell(\ell + 1)\,\hbar^2/2Mr^2\} + U(r)] \tag{4.9}$$

From Eq. (4.9), it may be observed that the sign of both the terms $\{(\ell(\ell + 1)\hbar^2/2Mr^2\}$ and $U(r)$ is the same. A potential with positive sign is derived from a repulsive force and, therefore, the first term in Eq. (5.9) also represents a potential that corresponds to a repulsive force. The first term, which has the dimensions of energy, is called the centrifugal potential because like the classical centrifugal force it has the tendency to pull the particle away from the centre of mass. Larger the value of the angular momentum ℓ more will be the centrifugal force.

Coming back to Eq. (4.4a), for the ground state of the system, which is the lowest energy state, ℓ should be zero. *It may, therefore, be said that for the ground state of a two-body system of spinless particles in a central potential, the orbital angular momentum of the relative motion of the particles is zero and the system is said to be in S state.* States of a system with relative angular momentum $\ell = 0, 1, 2, 3 \ldots$ are, respectively, designated as S, P, D, … states.

Finally, the Schrödinger equation for the ground state of a system of two spinless particles of reduced mass M, total energy W, and in a central potential $U(r)$, which depends only on r is

$$\frac{1}{\chi(r)}\frac{d}{dr}\left(r^2\frac{d}{dr}\chi(r)\right) + \frac{2Mr^2}{\hbar^2}[W - U(r)] = 0 \tag{4.10}$$

To simplify the calculations, we substitute

$$u(r) = r\chi(r) \tag{4.11}$$

The function $u(r)$ is called the reduced radial wave function. With this substitution, Eq. (4.10) reduces to

$$\left(\frac{d^2u(r)}{dr^2}\right) + \frac{2Mr^2}{\hbar^2}[W - U(r)]\,u(r) = 0 \tag{4.12}$$

Equation (4.12) is the *reduced radial equation* for the ground state of a system of two spinless particles of reduced mass M, total energy W, and potential energy $U(r)$, which is a function of r alone. Eq. (4.12) is quite general, the only conditions imposed so far are (i) the particles are spinless and (ii) that $U(r)$ is a spherically symmetric function of r. It may be remarked that in Eq. (4.12), if W is positive it refers to a system of two particles that are free and not bound. A positive value of $U(r)$ means a repulsive force between the two particles. A negative value of

W means a bound system and a negative value of $U(r)$ refers to a force of attraction between the particles.

Next, we consider the angular part of the wave function described by Eq. (4.4b)). If it is further assumed that the function $Y(\theta, \varphi)$ may be written as the multiplication of two independent functions $\Theta(\theta)$ and $\Phi(\varphi)$, which are functions only of θ and φ, respectively, then on substituting $Y(\theta, \varphi) = \Theta(\theta)\,\Phi(\varphi)$ in Eq. (4.4b)), we get

$$\sin^2\theta\left[\frac{1}{\Theta\sin\theta}\frac{d}{d\theta}\left(\sin\theta\frac{d\Theta}{d\theta}\right)+\ell(\ell+1)\right] = -\frac{1}{\Phi}\frac{d^2\Phi}{d\phi^2}$$

$$= m^2 \text{ (separation constant)} \tag{4.4c}$$

In Eq. (4.4c), the LHS and the RHS are independent of each other and are, therefore, equal to some constant of separation, for instance, m^2. Eq. (4.4c) gives two equations, one polar and the other azimuthal as

$$\left[\frac{1}{\Phi\sin\theta}\frac{d}{d\theta}\left(\sin\theta\frac{d\Phi}{d\theta}\right)+\ell(\ell+1)\right]-\frac{m^2}{\sin^2\theta}=0 \tag{4.4d}$$

And

$$\frac{d^2\Phi}{d\phi^2}+m^2\Phi=0 \tag{4.4e}$$

Equation (4.4d) is the well-known Legendre's equation. Being a second-order differential equation, it has two independent solutions. Except for the values of the separation constant $\ell(\ell+1) = 0, 2, 6 \ldots$ for which ℓ has the values 0, 1, 2, 3, ... both the solutions of Legendre's equation becomes infinite for $\cos\theta = \pm 1$ and, therefore, not acceptable. It can be shown that for $\ell = 0$ and for other positive integer values and for $\ell \geq |m|$, one of the solution of Legendre's equation remains finite, single valued, and continuous, that is a well-behaved function.

If $m = 0$, from Eq. (4.4e), $\Phi(\phi)$ becomes constant. Under these conditions, the acceptable Legendre's solutions are called Legendre's polynomials and are represented by $P_\ell(\cos\theta)$. The expressions for the first few Legendre's polynomials are as follows:

$$P_0(\cos\theta) = 1$$

$$P_1(\cos\theta) = \cos\theta;$$

$$P_2(\cos\theta) = \frac{1}{2}(3\cos^2\theta - 1) = \frac{1}{4}(3\cos 2\theta + 1)$$

$$P_3(\cos\theta) = \frac{1}{2}(5\cos^3\theta - 3\cos\theta) = \frac{1}{8}(5\cos 3\theta + 3\cos\theta)$$

The azimuthal Eq. (4.4e) has the solution

$$\Phi_m(\varphi) = e^{\pm m\varphi} \tag{4.4f}$$

As an increase in φ by 2π should bring back to the starting position in space and must give the same value of Φ, the allowed values of m, the magnetic quantum number, may be negative or positive integers.

In the case when the magnetic quantum number m is not zero but $|m| \leq \ell$, solution of Legendre's equation are called associate Legendre's polynomials and are represented by $P^{l}_{m}(\cos\theta)$.

$$P^{\ell}_{m}(\cos\theta) = \left[(1-\cos^2\theta)^{\frac{|m|}{2}}\right]\frac{d^{|m|}(P_{\ell}\cos\theta)}{d\cos\theta^{|m|}}$$

The angular part of the space wave function $Y = \Theta(\theta)\Phi(\varphi)$ depends on l, m, θ, and φ and may be written as

$$Y_{\ell m}(\theta, \varphi) = N_{\ell m}\, P^{m}_{\ell}(\cos\theta)\, e^{im\varphi}$$

Here, $N_{\ell m}$ is the normalization constant. $Y_{\ell m}(\theta, \varphi)$ are called the spherical harmonics.

4.3 SIMPLE QUANTUM MECHANICAL DESCRIPTION OF DEUTERON

Now, we move to the specific case of the deuteron and rewrite Eq. (4.12). In the case of deuteron, the reduced mass $M = \{(M_n M_p)/(M_n + M_p)\}$, where M_n and M_p are, respectively, the masses of neutron and proton. As the masses of neutron and proton are nearly equal, the reduced mass M is nearly equal to half the mass of a neutron or a proton. The deuteron is a bound system and, therefore, $W = -E_b$, where E_b is the binding energy of the ground state. As in deuteron the nuclear force between neutron and proton is attractive, we represent it by a negative square-well (also called rectangular-well) potential of depth V_0 and width b (figure 4.3).

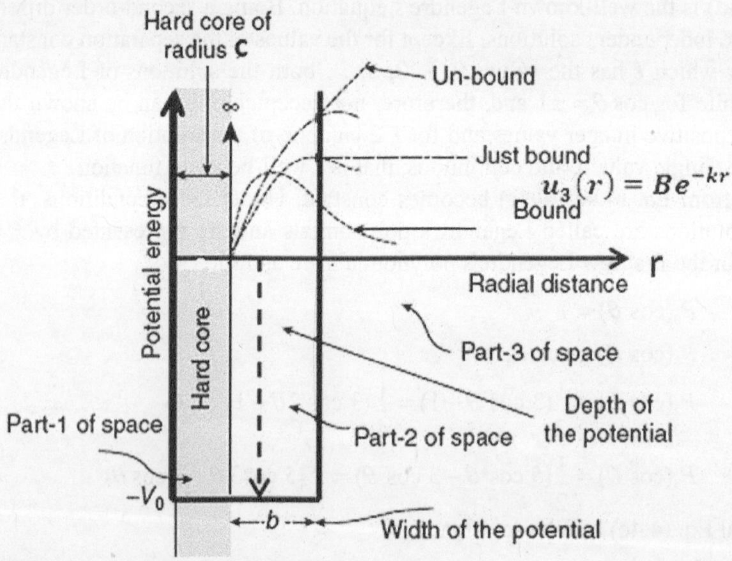

Figure 4.3 *Rectangular deuteron potential*

In figure 4.3, the distance c from the origin is the nuclear hard core. High-energy scattering experiments have indicated that in a nucleus two nucleons cannot come closer than c. The value of c is around 0.4 F. We chose a square-well potential to represent the deuteron as it is the simplest potential. The potential shown in figure 4.3 may be mathematically represented as

$$V = \infty \text{ for } r \le c$$

$$V = -V_0 \text{ for } c < r \le (b + c)$$

and $$V = 0 \text{ for } r > (c + b)$$

The Schrödinger equation for deuteron may be obtained by substituting the values of W and $U(r)$ in Eq. (4.12)

$$\frac{d^2u(r)}{dr^2} + \frac{2M}{\hbar^2}[(-E_b) - (-V_0)]u(r) = 0$$

Or $$\frac{d^2u(r)}{dr^2} + \frac{2M}{\hbar^2}[V_0 - E_b]u(r) = 0 \tag{4.13}$$

In Eq. (4.13), V_0 and E_b are positive numbers. Now, our task is to find the reduced radial wave function $u(r)$ that satisfies Eq. (4.13). The total space around the potential V may be divided into three parts: part 1 from $r = 0$ to c, part 2 from $r = c$ to $(c + b)$ and part 3 from $r = (c + b)$ and beyond. As the potential V is infinite between $r = 0$ and $r = c$, $u_1(r)$, the reduced wave function in part 1 is 0.

Or $$u_1(r) = 0$$

$u_2(r)$, reduced radial wave function in part 2 of the space, is the solution of the Eq. (4.13) in the limits r between c and $(c + b)$. The solution of Eq. (4.13), $u_2(r)$ in part 2 that vanishes at $r = c$ is

$$u_2(r) = A \sin K(r - c) \tag{4.14}$$

where A is a normalization constant and

$$K = \frac{1}{\hbar}\sqrt{2M(V_0 - E_b)} \tag{4.15}$$

In part 3 of space, Eq. (4.13) reduces to

$$\frac{d^2u(r)}{dr^2} + \frac{2M}{\hbar^2}[-E_b]u(r) = 0 \tag{4.16}$$

The solution $u_3(r)$ that satisfies Eq. (4.16) and vanishes at $r = \infty$ is

$$u_3(r) = Be^{-kr} \tag{4.17}$$

where B is a normalization constant and k is

$$k = \frac{1}{\hbar}\sqrt{2ME_b} \tag{4.18}$$

The solutions $u_2(r)$ and $u_3(r)$ should match at the boundary $r = (b + c)$. This requires

$$u_2(r)|_{r=b+c} = u_3(r)|_{r=b+c}$$

and $$\left[\frac{d}{dr}u_2(r)\right]_{r=c+b} = \left[\frac{d}{dr}u_3(r)\right]_{r=c+b}$$

The above-mentioned two conditions give

$$A \sin Kb = B\, e^{-k(c+b)} \tag{4.19}$$

And

$$AK\cos Kb = -kB\, e^{-k(c+b)} \tag{4.20}$$

To eliminate the normalization constants A and B, Eq. (4.20) is divided by Eq. (4.19) to get

$$K \cot Kb = -k \tag{4.21}$$

Equation (4.21) relates V_0, b, and E_b. It may be noticed that in Eq. (4.21) the central repulsive core c does not enter, and, therefore, the relation between the deuteron binding energy E_b, potential depth V_0, and the width of the potential b is independent of the hard core radius c.

A word about K and k, the wave numbers, it may be noted that $K = \frac{1}{\hbar}\sqrt{2M(V_0 - E_b)}$ is the reciprocal of the rationalized de Broglie wavelength of a particle of mass M and kinetic energy $(V_0 - E_b)$ and similarly, $k = \frac{1}{\hbar}\sqrt{2ME_b}$, represents the reciprocal of the rationalized wavelength of a particle of mass M and kinetic energy E_b. Hence, they are called wave numbers and are measured in inverse Fermi (F^{-1}). In general, a wave number for a particle of mass m and kinetic energy E_{kin} may be written as $|K_m| = \frac{1}{\hbar}\sqrt{2m\,|E_{kin}|} = \frac{1}{\hbar c}\sqrt{2mc^2\,|E_{kin}|}$. One can write simpler expressions for the wave number, depending on the system of units employed for mass m and E_{kin}. If both mass m and the kinetic energy E_{kin} are in MeV, then

$$K_m \text{ (in unit of inverse Fermi, F}^{-1}) = 0.007165\sqrt{\text{mass in MeV} \times \text{kinetic energy in MeV}}$$

$$= 0.2187\sqrt{\text{mass in mass unit u} \times \text{kinetic energy in MeV}}.$$

In Eq. (4.21), the value of $E_b = 2.225$ MeV is fixed. If this value of the binding energy is substituted, then Eq. (4.21) will give a relation between two variables V_0 and b. As such, a unique value of V_0 and/or of b cannot be determined from a single relation of Eq. (4.21). If the value of one of the two variables is fixed, the value of the other that satisfies Eq. (4.21) can be found. For example, if V_0 is fixed as 40 MeV, then b can be calculated to be 1.895 F. Also, $V_0 = 73$ MeV gives a value of $b = 1.337$ F. The relation $K\cot Kb = -k$ is plotted in figure 4.4. As may be seen from the figure any value of potential well depth V_0 with a corresponding value of the width b may satisfy the condition of matching the reduced radial wave functions at the boundary $r = c + b$. In conclusion, it may be stated that the *binding energy of the deuteron alone cannot uniquely determine the width and depth of the assumed square well potential*. We need to have some other property of the deuteron to uniquely determine both the depth and the width parameters of the square-well potential.

4.3.1 Bound, Just Bound and Un-bound States

The reduced radial wave function $u_3(r)$ in part 3 of space has the value $u_3(r) = Be^{-kr}$, which gives a bound state of the system if condition (4.21) is satisfied by V_0 and b. The wave functions $u_2(r) = A\sin K(r - c)$ and $u_3(r) = Be^{-kr}$ for the bound system are shown with solid lines in figure 4.5. The argument kr of the exponential term is related to the binding energy E_b through the relation

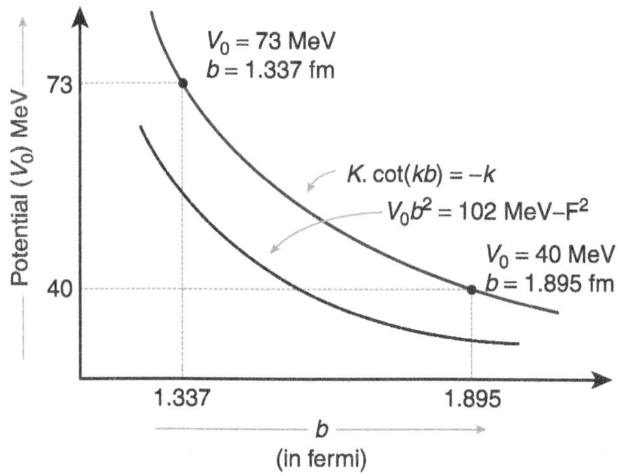

Figure 4.4 *A plot of KcotKb = –k*

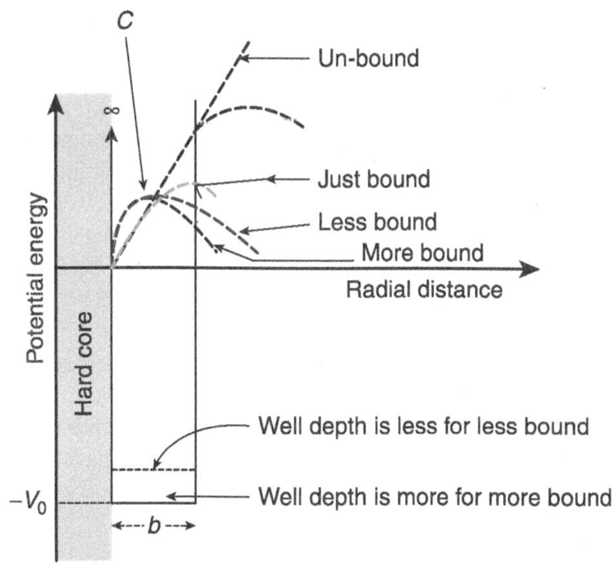

Figure 4.5 *Variation of $u_2(r)$, $u_3(r)$, and V_0 with the binding energy*

$k=(1/\hbar)[2ME_b]^{1/2}$. In case the system is just bound, that is $E_b=0$, k becomes zero, and $u_3(r)=Be^{-0}=B$, a constant. On substituting $E_b = 0$, K becomes equal to $(1/\hbar)[2MV_0]^{1/2}$. Now, $u_2(r)$, which is a sinusoidal function of Kb, must match $u_3(r)$ at the boundary of the well $r = (b + c)$. However, in the case of the limit of binding ($E_b = 0$), $u_3(r)$ is constant. The slope of the curve representing the function $\sin Kb$ is constant when $Kb = \pi/2$. Thus, for the limit of binding, we get

$$Kb = \pi/2; \quad \text{or} \quad (1/\hbar)\,[2MV_0]^{1/2}b = \pi/2 \quad \text{or} \quad V_0b^2 = (\pi/2)^2\,[\hbar^2/2M] \tag{4.22}$$

In the case of deuteron, the reduced mass $M = 0.504$ u. Substituting the values of π, $\hbar$, and M in Eq. (4.22) one obtains for the limit of binding

$$V_0b^2 = 102 \text{ MeV} - F^2 \tag{4.23}$$

The curve represented by Eq. (4.23) is also shown in figure 4.4 by dotted lines. Two points may be noted in figure 4.4.

1. For a given width b, the depth of the potential well that binds the two particles decreases as the magnitude of the binding energy decreases.
2. No bound state is possible if V_0b^2 is less than 102 MeV – F^2.

Figure 4.5 shows how the wave function $u_2(r)$ attains the peak much before the boundary of the well (at $c + b$) for strongly bound system, at the boundary for just bound system and does not reach the peak within the well for un-bound system. As may be inferred, the matching of the wave function $u_2(r)$ inside the well to the wave function $u_3(r)$ outside the well decides whether the system will be bound, just bound, or unbound. Larger the binding energy $|E_b|$ more rapidly $u_3(r)$ will go towards zero. To match the slope of $u_3(r)$ at the boundary $r = c + b$, the wave function $u_2(r)$ (which is a sinusoidal function) attains the peak much before the boundary and starts decreasing. In the case of just bound ($E_b = 0$), the outside wave function $u_3(r)$ becomes constant and its slope becomes zero. The inside wave function $u_2(r)$ is a sinusoidal function, which has a zero slope at the peak of the sinusoidal curve, therefore, to much the slope of $u_3(r)$ in just bound case, $u_2(r)$ attains peak at the well boundary $r = c + b$. In the case of the un-bound system, $u_3(r)$ increases with r and, therefore, $u_2(r)$ matches the slope of $u_3(r)$ before attaining the peak within the well. As the binding energy of the system decreases the depth of the potential well also decreases, because the depth of the well is the sum of the kinetic energy and the binding energy.

From the above-mentioned discussion, it is clear that a negative potential or a force of attraction between the two particles does not ensure that the system will be bound. It is the depth of the potential and the kinetic energy of motion that decides whether the system will be bound or unbound. We know that neutron–proton in singlet state $^1S_0$ with their spins pointing in opposite directions does not form a bound system. It does not mean that the force between neutron and proton in singlet state is not attractive. What it means is that the depth of the singlet potential is *just not enough* to bind the particles. Other scattering experiments have indicated that the singlet potential misses binding by only a few MeV.

4.3.2 Spin Dependence of Nuclear Force

Non-existence of a bound singlet $^1S_0$ state for neutron–proton system and the existence of a bound triplet $^3S_1$ state clearly show that the nuclear force between a neutron and a proton is spin dependent. The force is stronger when the spins are parallel and weaker when they are anti-parallel.

4.3.3 Tensor Nature of Nuclear Force

As has been shown earlier, the ground state of a system of two spinless particles in a central potential is an S-state. The ground state of deuteron is, therefore, also an S-state.

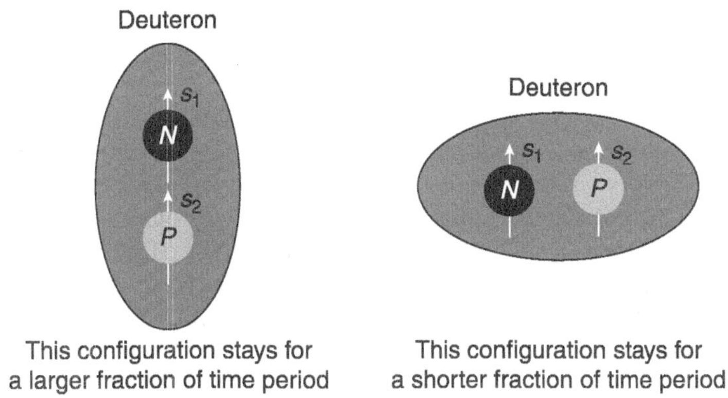

Figure 4.6 *During their motion the neutron and the proton stay in configuration (a) for a longer fraction of the time period than in configuration (b)*

However, S-state is spherically symmetric; as such, the charge distribution of deuteron should be spherically symmetric and hence its electric quadrupole moment should be zero. However, the shift in the energy of the lines of the rotational band spectrum of D_2 gas has shown that deuteron has a small positive quadrupole moment $Q = 2.86 \times 10^{-3}$ b. A positive value of Q means that the charge distribution is stretched along the spin axis, that is a prolate shape. This means that during their periodic motion about the centre of mass the neutron and the proton keep their spins parallel (so that $J = 1$) and stay one above the other for a larger fraction of the time period than the fraction of the time period they stay side by side. This is shown in figure 4.6, where the configuration (a) stays for a larger fraction of the time period and the configuration (b) for a smaller fraction of the time period. As a result, the time-averaged charge distribution of deuteron shows a prolate shape. The finite value of the quadrupole moment for deuteron shows the tensor nature of the n–p force. Tensor nature means that the force depends on the angles that the spins of the two particles make with the line joining the centres of the particles. For configuration (a), the angles are zero and for (b) they are of 90° and the force of attraction between the two nucleons are different for the two configurations.

Another hint about the presence of the tensor force in deuteron comes from the magnitude of the magnetic dipole moment μ_d of deuteron. In the ground state of the deuteron, the spins of the neutron and the proton always remain parallel to each other and if it is a pure S-state then the algebraic sum of the magnetic moments of neutron (μ_n) and proton (μ_p) should be equal to μ_d. However, the algebraic sum $\mu_n + \mu_p = -1.91350$ nm $+ 2.79275$ nm $= 0.87925$ nm, while the experimental value of μ_d is 0.857393 nm. The difference of about 0.2186 nm between the experimental and the expected value is much more than the errors of measurements and cannot be explained if deuteron is assumed to be a pure S-state. Further, it may also be remarked that the experimental value of μ_d is slightly less than the algebraic sum of μ_n and μ_p. We shall come back to this point a little later.

To explain the observed quadrupole moment and the magnetic moment of the deuteron, it is obvious to acknowledge that deuteron ground state is not a pure S-state. A mixture of S-state

($\ell = 0$) with some higher ℓ state that has non-spherical distribution and is compatible with the observed properties of the deuteron may explain the experimentally observed values for the quadrupole and the dipole moments of the deuteron. Let us discuss the properties of the higher ℓ states and whether or not they are compatible with the properties of the deuteron. It may be noted that both the quadrupole moment and the dipole moment are only slightly different from the value expected if it is assumed that the deuteron ground state is a pure S-state. It means that basically deuteron ground state is primarily an S-state and the other state responsible for the small finite values of Q and μ_d has only a small contribution. The parity of deuteron is even. The parity of a state with angular momentum ℓ is equal to $(-1)^\ell$. The parity of the S-state is even. Any other state that mixes with S-state must also have even parity so that the total parity of the system remains even. All states that have odd values of orbital angular momentum l will have negative or odd parity and, therefore, cannot mix with S-state. Only states of even l (=2, 4, 6) can mix with the S-state without violating the even parity requirements. It is also required that the mixed state must reproduce the angular momentum for the deuteron. Measured value of J for deuteron is 1. The total angular momentum J of the deuteron is given by the expression

$$J = \ell + s_p + s_n$$

where s_p and s_n are, respectively, the spins of the neutron and of proton each having the value ½ and the addition is quantum mechanical. The quantum mechanical sum of two angular momentums J_a and J_b may have values from $|J_a + J_b|$ to $|J_a - J_b|$ in steps of unity. For the S-state, $J = (0 + ½ + ½)$ may have two possible values 1 and 0. In case of deuteron, s_p and s_n are parallel and add up to give $J =$ 1 for the S-state. For the D-state, $J = (2 + ½ + ½)$ and can have values 3, 2, and 1. As such, D-state can mix with S-state as it may provide the required parity and the J value for deuteron. State F with orbital angular momentum $\ell = 4$, gives $J = 4 + ½ + ½$. Possible values of J in F-state are 5, 4, and 3. It may be observed that no odd ℓ-states and even ℓ-states with $\ell \geq 4$ could produce a state with total angular momentum $J = 1$. Thus, except the D-state, no other even ℓ-state can mix with S-state in case of the deuteron to reproduce the observed values of spin and parity of the deuteron.

It may also be seen that when the neutron and the proton are in D-state, the configuration is $^3D_1$, which means that the direction of the orbital angular momentum ℓ is opposite to the direction of the intrinsic spins of the neutron and proton. Further, the directions of the neutron and proton intrinsic spins remain parallel both in S-state and in D-state. When the system moves from S-state to D-state, the intrinsic spins flip in a direction opposite to the direction of ℓ (=2). When the deuteron system is in D-state, three components contribute to the magnetic moment μ_d.

$$\mu_d = \mu_n + \mu_p + \mu^\ell{}_p$$

Here, μ_n and μ_p are the intrinsic magnetic moments of neutron and proton and $\mu^\ell{}_p$ is the magnetic moment of proton due to its orbital motion. Orbital motion of neutron does not produce any magnetic moment. Now, μ_n is negative, μ_p is positive, and $\mu^\ell{}_p$ will also be negative as the direction of ℓ is opposite to the direction of the spin of the proton. As such, the resultant magnetic moment of deuteron in D-state will be less than the value of the magnetic moment in S-state. As the deuteron remains in D-state only for a very small fraction of the time period, the time-averaged value of μ_d is only slightly less than the value for the pure S-state.

In light of the discussion, the total wave function Ψ_d for deuteron may be written as a linear combination of the normalized wave functions Ψ_s and Ψ_D for the S- and D-states, respectively.

$$\Psi_d = A_s \Psi_s + A_D \Psi_D$$

The amplitude coefficients A_s and A_D satisfy the condition

$$A_s^2 + A_D^2 = 1$$

The magnitude of A_D and the radial dependence of Ψ_D decide the values of Q and μ_d. The radial dependence of Ψ_D in turn depends on the shape of the potential between the neutron and the proton. A_D^2 in the range of 3–6% for different shapes of the nucleon–nucleon potential reproduces the experimental value of deuteron quadrupole moment. In addition, A_D^2 of about 4% explains the experimental value of deuteron dipole moment.

4.3.4 Radius of Deuteron

Let us refer to $u_3(r)$ the reduced wave function outside the potential well that exponentially decreases with r as

$$U_3(r) = Be^{r/(1/d)}$$

The factor $(1/d)$ called the relaxation length, may be taken as the first approximation towards the radius of the deuteron.

$$r_d = (1/d) = [\hbar/(2ME_b)]$$

On substituting the values of M and E_b, we get $(1/d) = 4.31$ F. It is to be noted that the radius of deuteron as determined is much larger than the range of the nuclear force which is of the order of 2 F. This implies that for a substantial part of the time period neutron and proton in deuteron remain beyond the range of nuclear force. This is possible only in quantum systems where particles may tunnel out of the potential well, which is not possible in classical systems. Another point that may be noted is that the relaxation length is inversely proportional to the binding energy of the system. Smaller the binding energy, larger will be the relaxation length or the radius. Deuteron is a loosely bound system. In addition, it means that the depth of the deuteron potential well may be large, but it also has large kinetic energy of motion of nucleons. The difference of the potential depth and the kinetic energy gives the binding energy.

It may be recalled that the electromagnetic radius of deuteron is the distance of the proton from the centre of mass and is half the separation between neutron and proton = $r/2$. The mean square electromagnetic radius may be obtained from the following expression

$$(r_d^2)\text{avg} = (1/2)^2[\int_c^{b+c} r^2 u_2^2(r)dr + \int_{b+c}^{\infty} r^2 u_3^2(r)dr] \tag{4.24}$$

The value of r.m.s. electromagnetic radius of deuteron determined using Eq. (4.24) agrees well with the experimental value if the values of hard core radius $c = 0.4$ F, determined from high-energy scattering experiments and the r.m.s. radius of proton, obtained from electron scattering experiment = 0.8 F, are used. The values of A^2 and B^2 needed in these calculations may be obtained from the normalization condition

$$\int_c^{b+c} u_2^2(r)dr + \int_{c+b}^{\infty} u_3^2(r)dr = 1$$

Solved example S-4.1

Deuteron is a loosely bound system of two nucleons of nearly equal mass that may be represented by a square well of depth V_0 and width b. Show that to a good approximation $V_0b^2 = \left(\frac{\pi}{2}\right)^2\left(\frac{\hbar^2}{2M}\right)$, where M is the reduced mass of the system.

Solution: We have seen that for a two particle system of reduced mass M under a square-well potential of depth $-V_0$ and width b, the following relation (see Eq. (4.21)) holds good

$$K\cot Kb = -k.$$

Here, $K = \frac{1}{\hbar}\sqrt{2M(V_0 - E_b)}$; $k = \frac{1}{\hbar}\sqrt{2ME_b}$, and E_b is the binding energy of the system. As the binding energy of deuteron is only 2.22 MeV, and $2M$ is about 1 MeV, k is a small number. As such, $\cot Kb = -k/K =$ very small negative quantity. This means that Kb, the argument of $\cot Kb$, is around $\frac{\pi}{2}$, that is,

$$Kb \approx \frac{\pi}{2} \text{ and } (Kb)^2 = \left(\frac{\pi}{2}\right)^2 \text{ or } \left(\frac{1}{\hbar}\sqrt{2M(V_0 - E_b)}\right)^2 b^2 = \left(\frac{\pi}{2}\right)^2$$

Neglecting E_b in comparison with V_0, we get; $V_0b^2 = \left(\frac{\pi}{2}\right)^2\left(\frac{\hbar^2}{2M}\right)$.

Solved example S-4.2

Assuming that for deuteron the reduced radial wave function $u(r) = Be^{-kr}$ holds good from $r = 0$ to ∞, determine the value of constant B. In addition, find the probability that the separation between the neutron and proton inside deuteron exceeds a distance 'd'.

Solution. The normalization constant B may be determined from the condition that the probability of finding the system somewhere from $r = 0$ to $r = \infty$ is.equal to one. Therefore,

$$\int_0^\infty |u(r)|^2\, dr = 1 \text{ or } \int_0^\infty B^2 e^{-2kr}\, dr = 1 \text{ or } \frac{B^2}{2k} = 1.$$

Hence, $B = \sqrt{2k}$.

The probability P_d that the separation between neutron and proton is more than a certain distance (say) 'd' is equal to the value of the integral $\int_d^\infty |u_3(r)|^2\, dr$. Therefore,

$$P_d = \int_d^\infty |u_3(r)|^2\, dr = B^2 \int_d^\infty e^{-2kr}\, dr = 2k\left[\frac{1}{2k}\{[e^{-2kr}]_d^\infty\}\right] = [e^{-2kd}]$$
$$= [e^{-2kd}]$$

By substituting the values of k and d in appropriate units, both B and P_d may be calculated.

Solved example S-4.3

Calculate the percent of time the neutron and proton in deuteron remain outside the nuclear potential well.

Solution. The radial wave function inside the potential well $\psi_2(r)$ and beyond the potential well $\psi_3(r)$ is related to their respective reduced wave functions u_2 and u_3 by the following relations:

$$\psi_2(r) = u_2(r)/r \text{ and } \psi_3(r) = u_3(r)/r$$

The probability P_{out} that neutron and proton remains outside the nuclear potential is given by

$$P_{\text{out}} = \int_{(b+c)}^{\infty} |\psi_3|^2 dv = \int_{(b+c)}^{\infty} \frac{|u_3|^2}{r^2} 4\pi r^2 dr = 4\pi B^2 \int_{(b+c)}^{\infty} e^{-2kr} dr$$

$$= \frac{4\pi B^2}{2k} e^{-2k(b+c)} \tag{A}$$

As $u_2(b + c) = u_3(b + c)$, we have $A \sin K(r - c)|_{(b+c)} = Be^{-kr}|_{(b+c)}$, which gives

$$A \sin Kb = Be^{-k(b+c)} \tag{B}$$

However, because of the very low binding energy of deuteron it may be assumed that the wave function u_2 just reaches the maxima at boundary $r = (c + b)$, which means that $Kb = \frac{\pi}{2}$. Hence, it follows from Eq. (B)

$$A = Be^{-k(b+c)} \text{ and } e^{-2k(b+c)} = \frac{A^2}{B^2} \tag{C}$$

Substituting this value in Eq. (A), we get

$$P_{\text{out}} = \frac{4\pi B^2}{2k}\frac{A^2}{B^2} = \frac{2\pi A^2}{k} \tag{D}$$

To determine the value of A, we make use of the fact that the total probability of finding the deuteron somewhere in the space is 1.

$$\int_{c}^{(b+c)} |\Psi_2|^2 d\tau + \int_{(b+c)}^{\infty} |\Psi_3|^2 d\tau = 1 \text{ or } \int_{c}^{(b+c)} \frac{|u_2|^2}{r^2} 4\pi r^2 dr + \int_{(b+c)}^{\infty} \frac{|u_3|^2}{r^2} 4\pi r^2 dr = 1$$

or
$$A^2 \int_{c}^{(b+c)} |\sin K(r-c)|^2 \, dr + B^2 \int_{(c+b)}^{\infty} e^{-2kr} dr = \frac{1}{4\pi}$$

or
$$A^2 \int_{c}^{b+c} \frac{1}{2}(1-\cos 2K(r-c))dr + B^2 \int_{c+b}^{\infty} e^{-2kr} dr = \frac{1}{4\pi}$$

or
$$\frac{A^2}{2}\left[r - \frac{\sin 2K(r-c)}{2K}\right]_c^{c+b} + \frac{B^2}{2k} e^{-2k(c+b)} = \frac{1}{4\pi}$$

or
$$\frac{A^2}{2}\left[b - \frac{1}{2K}\sin 2Kb\right] + \frac{B^2}{2k} e^{-2k(b+c)} = \frac{1}{4\pi}$$

As $Kb \approx \frac{\pi}{2}$, $\sin 2Kb = 0$. In addition, $B^2 e^{-2k(b+c)} = A^2$.

$$A^2\left[b+\frac{1}{k}\right]=\frac{1}{2\pi}$$

or
$$A^2=\frac{k}{2\pi(kb+1)}A^2=\frac{k}{2\pi(kb+1)}.$$

Substituting this value of A^2 in Eq. (D) gives

$$P_{\text{out}}=\frac{1}{kb+1} \quad \text{(E)}$$

Now, $k=\frac{1}{\hbar}\sqrt{2ME_b}$ with $2M$ = nucleon mass ≈ 939 MeV; E_b = 2.22 MeV and $\hbar$ =0.658 × 10^{-21} MeV-s, considering b = 2 F. Substituting these values in Eq. (E) gives

P_{out} = 0.7. It means that for about 70% of the time neutron and proton remain outside the range of the nuclear field.

Exercise p-4.1: Show that the ground state of a system of two spinless particles under a central potential is an S-state.

Exercise p-4.2: Write Schrödinger equation for a system of two spinless particles and solve it for a central attractive potential in spherical polar coordinates.

Exercise p-4.3: Write notes on the dipole moment and the quadrupole moment of deuteron. Explain why the magnetic dipole moment of deuteron is less than the sum of the dipole moments of neutron and the proton.

Exercise p-4.4: It is known that the normalized wave function $\psi=\left(\frac{1}{r}\right)\left(\frac{\alpha}{2\pi}\right)e^{-\alpha r}$, where α^{-1} = 4.3F,. is a good approximate wave function for deuteron. Using this function, find the root mean separation between neutron and proton in deuteron.

Exercise p-4.5: Assuming that a system of two spinless particles can be represented by a square-well potential, discuss the condition (s) for producing a bound state.

Exercise p-4.6: Assuming deuteron to be a just bound system, calculate the depth of the square well potential that has a width 1.5 F.

[**Hint:** As it is assumed that deuteron is a just bound state, the wave function $u_2(r) = A \sin K(r-c)$ must peak at the boundary $r = (b + c)$, which means

$$K(r-c)\big|_{(c+b)}=\frac{\pi}{2} \text{ or } (Kb)^2=\frac{\pi^2}{4}$$

or
$$2M(V_0-E_b)=\frac{\pi^2\hbar^2}{4b^2} \text{ Or } V_0=\frac{\pi^2\hbar^2}{8Mb^2}+E_b \quad \text{(A)}$$

The important point is to put the quantities appearing in Eq. (A) in correct units. If we want our answer in MeV, then we may multiply by c^2 both in the numerator and denominator of the first term $V_0 = \frac{\pi^2(\hbar^2c^2)}{8(M c^2)b^2} + E_b$. Now, the value of $(\hbar c) = 197.32\,\text{MeV F}$ and the value of (Mc^2), the reduced mass of the neutron–proton system $\left(\approx \frac{940}{2} = 470\,\text{MeV}\right)$ and $E_b = 2.225$ MeV, may be substituted.

4.4 QUANTUM MECHANICAL APPROACH TO NUCLEON–NUCLEON SCATTERING

Scattering process serves to understand the structure of subatomic particles and the interactions between them. To initiate the process, a projectile is made to hit a target. As a result of the interaction between the projectile and the target, the projectile is either scattered or it may initiate reactions. In *elastic scattering*, the incident particle as well as the target remains in the *same state* as they were *initially*. Although the kinetic energies of both the incident and the target particles may change in elastic scattering, the total energy of the system and the state of excitation of the incident and the target nuclei does not change. In *inelastic* scattering, the state of excitation of one or of both, the incident particle and the target, changes. The de Broglie wavelength of the incident particle plays an important role in scattering processes. The rationalized de Broglie wavelength $ƛ$ for a particle of rest mass m_0 and kinetic energy E_{kin} is given by $ƛ = \frac{\hbar}{p}$. However, the total energy E of the particle, which is the sum of its kinetic and rest mass energy, is related to its kinetic energy and the rest mass m_0 by the relation

$$E^2 = p^2c^2 + m_0^2 c^4 \text{ or } (E_{\text{kin}} + m_0c^2)^2 = p^2c^2 + m_0^2 c^4$$

or

$$p = \frac{\sqrt{(E_{\text{kin}}^2 + 2m_0c^2E_{\text{kin}})}}{c}$$

$$\lambda = \frac{\hbar}{p} = \frac{\hbar c}{\sqrt{(E_{\text{kin}}^2 + 2m_0c^2E_{\text{kin}})}} \tag{4.25}$$

which reduces to

$$ƛ = \hbar/(2m_0E_{\text{kin}})^{1/2} \text{ when } m_oc^2 >> E_{\text{kin}} \tag{4.26a}$$

and

$$ƛ = (\hbar c)/E_{\text{kin}} \text{ when } m_0c^2 << E_{\text{kin}} \tag{4.26b}$$

Figure 4.7 shows the relation between the rationalized de Broglie wavelength, kinetic energy, and the momentum for different particles. It may be observed that with the increase of kinetic energy the deviation from the photon line vanishes for all particles, all particles behave like photons.

Most of the times in laboratory frame the incident particle approaches a stationary scattering centre as shown in figure 4.8a. Before scattering, the incident particle of mass M_1 is moving towards the stationary target particle of mass M_2 with velocity V. After scattering, the incident particle moves at angle θ with respect to the incident direction with a velocity V_1 and the target particle recoils at an angle ϕ with velocity V_2 (figure 4.9). The centre of mass of the two particles

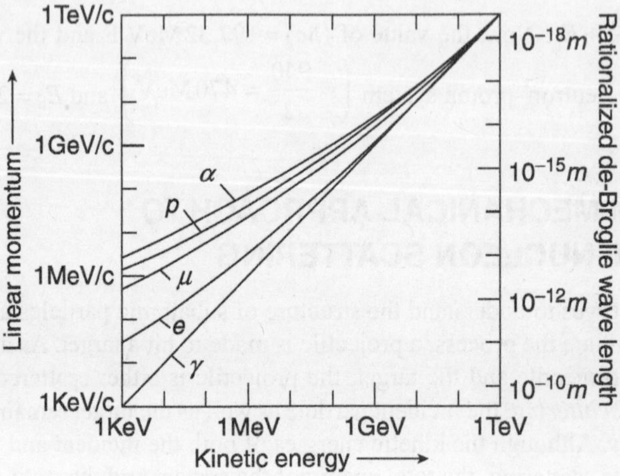

Figure 4.7 *Graphical representation of the kinetic energy, momentum, and rationalized de Broglie wavelength for different particles*

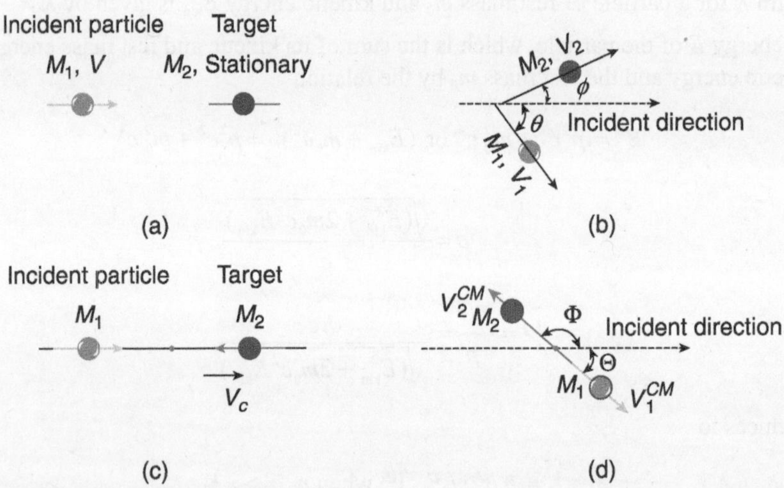

Figure 4.8 *Incident particle and the target (a) before interaction in laboratory frame, (b) after interaction in laboratory frame, (c) before elastic scattering in centre of mass frame, (d) after scattering in centre of mass frame*

moves with the velocity V_c and the reduced mass of the system is M_{red}. They are related to the initial parameters by the following relations:

$$M_{red} = \frac{M_1 M_2}{M_1 + M_2} \quad \text{and} \quad V_c = V\left(\frac{M_1}{M_1 + M_2}\right) = V\left(\frac{M_{red}}{M_2}\right) \tag{4.27}$$

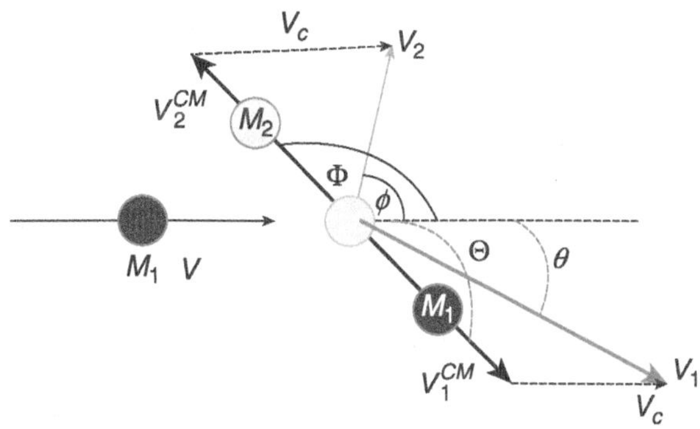

Figure 4.9 *Velocity triangles for the incident and the target particles*

In laboratory frame, the centre of mass moves with the velocity V_{cm} before, after, and during the scattering process.

In the centre of mass frame (CM frame), the incident and the target particles move in a straight line in opposite directions with equal momentum. Thus, the total linear momentum in the CM is zero. As shown in figure 4.8, the elastic scattering in CM frame corresponds to the rotation of the line of incidence. The angles Θ and Φ of CM frame are related to the angles θ and ϕ of laboratory frame by the relations

$$\Phi = 2\phi = (\pi - \Theta) \tag{4.28a}$$

and

$$\cot \theta = \frac{\left(\frac{M_1}{M_2}\right) + \cos \Theta}{\sin \Theta} \tag{4.28b}$$

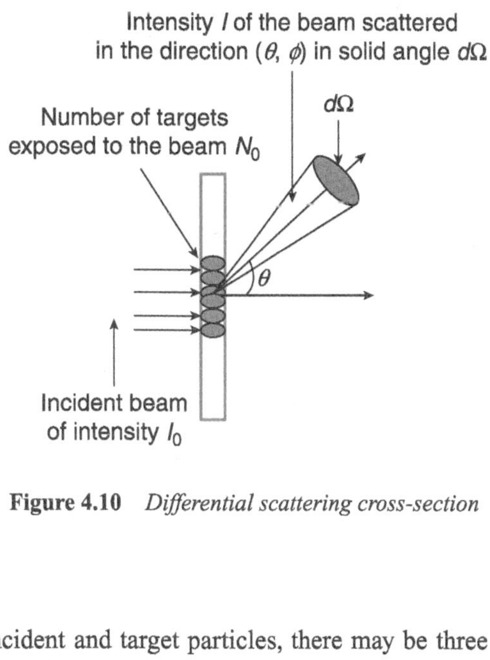

Figure 4.10 *Differential scattering cross-section*

Depending on the ratios of the masses of the incident and target particles, there may be three cases of interest

$$\text{If } M_2 >> M_1 \quad \text{then} \quad \theta \approx \Theta \tag{4.29a}$$

$$\text{If } M_2 = M_1 \quad \text{then} \quad \theta = \Theta/2 \tag{4.29b}$$

$$\text{If } M_2 << M_1 \quad \text{then} \quad \theta \approx \left(\frac{M_2}{M_1}\right) \sin \Theta \tag{4.29c}$$

The velocity triangles for the incident and the target particles are shown in figure 4.9. As may be seen in this figure,

$V_1 = V_c + V_1^{CM}$; $V_2 = V_c + V_2^{CM}$, where V_1 and V_2 are, respectively, the laboratory velocities and V_1^{CM}, V_2^{CM} the centre of mass velocities of particle M_1 and M_2. V_c is the velocity of the centre of mass.

The total kinetic energy available in the centre of mass frame, $(E_{kin})_{CM}$ is given by the relation

$$(E_{kin})_{CM} = \frac{1}{2} M_{red} V^2 = \frac{1}{2}\left(\frac{M_1 M_2}{M_1 + M_2}\right) V^2$$

The energy associated with the centre of mass is $\frac{1}{2}(M_1 + M_2)V_c^2$

If I_0 is the beam intensity of the projectiles, that is the number of incident particles crossing unit area held normal to the beam direction, per second and I the intensity of particles scattered in the direction (θ, ϕ) within solid angle $d\Omega$ and N_0 the number of target particles interacting with the beam, then

$$I \propto I_0 N_0 d\Omega$$

or

$$I = \frac{d\sigma(\theta,\phi)}{d\Omega}(I_0 N_0 d\Omega)$$

The constant of proportionality $\frac{d\sigma(\theta,\phi)}{d\Omega}$ is called the differential scattering cross-section. It is a measure of the probability of scattering in a given direction (θ, ϕ) per unit solid angle per target particle per unit time. The total cross-section σ is given as

$$\sigma = \int \frac{d\sigma(\theta,\phi)}{d\Omega} d\Omega$$

It may be shown that the scattering cross-section $\sigma(\theta)$ in laboratory frame is related to the cross-section in the centre-of-mass frame $\sigma(\Theta)$ according to the following relation:

$$\frac{d\sigma(\theta)}{d\Omega} = \frac{\left[1+\left(\frac{M_1}{M_2}\right)^2 + 2\left(\frac{M_1}{M_2}\right)\cos\Theta\right]^{3/2}}{|1+\frac{M_1}{M_2}\cos\Theta|}\frac{d\sigma(\Theta)}{d\Omega}.$$

The total and differential scattering cross-sections in the laboratory and the centre of mass frames are related to each other by the relation

$$\sigma = \int_0^{2\pi} d\phi \int_0^{\pi} \frac{d\sigma(\theta,\phi)}{d} \sin\theta\, d\theta = \int_0^{2\pi} d\Phi \int_0^{\pi} \frac{d\sigma(\Theta,\Phi)}{d\Omega} \sin\Theta\, d\Theta$$

The magnitude of the total scattering cross-section in laboratory and centre of mass frames is equal, as it gives the total number of scattering events per target particle per unit incident flux per unit time.

Solved example S-4.4

A particle of mass M_1 is elastically scattered by an stationary particle of mass M_2, which is projected at an angle ϕ, while the incident particle is scattered by an angle θ from the incident direction. Show that the ratio of the masses of the incident and target particles is given by $\frac{\sin(\theta+2\phi)}{\sin\theta}$.

Solution. Let Θ and Φ.be the scattering angles for the incident and the target particles in the CM frame. We now use the relations

$$\cot\theta = \frac{\left(\frac{M_1}{M_2}\right)+\cos\Theta}{\sin\Theta};\ \Theta=\pi-\Phi=\pi-(2\phi),\text{ as }\Phi=2\phi.$$

Therefore,

$$\sin\Theta=\sin(\pi-2\phi)=\sin 2\phi \text{ and } \cos\Theta=\cos(\pi-2\phi)=-\cos 2\phi$$

Hence,

$$\cot\theta = \frac{\cos\theta}{\sin\theta}=\frac{\left(\frac{M_1}{M_2}\right)+\cos\Theta}{\sin\Theta}=\frac{\left(\frac{M_1}{M_2}\right)-\cos 2\phi}{\sin 2\phi}$$

or

$$\frac{M_1}{M_2}\sin\theta-\sin\theta\cos 2\phi=\cos\theta\sin 2\phi;\text{ or }\frac{M_1}{M_2}\sin\theta=\sin(\theta+2\phi)$$

or

$$\frac{M_1}{M_2}=\frac{\sin(\theta+2\phi)}{\sin\theta}$$

Exercise p-4.7: A smooth sphere of mass M_1 moving with a velocity V collides elastically with a stationary similar sphere of mass $M_2\,(M_1 > M_2)$, gets scattered at an angle θ and deflects the target sphere by the angle ϕ in the laboratory frame. If θ_{CM} is the angle of scattering of the projectile in the CM-frame then show that

$$\tan\theta=\frac{\sin\theta_{CM}}{\left(\frac{M_1}{M_2}\right)+\cos\theta_{CM}}$$

Also show that in case $M_1 = 2M_2$, then $\theta=\phi=15^0$.

Solved example S-4.5

For the elastic scattering between an incident particle of mass m and a stationary target of mass $M\ (M < m)$ show that the sin θmax, where θmax, is the maximum angle of scattering in laboratory frame, is given by, $\sin\theta\text{max}=\frac{m}{M}$.

Solution. From Eq. (4.28b), we have

$$\tan\theta = \frac{\sin\Theta}{\left(\frac{m}{M} + \cos\Theta\right)} \tag{A}$$

Now, for maximum value of θ (we call it θ_{max}), $\frac{d\tan\theta}{d\theta}$ should be zero, which means that differential of Eq. (A) should be zero. This condition will give the maximum value of CM-angle Θ_{max}.

$$\frac{d\left[\frac{\sin\Theta}{\left(\frac{m}{M}+\cos\Theta\right)}\right]}{d\Theta} = 0; \text{ or } \frac{\cos\Theta}{\frac{m}{M}+\cos\Theta} + \frac{\sin\Theta^2}{\left(\frac{m}{M}+\cos\Theta\right)^2} = 0 \text{ or } \frac{\frac{m}{M}\cos\Theta + \cos\Theta^2 + \sin\theta^2}{\left(\frac{m}{M}+\cos\Theta\right)^2} = 0$$

The above-mentioned expression will be 0 when the numerator is 0, which gives

$$\cos\Theta_{max} = -\frac{M}{m} \text{ and } \sin\Theta_{max} = \frac{\sqrt{m^2 - M^2}}{m}.$$

When these values are substituted in Eq. (A), we get

$$\tan\theta_{max} = \frac{m}{\sqrt{M^2 - m^2}}$$

And
$$\sin\theta_{max} = \frac{m}{M}$$

Solved example S-4.6

Alpha particles of 5 MeV pass through a gas where they undergo elastic scattering with the nuclei of the gas. The maximum angle of scattering is found to be 30° and no more than that. Identify the gas. In addition, discuss the dependence of the maximum scattering angle on the incident alpha energy.

Solution. The first thing to be noted is that the problem is non-relativistic as the kinetic energy of alpha particle is much less than its rest mass energy. Hence, laws of classical scattering can be used. In solved example S-4.5, it is shown that

$$\sin\theta_{max} = \frac{\text{mass of the target}}{\text{mass of the incident particle}}.$$

Hence,

$$\text{Mass of target particle} = \text{mass of the incident particle} \times \sin\theta\text{max}.$$
$$= (\approx 4\text{ u}) \times \sin 30° = 2\text{ u}$$

Thus, the mass of the target gas nucleus is around 2 u, which is the mass of the deuteron. Hence, the gas is deuterium.

It may be noticed that θmax does not depend on the incident energy.

Exercise p-4.8: A particle of mass m is elastically scattered by a stationary particle of mass M ($M < m$). The target particle suffers a deflection of ϕ from the original direction of motion of the incident particle. Show that

$$\sin\phi_{max} = \left[\frac{m-M}{2m}\right]$$

Here, ϕ max is the maximum possible angle of scattering of the target particle.

Exercise p-4.9: A non relativistic alpha particle moving with the velocity v hits a stationary proton. Calculate the maximum velocity (in terms of v) that may be imparted to the proton.

Before proceeding further we once again recapitulate some definitions.

1. **Cross-section:** Cross-section for a given process is a measure of the probability of the occurrence of the process. It is defined in such a way that the dimension of cross-section comes out to be that of area. In case of nuclear events, cross-section may be defined as the interaction probability per unit target per unit incident flux. Cross-section is generally denoted by the symbol σ with a subscript or bracket denoting the process. Cross-section is a quantity that may be measured experimentally on one hand and may also be calculated theoretically on the other. As such, it is the meeting ground for experiments and theory. Unit of cross-section is barn, denoted by b. 1b = 10^{-28} m^2 or 10^{-24} cm^2.
2. **Flux:** In a beam of particles, the number of particles crossing the unit area held normal to the beam, per unit time is called the flux of the beam. Suppose that a beam of particles hits an area A held normal to the beam and dN particles passes through the area in time dt. Then, the flux $= \left(\frac{1}{A}\right)\frac{dN}{dt} = \left(\frac{dx}{A\,dx}\right)\frac{dN}{dt} = \left(\frac{dN}{A\,dx}\right)\frac{dx}{dt}$ = Number density of the particles $\left(\frac{dN}{A\,dx}\right) \times$ velocity of the particle $\frac{dx}{dt}$.
3. **Scattering rate W:** The number of particles scattered per unit time in solid angle $d\Omega$ is proportional to (i) the incident flux ϕ, (ii) number of independent scattering centres N_ϕ in the target, and (iii) the element of the solid angle $d\Omega$.

   or $$W \propto \phi N_\phi d\Omega$$

   The proportionality factor is called the differential scattering cross-section and is denoted by $\frac{d\sigma}{d\Omega}$.

   $$\frac{d\sigma}{d\Omega} = \frac{W}{\phi N_\phi d\Omega}$$

   The total number of particles scattered per unit time are given by

   $$W = \phi N_\phi \sigma$$

With the total scattering cross-section

$$\sigma = \int_0^{4\pi} \frac{d\sigma}{d\Omega} d\Omega.$$

In spherical polar coordinates, the element of solid angle dΩ is given by $\sin\theta\, d\theta d\phi$. Substituting this value in the above equation, we get

$$\sigma = \int_0^{2\pi} d\phi \int \frac{d\sigma(\theta,\phi)}{d\Omega} \sin\theta d\theta .$$

When the beam of the incident particles is not polarized, it means that the spins of the particles are randomly oriented or the particles are spinless, the differential cross-section is independent of φ and the above equation can be integrated over φ to get

$$\sigma = 2\pi \int_0^{\pi} \frac{d\sigma(\theta)}{d\Omega} \sin\theta d\theta$$

Proportionality factor between W and σ is called luminosity and is denoted by L

$$L = \varphi N_{\varphi}.$$

4.4.1 Neutron–proton Elastic Scattering

We now apply these considerations to the elastic scattering of neutron by proton. It is now possible to get collimated beam of neutrons starting from thermal energy, from nuclear reactors, to higher energies from particle accelerators. Although neutrons cannot be accelerated in an accelerator directly, nuclear reactions initiated by accelerated charged particle may be used to get high-energy neutrons. For example, the reaction $t(d, n)^4$He with threshold around 130 keV is often used to produce 14 MeV neutrons. Any hydrogenous material that has a large percentage of hydrogen may be used as the proton target. Large amount of data on (n, p) scattering cross-section over a wide range of energy is now available. To test the experimental data, we develop a simple quantum mechanical model for the elastic scattering of neutron and proton.

As usual, we consider the problem of n–p scattering in the centre of mass (CM) frame. In laboratory frame, a neutron with kinetic energy E_{kin} is approaching with velocity v towards a stationary proton, but in CM frame a particle of reduced mass M (= m/2), where m is the mass of neutron (which is taken equal to the mass of a proton) is approaching the centre of mass with kinetic energy $E = E_{kin}/2$. For simplicity, it is assumed that both the neutron and the proton are spinless and that their masses are equal. In laboratory frame, the incident neutron comes under the influence of potential $V(r)$ as soon as it reaches a distance from the proton, which is of the order of the range of the nuclear force. In the CM frame, the particle of reduced mass M comes under the influence of potential $V(r)$ when it is within the range of the nuclear force from the centre of mass. We now write the assumptions made so far

1. Neutron and protons are spinless.
2. Neutron and proton have same mass.

3. The interaction between neutron and proton may be represented by a potential $V(r)$.
4. $V(r)$ is spherically symmetric.
5. The interaction is non-relativistic

As a consequence of the above-mentioned assumptions,

1. The reduced mass of the system in CM coordinates is half the mass of neutron (or proton).
2. The kinetic energy in CM frame is half of the initial kinetic energy of the incident neutron.
3. The angles of scattering of the neutron and the proton in CM are twice the respective angles in laboratory frame ($\Theta_{CM} = 2\theta_{lab}$, $\Phi_{CM} = 2\varphi_{lab}$).
4. The space part of the wave function of the system $\Psi(r, \theta, \phi)$ may be written as the multiplication of three wave functions, each depending only one coordinate, $\Psi(r, \theta, \varphi) = \chi(r)\Theta(\theta)\Phi(\varphi)$.
5. The radial part of the wave function $\chi(r)$ may be written in terms of the reduced radial wave function $u_\ell(r) = \chi(r)/r$.

We consider the problem of scattering into two parts, initial state before scattering and the final state after scattering has taken place.

In the initial state, we consider a plane wave of particles of mass M, kinetic energy E moving towards the scattering centre, which is assumed to be in the direction of the z-axis. The wave function describing such a wave is

$$\Psi_{inc} = e^{ikz} \tag{4.30a}$$

where $k = \frac{1}{\hbar}(\sqrt{2ME})$ is the wave number of the incident particle and $E = ½MV^2$ its kinetic energy. If the value of $\hbar$ is put in the equation for k, we get $k = 0.2187\ (M|E|)^{1/2}\ F^{-1}$, where M is in unified mass unit u and the kinetic energy E is in MeV. Further, in the case of neutron–proton scattering the reduced mass M is 0.504 u, and therefore, $k = 0.1555\ (|E|)^{1/2}$.

The wave represented by Eq. (4.30a) has a probability density $\Psi\Psi^* = 1$, which means that the wave contains one particle per unit volume of space. To calculate the flux of the beam, we consider a cylinder of base area unity and length v (velocity of the beam particle) held in the beam with its base normal to the velocity of the beam particles. All particles inside this cylinder will cross the base of the cylinder in unit time. The number of particles in the cylinder is equal to the volume of the cylinder, because from the probability density we know that each unit volume of the space contains one particle. Hence, the particle flux of the wave is 'v' where v is the velocity of the particles in the wave.

The incident plane wave may be decomposed into imaginary cylindrical zones of radius $\lambda\!\!\!^-$, $2\lambda\!\!\!^-$, $3\lambda\!\!\!^-$, … $n\lambda\!\!\!^-$ with the target as the centre of these zones as shown in figure 4.11. In semi-classical term, the particles at the outer edge of the inner most zone have the angular momentum L_1 about the centre of mass, where $L_1 = Mv\lambda\!\!\!^-$, as the radius of the zone is $\lambda\!\!\!^-$ and $\lambda\!\!\!^-$ is perpendicular to the velocity v of the particle.

$$L_1 = Mv\lambda\!\!\!^- \quad \text{or} \quad L_1^2 = (Mv)^2\lambda\!\!\!^{-2} = (Mv)^2\{\hbar/Mv\}^2 = \hbar^2 \tag{4.30b}$$

$$\text{However, } L_1^2 = \ell(\ell+1)\ \hbar^2 \tag{4.30c}$$

Equating (4.30a) and (4.30b), we get $\ell = 0$ for the particles in the inner most zone. Similarly, it can be argued that the angular momentum of particles in the second zone of outer radius $2\lambda\!\!\!^-$ and

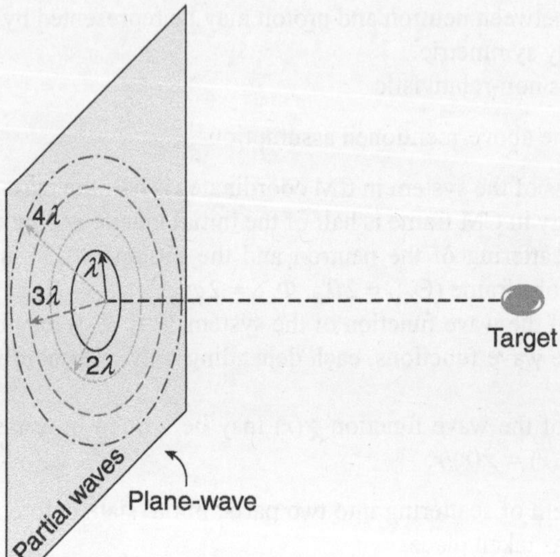

Figure 4.11 *Decomposition of the plane wave into zones of different orbital angular momentum*

the inner radius $\lambda\!\!\!^{-}$ contains particles that have orbital angular momentum $\ell = 1$, the third zone with outer radius $3\lambda\!\!\!^{-}$ and inner radius $2\lambda\!\!\!^{-}$ contain particles that have orbital angular momentum $\ell = 3$ and so on. In this way, the incident plane wave may be considered as a wave having particles of orbital angular momentum 0, 1, 2, … with respect to the target in different zones. The inner most zone having particles of $\ell = 0$ is called the S-wave, the next P-wave, and so on. It may, however, be emphasized that the size of each zone depends on $\lambda\!\!\!^{-}$ which in turn depends on the energy of the particle. As the incident energy increases, the size of the zone decreases. Each component of the incident wave with a fixed value of ℓ is called a partial wave. One advantage of this decomposition of the incident wave into partial waves is that only those partial waves that fall within the range of the nuclear force around the target will be affected by the target potential and other partial waves of higher orbital angular momentum will move out unaffected. Further, the orbital angular momentum is conserved in elastic scattering as no torques are present. As the energy of the incident beam increases, more and more partial waves of higher orbital angular momentum come within the range of the nuclear field of the target and are, therefore, affected. At lower incident energies, only first few partial waves interact with the target. For example, for 10 MeV neutrons in lab, the CM energy is 5 MeV and the rationalized de Broglie wavelength $\lambda\!\!\!^{-}$ for 5 MeV neutrons is 2.8735 F. It means that up to a distance of 2.8735 F from the centre of the target only S ($\ell = 0$) partial waves will extend. Assuming the range of the nuclear force as 2 F, it is clear that only S-waves will contribute to the elastic scattering at 10 MeV incident and 5 MeV CM-energy. Partial waves of higher ℓ values will not come in the range of the nuclear force and, therefore, will not be affected.

In spherical polar coordinates, the incident wave Ψ_{inc} may be written as

$$\Psi_{\text{inc}} = e^{ikz} = e^{ikr\cos\upsilon} = \sum_{\ell=0}^{\ell=\infty} B_\ell(r) Y_{\ell,0}(\theta) \tag{4.31}$$

The function $B_\ell(r)$ gives the radial dependence of the incident wave and the spherical harmonic function $Y_{\ell,0}(\theta)$, the angular dependence for each value of the orbital angular momentum ℓ. As such, the probability density $\Psi\Psi^*$ depends on $B^2_\ell(r)Y^2_{\ell,0}(\theta)$. The radial function $B_1(r)$ is related to the spherical Bessel function $j_\ell(kr)$ through the following equation:

$$B_\ell(r) = i^1\{4(2\ell + 1)\}^{1/2} j_\ell(kr) \tag{4.31a}$$

Figure 4.12 shows the variation of $|B_\ell(r)|^2$ and $Y^2_{\ell,0}(q)$ for $\ell = 0$ and 1. Now, for 5 MeV neutron $k = 0.348$ F^{-1} and if it is assumed that the range of nuclear field is 2 F, then $kr = 0.348 \times 2 = 0.696$. As may be observed from figure 4.12a, mostly S-partial wave will contribute to elastic scattering, a conclusion arrived earlier on semi-classical arguments. A small contribution may also come from the P-wave ($\ell = 1$) but it will be very small. Further, it is known that interaction in odd ℓ-states is comparatively much weaker than for the even ℓ-states.

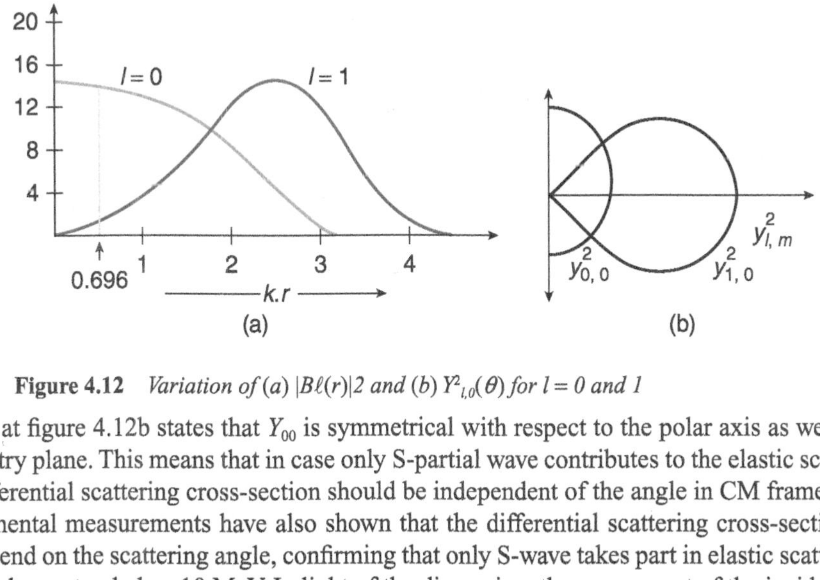

Figure 4.12 *Variation of (a) |Bℓ(r)|2 and (b) $Y^2_{1,0}(\theta)$ for l = 0 and 1*

A look at figure 4.12b states that Y_{00} is symmetrical with respect to the polar axis as well as the symmetry plane. This means that in case only S-partial wave contributes to the elastic scattering, the differential scattering cross-section should be independent of the angle in CM frame. Actual experimental measurements have also shown that the differential scattering cross-section does not depend on the scattering angle, confirming that only S-wave takes part in elastic scattering of neutron by proton below 10 MeV. In light of the discussion, the component of the incident wave that is going to be affected by the scattering potential will be the S-partial wave denoted by ψ_0^{inc} and is given by

$$\psi_0^{\text{inc}} = B_0(r)Y_{0,0}(\theta) = \frac{\sin kr}{kr} = \frac{e^{ikr} - e^{-ikr}}{2i\,kr} \tag{4.31b}$$

Equation (4.31b) represents two waves: $\frac{e^{ikr}}{2ikr}$ that is an outgoing wave (a wave going away from the scattering centre) and $\frac{e^{-ikr}}{2ikr}$ that represents a wave moving towards the scattering centre. As we are considering elastic scattering, the scattering potential can only change the phase of the outgoing wave contained in ψ_0^{inc}. Suppose as a result of scattering the phase of the outgoing wave is changed by $2\delta_0$, after scattering, the ψ_0^{inc} becomes Ψ'_0 given by

$$\Psi'_0 = \frac{e^{i(kr+2\delta 0)} - e^{-ikr}}{2ikr} \tag{4.31c}$$

The effect of scattering is contained in the difference $(\Psi'_0 - \psi_0^{\text{inc}})$

$$(\Psi'_0 - \psi_0^{\text{inc}}) = \Psi'' = \left(\frac{e^{i(kr+2\delta 0)} - e^{-ikr}}{2ikr}\right) - \left(\frac{e^{ikr} - e^{-ikr}}{2ikr}\right) = \left[\frac{e^{i\delta 0}\, e^{i(kr+\delta 0)} - e^{-i\delta 0}\, e^{i(kr+\delta 0)}}{2ikr}\right]$$

$$= \left(\frac{e^{i\delta 0} - e^{-i\delta 0}}{2i}\right)\left(\frac{e^{i(kr+\delta 0)}}{kr}\right) = \sin\delta_0 \left(\frac{e^{i(kr+\delta 0)}}{kr}\right) \tag{4.32}$$

At a large distance from the scattering centre and after the scattering has taken place, there will be two waves, Ψ^{inc} and a wave Ψ'' represented by Eq. (4.32) that contains the net effect of scattering. Figure 4.13 shows the incident plane wave and the scattered spherical wave after the scattering.

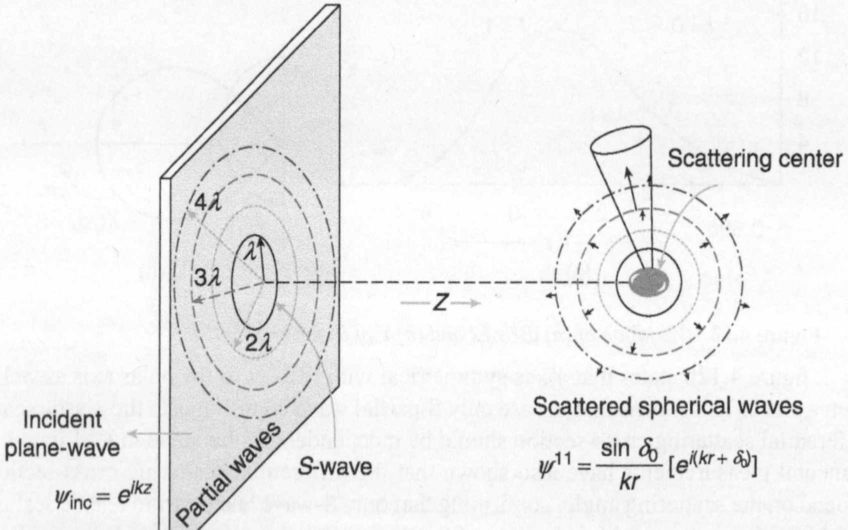

Figure 4.13 *Pictorial representation of the incident plane wave and the scattered spherical wave moving out from the scattering centre*

Equation (4.32) represents a wave Ψ'' moving away from the centre of scattering with an amplitude $(\sin\delta_0/kr)$. As a result of elastic scattering by the potential around the target (CM) three changes have taken place:

1. The S-partial wave of the incident plane wave Ψ^{inc} has been scattered back with same ℓ (=0) and velocity v as a spherical wave Ψ''.

2. The phase of the incident S-partial wave has been shifted by $2\delta_0$.
3. The particles contained in the S-partial wave have been redistributed into the scattered spherical out going wave Ψ''.

To find out the number of particles scattered per unit time N^{sc}, we enclose the target (which in this case is the centre of mass) with an imaginary sphere of radius r and calculate the particles that come out of the surface of the sphere per unit time. We know that $\int \Psi'' \Psi''^*$ gives the number density, which on multiplying with the velocity v gives the number of particles per unit area per unit time. As such $\{\int \Psi'' . \Psi''^* dv\} v(4\pi r^2)$, where $(4\pi r^2)$ is the surface area of the sphere of radius r, gives the number of particles scattered per unit time N^{sc}.

$$N^{sc} = (\sin\delta_0/kr)^2 4\pi r^2 v = \frac{4\pi \sin\delta_0^2}{k^2} v$$

Scattering cross-section

$$\sigma_0 = (N^{sc}/\text{incident flux } v) = (4\pi \sin^2\delta_0/k^2) \tag{4.33}$$

Experimental measurement of differential scattering cross section $\frac{d\sigma}{d\Omega}$ has shown that it is isotropic in the centre of mass, as expected. In light of this

$$\frac{d\sigma_0}{d\Omega} = \frac{\sigma_0}{4\pi} = \frac{\sin^2\delta_0}{k^2} \tag{4.34}$$

Quantities in Eq. (4.34) are in CM system. We now want to write these quantities in laboratory system, for which, we use the relations for two particles of same mass

$$\theta_{CM} = 2\theta_{ab} \text{ and } d\Omega_{CM} = 2 \ \sin\theta_{CM}\, d\theta_{CM} = 2\ [2\sin\theta_{lab} \cos\theta_{lab}]2\, d\theta_{lab}$$

$$\text{And } d\Omega_{lab} = 2 \sin\theta_{ab}\, d\theta_{lab}$$

Now,

$$\frac{d\sigma_0}{d\Omega_{CM}} = \frac{d\Omega_{CM}}{d\Omega_{lab}} = \left(\frac{\sigma_0}{4\pi}\right)(4\cos\theta_{lab}) = \frac{\sigma_0 \cos\theta_{\,lab}}{\pi} \tag{4.35}$$

It may be remembered that the maximum extension of θ_{lab} is between 0 and $\pi/2$. The differential cross-section in laboratory system will be zero for θ_{lab} beyond $\pi/2$ and up to π. In addition, the total scattering cross-section will be same both in CM frame and laboratory system.

So far we have considered only the S-wave scattering. However, at energies greater than 15 MeV, higher partial waves are also affected. It can be shown that the total scattering cross-section in the general case may be written as

$$\sigma = \sum_0^l \sigma_l \quad \text{and} \quad \sigma_\ell = \frac{4\pi\,(2l+1)\,\sin\delta_l^2}{k^2} \tag{4.36}$$

Here, σ_ℓ is the scattering cross-section for the ℓth partial wave and δ_ℓ the corresponding phase shift.

4.4.2 Triplet and Singlet Cross-sections

In n–p scattering with an unpolarized neutron beam and unpolarized proton target, it is equally likely that the incident neutron and the target proton have their spins parallel or anti-parallel, that is they interact in triplet or singlet state. In case of the triplet state, there are three possible sub-states, and in the case of singlet interaction there is only one sub-state. According to statistical quantum mechanics, each sub-state of a system is equally likely. In this case, in total, there are four sub-states (three of triplet and one of singlet) and they are all equally likely. Therefore, the statistical weight of the triplet state is $\frac{3}{4}$ and that of the singlet state $\frac{1}{4}$. The total S-wave scattering cross-section σ_0 will, therefore, be the sum of the S-wave scattering cross-sections for the triplet and the singlet states multiplied by their statistical weights.

$$\sigma_0 = \frac{3}{4}\,\sigma_0^{\text{trip}} + \frac{1}{4}\,\sigma_0^{\text{sing}} \tag{4.37}$$

Here, σ_0^{trip} and σ_0^{sing} are, respectively, the triplet and singlet state S-wave scattering cross-sections. In the study of deuteron, it has been found that singlet state is unbound and, therefore, it is likely that the scattering cross-section for the singlet state may be different from that for the triplet state.

4.4.3 Solving Schrödinger Equations for Scattering Problem

In section 4.4.1, it was shown that for energies less than 10 MeV or so only the partial wave S, of the incident plane wave, interacts with the scattering centre and that the phase of the wave is shifted by $2\delta_0$. In the case of deuteron, it was found that quantum mechanically the neutron–proton system can be represented by a square well of depth $-V_0$, width b, and binding energy $-E_b$ along with a hard repulsive core of width c. In this section, attempt will be made to develop a similar quantum mechanical model for the scattering and to correlate the magnitude of the phase shift $2\delta_0$ with the parameters of the system. The main difference between the scattering and the deuteron problem is that in the case of scattering, the system is not bound, which means the energy W is positive and not $-E_b$. The Schrödinger equation for the reduced radial wave function $u(r)$ in case of scattering may be obtained by replacing $-E_b$ by +W in Eq. (4.13) of the deuteron problem.

$$\frac{d^2u(r)}{dr^2} + \left(\frac{2M}{\hbar^2}\right)[V_0 + W]u(r) = 0 \tag{4.38}$$

The boundary conditions on potential $V(r)$ remains same, that is
$V(r)$ is spherically symmetric and

$$V = \infty \text{ for } r \le c$$
$$V = -V_0 \text{ for } c < r < (c + b)$$
$$V = 0 \text{ for } r > b + c$$

The space is again divided into three parts: $r = 0$ to $r = c$; $r = c$ to $r = (b + c)$; and $r > (b + c)$.

The reduced radial wave function $u_1(r)$ will be zero in the first part of the space because of the infinite high repulsive core. The reduced wave functions $u_2(r)$ and $u_3(r)$ for the second and the third parts of the space are, respectively, given by the following relations:

$$u_2(r) = A\sin K\,(r - c) \tag{4.39a}$$

where wave number

$$K = \frac{1}{\hbar}\sqrt{2M(W+V_0)} \tag{4.39b}$$

and

$$u_3(r) = B\sin(kr + \delta_0) \tag{4.40a}$$

With wave number

$$k = \frac{1}{\hbar}\sqrt{2MW} \tag{4.40b}$$

To match the two wave functions in (4.39a) and (4.40a) at the boundary $r = b + c$, we equate $\{1/u(r)\}$ (du/dr) at the boundary to get

$$K\cot(Kb) = k\cot[k(b+c) + \delta_0] \tag{4.41}$$

The value of phase shift δ_0 may be obtained from Eq. (4.41) if the values of V_0, W, b, and c are substituted. Once δ_0 is known, the S-wave scattering cross-section σ_0 may be determined from Eq. (4.33). When values of triplet potential parameters V_0 and b that reproduces the binding energy and the r.m.s. radius of deuteron were used to calculate σ_0, it was found that for energies $W > 5$ MeV, the experimental σ_0 data matched well with the theoretically calculated values. However, for lower incident energies $W < 5$ MeV, the theoretical predictions were much smaller than the experimental values. The large discrepancy in calculated and experimental values of σ_0 could not be assigned to the shape of the potential $V(r)$ used for the calculations. As a matter of fact, it has been found that σ_0 does not depend much on the shape and width of the potential well as long it reproduces the binding energy and r.m.s. radius of deuteron. Theoretical calculations have also shown that the presence of tensor force affects σ_0 at the most by 1%. Theoretical calculations for σ_0 referred earlier have been done only for the triplet state, since triplet potential-well parameters of deuteron have been used. The singlet interaction also contributes to scattering, and if at lower energies σ_0^{sing}, which has not been included in calculations, is much larger than σ_0^{trip} then the discrepancy between the calculated and experimental values of S-wave scattering cross-sections at lower energies may be explained.

4.4.4 Zero Energy Limit and Scattering Length

Neutrons of extremely low energy ≈0.0008 eV called 'cold neutrons' may be obtained by the velocity selector techniques. Detailed experiments of coherent and incoherent scattering of cold and thermal neutrons by hydrogen molecules have been carried out to get information regarding the singlet-state scattering. Before discussing the results of these experiments, we consider the modification that takes place in the reduced radial wave function $u_3(r)$ when the energy W approaches zero.

In the limit W going to zero, the wave number k goes to zero

$$\text{Limit } k \to 0,\ u_3(r) = \text{Limit } k \to 0\ B\sin(kr + \delta_0) = \text{Limit } k \to 0\ B\sin k(r + \delta_0/k) \tag{4.42}$$

It is clear from Eq. (4.42) that when k approaches zero, δ_0 should also approach zero to keep (δ_0/k) finite. In Eq. (4.42) the argument of sin is very small in the limit $k \to 0$ and the sin may be replaced by its argument. (If θ is small $\sin\theta = \theta$).

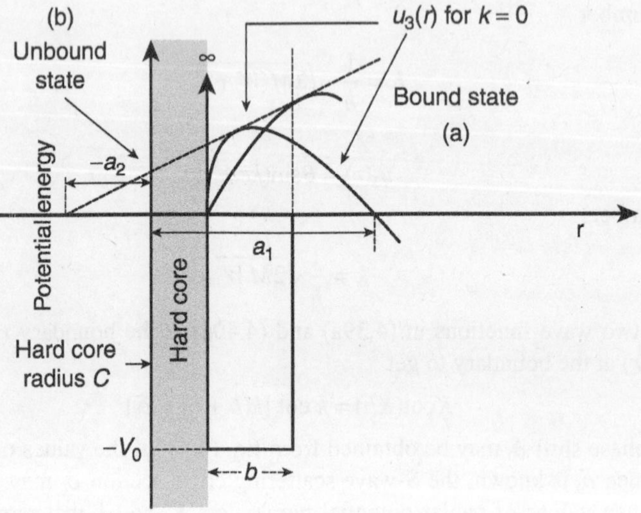

Figure 4.14 *In the limit W→0, k→0 and u3(r) degenerate into a straight line. Case (a) the well depth $-V_0$ is deep enough to produce bound state and a positive scattering length a_1. Case (b) a well not deep enough to produce a bound state has a negative scattering length $-a_2$*

Hence,

$$\text{Limit } k \to 0 \; u_3(r) = (B/k)\,(r + \delta_0/k) \tag{4.43}$$

Equation (4.43) is the equation of a straight line that crosses the r axis at $r = \delta_0/k$.

We thus see that the reduced wave function $u_3(r)$ degenerates into a straight line as shown in figure 4.14. In case the well depth is deep enough to produce the bound state, the wave function $u_2(r)$ attains the maximum value within the well and has a negative slope at the well boundary $r = b + c$. The wave function $u_3(r)$, which in the zero energy limit becomes a straight line, joins $u_2(r)$ and makes an intercept $+\,a_1$ on the r axis. This intercept is called the scattering length. In the case when the well is not deep enough to produce a bound state, the wave function $u_2(r)$ is still rising and has a positive slope at the well boundary. The straight line wave function $u_3(r)$ in this case (b) joins the rising $u_2(r)$ and makes a negative intercept $-\,a_2$ on the r-axis. Thus, the scattering lengths have different signs for the bound and unbound states. Now,

$$\delta_{0\text{trip}} = ka_1 \text{ and } \delta_{0\text{sing}} = -ka_2$$

where $\delta_{0\text{trip}}$ and $\delta_{0\text{sing}}$ are, respectively, the scattering lengths for the bound (triplet) and the unbound (singlet) states.

From Eq. (4.33), we have

$$\sigma_0 = \frac{4\pi \sin{\delta_0}^2}{k^2}.$$

However, for $k \to 0$

$$\sigma_0 = \frac{4\pi{\delta_0}^2}{k^2}$$

And $$\sigma_0^{\text{trip}} = 4\,a_1^2;\ \sigma_0^{\text{sing}} = 4\,a_2^2 \quad (4.44)$$

Also, $$\sigma_0 = \tfrac{3}{4}\,\sigma_0^{\text{trip}} + \tfrac{1}{4}\,\sigma_0^{\text{sing}} = 3\,a_1^2 + a_2^2 \quad (4.45)$$

If the measured value of σ_0 (=20.3 b) is used in Eq. (4.45), we get a relationship between the squares of the scattering lengths a_1 and a_2. One cannot determine the signs and the magnitudes of the scattering lengths from this relationship. To determine the sign and the values of individual scattering lengths, we require another relation between them. Scattering data of cold neutrons with the hydrogen molecule provide the desired relationship. The intra-nuclear separation in hydrogen molecule is of the order of 0.8×10^{-8} cm. Cold neutrons with de Broglie wave length of about 10×10^{-8} cm undergo coherent scattering with the two protons in the hydrogen molecule. In the case of coherent scattering, which is diffraction like phenomenon, the scattering cross-section depends on the weighted algebraic sum of the scattering lengths $[1/4(3a_1 + a_2)]$ and not on the square of them. The results of coherent scattering of cold and thermal neutrons with hydrogen molecules along with the neutron–proton scattering data at energies below 5 MeV have given the following values for the scattering lengths.

a_1 (triplet scattering length) = 5.38 F and a_2 (singlet scattering length) = −23.7 F

When one calculates the depth of the square-well potential that may give the scattering length of −23.7 F, keeping the width same as for the deuteron well, we get $V_0^{\text{sing}} \approx 54$ MeV. A well of depth −57 MeV gives a just-bound state. It can, therefore, be said that the singlet n–p state misses binding by only 4 MeV.

4.4.5 Shape Independence and Effective Range Theory

Equation (4.41) gives the relationship between K, k, and the phase shift δ_0 for the square-well potential. The rectangular-well parameters V_0, b, and the incident neutron energy W appear in this relation through K and k.

$$K \cot (Kb) = k \cot [k\,(b + c) + \delta_0] \quad (4.41)$$

Once the phase shift is known from Eq. (4.41), the scattering cross-section can be calculated using the expression

$$\sigma_0 = \frac{4\pi \sin^2 \delta_0}{k^2}$$

It may, however, be noted that if instead of a rectangular well some other shape of the well is considered, then another equation like Eq. (4.41) will have to be developed to determine the phase shift. Further, if phase shift is required at different energies, generally numerical integration has to be carried out. In nutshell, it may be stated that the derivations done so far depend on the shape of the potential used to represent the interaction between the incident and the scattering particle. However, the method to be developed now will show that phase shift can be calculated, without much error, by using two parameters, the scattering length and the effective range for wells of any shape.

Figure 4.15a shows the radial wave function $u(r)$ both inside and outside the potential well. A new function $v(r)$ is defined in such a way that it is equal to $u(r)$ outside the potential well;

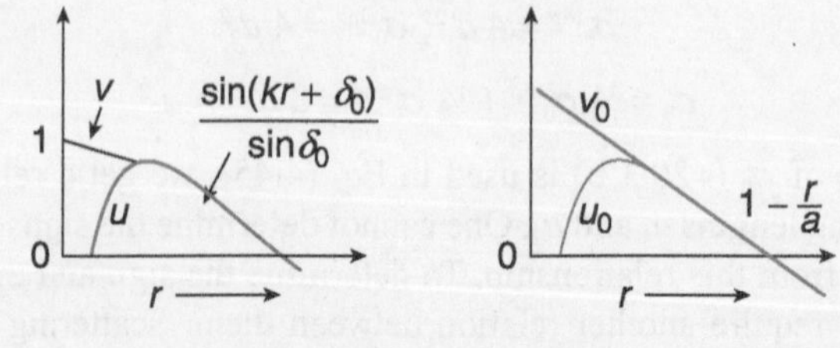

Figure 4.15 *The radial wave functions u, u_0, v, and v_0 in effective range theory*

however, it extrapolates $u(r)$ within the well up to $r = 0$. Further, $v(r)$ is normalized such that it has a value 1 at $r = 0$. As $u(r)$ outside the well goes as $\sin(kr + \delta_0)$

$$v(r) = \frac{\sin(kr + \delta_0)}{\sin \delta_0} \tag{4.46}$$

The reduced radial wave function $u(r)$ is the solution of the Schrödinger equation

$$-\frac{\hbar^2}{2M}\frac{d^2u}{dr^2} + V(r)u(r) = W\,u(r)$$

or

$$-\frac{d^2u}{dr^2} + \left(\frac{2M}{\hbar^2}\right)V(r)u(r) = (2M/\hbar^2)W\,u(r) = k^2u(r) \tag{4.47}$$

Similarly,

$$-\frac{d^2u_0}{dr^2} + \frac{2M}{\hbar^2}V(r)u_0 = 0 \tag{4.48}$$

We multiply Eq. (4.47) by u_0 and Eq. (4.48) by u and on subtracting, we get

$$u\frac{d^2u_0}{dr^2} - u_0\frac{d^2u}{dr^2} = k^2\,u\,u_0$$

This equation may also be written as

$$\frac{d}{dr}\left(u\frac{du_0}{dr} - u_0\frac{du}{dr}\right) = k^2\,u\,u_0 \tag{4.49}$$

Similarly for the function $v(r)$, one may write

$$\frac{d}{dr}\left(v\frac{dv_0}{dr} - v_0\frac{dv}{dr}\right) = k^2\,v\,v_0 \tag{4.50}$$

We subtract Eq. (4.49) from (4.50) and integrate from 0 to infinity to get

$$\int_0^\infty \frac{d}{dr}\left[v\frac{dv_0}{dr} - v_0\frac{dv}{dr} - u\frac{du_0}{dr} + u_0\frac{du}{dr}\right]dr = k^2\int_0^\infty (vv_0 - uu_0)\,dr \tag{4.51}$$

Now, in Eq. (4.51), $v = v_0 = 1$ for $r = 0$, and outside the potential $v = v_0$ and $u = u_0$, therefore, we get

$$-\left(\frac{dv_0}{dr} - \frac{dv}{dr}\right)_{\text{at } r=0} = k^2 \int_0^\infty (vv_0 - uu_0)\,dr \tag{4.52}$$

However, $[d/dr(v_0)]_{\text{at } r=0} = -1/a$ and $[d/dr(v)]_{\text{at } r=0} = k\cot\delta_0$. Substituting these values in Eq. (4.52) we get

$$k\cot\delta_0 + 1/a = k^2 \int_0^\infty (vv_0 - uu_0)\,dr \tag{4.53}$$

No approximation has been made in deriving Eq. (4.53). We now assume that at moderate energies no serious error is introduced if v is replaced by v_0 and u by u_0 in Eq. (4.53), so that the equation becomes

$$k\cot\delta_0 + 1/a = k^2 \int_0^\infty (v_0^2 - u_0^2)\,dr \tag{4.54}$$

We now define a quantity called effective range r_0 as follows:

$$r_0 = 2\int_0^\infty (v_0^2 - u_0^2)\,dr \tag{4.55}$$

The effective range r_0 defined as in Eq. (4.55) is nearly equal to the radius where the potential becomes small and that is why it is called the effective range. By substituting the value of effective range in Eq. (4.54), we get

$$\cot\delta_0 = -\frac{1}{ka} + \frac{1}{2}\,r_0\,k \tag{4.56}$$

Equation (4.56) gives the phase shift in terms of two parameters: the scattering length and the effective range under the so-called shape-independent approximation. S-wave scattering cross-section σ_0 may now be written as

$$\sigma_0 = \frac{4\pi a^2}{\left[1 - \frac{1}{2}\,a r_0 k^2\right]^2 + a^2k^2} \tag{4.57}$$

To determine the total S-wave scattering cross-section, which is the weighted sum of the triplet and singlet state S-wave cross-sections (Eq. (4.45)), we need the value of four parameters: a^{trip}, a^{sing}, r_0^{trip}, and r_0^{sing}. The values of the triplet and singlet scattering lengths have already been determined from the deuteron analysis and from the incoherent and coherent n–p scattering cross-sections. Triplet effective range r_0^{trip} can again be determined from the deuteron analysis. It comes out to be

$$r_0^{\text{trip}} = \frac{2}{k^2}\left[k - \frac{1}{a^{\text{trip}}}\right] \tag{4.58}$$

The singlet effective range r_0^{sing} may now be treated as a free parameter and the experimental data on n–p elastic scattering cross-section at different energies may be fitted to deduce the singlet effective range. A value of $r_0^{sing} = 2.4$ F is found to reproduce the experimental n–p scattering cross-section data from 0.02 MeV to 20 MeV neutron laboratory energy extremely well. The simple quantum mechanical approach to the deuteron and the n–p scattering with shape-independent approximation gives the following parameters:

$$a^{trip} = 5.38 \text{ F},\ a^{sing} = -23.7 \text{ F},\ r_0^{trip} = 1.70 \text{ F},\ r_0^{sing} = 2.4 \text{ F} \tag{4.59}$$

4.4.6 Proton–Proton Elastic Scattering at Low Energies

The p–p elastic scattering differs from the n–p elastic scattering on the following counts:

1. **Coulomb scattering:** In p–p scattering, Coulomb interaction between protons is always present along with nuclear interaction, which acts only at short distances of the order of 2 F. Even at energies where only S-wave nuclear interaction is likely, long-range Coulomb interaction affects the phases of partial waves of higher orbital angular momentum ℓ. As a result, the total S-wave cross-section does not provide the desired information about phase shift. In general, measurement of differential scattering cross-section at different scattering angles is made and the contribution of Coulomb scattering, calculated theoretically, is subtracted from the measured data to get the nuclear scattering cross-section.
2. **Identical particles:** The fact that the target and the incident particles are identical and cannot be distinguished from each other affects the process in two ways. First, the scattering events in which the incident proton is scattered at angle θ and at $\pi - \theta$ cannot be distinguished. Second, Pauli exclusion principle excludes some states, in particular, at low energies only singlet state contributes to scattering, as the triplet state with identical (or parallel) spins is forbidden.
3. **Vacuum polarization:** According to Dirac's hypothesis, the vacuum is filled with states of negative energy electrons. The wave functions of these states get distorted in the immediate neighbourhood of the electrostatic charge. This gives rise to what is called the 'vacuum polarization'. Although the effect of vacuum polarization is only a fraction of a percent in case of low energy p–p scattering but it needs to be included for the sake of completeness.

Like the n–p scattering data, the p–p elastic scattering data may be well reproduced by a hard core of width 0.4 F, a square-well potential of about 40 MeV depth and 2 F width with a small positive Coulomb potential starting from the boundary of the potential well and decreasing as $1/r$. A good matching of the experimental data on deuteron and of n–p and p–p elastic scattering proves the success of the quantum mechanical approach to these systems. Further, these studies also establish the charge independence, spin dependence, and tensor nature of nucleon–nucleon force. However, the exact shape, the depth, and the width of the nucleon–nucleon potential cannot be determined uniquely from these studies.

4.4.7 n–p and p–p Elastic Scattering at High Energies

At higher energies for both p–p and n–p scattering, partial waves of higher ℓ values participate in scattering. The tensor force that mixes states of different ℓ values also becomes significant.

As a result, experiments on the measurement of differential cross-sections at various scattering angles do not suffice. Additional experiments on determining the polarization produced as a result of scattering have to be done. Large number of such experiments has been carried out to find out the phase shift parameters that are smooth functions of the incident energy. The computer analysis of the large amount of experimental data has yielded the desired phase-shift parameters that may reproduce the experimental data in the given range of incident energy. This shows that the partial wave analysis approach for the nucleon–nucleon elastic scattering is valid.

Solved example S-4.7

Show that the differential cross-section $\frac{d\sigma(\theta)}{d\Omega}$ for the scattering of proton with proton in laboratory system (LS) is related to the corresponding centre of mass differential scattering cross-section $\frac{d\sigma(\theta)}{d\Omega}$ by the relation $\frac{d\sigma(\theta)}{d\Omega} = 4\cos\frac{\Theta}{2}\frac{d\sigma(\Theta)}{d\Omega}$.

Solution. We use the transformation formula

$$\frac{d\sigma(\theta)}{d\Omega} = \frac{\left[1+\left(\frac{M_1}{M_2}\right)^2 + 2\left(\frac{M_1}{M_2}\right)\cos\Theta\right]^{3/2}}{|1+\frac{M_1}{M_2}\cos\Theta\,|}\frac{d\sigma(\Theta)}{d\Omega}$$

As in the problem scattering is of proton with proton, $(M_1/M_2) = 1$ and therefore,

$$\frac{d\sigma(\theta)}{d\Omega} = (2^{3/2})\left[(1+\cos\Theta)^{1/2}\frac{d\sigma(\Theta)}{d\Omega}\right] = (2^{3/2})\left[(\{2\cos^2\Theta/2)^{1/2}\right]\frac{d\sigma(\Theta)}{d\Omega}$$

Or
$$\frac{d\sigma(\theta)}{d\Omega} = 4\cos(\Theta/2)\frac{d\sigma(\Theta)}{d\Omega}$$

Exercise p-4.10: An alpha particle moving with a velocity of 1×10^3 m/s hits a stationary proton and gets scattered to 10° from its original direction of motion. Calculate the angle of scattering of proton.

Exercise p-4.11: What is partial wave analysis and what is its advantage?

Solved example S-4.8

Neutrons of energy E_n are made to hit protons at rest. If σ_{np} is the total n–p scattering cross-section at low energy, which is assumed to be isotropic in the centre of mass, then show that in the laboratory frame the proton energy distribution is given by

$$\frac{d\sigma_{proton}}{dE_{proton}} = \frac{\sigma_{np}}{E_n} = \text{constant}$$

Solution. We assume that the neutron and proton have same mass so that the reduced mass of the system is ½ the mass of neutron or proton. Let us assume that the velocity of the incident neutron is V_0, so that the velocity of the centre of mass V_c is $V_0/2$. The proton was initially at rest. However, in centre of mass it was moving with the velocity V_c before collision and also after collision. Let Φ be the angle in CMS that proton makes with the direction of motion of the incident neutron. As shown in the figure, the laboratory velocity of the proton V_p may be determined from the velocity triangle as

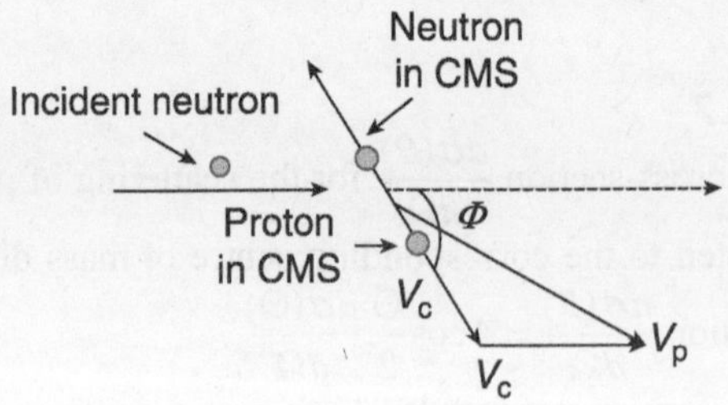

$$V_p^{\,2} = V_c^{\,2} + V_c^{\,2} + 2V_c^{\,2}\cos\Phi = 2V_c^2(1+\cos\Phi) = \frac{V_0^2}{2}(1+\cos\Phi)$$

Therefore, the kinetic energy of proton in LS is

$$E_p = \frac{E_n}{2}(1+\cos\Phi)$$

We differentiate this expression to get,

$$dE_p = -\frac{E_n}{2}\sin\Phi\, d\Phi = -E_n\frac{d\Phi_{CM}}{4p}\, dE_p = -\frac{E_n}{2}\sin\Phi\, d\Phi = -E_n\frac{d\Phi_{CM}}{4p}$$

The negative sign in the above-mentioned expression simply means that when E_p increases Φ decreases.

Or $$\frac{dE_p}{d\Omega_{CM}} = \frac{E_n}{4\pi} \text{ or } \frac{d\Omega_{CM}}{dE_p} = \frac{4\pi}{E_n} \qquad \text{(A)}$$

It is given that the differential scattering cross-section for protons is isotropic in centre of mass, that is

$$\frac{d\sigma_p}{d\Omega_{CM}} = \frac{\sigma_{np}}{4\pi} \qquad \text{(B)}$$

Multiplying Eqs. (A) and (B), we get

$$\frac{d\sigma_p}{dE_p} = \frac{\sigma_{np}}{E_n} = \text{constant}$$

Solved example S-4.9

Particles of mass m are elastically scattered by target nuclei of mass M, initially at rest. Assuming that the scattering in centre of mass is isotropic, show that the distribution of the target nuclei in laboratory system has $\cos\phi$ dependence.

Solution. Let us assume that the total scattering cross-section is σ. Now, the differential scattering cross-section for the target particles in CMS is isotropic; therefore,

$$\frac{d\sigma(M,\Phi)}{d\Omega_{\text{CM}}} = \frac{\sigma}{4\pi}$$

Since total cross-section is same for both the laboratory and the CM systems, we have

$$\frac{d\Phi(M,\phi)}{d\Omega_{\text{LS}}} \sin\phi\, d\phi = \frac{d\sigma(M,\Phi)}{d\Omega_{\text{CM}}} \sin\Phi\, d\Phi$$

Or $$\frac{d\sigma(M,\phi)}{d\Omega_{\text{LS}}} = \frac{d\sigma(M,\Phi)}{d\Omega_{\text{CM}}} \sin\Phi\, d\Phi / \sin\phi\, d\phi \quad \text{(A)}$$

However, the scattering angle of the target particle in CMS $\Phi = 2\phi$ the scattering angle in LS. Substituting this value of $\Phi = 2\phi$, and $d\Phi = 2d\phi$, we get.

$$\frac{d\sigma(M,\phi)}{d\Omega_{LS}} = \frac{d\sigma(M,\Phi)}{d\Omega_{CM}} [2\sin\phi\cos\phi\, 2d\phi / \sin\phi\, d\phi]$$

$$= \frac{d\sigma(M,\Phi)}{d\Omega_{CM}} [4\cos\phi] = \frac{\sigma}{4\pi}[4\cos\phi] = \frac{\sigma}{\pi}[\cos\phi]$$

Thus, the differential cross-section of target particles in laboratory frame depends on $\cos\phi$.

Exercise p-4.12: Describe the quantum mechanical approach to n–p scattering at energies below 10 MeV.

Solved example S-4.10

Particles of negligible size are projected on a hard, fixed sphere of infinite mass and radius r_0. The incident particles are elastically scattered with the condition that in each encounter the angle of reflection is equal to the angle of incidence of the particle. Calculate the differential scattering cross-section, investigate its nature and hence find the total scattering cross-section.

Solution. As shown in figure S-4.10, a beam of intensity I_0 is moving from left to right on to a fixed heavy sphere of radius r_0. All particles of the beam moving with impact parameter a and $(a + da)$ will hit the sphere and will get scattered by an angle θ, in a solid angle $d\Omega = \sin\theta\, d\theta\, d\varphi$. As there is azimuthal symmetry integration over $d\varphi$ will give 2π. Hence, $d\Omega = 2\Leftarrow \sin\theta\, d\theta$. The number N of particles with impact parameter between a and $(a + da)$ may be calculated from the area of the annular ring and the incident intensity I_0.

$N = I_0 \times$ area of the annular ring with inner radius a and outer radius $(a + da)$

Or $$N = I_0 \times [\pi a^2 - \pi(a + da)^2] = I_0 \times (2\pi a\, da) \quad \text{(A)}$$

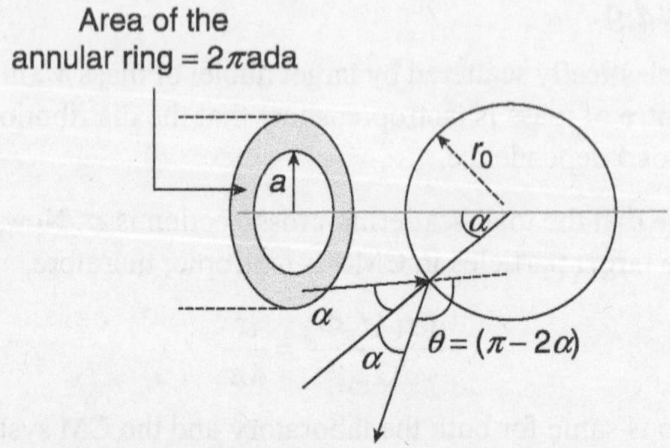

Figure S-4.10 *Elastic scattering of particles with impact parameter a and (a+da) by a hard sphere.*

Let $\left(\dfrac{d\sigma(\theta)}{d\Omega}\right)$ denote the differential scattering cross-section in direction θ. Then,

$$I_0\left(\frac{d\sigma(\theta)}{d\Omega}\right)d\Omega = I_0\left(\frac{d\sigma(\theta)}{d\Omega}\right)(2\pi \sin\theta\, d\theta) = N = I_0\, 2\pi\, a da$$

Or
$$\left(\frac{d\sigma(\theta)}{d\Omega}\right) = \frac{ada}{\sin\theta\, d\theta} \tag{B}$$

From figure S-4.10, it may be seen that angle of incidence α is equal to the angle of reflection, and hence, the angle of scattering $\theta = (\pi - 2\alpha)$ and the impact parameter

$$a = r_0 \sin\alpha;\ \text{therefore,}\ da = r_0 \cos\alpha\, d\alpha \tag{C}$$

$$\text{Also, } \theta = (\pi - 2\alpha) \text{ and hence, } d\theta = -2\, d\alpha \tag{D}$$

The negative sign only implies that when θ increases α decreases.
Further,

$$\sin\theta = \sin(\pi - 2\alpha) = \sin 2\alpha = 2\sin\alpha\cos\alpha \tag{E}$$

Substituting the values of a, da, $\sin\theta$, and $d\sin\theta$ from Eqs. (C)–(E) in (B), we get

$$\left(\frac{d\sigma(\theta)}{d\Omega}\right) = \frac{r_0^2}{4} \tag{F}$$

Equation (F) tells that the differential scattering cross section does not depend on the angle of scattering θ, which means that the distribution of scattered particles in laboratory is isotropic. The total cross-section can be obtained by integrating the following expression:

$\sigma_{total} = \int_{\phi=0,\theta=0}^{\phi=2\pi,\theta=\pi}\left(\dfrac{d\sigma(\theta)}{d\Omega}\right)\sin\theta\, d\theta\, d\phi \sigma_{total} = \int_{\phi=0,\theta=0}^{\phi=2\pi,\theta=\pi}\left(\dfrac{d\sigma(\theta)}{d\Omega}\right)\sin\theta\, d\theta\, d\phi = \pi r_0^2$. It may thus be observed that the total scattering cross-section is equal to the geometric cross-section of the sphere πr_0^2.

Exercise p-4.14: 10 MeV neutrons get scattered by stationary proton. What maximum angular momentum may be imparted to the n–p system? Consider the classical radius of proton to be 0.8 F.

Solved example S-4.11

A beam of particles of mass m is represented by the wave function $\psi(x) = Ae^{ikx} + Be^{-ikx}$. Find the flux of the beam.

Solution. The flux is given by

$$\text{Flux} = \left(\frac{\hbar}{2im}\right)\left[\psi^* \frac{d\psi}{dx} - \left(\frac{d\psi^*}{dx}\right)\psi\right] =$$

$$\left(\frac{\hbar}{2im}\right)\left[\{Ae^{-ikx} + Be^{ikx}\}ik\{Ae^{ikx} - Be^{-ikr}\} + ik\{Ae^{-ikx} - Be^{ikx}\}\{Ae^{ikx} + Be^{-ikx}\}\right]$$

$$= \left(\frac{\hbar k}{m}\right)\{A^2 - B^2\}$$

The required flux is $\left(\frac{\hbar k}{m}\right)\{A^2 - B^2\}$.

Exercise p-4.15: Define scattering length and the effective range and discuss their significance.

Exercise p-4.16: What conclusions regarding the nature of nucleon–nucleon force can be drawn from the study of the deuteron and the nucleon–nucleon scattering? Explain your answer.

4.5 ONE-DIMENSIONAL RECTANGULAR BARRIER TRANSMISSION

In nuclear physics, we often encounter situations when charged nuclear particles, such as alpha particles or heavier ions are projected towards a stationary nucleus. The incident particles face a potential barrier of increasing magnitude as it approaches the target. In classical physics, the incident particles are reflected back by the barrier if their energy is less than the barrier height and no incident particle can ever reach the target. However, when the problem is treated quantum mechanically, a small number of incident particles of energy less than the barrier height are able to transmit through the barrier to reach the target. This is called quantum mechanical tunnelling. Quantum mechanical tunnelling has explained a large number of nuclear and other phenomena, such as the alpha decay and sub-barrier fusion of heavy ions, which had no explanation in classical physics. We shall outline the elementary quantum mechanical treatment of transmission through a one-dimensional barrier.

A rectangular barrier of (potential) height U and width 'a' is shown in figure 4.16. A beam of mono-energetic particles of energy W (which in the present case is equal to the kinetic energy E of the particles of the beam, that is ($W = E$) is made to fall on the barrier from left to right in the direction of the positive z-axis. It is assumed that W (or E) is less than the height U of the barrier. As shown in figure 4.16, the space around the rectangular barrier may be divided into three regions I, II, and III. In region I ($z < a$), there will be the incident wave and a part of the incident wave that undergo reflection at the boundary AB of the potential barrier. Thus, in region I, there will be two waves one moving in the positive z-direction and another wave moving in the negative z-direction. However, the amplitudes of the two waves will be different. In region II (z between A and D) also there will be two waves one moving in the positive z-direction and the other, the wave reflected from the other boundary CD of the barrier, moving in the negative z-direction. There will be only one wave moving in the positive z-direction in region III ($z > a$), as there is free space beyond CD and, therefore, no chance of any further reflection.

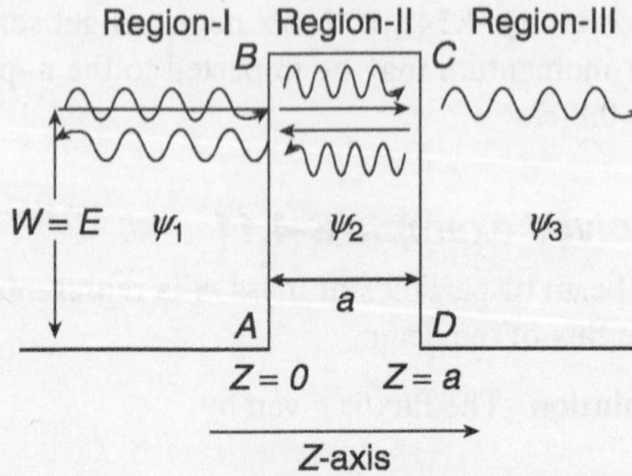

Figure 4.16 *Rectangular potential barrier. The total space is divided into three parts: Region I, Region II, and Region III. In regions I and II, there are reflected as well as components of the incident waves. However, in region III there is only an outgoing wave, and no reflected wave*

The wave number k_1 of the incident particle in region I is given by

$$k_1 = \frac{\sqrt{2MW}}{\hbar} = \frac{\sqrt{2ME}}{\hbar} \tag{4.60a}$$

where M is the mass of the incident particle and $W = E = \frac{1}{2}Mv^2$ is the kinetic energy of the particle. The space part of the incident plane wave in region I may be written as

$$\psi_{1i} = e^{ik_1 z} \tag{4.60b}$$

Here, the amplitude of the wave is taken as unity. The wave ψ_{1r} reflected from the barrier boundary AB and moving in the negative z-direction may be given as

$$\psi_{1r} = Ae^{-ik_1 z} \tag{4.60c}$$

The amplitude A of the reflected wave will obviously be smaller than 1. The total wave function in region I is which is denoted by ψ_1; therefore, the sum of ψ_{1i} and ψ_{1r}

$$\psi_1 = e^{ik_1 z} + Ae^{-ik_1 z} \tag{4.61}$$

The wave number k_2' in region II is given by

$$k_2' = \frac{\sqrt{2M(W-U)}}{\hbar} = \frac{i\sqrt{2M(U-W)}}{\hbar} = i\,k_2 \tag{4.62a}$$

Here,

$$k_2 = \frac{\sqrt{2M(U-W)}}{\hbar} \text{ is a real number.} \tag{4.62b}$$

There are two waves in region II: one is the component of the incident wave that has penetrated inside and the other is the wave reflected from boundary CD. If B and C are, respectively, the amplitudes of the two waves, then the total wave function ψ_2 in region II may be written as

$$\psi_2 = Ce^{ik_2'z} + Be^{-ik_2'z}$$

$$= Ce^{-k_2 z} + Be^{k_2 z} \tag{4.63}$$

In region III, potential U is zero-like region I and the wave number k_3 is same as k_1.

$$k_3 = k_1 = \frac{\sqrt{2MW}}{\hbar} = \frac{\sqrt{2ME}}{\hbar} \tag{4.64}$$

Also there is only one wave, a wave moving in positive z direction in region III and no reflected wave. As such, the wave function ψ_3 in region III with amplitude A_3 may be written as

$$\psi_3 = D\,e^{-ik_3 z} \tag{4.65}$$

The wave functions ψ_1, ψ_2, and ψ_3 satisfy the Schrödinger equations appropriate to regions I, II, and III, respectively. The one-dimensional Schrödinger equation for a particle of total energy W and potential energy U, appropriate for region II is

$$-\frac{\hbar^2}{2M}\frac{\partial^2\psi}{\partial z^2} + (U - W)\psi = 0 \tag{4.66}$$

Schrödinger equations for regions I and II may be obtained by putting $U = 0$ in Eq. (4.66)

Wave functions ψ_1, ψ_2, and ψ_3 must satisfy the following boundary conditions:

$$\psi_1 = \psi_2 \text{ and } \frac{d\psi_1}{dz} = \frac{d\psi_2}{dz} \text{ at } z = 0$$

This gives

$$1 + A = B + C \text{ and } i\,k_1(1 - A) = k_2(B - C) \tag{4.67a}$$

And

$$\psi_2 = \psi_3 \text{ and } \frac{d\psi_2}{dz} = \frac{d\psi_3}{dz} \text{ at } z = a$$

This gives

$$B\,e^{k_2 a} + C\,e^{-k_2 a} = D\,e^{ik_1 a} \tag{4.67b}$$

And

$$B\,k_2\,(e^{k_2 a}) - C\,k_2\,e^{-k_2 a} = i\,k_1\,D\,e^{ik_1 a} \tag{4.67c}$$

From Eqs. (4.67a–c), it is easy to get

$$B = \frac{2i\,k_1(k_2 + i\,k_1)}{[(k_2 + i\,k_1)^2 - e^{2k_2 a}(k_2 - i\,k_1)^2]} \tag{4.68a}$$

$$C = \frac{2ik_1(k_2 - ik_1)e^{2k_2a}}{[(k_2 + ik_1)^2 - e^{2k_2a}(k_2 - ik_1)^2]} \tag{4.68b}$$

$$D = \frac{4k_1k_2e^{-ik_1a}}{(k_2 + ik_1)^2 e^{-k_2a} - e^{k_2a}(ik_1 - k_2)^2} \tag{4.68c}$$

Here k_1 and k_2 are real numbers.

Further, the density of incident particles (i.e. the number of incident particles per unit volume) $|\psi_1|^2$ and of the transmitted particles $|\psi_3|^2$ is equal to 1 and $|D|^2$, respectively. Also, the flux that is equal to the particle density times the velocity is v and $|D|^2\, v$, respectively, for the incident and the transmitted waves. Here, v is the velocity of the incident particle and the particles coming out on the other side of the barrier after tunnelling. The transmission coefficient or barrier transmission factor τ is defined as the ratio of the transmitted to the incident flux and in the present case may be given by

$$\tau = \frac{|D|^2 v}{v} = |D|^2 = D^* D^* \tag{4.69a}$$

$$\tau = \frac{16k_1^2 k_2^2}{((k_1^2 + k_2^2)(e^{2k_2a} + e^{-2k_2a})) - 2\{k_2^4 - 6k_2^2k_1^2 + k_1^4\}} \tag{4.69b}$$

Using the identity $\left[\sin h^2\theta = \frac{1}{4}\{e^{2\theta} + e^{-2\theta}\} - \frac{1}{2}\right]$, it may be shown that

$$\tau = \frac{4k_1^2 k_2^2}{(k_1^2 + k_2^2)^2 \sin h^2(k_2a) + 4(k_1k_2)^2} \tag{4.69c}$$

Substituting

$$k_1^2k_2^2 = \frac{4m^2E(U-E)}{\hbar^4} \text{ and } (k_1^2 + k_2^2) = \frac{2mU}{\hbar^4}$$

we may get

$$\tau = \left[1 + \frac{U^2}{4E(U-E)} \sin h^2 k_2 a\right]^{-1} \tag{4.69d}$$

The potential barrier is said to be thick if $k_2a >> 1$. In that case, the term $\sin h^2 k_2 a$ reduces to $\frac{1}{4}e^{2k_2a}$ and, therefore, the transmission coefficient for a thick barrier may be given as

$$\tau_{\text{thick}} = 16\frac{E}{U}\left(\frac{E-U}{U}\right)e^{-2k_2a} \tag{4.70}$$

$$= f\, e^{-2\gamma} \tag{4.71}$$

where the energy-dependent term f is of the order of unity and $\gamma = k_2 a$ is a dimensionless quantity, the value of which depends on the kinetic energy of the incident particle and the height of the rectangular potential barrier. For a thick rectangular barrier, we may write

$$\tau_{\text{thick}} \approx e^{-2\gamma} \tag{4.72}$$

It follows from Eq. (4.72) that the barrier transmission coefficient for thick barrier decreases exponentially with γ, which in turn depends on the difference of the barrier height U and the kinetic energy E and the width 'a' of the barrier.

In general, transmission coefficient also depends on the orbital angular momentum ℓ of the transmitted particle but in the present simple case we have considered only S-wave transmission (for which $\ell = 0$).

The coefficient of reflection R may be obtained from the amplitude A of the wave reflected by the barrier in part I of the space.

$$R = |A|^2 = \frac{(k_1^2 + k_2^2)^2 \sinh^2(k_2 a)}{(k_1^2 + k_2^2)^2 \sinh^2(k_2 a) + 4(k_1 k_2)^2} \tag{4.73}$$

It is obvious that $R = 1 - \tau$.

Solved example S-4.12

Calculate the barrier transmission coefficient for the case $E > U$ and determine the condition for total transmission.

Solution. We refer to Eq. (4.69c) that gives the transmission coefficient as,

$$\tau = \frac{4k_1^2 k_2^2}{(k_1^2 + k_2^2)^2 \sinh^2(k_2 a) + 4(k_1 k_2)^2}$$

However, when $U > E$, $k_2 k_2$ becomes imaginary and $\sinh(k_2 a)$ becomes $i \sin(k_2 a)$. Substituting the values of k_1^2 and k_2^2, we get

$$\tau_{(E>U)} = \frac{1}{1 + \dfrac{U^2 \sin^2(k_2 a)}{4E(E-U)}}$$

Figure S-4.12 shows the variation of the transmission coefficient with the ratio of the kinetic energy to the barrier height.

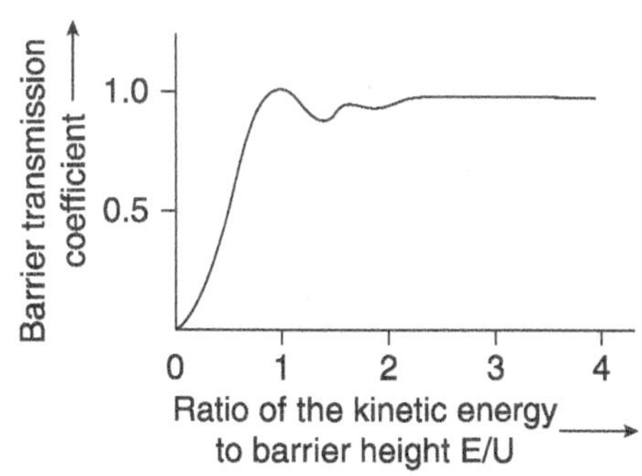

Figure S-4.12 *Variation of the transmission coefficient with the ratio of the particle energy to the barrier height.*

Exercise p-4.17: A particle of mass m and kinetic energy E is moving in a region where there is a potential step: $V(x) = 0$, for $x < 0$; and $V(x) = V$, for $x > 0$. Discuss the transmission of the particle through the barrier for the two cases: (1) $V > E$ and (2) $V < E$.

[Hint: Case (1), when $V > E$]: In this case, there will be two waves one incident wave $\psi_{i=Ae^{ik_1x}}$ and the other reflected wave $\psi_r = Be^{-ik_1x}$, where $k_1 = \dfrac{\sqrt{2mE}}{\hbar}$. The complete wave in the region $x < 0$ is $\psi = Ae^{ik_1x} + Be^{-ik_1x}$.

In the region $x > 0$, there will be a transmitted wave $\psi_t = Ce^{-k_2x}$, only as there will be no reflected wave. Here, $k_2 = \frac{\sqrt{2m(V-E)}}{\hbar}$. Relations between the amplitudes A, B, and C may be obtained by matching the wave functions ψ and ψ_t and their derivatives at $x = 0$.

Case (2), when $E > V$: In this case, the total wave in the region $x < 0$ will still have two components ψ_i and ψ_r as in the previous case. However, in the region $x > 0$, there will still be one wave moving in the positive x direction given by $\psi_{(x>0)} = De^{ik_2x}$. Again, the relations between the amplitudes A, B, and D may be found by the conditions of matching the wave functions and their derivatives at $x = 0$. Figure p-4.17 shows the situations in the two cases.

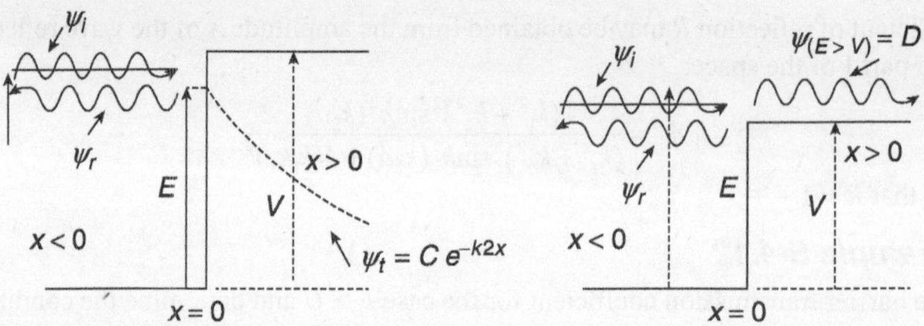

Figure p-4.17

Multiple Choice Questions

Note: Some of the problems may have more than one correct alternative. All correct alternatives must be ticked in such cases for complete answer.

Exercise M-4.1: The barrier transmission probability for particles of kinetic energy $E > U$ and $k_2 = \frac{\sqrt{2M(E-U)}}{\hbar}$ is

(a) $\left\{1+\frac{U^2\sin^2 k_2a}{4E(E-U)}\right\}^{-2}$ (b) $\left\{1+\frac{U^2\sin^2 k_2a}{4E(E-U)}\right\}^{-1}$

(c) $\left\{1-\frac{U^2\sin^2 k_2a}{4E(E-U)}\right\}^{-2}$ (d) $\left\{1-\frac{U^2\sin^2 k_2a}{4E(E-U)}\right\}^{-1}$

Exercise M-4.2: The reduced radial wave function for a system of two spinless particles described by a rectangular potential well attains the peak value at the boundary of the well. The two particle system is

(a) strongly bound
(b) just bound
(c) not bound
(d) whether bound or not depends on the depth of the potential well

Exercise M-4.3: If the total n–p force is spin-independent then the ratio of cross-sections for ortho and para scattering should be

(a) 3:1 (b) 1:3 (c) 3:4 (d) 1:1

Exercise M-4.4: The negative value of the scattering length indicates scattering from a

(a) bound state
(b) just bound state
(c) un-bound state
(d) state of mixed configurations

Exercise M-4.5: The parameters 'a' and 'r_0' of shape-independent approximation

(a) determine the shape of the nuclear potential
(b) determine the depth of the potential
(c) determine the width of the potential
(d) fix the depth and the width of the potential for a chosen shape of the potential

Exercise M-4.6: Deuteron is an admixture of S- and D-states. When in D-state, the spin J and the orbital angular momentum ℓ are

(a) parallel
(b) anti-parallel
(c) orthogonal
(d) may be parallel or orthogonal

Exercise M-4.7: A beam of mono-energetic particles of mass m, kinetic energy E_0, and intensity I_0, is made to fall on a potential barrier of height $U > E_0$. Particles with intensity I and energy E undergo tunnelling, then

(a) $I < I_0$ and $E < E_0$ (b) $I < I_0$ and $E = E_0$ (c) $I = I_0$ and $E = E_0$ (d) $I/I_0 = E/E_0$

Exercise M-4.8: Which state of n–p system has negative scattering length?

(a) singlet
(b) triplet
(c) both singlet and triplet
(d) none of the singlet and triplet state

Exercise M-4.9: The wave of mono-energetic particles of energy E tunnelled through a potential step of height U decays exponentially with distance. The energy E and the height of potential step U are related by

(a) $E = U$
(b) $E > U$
(c) $E < U$
(d) there is no definite relation between E and U

Exercise M-4.10: Quantum mechanical analysis of deuteron has indicated that deuteron is a mixed state of about 96% S-state and about 4% D-states. Imagine that the contribution of D-state in deuteron is 20%. What will happen to the magnetic dipole moment μ and the electric quadrupole moment Q of deuteron?

(a) μ will increase, Q will increase
(b) μ will increase, Q will decrease
(c) μ will decrease, Q will increase
(d) no change will take place in the magnitudes of μ and Q

5

Characteristics of Stable Nuclei and Nuclear Models

5.1 SYSTEMATIC TRENDS IN STABLE AND LONG-LIVED NUCLIDES

Studies of systematic trends observed in the naturally occurring nuclides are of considerable importance as they provide insight into the nature of nuclear forces and also because any good model for the nucleus should be able to reproduce most of them. Abundance of different elements in meteorites, earth's crust, hydrosphere, and atmosphere and the distribution of isotopes/isotones throw light on the synthesis of nuclei as well as on the nature of nuclear force. In this regard, it is worth noting that the distribution of elements in terrestrial bodies is essentially identical to those in non-terrestrial bodies such as meteorites. Further, no new stable element could be synthesized through transmutation or otherwise, which has not already existed. Some important systematic trends observed in the stable and long-lived radioactive nuclei are as follows:

1. The number of β stable and long-lived nuclides in nature is largest for even–even nuclides. The number of stable and long-lived even–odd and odd–even nuclides is nearly equal while there are only four stable and five long-lived nuclei that contain odd number of both protons and neutrons. Details of stable and long-lived isotopes of nuclides are given in table 5.1, and figure 5.1 shows the chart of stable nuclides as a function of their proton (Z) and neutron (N) numbers. Abundance of naturally occurring nuclides is shown in figure 5.2.
2. The spin or the intrinsic angular momentum J in the ground state of all even–even nuclides is zero, for odd–odd nuclides is integer $1\hbar, 2\hbar, 3\hbar, \ldots$ and for even–odd or odd–even (odd A nuclides) nuclides half integer, $1/2\hbar, 3/2\hbar, \ldots$ Further, the mirror nuclei have identical J values.

Table 5.1 *Number of stable nuclides for different combinations of N and Z*

Proton number	Neutron number	β stable	Half life > 10^9 years
Even	Even	155	11
Even	Odd	53	3
Odd	Even	50	3
Odd	Odd	4	5

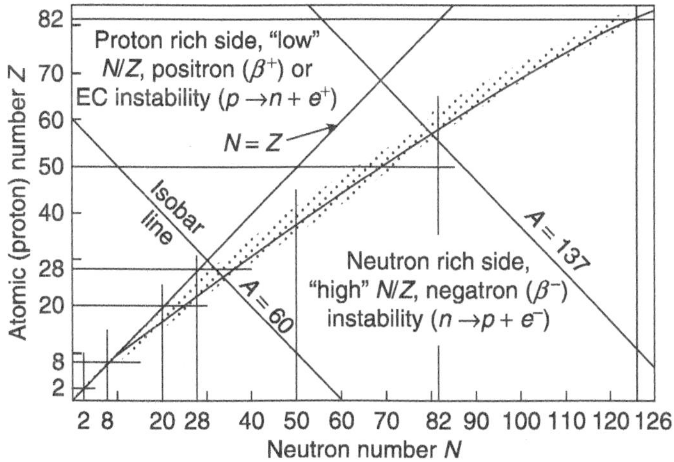

Figure 5.1 *Chart of stable nuclides as a function of their proton (Z) and neutron (N) numbers*

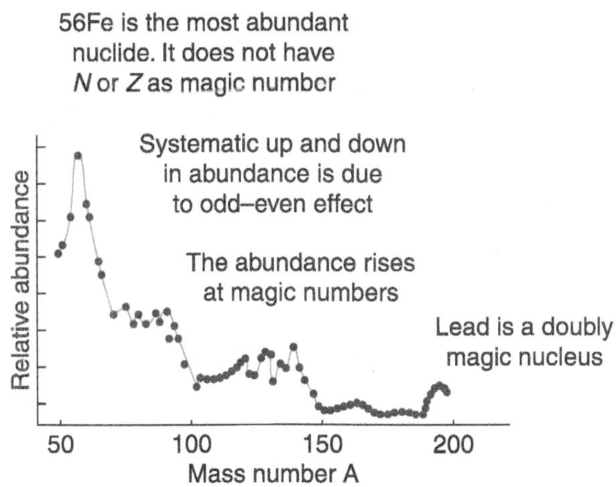

Figure 5.2 *Abundance of naturally occurring nuclides*

3. Stable nuclides lie in a narrow band on N versus Z graph (see figure 5.1). However, in general, there are two or even three stable nuclides for a given value of A if it is even and only one if A is odd. The relative frequency distribution of isotopes (same Z) and of isotones (same N) is similar. When masses of nuclides for a given value of A are plotted against the proton number Z, the most stable isobar lies at the bottom of the nuclear energy surface (figure 5.3). All heavier isobars decay into the stable members by β^-, β^+, or electron

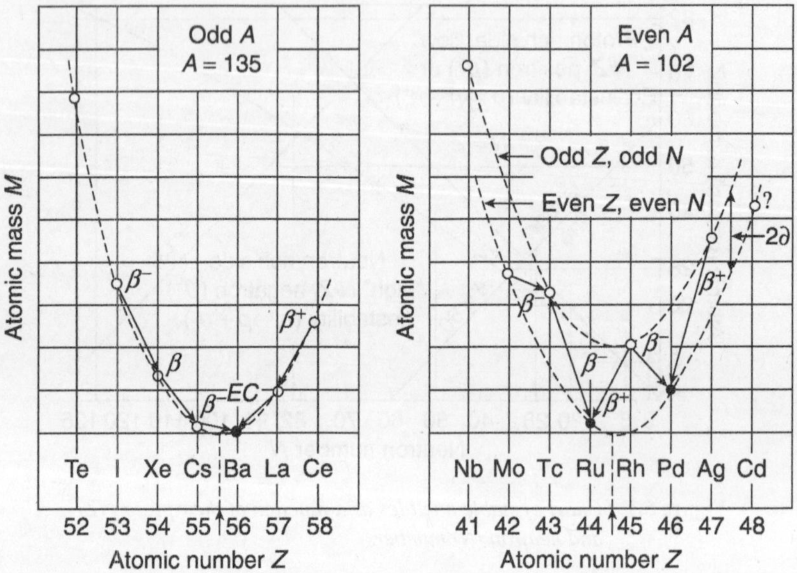

Figure 5.3 *Isobaric parabolas for (a) odd-A case and (b) even-A case*

capture decays. In case of odd-A nuclides there is in general, one most stable isobar, while in case of even-A there are two most stable members belonging to even *N* and even *Z*.

4. The fact that the radii of nuclei are proportional to the cube root of their atomic mass number *A* indicates a substantially constant density of nuclear matter.
5. The neutron excess $(N - Z)$ for stable nuclei varies as $A^{5/3}$.
6. Except for very light nuclei, the binding energy per nucleon B/A is nearly constant.
7. There is a strong tendency for two neutrons or two protons to pair in a nucleus. In case of an odd–odd nucleus, *A* is even and the neighbouring nuclei are even–even. The mass of the even–even neighbour nuclide is smaller due to the pairing energy and, therefore, the odd–odd nucleus decays into even–even neighbouring nucleus. The only exceptions are ${}^{2}_{1}H$, ${}^{6}_{3}Li$, ${}^{10}_{5}B$, and ${}^{14}_{7}N$. In case of these four nuclides, symmetry energy outplays the pairing energy.
8. Discontinuities in binding energies and in separation energies of the last nucleon have been observed at specific neutron and proton numbers (figure 5.4). Although B/A is substantially constant the total binding energy of nuclides shows discontinuities at Z or $N = 20, 28, 50, 80$, and 126. Similarly, the separation energies for the last neutron or proton show sudden jumps for N or $Z = 2, 8, 20, 28, 50, 82$, and 126. These numbers are called magic numbers. Nuclei with *N* or Z equal to one of these values are called magic nuclei. If both *N* and *Z* are magic, the nucleus is doubly magic. ${}^{208}_{82}Pb$ is a doubly magic nucleus. Special properties such as large binding energy, large number of isotopes, large number of isotones, and large separation energy for the last nucleon are associated with magic nuclei (figure 5.5).
9. The frequency of occurrence of isomeric states suddenly increases when *N* or *Z* lies between 40 and 50 and again between 70 and 80. These regions are called the islands of isomerism.

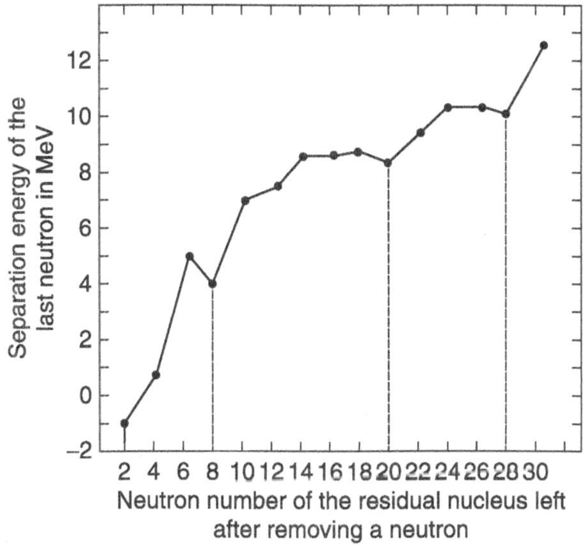

Figure 5.4 *Sharp change in the neutron separation energy at magic numbers*

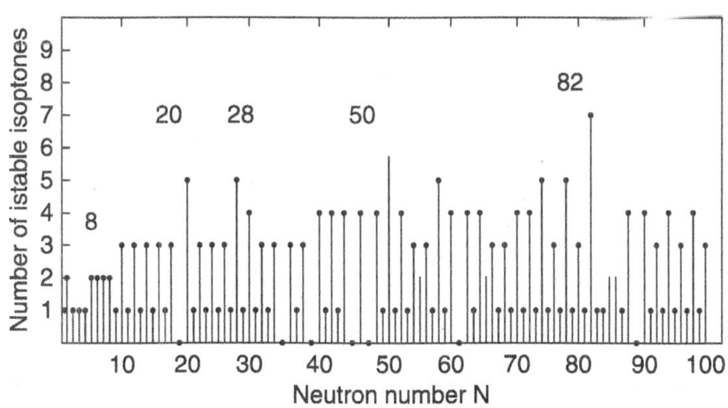

Figure 5.5 *The number of stable isotones increases at magic numbers*

10. Most of the nuclei with $A > 140$ decay by alpha emission, though they are stable against a neutron or a proton decay. The half-life of the alpha emitter depends very strongly on the energy of the emitted alpha particle. A slight change in the energy of the emitted alpha particle, for instance, from 5 MeV to 8 MeV changes the half life by a factor of 10^{20}. Alpha decay energy follows a systematic trend with the decay constant as shown in Geiger–Nuttall plot (figure 5.6)
11. The magnetic dipole moment μ of the nuclei is found to depend on their spin J and whether the neutron and/or proton number is even or odd (see table 5.2)

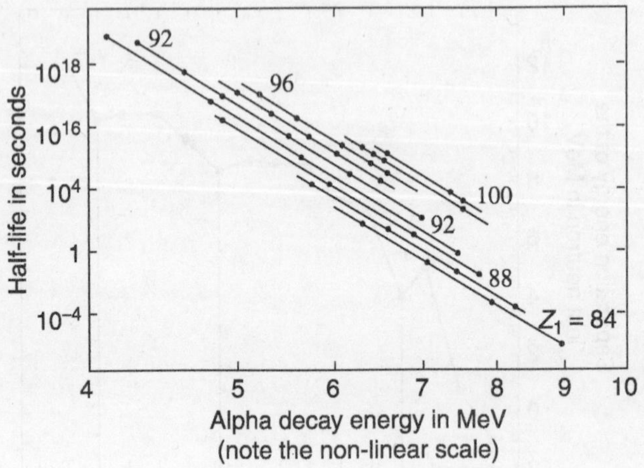

Figure 5.6 *Geiger–Nuttall plot*

Table 5.2 *Dependence of magnetic moment on proton and neutron numbers*

Proton number Z	Neutron number N	Spin J	Magnetic moment μ
Even	Odd	Half integer	Usually small and often negative
Odd	Even	Half integer	Usually positive and large
Even	Even	Zero	Intermediate
Odd	Odd	Integer	Usually positive non–zero

For odd-A nuclei, the magnetic moments are found to lie near the Schmidt lines (see figure 5.7(a) and (b)). Schmidt lines give the magnetic moments assuming that the spin and the magnetic moment of the nucleus are due to the last unpaired nucleon. If the orbital angular moment of the last nucleon is ℓ, then the total angular moment of the nucleon can have two possible values $j = \ell + 1/2$ and $j = \ell - 1/2$. Corresponding to these two values, there are two Schmidt lines for each case when the odd particle is proton or neutron.

12. The electric quadrupole moment of nuclei show typical behaviour with the number of nucleons in the nucleus. When the number of nucleons reaches 2, 8, 20, 28, 50, 82, and 126 it tends to be zero or very small indicating that near the magic number the nucleus tends to become spherical.
13. Some heavy nuclei with odd A undergo fission by thermal neutrons.
14. The spacing between the excited levels of a nucleus decreases with the increase in the excitation energy.
15. Nuclei heavier than ^{238}U do not exist in nature.

It is obvious that these trends are the outcome of the properties of the nuclear force. As the exact nature of nuclear force is not known, models for the nucleus, based on some assumptions about the nature of the nuclear force, are developed. A good model should reproduce most of the observed trends. It is, however, neither possible nor expected that a single model will be able to

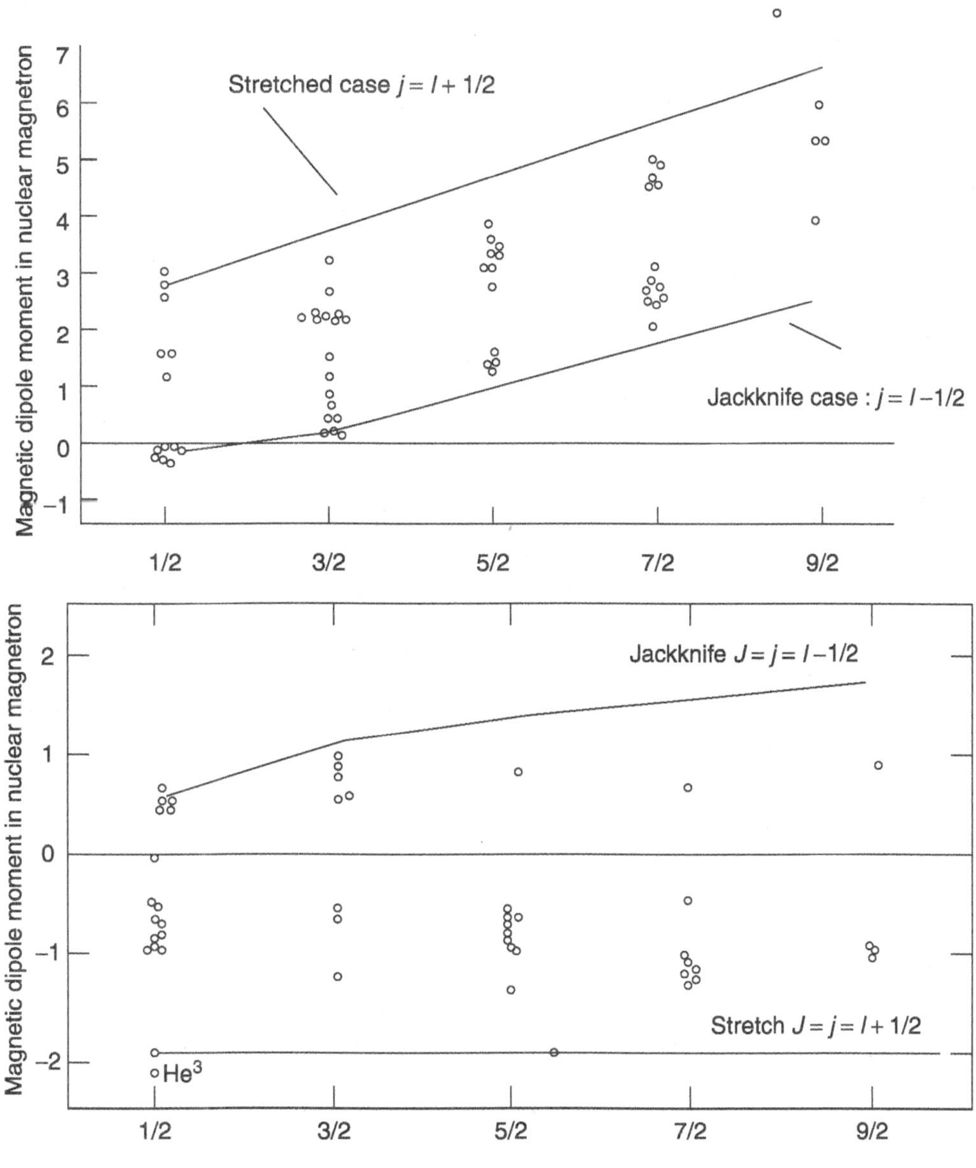

Figure 5.7 *Schmidt lines for (a) odd proton and (b) for odd neutron nuclides*

explain all the observed trends. Two models, namely the *shell model* and the *liquid drop model* are of special significance as the former successfully explains the special properties associated with magic numbers and the latter the fission of heavy nuclides. *Fermi gas model*, in addition, successfully explains the dependence of average nucleon binding energy in case of heavy nuclei, on $\frac{(N-Z)^2}{A}$.

5.2 SHELL MODEL

Nuclear shell model has developed out of the independent particle (or individual particle) model proposed by Bartlett, Guggenheim, Elsasser, and others. Considering analogy from the Bohr's model of the atom, it was assumed that nucleons inside the nucleus move independent of each other in a central potential. However, unlike the atom where the nucleus of the atom provides the central potential, there is no such agency to provide the central potential in case of the nucleus. It was, therefore, assumed that all the other $(A - 1)$ nucleons jointly produce a central field. In this field, which may be represented by a short-range central attractive potential, nucleons are assumed to move independent of each other in orbits of well-defined energy and orbital angular momentum. It was further assumed that the potential is same for all values of the orbital angular momentum ℓ. This is called the independent or individual particle model. One big problem encountered with the individual particle approach was the validity of the assumption that nucleons move independent of each other. Independent motion means either no or weak interaction between nucleons. However, nucleon–nucleon scattering and nuclear reactions suggest a strong interaction between nucleons. The problem was overcome by invoking Pauli exclusion principle. It was argued that in the ground state or even at small excitation energies, all the low-lying energy states of nucleons are already occupied. Nucleons being fermions cannot go to an already occupied state if they interact with each other. If interaction between the nucleons does take place then they must go to some vacant higher energy state. As such, in spite of being highly interactive, exclusion principle restricts interactions between the nucleons.

Assuming that the average central potential produced by all other nucleons is represented by a square well of depth V_0 and width r_0, the space part of the wave function $\Psi(r, \theta, \varphi)$ may be written as

$$\Psi(r, \theta, \varphi) = \chi_\ell(r)\, Y^\ell{}_m(\theta, \varphi) \tag{5.1}$$

where the radial wave function $\chi_\ell(r)$ is the solution of the radial equation

$$\frac{1}{r^2}\frac{d}{dr}\left(r^2\frac{d\chi_\ell}{dr}\right)+\frac{2M}{\hbar^2}\left[W+V_0-\frac{\ell(\ell+1)\hbar^2}{2Mr^2}\right]\chi_\ell=0 \tag{5.2}$$

Here, W is the energy of the nucleon and M the reduced mass. In case the nucleus is heavy, the reduced mass M is equal to the mass of the nucleon. The boundary condition on the wave function $\chi_\ell(r)$ is that it should be finite for all values of r. Spherical Bessel functions $j_\ell(kr)$ fulfil this requirement.

$$\chi_\ell(r) = j_\ell(kr) = \left(-\frac{r}{k}\right)^\ell \left(\frac{1}{r}\frac{d}{dr}\right)^\ell \left(\frac{\sin kr}{kr}\right) \tag{5.3}$$

where $k = \frac{1}{\hbar}\sqrt{2MW}$ (5.4)

The spherical Bessel functions for $\ell = 0, 1, 2, 3$, and 4 are plotted in figure 5.8.

Another boundary condition in the present case is that the wave function $\chi_\ell(r)$ should also go to zero at the boundary of the well $r = r_0$. As may be observed from figure 5.8, Bessel's function for each ℓ goes to zero for several different values of kr. For example, $j_0(kr)$ passes through zero for

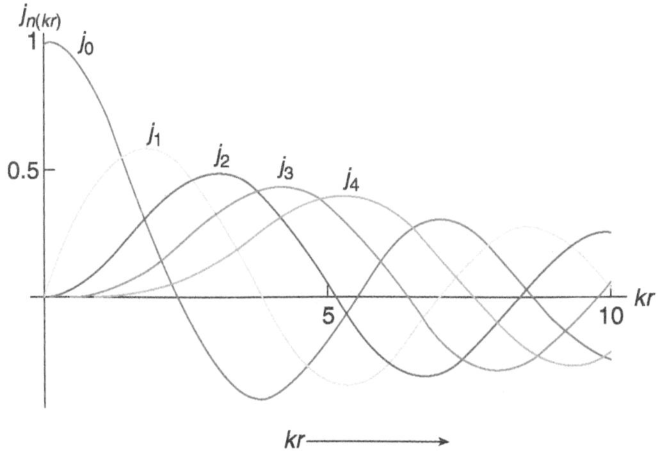

Figure 5.8 *Plot of $J_\ell(kr)$ for $\ell = 0, 1, 2, 3,$ and 4*

$(kr) = 3.14$; 6.28; 9.42; and so on, $j_1(kr)$ goes to zero at $kr = 4.48, 7.73, 10.90, \ldots$; $j_2(kr)$ becomes zero for $(kr) = 5.76, 9.09, \ldots$. Each zero crossing of $j_\ell(kr)$ gives an energy level for that ℓ value. The energy of the level is proportional to the square of the (kr) value that gives the zero of $j_\ell(kr)$. As zero crossings of $j_\ell(kr)$ for a given ℓ are not equidistant, the successive levels for that ℓ are also not equidistant. Further, between the two successive levels of a given ℓ, level from some different ℓ may also be present. Various different shapes for the central potential including harmonic oscillator, infinite depth rectangular well, well with rounded edges, and so on have been used but the sequence of levels and their relative separations are not found to be very sensitive to the shape of the well. Each level for a given value of ℓ degenerates with respect to the magnetic quantum number m_ℓ as it does not enter into the radial equation. As each ℓ value has a multiplicity of $(2\ell + 1)$, a maximum of $(2\ell + 1)$ neutrons and $(2\ell + 1)$ protons can be accommodated in that level if nucleons are considered as spinless particles. In case the spin of the nucleon is taken as $1/2\hbar$, which can have two possible orientations, the maximum number of $2\,(2\ell + 1)$ nucleons of each type may be accommodated in each state of the given ℓ. The order of levels for a finite square well along with the maximum number of nucleons of a given kind that may be accommodated in the level (or state), called the occupation number, are shown on the left hand side in figure 5.9. The total number of nucleons of a given kind that may be filled in the given level and in all other levels of lower energy is also shown by numbers in circles.

As may be observed from figure 5.9, this sequence of levels reproduces the closure of shells at nucleon numbers (N or Z) = 2, 8, 18, and 20 as after filling of the levels 1s, 1p, 1d, and 2s there are relatively large jumps in the energy to the next level, which is characteristic of the shell closure. Thus, lower magic numbers are explained by the independent particle model. However, the higher shell closures at N or $Z = 34, 58, 68, \ldots$ do not tally with the empirically known magic numbers 50, 82, and 126.

5.2.1 Spin–Orbit Coupling

To reproduce the higher magic numbers (50, 82, and 126) Mayer and Haxel, Jensen and Suess, independently made the ad-hoc assumption of strong spin–orbit coupling for nucleons. Accordingly, for

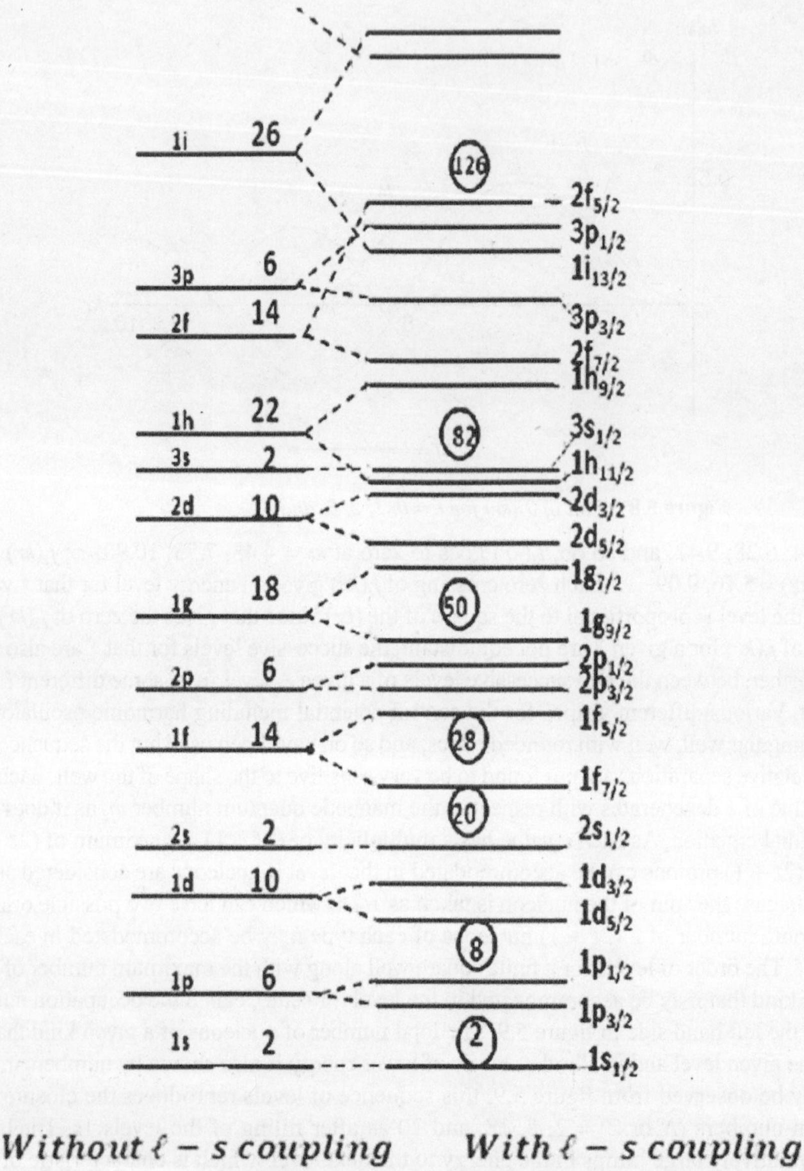

Figure 5.9 *Order of levels for a square well of finite depth and the maximum number of nucleons of one kind that may be accommodated in the level*

the given ℓ the energy of the $j = \ell + 1/2$ state may be quite different from the energy for the $j = \ell - 1/2$ state. Apart from the assumptions already made in the independent particle model, following additional assumptions have been made to reproduce the higher magic numbers (N or $Z = 50$, 82, and 126), the spin J, parity π, and the magnetic dipole moment μ of the nucleus.

1. For a fixed ℓ, the $j = \ell + 1/2$ state is more tightly bound than the $j = \ell - 1/2$ state.
2. The energy separation ΔE between the $j = \ell + 1/2$ and $j = \ell - 1/2$ states increases with ℓ and $\Delta E \propto \frac{(2\ell+1)}{A^{2/3}}$.
3. An even number of identical nucleons having the same ℓ and j will always couple to give even parity, zero total angular momentum J, and zero magnetic moment.
4. An odd number of identical nucleons having same ℓ and j will always couple to give odd parity if ℓ is odd and even parity if ℓ is even, a total angular momentum J equal to the j of the odd unpaired nucleon and the magnetic moment μ equal to the magnetic moment of the unpaired nucleon in state j.
5. Additional binding energy δ, called the pairing energy, is associated with the double occupancy of any state of a given ℓ and j by two identical nucleons. The pairing energy δ is largest for the state with largest j and is proportional to $\frac{(2j+1)}{A}$.

If the spin orbit interaction is assumed to be of the form

$$V_{s\ell} = v(r)\,(s\cdot\ell) \tag{5.5}$$

where $v(r)$ is potential function like the harmonic oscillator or a square well and s and ℓ are, respectively, the spin and orbital angular momentum vectors, the assumed strong coupling ℓ and s combine to give a resultant j given by

$$j(j+1) = \ell\,(\ell+1) + s\,(s+1) + 2\,(s\cdot\ell) \tag{5.6}$$

The value of the dot product $(s\cdot\ell)$ may be found from Eq. (5.6). Now, as $s = 1/2$, there are two possible values $j = \ell + 1/2$, called the stretched case and the $j = \ell - 1/2$ termed as jack-knife case. Using two values of j, the dot product $(s\cdot\ell)$ for the stretched and jack-knife case may be found as

$$(s\cdot\ell)_{\text{str}} = 1/2\ \ell;\ \text{and}\ (s\cdot\ell)_{\text{jack}} = -1/2\ (\ell+1)\ \text{excluding}\ \ell = 0$$

If these values of $(s\cdot\ell)$ are put in Eq. (5.5), we may get the difference in the energy ΔE of the two j states as

$$\Delta E = 1/2\ v(r)\,(2\ell+1) \tag{5.7}$$

Eq. (5.7) justifies the assumption 1 made earlier. Relativistic theories predict a spin orbit term of the correct sign as required but the strength of the relativistic spin orbit term is much less than what is required to produce the correct separation of the levels.

The splitting of levels under the strong ℓ–s coupling scheme is shown on the right hand side of figure 5.9.

Shell model states are designated by the principle quantum number (that gives the number of the zero crossings of $j_\ell(kr)$ for a given ℓ), the symbol specifying the value of the orbital angular moment ℓ (s for $\ell = 0$, p for $\ell = 1$, d for $\ell = 2$, f for $\ell = 3$, and so on) and the j-value ($= \ell \pm 1/2$). Typical examples are $1s_{1/2}$ and $2p_{3/2}$. As is obvious, the lower energy shell states are filed first and once the state is completely filled with the maximum number of nucleons that it can accommodate, the next nucleon goes in the state of next higher energy.

The groupings of shell model states give rise to shells. Depending on the jump in the energy at the discontinuity, there are six major and two minor shells. These are shown in figure 5.9. Grouping of shell may be written as follows:

$$\left[1s_{\frac{1}{2}}\right]\overset{2}{\uparrow}\left[1p_{\frac{3}{2}}, 1p_{\frac{1}{2}}\right]\overset{8}{\uparrow}\left[1d_{\frac{3}{2}}, 2s_{\frac{1}{2}}, 1d_{\frac{s}{2}}\right]\overset{20}{\uparrow}\left[1f_{\frac{7}{2}}\right]\overset{28}{\uparrow}\left[2p_{\frac{s}{2}}, 1f_{\frac{3}{2}}, 2p_{\frac{1}{2}}, 1g_{\frac{9}{2}}\right]\overset{50}{\uparrow}\left[1g_{\frac{7}{2}}, 2d_{\frac{3}{2}}, 2d_{\frac{3}{2}}, 3s_{\frac{1}{2}}, 1h_{\frac{11}{2}}\right]\overset{82}{\uparrow}\ldots\ldots$$

5.2.2 Prediction of the Spin and the Parity of the Nucleus

Apart from explaining the occurrence of magic numbers, a more stringent test of the shell model is its power of predicting the angular momentum (or spin J) and the parity of the nucleus.

1. In case of the even–even nucleus, the shell model predicts $J^{\pi} = 0^{+}$. There is no exception to this rule.
2. In case of the even–odd or odd–even nucleus, the nuclear spin is equal to the j of the unpaired particle and the parity is $= (-1)^{\ell}$, where ℓ is the orbital angular momentum of the unpaired particle. There are some exceptions to this rule.
3. For the odd–odd nucleus, the angular momentum J is an integer equal to the quantum mechanical sum of the j_n and j_p, which are, respectively, the j values for the odd unpaired neutron and the proton.

$$J = j_n + j_p$$

and the value of J lies between

$$|j_{n-j_p}| \leq J \leq j_n + j_p.$$

The parity of the nucleus is given by $\pi = (-1)^{(\ell_p + \ell_n)}$, where ℓ_n and ℓ_p are the orbital angular momentum of the neutron and the proton, respectively.

4. In the case where the number of neutrons (or protons) is one less than the next shell closure and the number of protons (or neutron) are even, then, the spin and parity of the nucleus is predicted by the so-called 'hole' in the next shell. For example, in the case of $^{39}_{19}K$ there are even number of (20) neutrons, which contribute zero spin and even parity. Now, 19 protons are one less than the number 20, one short of the next shell closure. This may be considered as if there is a hole in the next state shell that closes with $1d_{3/2}$. The spin and parity of $^{39}_{19}K$ is decided by the proton hole in $1d_{3/2}$ state. Therefore, J^{π} of $^{39}_{19}K$ is $3/2^{+}$. In a similar way, the spin and parity of $^{39}_{20}Ca$ may be assigned to the neutron hole in $1d_{3/2}$ shell as $3/2^{+}$.

The exact ordering of energy states within a shell is somewhat flexible. It is because some levels lie so close to each other that their ordering may change from nucleus to nucleus. For example, let us consider the case of the two isotopes of antimony $^{123}{}_{51}Sb$ and $^{121}{}_{51}Sb$, with 51 protons and, respectively, 72 and 70 neutrons. As the number of neutrons is even in both the isotopes, they contribute zero angular momentum and even parity to the nucleus. The spin, parity, and the magnetic moment of these two isotopes are decided by the 51st odd proton. The 51st proton, according to the shell model should go to $g_{7/2}$ level and, therefore, both the nuclides should have $J = 7/2$, and even parity (for g-state $\ell = 4$ and parity $\pi = (-1)^{\ell}$. The spin and parity of $^{123}{}_{51}Sb$ are actually $7/2^{+}$, as predicted by the shell model, but for $^{121}{}_{51}Sb$ they are $5/2^{+}$ suggesting that in case of ^{121}Sb the 51st proton goes to $d_{5/2}$ level. As such, there appears to be a crossover of $g_{7/2}$ and $d_{5/2}$

levels in these two isotopes of the same element. Actual ordering of levels may also depend on the number of neutrons in the nucleus.

According to the shell model level scheme, the fifth major shells closes with the state $h_{11/2}$. This state may accommodate a maximum of 12, $(2j + 1)$ neutrons or protons. For any nucleus that has odd number of neutrons or protons between 71 and 81, the odd nucleon should have gone to $h_{11/2}$ state and its spin should have been $J = 11/2$. However, no nucleus is found to have a spin of 11/2. The natural inference is that there is a crossover of $s_{1/2}$ and $h_{11/2}$ levels.

Let us now consider the case of $^{123}{}_{52}$Te nucleus. It has 71 neutrons and 52 protons. The odd neutron in this case decides the spin, parity, and the magnetic moment of the nucleus. In any case, whether there is a crossover between the $s_{1/2}$ and $h_{11/2}$ states or no crossover, the odd neutron must go to the $h_{11/2}$ state. It is because up to 68 neutrons get accommodated up to $2d_{3/2}$ level. Out of the remaining three neutrons, even if $s_{1/2}$ state is lower in energy than the $h_{11/2}$ state, it can accommodate only a maximum of two neutrons. As such, the remaining one neutron must go to the $h_{11/2}$ level. In the other case, when there is no crossover, all the three neutrons must go to the $h_{11/2}$ state. As such the spin, parity, and magnetic moment of ^{123}Te should have the values characteristic of the state $h_{11/2}$. However, the spin, parity, and magnetic moment of ^{123}Te ($½^+$, $\mu = -0.74$ nm) are typical of the $s_{1/2}$ state. Non-existence of any nucleus with spin 11/2 and the cases like that of ^{123}Te suggests that the state $h_{11/2}$ always robs a nucleon from the $s_{1/2}$ state, to make the number of nucleons even. This is attributed to the pairing energy δ. As pairing energy is larger for a higher ℓ state, $h_{11/2}$ state pulls a nucleon from the neighbouring state whenever the number of nucleons is odd in it.

Although the major force of attraction between the nucleons is derived from the central potential and from the potential of spin–orbit coupling, there is a so-called residual interaction between nucleons in the same state. This residual interaction becomes more visible if the state is only partly filled. As per assumption 3 made earlier, all completely filled states contribute zero angular momentum and positive parity to the nucleus; the angular momentum and the parity of the ground state of the nucleus are determined by the residual interaction. Pairing energy δ, which has values of the order of 1–2 MeV, forms the most important component of the residual interaction. The origin of the pairing energy may be the large overlap of the wave functions of the identical particles in the same state when they have opposite magnetic quantum numbers m_j.

Let us test the power of predicting the ground-state angular momentum and parity of the shell model by taking the example of an odd–odd nucleus. $^{38}{}_{17}$Cl nucleus has 17 protons and 21 neutrons. The angular momentum J and the parity π of this nucleus, according to the shell model, are due to the last unpaired (i.e. 21st) neutron and 17th proton. According to the shell model level scheme, the proton should go to the level $1f_{7/2}$ and the neutron to the level $1d_{3/2}$. For level f, the orbital angular momentum ℓ is 3. In case of f level, the $j = 7/2$ state is formed when $j = \ell + 1/2$, which means the orbital angular moment ℓ of the proton is in the same direction as its spin (1/2). For the odd neutron, which is in $d_{3/2}$ level, the $j = 3/2$ indicates that the orbital angular moment ℓ (=2) and the spin 1/2 are in the opposite direction ($j = \ell - 1/2$). As the combination of a neutron and a proton favours parallel spins, the orbital angular momentums of proton and neutron are opposite to each other. Therefore, the orbital angular momentum J of the nucleus ^{38}Cl is the sum of two parallel spins (1/2 + 1/2) + 3 (the orbital angular momentum of proton) – 2 (the orbital angular momentum of neutron which is opposite to the spins) = 2. Thus, according to shell model, the spin of ^{38}Cl, $J = 2$. Next, let us consider the parity π of ^{38}Cl, which will be equal to the multiplication of the parities of the 17th proton and the 21st neutron. $\pi = (\pi)^P\,(\pi)^N = (-1)^{\ell_p}\,(-1)^{\ell_n}$, where

ℓ_p and ℓ_n are, respectively, the orbital angular momentum of proton and the neutron. However, $\ell_p = 3$ and $\ell_n = 2$. As such $(\pi)^P$ = negative and $(\pi)^N$ = even or positive. Hence, according to shell model π = negative or odd. We, thus, see that the shell model predicts spin 2 and parity odd for ^{38}Cl nucleus. The predicted value agrees with the experimental value of the spin and parity of ^{38}Cl.

5.2.3 Quadrupole Moment and Shell Model

Figure 5.10a shows the measured value of nuclear quadrupole moments as a function of the proton number of the nucleus. It may be noticed that the quadrupole moment is zero for nuclei at major proton magic numbers. Nuclear configurations that have one extra neutron or proton above the closed shell appears to have an oblate shape, while those with a neutron or proton hole in the shell as a prolate ellipsoid. Shell model does not really correctly predict the quadrupole moments of nuclei, particularly for nuclei far away from the closed shells. In case of the nuclei near the closed shell, detailed shell model calculations give the following expression for the quadrupole moment Q:

$$Q = -\frac{(2j-1)}{2(j+1)}\left(\frac{3}{5}R^2\right),$$ where R is the radius of the nucleus given by $r_0 A^{1/3}$.

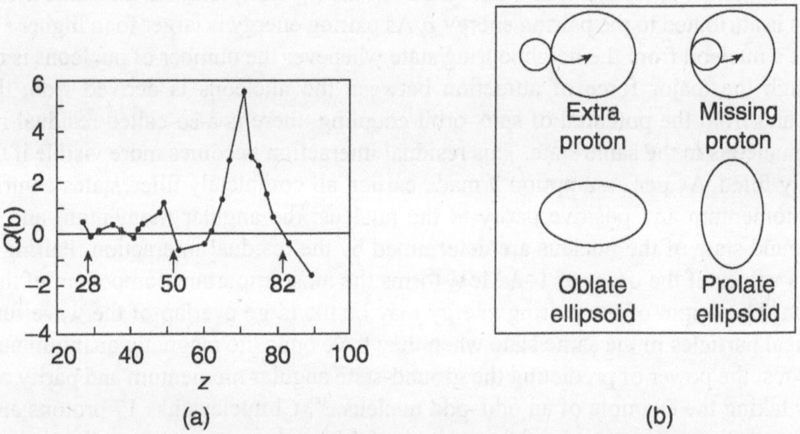

Figure 5.10 *(a) The measured quadrupole moments as a function of the atomic number Z. (b) Nuclei with an extra neutron or proton than the closed shell behave like an oblate ellipsoid and those with a nucleon hole in the shell like a prolate ellipsoid*

5.2.4 Prediction of Low-lying Excited States

The shell model is partially successful in accounting for the measured excited states of odd A nuclei. The excited states in the shell model may be generated by exciting the odd nucleon in the unfilled shell to the available higher energy states. As the number of possible states available for excitation increases with the excitation energy, the density of excited levels increases rapidly with energy. This rapid rise of level density with the excitation energy has been confirmed by experiments.

In general, it is expected that an excited state permitted by the shell model will have definite excitation energy. However, often more than one excited states with same characteristic (spin j and parity) but at slightly different energies have been observed experimentally. This multiplicity of levels is termed as 'fragmentation'. The fragmentation of levels is not easily explained by the shell model. It appears that some excited states are formed by the excitation of nucleons from the filled shells called the core. A simple shell model cannot explain the excitation of nucleons from the core.

In general, it may be stated that the shell model is good in predicting the spin, the parity, and the low-lying excited states only for the nuclei that are near the closed shells. As the simple shell model assumes a central potential, it can be used for spherical nuclei. The characteristics of distorted nuclei that have large quadrupole moment are not expected to be correctly explained by the shell model. As the energy of residual interaction, the pairing energy, is quite large of the order of separation between levels, the exact sequence of shell model states is flexible.

5.2.5 Prediction of Magnetic Moment: Schmidt Lines

According to the shell model, the unpaired nucleon in the odd-A nucleus decides the spin and the parity of the nucleus. In this model, it is, therefore, expected that the magnetic moment of the odd-A nucleus is also decided by the unpaired nucleon. Except for the nucleon in the $s_{1/2}$ state, the unpaired nucleon in any other shell model state possesses an orbital angular momentum ℓ and the intrinsic spin angular momentum s that couple to give a resultant angular momentum j. The resultant magnetic moment of the unpaired nucleon μ_j is the vector sum of the magnetic moments due to the intrinsic spin μ_s and the magnetic moment due to the orbital motion μ_ℓ. The spin and orbital magnetic moments are related to the respective angular momentum s and ℓ through the g-factors:

$\mu^p{}_s = g^p{}_s\, s$, where the g-factor for proton corresponding to its intrinsic spin $g^p{}_s = 5.5855$.

$\mu^n{}_s = g^n{}_s\, s$, where the g-factor for neutron corresponding to its intrinsic spin $g^n{}_s = -3.8270$.

$\mu^p{}_\ell = g^p{}_\ell\, \ell$, where the g-factor for proton corresponding to its orbital motion $g^p{}_\ell = 1$.

$\mu^n{}_\ell = g^n{}_\ell\, \ell$, where the g-factor for neutron corresponding to its orbital motion $g^n{}_\ell = 0$.

To measure the magnetic moment, the nucleus has to be put in some external magnetic field. Let us assume that the nucleus is put in an external magnetic field B in z-direction as shown in figure 5.11.

Now, s and ℓ do not have sharp components in the direction of the external magnetic field and only j has sharp component along z-axis. As such, the component of the resultant magnetic moment μ_j in the z-direction may be written as:

$$\mu_j = g_j \text{ (absolute magnitude of } j) = g_s \text{ (absolute magnitude of } s) \cos(s, j) + g_\ell \text{ (absolute magnitude of } \ell) \cos(\ell, j) \quad (5.8)$$

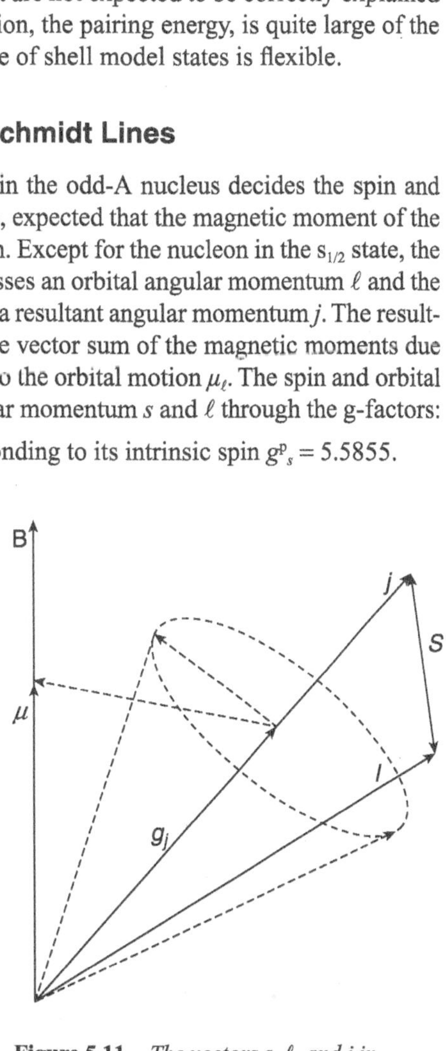

Figure 5.11 *The vectors s, ℓ, and j in external magnetic field*

$$= g_j\,[j(j+1)]^{1/2} = g_s\,[s\,(s+1)]^{1/2}\cos(s,j) + g_\ell\,[\ell\,(\ell+1)]^{1/2}\cos(\ell,j) \tag{5.9}$$

where cos (*s*, *j*) and cos (ℓ, *j*) are, respectively, the cosines of the angles between the vectors *s* and *j* and ℓ and *j*. Using the rule of triangles, these angles may be written as

$$\cos(s,j) = \frac{j(j+1)+s(s+1)-\ell(\ell+1)}{2\sqrt{s(s+1)j(j+1)}} \text{ and}$$

$$\cos(\ell,j) = \frac{j(j+1)-s(s+1)+\ell(\ell+1)}{2\sqrt{\ell(\ell+1)j(j+1)}} \tag{5.10}$$

Substituting these values of cosines in Eq. (5.9), we get

$$g_j = g_s\,\frac{j(j+1)+s(s+1)-\ell(\ell+1)}{2j(j+1)} + g_\ell\,\frac{j(j+1)+\ell(\ell+1)-s(s+1)}{2j(j+1)} \tag{5.11}$$

Now, in Eq. (5.11), $s = 1/2$ and we write the value of ℓ in terms of *j*. For the unpaired nucleon, ℓ can have two possible values $\ell = j - 1/2$ (called the stretched case) or $j + 1/2$ (called the jack-knife case). As such, there will be two values of g_j, $(g_j)_{\text{str}}$ for the stretched case and $(g_j)_{\text{jack}}$ for the jack-knife case. They are obtained by substituting the respective value of ℓ in Eq. (5.11)

$$(g_j)_{\text{str}} = \frac{1}{2j}\,g_s + \frac{(2j-1)}{2j}\,g_\ell \text{ and } (g_j)_{\text{jack}} = \frac{\left[-\frac{1}{2}\,g_s + \left(j+\frac{3}{2}\right)g_\ell\right]}{j(j+1)} \tag{5.12}$$

The magnetic moment μ_j can be calculated for the stretched and jack-knife cases by substituting the respective values of *g* in Eq. (5.9). Further, the values of g^n_ℓ and g^n_s have to be used for the odd neutron and g^p_s and g^p_ℓ for the odd proton case. After substituting the appropriate values for g_ℓ, we get

For odd proton: $\ell = j + 1/2$ $\quad \mu = \frac{1}{2}\,\frac{j}{j+1}\,(2j+3-g^p_s)$ (5.13)

$\ell = j - 1/2$ $\quad \mu = \frac{1}{2}\,(2j-3+g^p_s)$ (5.14)

For odd neutron: $\ell = j + 1/2$ $\quad \mu = -\frac{1}{2}\,\frac{j}{j+1}\,g^n_s$ (5.15)

$\ell = j - 1/2$ $\quad \mu = \frac{1}{2}\,g^n_s$ (5.16)

Eqs. (5.13) and (5.14) define the Schmidt lines for the odd proton and for the odd neutron cases. These are plotted in figure 5.7. The experimental magnetic moments for odd *Z* and odd *N* nuclides are also shown in these figures. Although the measured magnetic moments do not fall on Schmidt lines, they lie between the two lines. In addition, each measured value is closer either to the stretched or to the jack-knife Schmidt line. There appears to be a tendency in the measured values of the magnetic moments of odd-A nuclides that they fall into two distinct groups, each parallel to the Schmidt line. These groups, called Schmidt groups, are assigned to the $j = \ell + 1/2$ and $j = \ell - 1/2$ cases.

The value of the spin J of the odd-A nucleus along with the measured value of the magnetic moment helps to determine the orbital angular moment ℓ of the odd nucleon, for example, ${}^{59}_{27}Co$ has $J = 7/2$ and $\mu = 4.64$ nm. The measured value of the magnetic moment, which is large, puts this nucleus near the $j = \ell + 1/2$ Schmidt line in figure 5.7. Clearly, the ℓ of the unpaired proton in ${}^{59}_{27}Co$ is 3.

Measured magnetic moments of odd-A nuclides with one nucleon more than the magic numbers do not show sharp changes or very different value compared with the neighbouring nuclei. This indicates that the magic core may not be perfectly spherical and may contribute to the magnetic moment. The possibility of non-spherical core is also pointed by the measured quadrupole moments. Polarization of the core by the odd nucleon is a distinct possibility in this regard.

Solved example S-5.1

Calculate the energy gap between the $1p_{1/2}$ and $1d_{5/2}$ neutron shells, given that the total binding energies of ${}^{15}O$, ${}^{16}O$, and ${}^{17}O$ are respectively, 111.9556 MeV, 127.6193 MeV, and 131.7627 MeV.

Solution. According to the shell model eight protons in all the three isotopes of oxygen contribute zero spin and even parity. Seven neutrons in ${}^{15}O$ have the configuration $\left[(1s_{1/2})^2, (1p_{3/2})^4, (1p_{1/2})^1\right]$. The one unpaired neutron is in $((1p_{1/2}))$ state. The configuration of ${}^{16}O$ is $\left[(1s_{1/2})^2, (1p_{3/2})^4, (1p_{1/2})^2\right]$. The configuration of ${}^{17}O$ is $\left[(1s_{1/2})^2, (1p_{3/2})^4, (1p_{1/2})^2, (1d_{5/2})^1\right]$. We denote by $M(x)c^2$ the mass energy and by $B(x)$, the binding energy of nucleus x, where x can be nuclei ${}^{15}O$, ${}^{16}O$, and ${}^{17}O$. The mass energies of neutron and proton are represented by M_nc^2 and M_pc^2. The energy diagram of the three isotopes is shown in figure S-5.1. As is evident from the figure, the energy difference between $1p_{1/2}$ and $1d_{3/2}$ levels $\Delta E = \Delta E_1 + \Delta E_2$. We now calculate the mass energy for three isotopes.

$$M({}^{15}O)c^2 = 8M_pc^2 + 7M_nc^2 - B({}^{15}O);$$
$$M({}^{16}O) = 8M_pc^2 + 8M_nc^2 - B({}^{16}O) \text{ and}$$
$$M({}^{17}O)c^2 = 8M_pc^2 + 9M_nc^2 - B({}^{17}O)$$

Hence,

$$\Delta E_1 = M({}^{17}O)c^2 - M({}^{16}O)c^2 = M_nc^2 - B({}^{17}O) + B({}^{16}O) \quad \text{(A)}$$
$$\Delta E_2 = M({}^{15}O)c^2 - M({}^{16}O)c^2 = -M_nc^2 - B({}^{15}O) + B({}^{16}O) \quad \text{(B)}$$

Adding (A) and (B), we get

$$\Delta E = \Delta E_1 + \Delta E_2 = 2B({}^{16}O) - \{B({}^{17}O) + B({}^{15}O)\}$$
$$= 2(127.6193) - (131.7627 + 111.9556) \text{ MeV}$$

Ans: The energy difference between the d5/2 and p1/2 states is 11.5203 MeV.

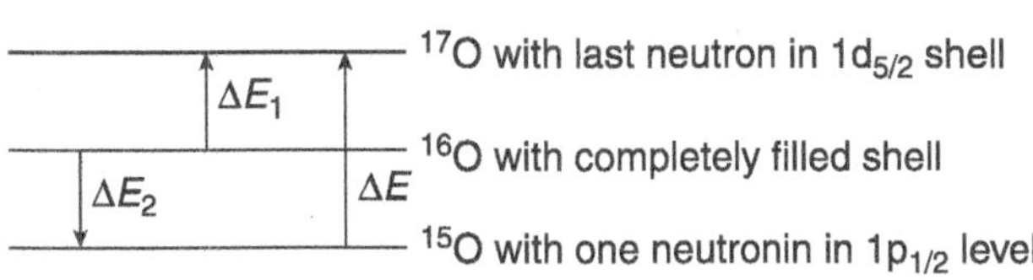

Figure S-5.1

Solved example S-5.2

Write down the shell model predictions for the neutron and proton configurations and the spin and parity for the ground states of the following nuclides, ${}^{3}_{2}\text{He}$, ${}^{20}_{10}\text{Ne}$, ${}^{30}_{14}\text{Si}$ and ${}^{41}_{21}\text{Sc}$.

Solution. Shell model predictions of neutron, proton configurations, and spin and parity are as follows:

Nuclide	Neutron configuration	Proton configuration	J^π
${}^{3}_{2}\text{He}$	$(1s_{1/2})^1$	$(1s_{1/2})^2$	$\frac{1^+}{2}$
${}^{20}_{10}\text{Ne}$	$(1s_{1/2})^2(1p_{3/2})^4(1p_{1/2})^2\ (1d_{5/2})^2$	$(1s_{1/2})^2(1p_{3/2})^4(1p_{1/2})^2(1d_{5/2})^2$	0^+
${}^{30}_{14}\text{Si}$	$(1s_{1/2})^2(1p_{3/2})^4(1p_{1/2})^2(1d_{5/2})^6(2s_{1/2})^2$	$(1s_{1/2})^2(1p_{3/2})^4(1p_{1/2})^2(1d_{5/2})^6$	0^+
${}^{41}_{21}\text{Sc}$	$(1s_{1/2})^2(1p_{3/2})^4(1p_{1/2})^2(1d_{5/2})^6(2s_{1/2})^2(1d_{3/2})^4$	$(1s_{1/2})^2(1p_{3/2})^4(1p_{1/2})^2(1d_{5/2})^6$ $(2s_{1/2})^2(1d_{3/2})^4[1f_{7/2}]^1$	$\frac{7^-}{2}$

Exercise p-5.1: What are the basic assumptions of the independent particle model and how they be justified? Determine the spin and parity of ${}^{13}_{6}\text{C}$ and ${}^{15}_{6}\text{C}$.

Exercise p-5.2: What are the additional assumptions made in the shell model? Discuss the possible cause for (a) the difference in the energy of $j + 1/2$ and $j - 1/2$ states (b) the pairing energy.

Exercise p-5.3: List the basic characteristics associated with the naturally occurring stable nuclides, and point out which can be explained by the shell model.

Exercise p-5.4: Discuss the spin, parity, and the magnetic moment of the following nuclides in light of the shell model: ${}^{7}{}_{3}\text{Li}$, ${}^{14}{}_{7}\text{N}$, and ${}^{41}{}_{20}\text{Ca}$.

Exercise p-5. 5: What are Schmidt lines? Derive expressions for Schmidt lines.

5.3 LIQUID DROP MODEL

Theoretical nuclear models encounter two principal problems:

1. There is no exact mathematical expression that accounts for the nuclear force, unlike the atomic case, for which the electromagnetic force is well-defined by Coulomb's Law.
2. There is no mathematical solution to the many-body problem, a limitation shared by both nuclear and atomic systems.

Fortunately, the power of the computer permits calculations that use sophisticated approximations to minimize these problems and provide increasingly accurate models for describing nuclear and atomic properties.

The starting point for theoretical models of the nucleus treats the problem from two divergent perspectives. At the macroscopic extreme is the *liquid drop model*, which examines the global properties of nuclei, such as energetic, binding energies, sizes, shapes, and nucleon distributions. This model assumes that *all nucleons are alike* (other than charge). In contrast, the *shell model* is designed to account for the quantum properties of nuclei such as spins, quantum states, magnetic moments, and magic numbers. The basic assumption of the shell model is that *all nucleons are different*, that is nucleons are fermions and must occupy different quantum states, as is the case for atoms. In a way, the liquid drop model is the antithesis of shell model. In liquid drop model, one assumes strong, short-range interaction between nucleons against the assumption of no interaction between nucleons and completely independent motion of each nucleon in a central field in the shell model.

The idealized goal of theoretical nuclear physics is to combine these two concepts into a unified model that may describe both the macroscopic and microscopic aspects of nuclear matter in a single comprehensive framework.

5.3.1 Basis of the Liquid Drop Model: Similarity Between a Liquid Drop and the Nucleus

The basis of the liquid drop model are the following similarities between the drop of an incompressible liquid and the nucleus.

1. Like a drops of a liquid, the density of nuclear matter in different nuclei is constant.
2. High-energy electron scattering experiments have shown that the interior of all nuclei, big or small, is identical with regard to the charge density and the nucleon density. As such, the big and small nuclei are like the big and small drops of a liquid.
3. Like a liquid drop, the nuclei also show surface effects that are analogous to the surface tension in liquids.
4. The near-constant binding energy per nucleon (B/A) may be compared to the constant latent heat of evaporation of a liquid.
5. The nucleon–nucleon force is a short-range force like the intra-molecular force in liquid drops.
6. All nuclei are nearly spherical like the drops of a liquid.
7. Two small liquid drops combine to make a big drop and a big liquid drop breaks into small drops. This may be compared to the fusion and fission of nuclei.

5.3.2 Differences Between a Liquid Drop and a Nucleus

In spite of so many similarities between a liquid drop and the nucleus, there are also strong differences between the two. They are as follows:

1. Unlike the liquid drop, there are positively charged protons in the nucleus.
2. The number of nucleons in the heaviest stable nucleus (<270) is much smaller than the molecules in an average liquid drop ($\approx 10^{23}$).
3. Unlike a liquid drop, a nucleus contains two components: neutrons and protons.

4. Much larger fraction of nucleons is at the nuclear surface as compared to the molecules in a quid drop.
5. No microscopic effects, such as the shell effect and magic numbers that are observed in nuclei, occur in liquid drops.

5.3.3 Weizsacker's Semi-empirical Mass or Binding Energy Formula

Based on the above-mentioned considerations, Bethe and Weizsacker developed a formula based partially on theory and partially on empirical data (and hence called the semi-empirical) for the mass or the binding energy of a nucleus.

The mass $M(Z, N)$ of a nucleus ${}^{A}_{Z}X$ having Z protons, N neutrons, and A (=Z + N) number of total nucleons may be written as

$$M(Z, N) = Z\,M_p + N\,M_n - B(Z, N)/c^2 \tag{5.17}$$

where M_p and M_n are, respectively, the masses of proton and neutron and $B(Z, N)$ the binding energy of the nucleus ${}^{A}_{Z}X$.

5.3.3.1 The volume energy

The main contribution to the binding energy comes from the strong nucleon–nucleon attractive force. The nucleon–nucleon attractive potential is estimated to be of the order of 40 MeV. In a nucleus with A nucleons and assuming that nucleon–nucleon force is charge independent, the total binding energy of the nucleus should have been proportional to the number of pairs of nucleons, that is $A\,(A-1)/2$. However, a given nucleon does not attract all other nucleons because of the short range (of the order of 2 F) of nucleon–nucleon force. As a result, all pairs of nucleons do not contribute to the binding energy, just like the molecules in a drop of liquid. As such, the binding energy is proportional to A and not to $A\,(A-1)$. The volume V of the nucleus ${}^{A}_{Z}X$ may be written as $V = 4/3\pi R^3$ and $R = r_o A^{1/3}$, $V = 4/3\pi r_o^{\,3} A$ = constant × A, where r_0 is the unit nuclear radius. As the atomic mass number A is proportional to the nuclear volume V, the binding energy due to the attractive nucleon–nucleon force, denoted by B_v, which is proportional to A, is called the volume energy.

$$B_v = a_v\,A \tag{5.18}$$

This is the only and dominating attractive term that will hold in case of the idealized neutron star of infinite dimensions. A neutron star consists of only neutrons packed together.

In case of the real nuclei, the finite size of the nucleus and the presence of protons contribute repulsive terms and diminish the binding energy. The constant a_v, which will ultimately be determined from the experimental data, may be roughly estimated. The binding energy of a nucleon in a nucleus is the difference between its potential energy and its kinetic energy. According to the Fermi gas model of the nucleus, the potential energy of each nucleon is around 40 MeV and the average kinetic energy of the nucleon is $\approx \frac{3}{5}\,\varepsilon_f$, where ε_f is the Fermi energy, the kinetic energy of the most energetic nucleon. Estimated value of Fermi energy is around 33 MeV. The binding energy of each nucleon then comes as $(40 - 3/5 \times 33) \approx 20$. If a nucleus has A nucleons, the total binding energy of the nucleus will be $20\,A$, giving $a_v \approx 20$. As will turn out, it is not a bad estimate for a_v, the empirical value for which is found to be 17.

5.3.3.2 The surface energy B_s

The nucleons at the surface of the nucleus are not surrounded by other nucleons on all sides as are the nucleons deep inside the volume. As such, these surface nucleons do not contribute the same amount of binding as is provided by the inner nucleons. Eq. (5.18) has been derived assuming that all A nucleons contribute the same amount of binding. Therefore, an amount of energy proportional to the number of surface nucleons has to be subtracted, on account of surface effect. The surface area S of a sphere of radius R is $4\pi R^2$, and for the nucleus $R = r_0 A^{1/3}$, the surface area of a nucleus with A nucleons $S = 4\pi r_0^2 A^{2/3}$. As the number of nucleons at the surface are proportional to the surface area S, the loss in the binding energy.

$$B_s = -a_s\, A^{2/3} \tag{5.19}$$

The negative sign indicates a loss in energy. In a small nucleus, larger fraction of total nucleons is at the surface and contributes a smaller binding energy. This is reflected in the B/A curve that falls for lower values of A.

5.3.3.3 Coulomb energy B_c

Mutual repulsions between protons give a negative contribution to the binding energy. The electrostatic energy of sphere of radius R having uniformly distributed Z protons is given by

$$B_c = -\frac{3}{5}\frac{1}{4\pi\varepsilon_0}\frac{Z(Z-1)e^2}{R} = -\frac{3}{5}\frac{1}{4\pi\varepsilon_0}\frac{Z(Z-1)e^2}{r_0\ A^{1/3}} = -a_c Z^2 A^{-1/3} \tag{5.20a}$$

where the coefficient $a_c = \dfrac{3}{5}\dfrac{1}{4\pi\varepsilon_0}\dfrac{e^2}{r_0}$ (5.20b)

The constant a_c can be calculated by substituting the values of the constants e, r_0, $4\pi\varepsilon_0$, and so on. The calculations give $a_c = 0.691$ MeV.

The effect of Coulomb energy is more in heavy nuclei that contain larger number of protons. It is the Coulomb energy that ultimately decides the stability of heavy nuclei.

5.3.3.4 Symmetry energy B_{sym}

In simple quantum mechanical approach, a nucleus is represented by a rectangular potential well of depth $-V_0$ in which nucleons are filled up to some highest (kinetic) energy state E. The difference between E and the top of the potential well gives the binding energy of the system. It is, therefore, obvious that the binding energy of the nucleus will be less if E (the highest energy) is large. In the case of the nucleus, the energy E is least or minimum when it contains equal numbers of neutrons and protons. The neutrons and the protons in a nucleus, which is a quantum entity, fill into two separate but almost identical systems of levels, each level accommodating a maximum two nucleons of a given type. The levels up to a certain energy E are filled when the number of neutrons and protons is equal as shown in figure 5.12. However, if the number of protons (or of neutrons) is more than the extra protons will go to levels of higher kinetic energy. As such, the binding energy, which is the difference of the potential energy and the kinetic energy, will decrease. Similarly, if the number of neutrons is more, the additional neutrons will occupy levels of higher kinetic energy, lowering the binding energy.

Raoult's law in chemistry states the same fact. According to Raoult's law, the energy of a two-component system with non-polar molecules is at minimum when the concentration of the two components is same. The vapour pressure of a system of two liquids is also at minimum and hence the system is most stable when the two liquids are in equal concentration. In case of the nucleus, the light stable nuclei have equal number of neutrons and protons. Had there been no repulsion between protons, the stable heavy nuclei would have preferred equal number of neutrons and protons. The symmetry energy B_{sym} may be written as

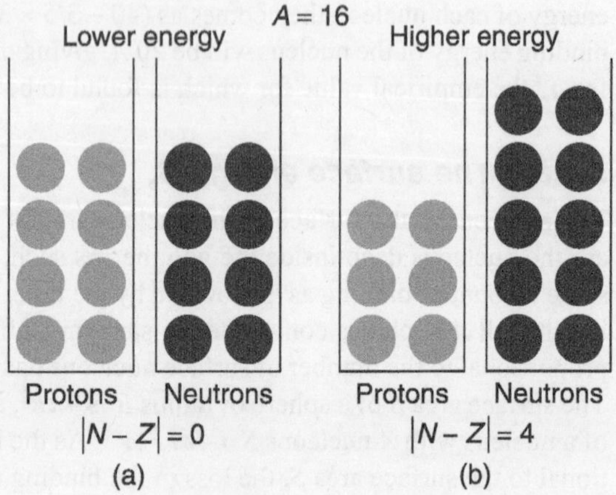

Figure 5.12 *Filling of energy levels by neutrons and protons in a nucleus (a) when their number is equal (b) when the number of one is more than the other*

$$B_{sym} = -a_{sym}\,(N - Z)^2/A = -a_{sym}\,\frac{(A-2Z)^2}{A} = -a_{sym}\,\frac{(N-Z)^2}{A} \tag{5.21}$$

The negative sign indicates that the symmetry energy diminishes binding and the square of $(N - Z)$ makes it symmetrical to the difference between the number of neutrons and protons. Symmetry energy plays a big role in supernova explosions where the neutron proton ratio may diverge from unity.

5.3.3.5 Pairing energy δ

To include the observed additional stability associated with the pairing of nucleons, a term δ is added to the binding energy expression. Conventionally, $\delta = 0$ for odd-A nuclei, positive for even–even nuclei and negative for odd–odd nuclei. Fermi empirically suggested the following form for δ

$$\delta = a_p\,\frac{1}{A^{3/4}} \tag{5.22}$$

With the inclusion of the above-mentioned terms, the final expression for the binding energy becomes

$$B(Z, N) = a_vA - a_sA^{2/3} - a_cZ^2A^{-1/3} - a_{sym}\,\frac{(A-2Z)^2}{A} \mp \delta \tag{5.23}$$

Here, the pairing energy δ is defined by Eq. (5.22). By substituting the above-mentioned value of the binding energy in Eq. (5.17), masses of nuclei may be calculated. Eq. (5.23), in general, gives nuclear masses within 5%. Further, correction terms may be added to the equation to include strong binding at magic numbers, and for other effects. Binding energy equations up to 250 parameters, reproducing masses within 0.1% have been used.

5.3.3.6 Values of coefficients

The value of the coefficients a_v, a_s, a_c, and so on can be determined by fitting the binding energy equation/mass equation to the known masses. As different authors use different additional correction terms in the binding energy expression, it becomes difficult to compare the values obtained by different authors. Segree who has made a detailed analysis used additional correction terms for the skin of the nucleus and for the shell effects. The values of the coefficients obtained by least square fitting of about 300 masses are a_v = 15.8 MeV, a_s = 18.3 MeV, a_c = 0.714 MeV, a_{sym} = 23.2 MeV, and a_p = 12 MeV.

Another set of the coefficients obtained by Wapstra and often used for calculations gives a_v = 14.1 MeV, a_s = 13.0 MeV, a_c = 0.596 MeV, a_{sym} = 19.0 MeV, δeven–even = –33.5 MeV, δodd–odd = +33.5, δodd-A = 0.

5.3.3.7 Isobaric parabolas

From Eq. (5.17) and Eq. (5.23), the mass of the nucleus $M(Z, N) = M(Z, A)$ may be written as

$$M(Z, A) = Z(M_p - M_n) + A(M_n - a_v) + a_s A^{2/3} + a_c Z^2/A^{1/3} + a_{sym} A + a_{sym} 4 Z^2/A + a_{sym} 4Z \mp \delta \quad (5.24)$$

Eq. (5.24) may be written as

$$M(Z, A) = \alpha A + \beta Z + \gamma Z^2 \mp \delta \quad (5.25)$$

where $\alpha = M_n - (a_v - a_{sym} - a_s A^{-1/3})$, $\beta = 4a_{sym} - (M_n - M_p)$ and $\gamma = (4a_{sym} A^{-1} + a_c A^{-1/3})$

For a constant A, Eq. (5.25) is the equation of the parabola. The coefficients α, β, and γ have dimensions of energy or mass. If A is odd, the paring energy δ is zero and therefore Eq. (5.25) represents only one parabola. In the case of even-A, Eq. (5.25) gives two parabolas one for even–even nuclei and the other for odd–odd nuclei.

The charge of the most stable isobar for a given parabola may be obtained by differentiating Eq. (5.25) with respect to Z and considering it equal to zero. In case of odd-A, generally there is only one stable isobar and all other heavier isobars decay into the most stable isobar either by β^+, β^-, or electron capture. In case of even-A, the decay of the heavier isobar leads to an isobar on the

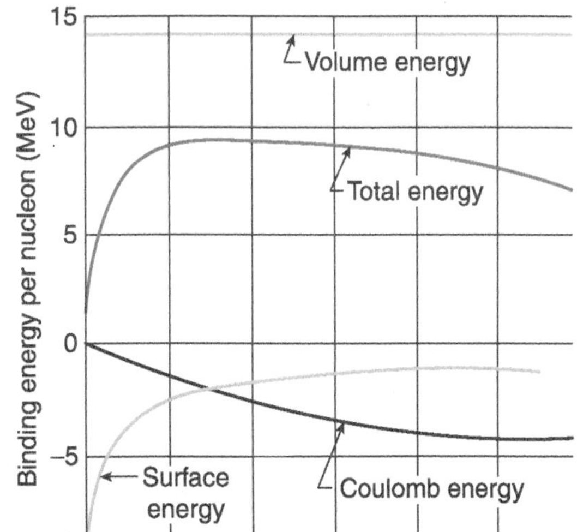

Figure 5.13 *Graph showing relative contributions of various terms of the semi-empirical binding energy formula*

other parabola. In general, in case of even–even nuclei there are at least two and sometimes even three most stable isobars. This is shown in figure 5.3.

The relative contribution of the various terms to the binding energy is shown in figure 5.13. As may be seen from the graph, at low A symmetry and Coulomb energies are not very important and essentially the surface term reduces the binding energy. Decrease in surface energy and increase in Coulomb energy results in the maximum value of B/A at around $A \approx 60$. With further increase in A, the surface term decreases but both the Coulomb and symmetry energies increase to reduce the binding energy.

5.3.4 Liquid Drop Model and Nuclear Fission

The liquid drop model could not explain the excited states of nuclei and the close packing of excited levels at higher energies. Assuming the nucleus to be a drop of an incompressible liquid, the excitation energy can be visualized as the surface vibrations that periodically deform the drop. The energy of these periodic deformations depends on the surface energy that in turn depends on the coefficient a_s of the surface energy term. Calculations have shown that the deformation energy for the least energetic mode of surface vibration is much larger than the excitation energy of low-lying excited levels of nuclei. However, the deformation of the liquid drop leads directly to the possibility of breaking of the nucleus in two halves, that is, to fission.

5.3.4.1 *Energy released in symmetric fission*

The energy released when a heavy nucleus $M(Z, A)$ divides into two equal halves $2M\,(Z/2, A/2)$ is equal to the difference in their binding energies. This difference in binding energy ΔE can be calculated from the semi-empirical binding energy formula (5.23). If the pairing term δ is dropped, the energy difference ΔE comes out to be

$$\Delta E = M(Z, A) - 2M\,(Z/2, A/2) = a_s A^{2/3}\,(1 - 2^{1/3}) + a_c\,(z^2/A^{1/3})\,(1 - 1/2^{2/3})$$

Or

$$\Delta E = -0.260\,a_s A^{2/3} + 0.370\,a_c \frac{Z^2}{A^{1/3}} \tag{5.26a}$$

$$= -\Delta E_{\text{surface}} + \Delta E_{\text{Coulomb}}$$

Equation (5.26a) shows that the change in the surface energy and the change in the Coulomb energy are of opposite signs, which means that while the surface energy tries to hold the two parts together, the Coulomb energy opposes it and tries to break the nucleus (Z, A) into two equal parts $(Z/2, A/2)$. For light nuclides, the surface energy term dominates over the Coulomb energy, the energy released is negative and the spontaneous breaking of the nucleus into two equal parts is energetically not possible. As the size and the charge of the nucleus increases Coulomb energy increases rapidly and the limit of stability (from only the energy considerations) reaches when $|\Delta E_{\text{surface}}|$ becomes equal to $|\Delta E_{\text{Coulomb}}|$ or $\Delta E = 0$. It gives

$$\frac{Z^2}{A} = \frac{0.260\,a_s}{0.370\,a_c} = \frac{0.26 \times 13}{0.37 \times 0.596} = 15.32 \tag{5.26b}$$

According to Eq. (5.26b), the energy released in the symmetric fission will be zero for $Z^2/A = 15$ or 16 and the energy released will be positive if Z^2/A for the nucleus is larger than 16.

For nuclides heavier than ${}^{83}_{36}Kr$, Z^2/A is larger than 16 and, therefore, energetically they may undergo self-symmetric fission. This is, however, against the observations. No nucleus almost up to Bismuth shows self-fission.

5.3.4.2 Fission barrier

Heavier nuclei such as uranium show very slow self-fission. The reason why nuclides for which self-fission is energetically possible $\left(\frac{Z^2}{A} > 16\right)$ do not undergo self-fission lies in the fission barrier. The two fission fragments of equal charge and mass formed as a result of symmetric self-fission face a Coulomb barrier as they depart. The concept of fission barrier can be better visualized if we look at the reverse process that is the fusion of two nuclei of charge $Z/2$ and mass number $A/2$. Let us assume that a nucleus $(Z/2, A/2)$ is projected towards another stationary target nucleus $(Z/2, A/2)$. As shown in figure 5.14, the potential energy of the incident nucleus increases as it approaches the target nucleus. Assuming that the target and the incident nuclei are both spheres of radius $R = r_0(A/2)^{1/3}$ and remain so all through, the Coulomb potential energy of the incident nucleus $U(r) = \frac{\left(\frac{Ze}{2}\right)^2}{4\pi\epsilon_0 r}$ increases with the decrease in the relative separation r as shown in figure 5.14. As the incident nucleus reaches within the range of the nuclear force (say within 2F) of the target nucleus, the potential energy $U(r)$ starts diminishing due to the attractive nuclear force. When the incident nucleus just touches the target nucleus that is $r = 2R$, the Coulomb potential energy in the absence of the nuclear force is given by $U(2R) = \frac{\left(\frac{Ze}{2}\right)^2}{4\pi\epsilon_0 2R}$ but the actual potential energy will be less than this value because of the increasing and strong attractive nuclear force. On further reducing the separation r, the two nuclei come under strong nuclear force and amalgamate to make the nucleus (Z, A). The energy that the quasi-nucleus $(Z/2, A/2)$ has within the nucleus (Z, A) is ΔE given by Eq. (5.26a).

In the case of the symmetric fission of the nucleus (Z, A), the two fission fragments will again face the potential barrier shown in figure 5.14 when they depart. Classically, if the decay energy ΔE is equal or larger than the maximum value of the potential barrier, self-fission will take place otherwise not. However, quantum mechanically self-fission may take place even if ΔE is smaller than the maximum barrier by the process of tunnelling. Substituting the value of $\left(\frac{e^2}{r_0}\right)$ in terms of a_c from Eq. (5.20b) in $U(2R)$ becomes

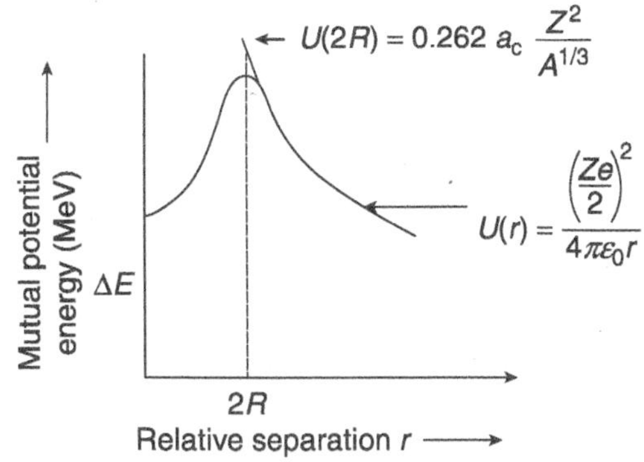

Figure 5.14 *Potential barrier between the two identical nuclei (Z/2, A/2)*

$$U(2R) = 0.262\, a_c \frac{Z^2}{A^{1/3}} \tag{5.26c}$$

The upper limit for self-fission may be obtained with the condition $\Delta E \geq U(2R)$. On substituting the values of $U(2R)$ from Eq. (5.26c) and ΔE from Eq. (5.26a), we get

$$\frac{Z^2}{A} = 2.4\frac{a_s}{a_c} = 2.4 \times \frac{13}{0.596} = 52.35 \text{ or } 53 \tag{5.27}$$

Here, Wapstra's values of the coefficients have been used. Eq. (5.27) gives the condition for the limit of stability against symmetric self-fission in which a nucleus spontaneously divides into two equal parts. The variation of the decay energy ΔE and the fission barrier for light, medium, heavy, and very heavy nuclei are shown in figure 5.15. As may be observed from these figures, self-fission is possible only in sufficiently heavy nuclei for which the decay energy is comparable to the barrier height.

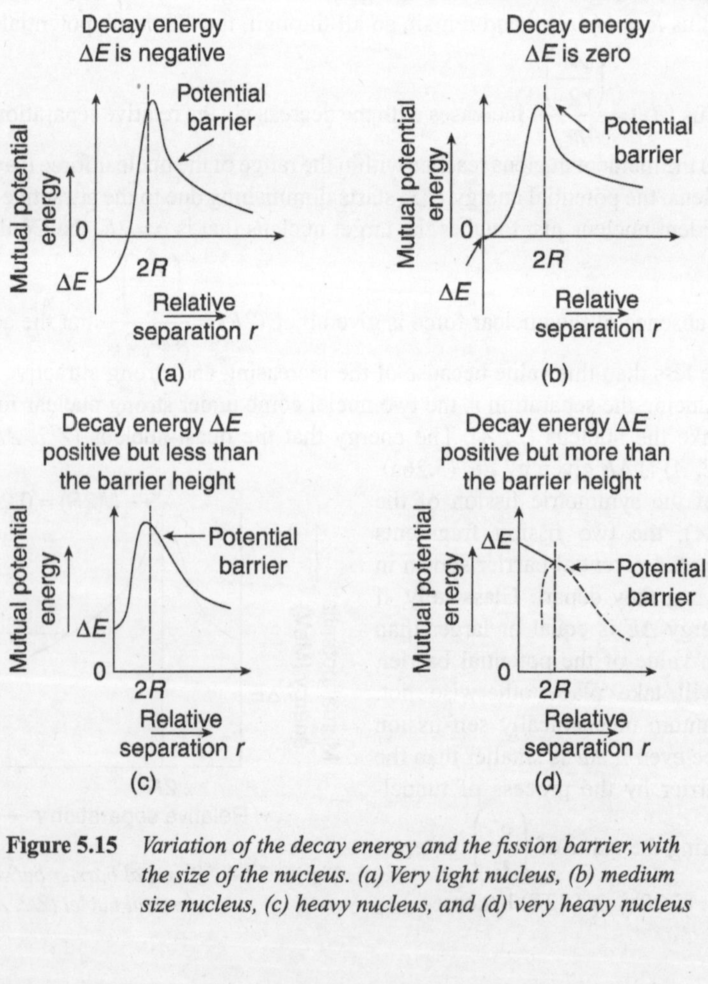

Figure 5.15 *Variation of the decay energy and the fission barrier, with the size of the nucleus. (a) Very light nucleus, (b) medium size nucleus, (c) heavy nucleus, and (d) very heavy nucleus*

5.3.4.3 Shape oscillations

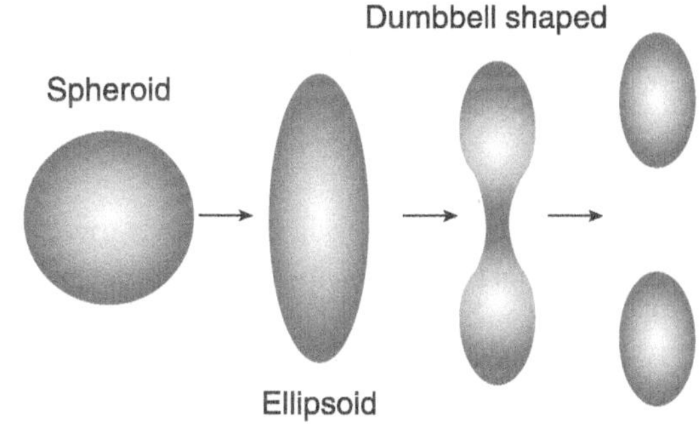

Figure 5.16 *Shape oscillations of the liquid drop*

Bohr and Wheeler considered fission from the perspective of the oscillations in the shape of the nuclear liquid drop. They argued that if the nuclear liquid drop is deformed, for instance, from a spherical shape to an ellipsoidal shape, the Coulomb energy of the drop will reduce as the protons in the ellipsoidal shape will be farther apart while the surface energy will be increased on account of the increase in the surface area. Thus, there is a constant struggle between the disruptive Coulomb force and the force due to the surface tension that tries to bring back the distorted liquid drop to the spherical shape of least surface area. For small oscillations, the surface tension force will over ride the Coulomb force; however, for a critically larger deformation, the Coulomb force becomes just equal to the restoring surface force. For any deformation beyond that critical value, the single nuclear drop will break into two equal halves. Shape oscillations are shown in figure 5.16.

They found that the disruptive Coulomb force wins over the surface force when

$$\frac{Z^2}{A} \geq 2\frac{a_s}{a_c} = 44 \tag{5.28a}$$

Z^2/A is called the fissionability parameter. Bohr and Wheeler's estimate of the fissionability parameter are considered better as they are based on the shape deformations of the liquid drop.

5.3.4.4 Induced fission

In principle, fission in any nucleus irrespective of its size can be initiated if the energy difference between the maximum fission barrier and the decay energy (called the activation energy) is supplied to the nucleus by some external source. This may be done in several ways. For example, gamma rays of about 8 MeV initiate the fission of ${}^{238}_{92}U$ nucleus. This is called photofission and is represented by (γ, f). The photofission of uranium-238 also shows that the difference between the maximum fission barrier and the decay energy in case of ^{238}U is only about 8 MeV or less. Thermal neutrons of negligible kinetic energy ($\approx$0.025 eV) when absorbed by the nucleus ^{235}U produce fission. A nuclear particle when absorbed by a nucleus increases the energy of the nucleus by the amount of its separation energy from the composite nucleus. The separation energy of a neutron from the composite nucleus ^{236}U is about 7 MeV. As such, on absorption of a neutron the internal energy of the nucleus will be increased by about 7 MeV. The pairing energy δ also plays an important role in the case of fission by thermal neutrons. With the absorption of a neutron, the odd-A nucleus ^{235}U becomes an even–even nucleus ^{236}U. ^{236}U is more tightly bound by an amount of energy 2δ as compared with ^{235}U. As δ in this mass region is of the order of

0.5 MeV, another 1 MeV of excitation energy is provided by the pairing effect. Thus, a total of about 8 MeV of additional excitation energy becomes available to the nucleus ^{236}U by the absorption of a thermal neutron. This is enough to overcome the fission barrier and initiate fission. Fission initiated by thermal neutrons has acquired great importance as a source of nuclear energy as well as of nuclear weapons.

Assuming that the maximum fission barrier is given by the Coulomb potential $U(2R)$ when the two identical nuclei ($Z/2$, $A/2$) are just touching each other, we may write

$$U(2\text{R}) = \frac{\left(\frac{Ze}{2}\right)^2}{4\pi\epsilon_0\, 2R} = \frac{(Ze)^2}{8\,(4\pi\epsilon_0)r_0\left(\frac{A}{2}\right)^{1/3}} = 0.15\,\frac{Z^2}{A^{1/3}} \qquad (5.28\text{b})$$

And the energy released in symmetrical fission ΔE from Eq. (5.26a)

$$\Delta E = -0.260\, a_s A^{2/3} + 0.370\, a_c \frac{Z^2}{A^{1/3}}$$

When values of a_s (=13.0) and a_c (=0.596) are substituted in the previous equation we get

$$\Delta E = -3.38 A^{2/3} + 0.22 \frac{Z^2}{A^{1/3}} \qquad (5.28\text{c})$$

The difference $[U(2R) - \Delta E]$ is equal to the activation energy E_a. Figure 5.17 shows the variation of $U(2R)$ and ΔE with the mass number A. As may be seen from this figure, the activation energy becomes zero for $A \approx 250$ and the spontaneous fission takes over.

Nuclei that may undergo fission by thermal neutrons (which are available in the atmosphere in large number) are called fissile materials. ^{233}U, ^{235}U, and ^{239}Pu are important fissile materials. Other nuclides such as ^{232}Th and ^{238}U, which may be converted by some nuclear reactions into a fissile material, are called fertile materials.

In the discussion we have made so far, we considered only the symmetric fission. However, the symmetric fission is rather rear. Fission initiated by thermal neutrons actually is asymmetric, that is the fission fragments are not of equal size. It is not only that the fission fragments are not always the same but there is also a distribution of the mass of the fission fragments. However, as the energy of the incident neutron is increased, the fission starts becoming more and more symmetric. Fission produced by neutrons of about 20 MeV in ^{238}U is symmetric to a large extent.

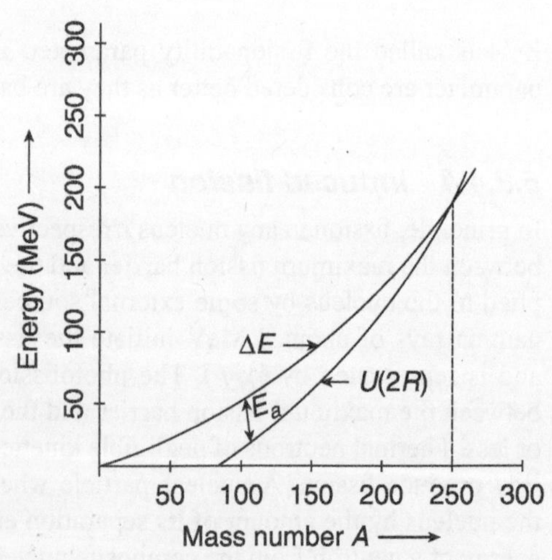

Figure 5.17 *Variation of decay energy ΔE, fission barrier height U(2R) and the activation energy Ea with the mass number A*

The liquid drop model of the nucleus has been quite successful in driving the binding energies of almost all nuclei, of course with some additional corrections. The liquid drop model is also quite successful in explaining nuclear fission.

Exercise p-5.6: Calculate the binding energy of ${}^{12}_{6}C, {}^{40}_{20}Ca,$ and ${}^{238}_{92}U$ nuclei using semi-empirical formula with Wapstra's values of coefficients and compare them with the experimental values. Explain the discrepancies between the experimental and the calculated values.

Exercise p-5.7: Discuss the origin of the surface and the asymmetry terms of the binding energy formula. Neglecting the pairing energy term of the mass formula, calculate the charge of the most stable isobar with mass number $A = 125$.

[**Hint:** We know from Eq. (5.25) that the mass of the nucleus (Z, A) may be written as

$$M(Z, A) = \alpha A + \beta Z + \gamma Z^2 \mp \delta$$

where $\beta = 4a_{sym} - (M_n - M_p)$ and $\gamma = (4a_{sym}A^{-1} + a_cA^{-1/3})$].

Neglecting the pairing energy term δ and differentiating $M (Z, A)$ with respect to Z and substituting $\frac{dM}{dZ} = 0$ to get the value of the most stable isobar, we get

$$\beta + 2Z\gamma = 0 \text{ or } Z \text{ (for most stable isobar)} = -\frac{\beta}{2\gamma}$$

$$= -\frac{4a_{sym} - (M_n - M_p)}{2(4a_{sym}A^{-1} + a_cA^{-1/3})}$$

The negative sign in the above-mentioned equation simply means that the maximum stability is associated with this value of Z. Neglecting the neutron–proton mass difference, the relation reduces to

$$Z = \frac{A}{2\left(1 + \frac{a_c}{4a_{sym}}A^{2/3}\right)} = \frac{A}{2 + \frac{0.596}{2\times 19.0}A^{2/3}} = \frac{A}{2 + 0.01568A^{2/3}} \quad \text{(p-5.7a)}$$

$$\text{For } A = 125, \text{ we get } Z = \frac{125}{2 + 0.01568\,(125)^{2/3}} = 52.26 = 53.$$

Exercise p-5.8: In what respect is a nucleus similar to and different from a liquid drop? Show that the ratio $\frac{Z}{A}$ for the most stable isobar for a heavy nucleus is around 0.4 and for light nuclei around 0.5.

[**Hint:** It follows from Eq. (p-5.7a) derived in Exercise p-5.7.]

Exercise p-5.9: Discuss how the liquid drop model can account for the isobaric mass parabolas. The binding energies of the mirror nuclei ${}^{41}_{21}Sc$ and ${}^{41}_{20}Ca$ are, respectively, 343.14 and 350.42 MeV. Calculate the radius of the nucleus with $A = 41$.

[**Hint:** The binding energies of the two isobars are different because of the two facts. (i) The Coulomb or electrostatic energy of ${}^{41}_{21}Sc$ is more than ${}^{41}_{20}Ca$ as it has one more proton but its mass energy is less than the mass energy of ${}^{41}_{20}$Ca as proton has a lesser mass than a neutron. Neutron–proton mass difference $\Delta m = 1.29$ MeV. As such, the difference in the binding energies of $B\,({}^{41}_{21}\text{Sc})$ and $B\,({}^{41}_{20}\text{Ca})$ may be given as

$$B\,({}^{41}_{21}\text{Sc}) - B\,({}^{41}_{20}\text{Ca}) = \text{Difference in the electrostatic energies} - 1.29\text{ MeV}$$

Or $(350.420 - 343.143) + 1.29 =$ Difference in the electrostatic energies

$$= \frac{3}{5}\left[\frac{(21 \times 20 - 20 \times 19)e^2}{4\pi\varepsilon_0\, r}\right]$$

$$\text{Or } 7.227 + 1.29 = 8.517 = \frac{24}{4\pi\varepsilon_0 r}.$$

$$\text{Hence, } r = \frac{e^2}{4\pi\varepsilon_0}\left(\frac{24}{8.517}\right) = 1.44 \times 2.82 = 4.06\text{F}].$$

Exercise p-5.10: Discuss the role of the various nuclear energy components in fission and explain why nuclei with positive decay energy do not undergo symmetric self-fission. Why only odd-A nuclei such as ^{233}U, ^{235}U, and ^{239}Pu undergo fission by thermal neutrons?

Exercise p-5.11: What conditions must be fulfilled for the spontaneous symmetric fission of a nucleus? The masses of nuclei ${}^{27}_{13}$Al and ${}^{27}_{14}$Si are, respectively, 26.9815 u and 26.9867 u. Using these data, calculate the value of the Coulomb energy coefficient of the semi-empirical mass formula.

$$\textbf{[Hint: } (26.9867 - 26.9815) \times 931.48 + 1.29 = a_c\left(\frac{(Z^2 - (Z-1)^2)}{A^{1/3}}\right)]$$

Exercise p-5.12: Discuss the limitations of the liquid drop model.

Exercise p-5.13: Using semi-empirical formula for binding energy, determine the maximum value of the charge Z for a fixed value of A and hence calculate the atomic number Z for the most stable isobar.

5.4 COLLECTIVE MODELS OF THE NUCLEUS

On one hand, the shell model of the nucleus could not satisfactorily explain the low-lying excited states of the nucleus. The liquid drop model on the other hand does not consider an individual nucleon in the nucleus and treat nucleus as uniform incompressible fluid. The energies associated with the shape oscillations in liquid drop model are much too large than the excitation energies of low-lying nuclear states.

5.4.1 Rotational Bands

There is close resemblance between the energy spectra of a nucleus and that of a molecule. Nucleon excited states in the nuclear case are analogous to the electronic excitations in case of the molecule. The low-lying excited bands that have characteristics of vibrational and rotational spectra have been observed in case of even–even heavy nuclei. These bands are similar to the rotational and vibrational bands observed in case of molecules. The rotational bands are characterized with energy levels given by $E = \frac{J(J+1)\hbar^2}{2\Theta}$, where Θ is the moment of inertia of the nucleus and J the spin of the level while the vibrational bands that generally lie higher in energy are characterized by equidistant levels corresponding to harmonic oscillator energy states $E = n\omega\hbar$.

These observations point towards some kind of collective motion of many nucleons together in the nucleus. The rotational and vibrational bands are produced by the rotation and vibration of the nuclear surface. These nuclear rotations and vibrations are quite different from the rigid body rotations and vibrations. Nuclear rotation essentially is a rotation of the deformed nuclear surface containing A nucleons. Similarly, nuclear vibrations are made up of periodic deformations in nuclear surface. In both cases, orderly displacements of many nucleons are involved and, therefore, they are classified under collective models.

5.4.2 Vibrational States

A vibrating nuclear surface may be described quite generally by an expansion in spherical harmonics with time-dependent shape parameters as coefficients

$$R(\theta,\varphi,t) = R_0\left(1+\sum_{\lambda=0}^{\infty}\sum_{\mu=-\lambda}^{\lambda}\alpha^*_{\lambda\mu}(t)Y_{\lambda\mu}(\theta,\varphi)\right) \tag{5.29}$$

where $R(\theta, \varphi, t)$ is the radius of the nucleus in direction (θ, φ) at time t, and R_0 is the radius of the spherical nucleus that is realized when all $\alpha_{\lambda\mu}$ are zero. The time-dependent amplitude $\alpha\lambda_\mu$ (t) describes the vibrations of the nucleus with different multipolarities around the ground state and are, therefore, called collective coordinates.

5.4.2.1 Types of multipole deformations

Monopole mode, $\lambda = 0$: As $Y_{0,0} = 1/4\pi$, is constant, the non-zero value of α_{00} corresponds to the change of the radius of the nucleus. This leads to the so-called breathing mode. In view of the negligible compressibility of the nuclear matter, the energy associated with the breathing mode is quite large and therefore these modes are not relevant for low-lying excited states (figure 5.18).

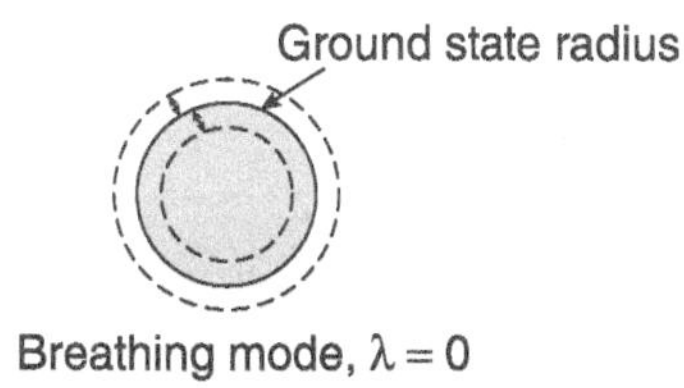

Figure 5.18 *Breathing mode of nuclear vibration*

Dipole mode, $\lambda = 1$, $Y_{1,0} \sim \cos\theta$: The dipole mode does not really correspond to the deformation of the nuclear surface but only refers to the shifting of the centre of mass or to the translation of the nucleus from one point to another in space. As translational motion does not change the energy of the nucleus, the dipole mode does not produce excited states.

Quadrupole mode, $\lambda = 2$: It is the most important deformation and produces important low-lying excited vibrational nuclear states.

Octupole mode, $\lambda = 3$: This is the fundamental asymmetric deformation mode and produces low-lying excited states of negative parity.

Hexadecapole mode, $\lambda = 4$: This mode involves excited states of highest angular momentum. Generally pure hexadecapole excitations are not observed. However, admixtures of dipole and hexadecapole excitations have been observed (figure 5.19).

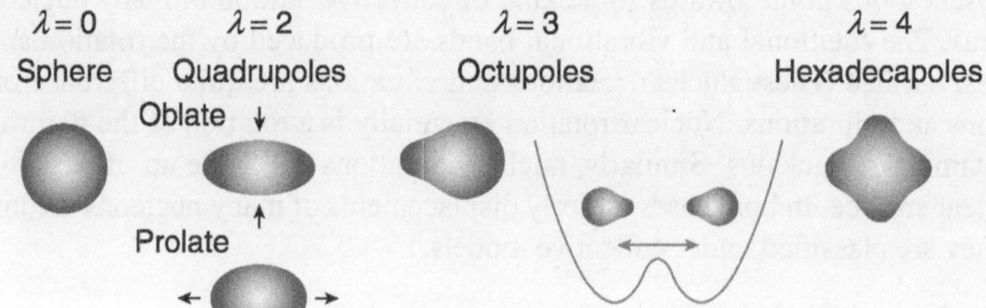

Figure 5.19 *Dipole and other higher modes of nuclear vibrations*

The lowest quantum of collective vibrational excitations is called a phonon. It has a quadrupole type of deformation. It carries two units of angular momentum and a positive or even parity. In even–even nuclei with small deformation, the lowest excited state is generally the one phonon vibrational state of $J = 2^+$. However, in case of nuclei with large deformation, the lowest excited state is rotational state with $J = 2^+$. The pure vibrational spectrum of slightly deformed medium mass nuclei has degenerate states for two or higher phonon excitations. However, in the experimental spectrum, the degeneracy of the states is removed, most likely by the admixture of higher order vibrations (figure 5.20).

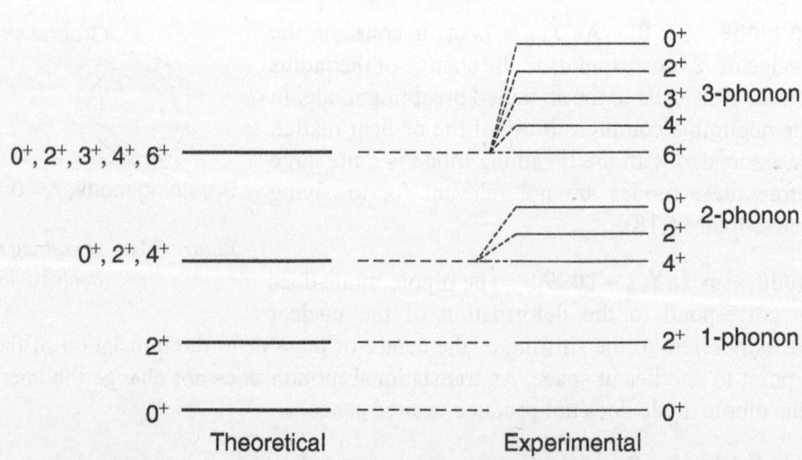

Figure 5.20 *Vibrational level diagram for a medium mass even-even nucleus*

5.4.3 Rotating Nuclei

The Hamiltonian for a rigid rotor is given as

$$H = \sum_{i=1}^{3} \frac{J'^2_i}{2\Theta} \tag{5.30}$$

where Θ is the moment of inertia and J' the angular momentum with respect to the body fixed axis while J refers to the rotation about a space fixed or stationary axis. If the nucleus is symmetric with respect to the Z-axis, J'_z is zero. Hence, Eq. (5.30) may be written as

$$H = \frac{J'^2_x + J'^2_y}{2\Theta_{xy}} \tag{5.31}$$

The quantum number for the rotor will be generated from space fixed J^2 and J_z. As the energy of the nucleus does not depend on its orientation in space $j^2 = J'^2$. This leads to

$$H = \frac{J^2}{2\Theta} \tag{5.32}$$

Treating H and J as operators and writing their quantized values, we get

$$E = \frac{J(J+1)\hbar^2}{2\Theta} \quad J = 0, 1, 2, \ldots \tag{5.33}$$

However, in case of the even–even nuclei, the symmetry of the system restricts the J values to 0, 2, 4 ….

The experimentally observed rotational spectra for even–even nuclei have given the nuclear moment of inertia Θ much smaller than the rigid body value. This may be due to the dragging of the nucleons with the surface in the rotational motion. Further, the ratio $\Theta/\Theta_{\text{rigid}}$ is found to increase with the deformation of the nucleus. Although the expression (5.33) generally reproduces the measured energies of rotational excited states, there are often systematic differences in the measured energies of the rotational states and the ones predicted by (5.33). This difference has been attributed to the increase in the moment of inertia with the increase in the angular momentum J. When correction for this is included the expression (5.33) gets modified to

$$E = \frac{J(J+1)\hbar^2}{2\Theta} - CJ^2\,(J+1)^2 \tag{5.34}$$

With two adjustable parameters Θ and C, it is now possible to fit the rotational level data for even–even nuclei to a much better extent (figure 5.21).

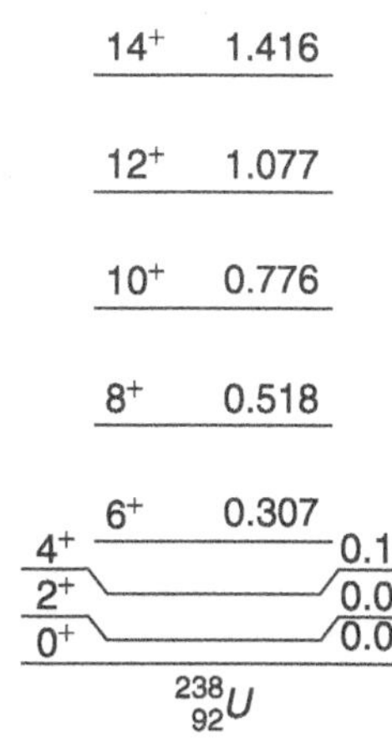

Figure 5.21 *Schematic rotational levels of* $^{238}_{92}U$

Exercise p-5.14: Discuss the origin of rotational and vibrational spectra of nuclei. Explain why the monopole and dipole vibrational modes are not observed. Outline the important characteristics of rotational levels of an even–even nucleus.

5.5 FERMI GAS MODEL OF THE NUCLEUS

Fermi-gas may be defined as an assembly of quasi-free particles that obey Fermi–Dirac statistics. A particle with binding energy much smaller than its kinetic energy is a quasi-free particle. Free electrons in metals fulfil both these conditions, they are fermions and their binding energy with the positive ion lattice is much smaller than their kinetic energies. As such, free electrons in metals are often treated as a Fermi gas. Extending this analogy, Fermi gas model of the nucleus has been developed. As is known, the average binding energy of nucleons ($\approx 7-8\,\text{MeV}$) in a nucleus is much smaller than their average kinetic energy of about 33 MeV. The Fermi gas model of the nucleus assumes protons and neutrons, which are fermions as two components of a Fermi gas. It is further assumed that these fermions move in an average potential that is generated from the mutual interactions between nucleons. Since nucleons are particles of half-integer spin, they follow Fermi–Dirac statistics. As such, no two identical nucleons can occupy the same energy state. In view of the $1/2\hbar$ spin of the nucleons, two nucleons of the same kind one with spin up and the other with spin down may be accommodated in each individual level. For simplicity, it is assumed that protons and neutrons in the nucleus move in two separate square-well potentials of same radius. As has already been stated, the average potential is generated from the mutual interactions of nucleons. From phase space (an imaginary six-dimensional space made up of three spatial components and three momentum components) considerations, the number of energy states dn that a nucleon in a volume V and in a momentum interval of p and $p + dp$ can have is given by

$$dn = \frac{4\pi p^2 dp}{(2\pi\hbar)^3} V \tag{5.35}$$

Assuming that the nucleus is in ground state, all states up to a maximum momentum p_F are filled. The energy level corresponding to the maximum momentum p_F is called the Fermi level and p_F the Fermi momentum. On integration, Eq. (5.35) yields

$$n = \frac{Vp_F^3}{6\pi^2\hbar^3} \tag{5.36}$$

Here, n is the number of energy sates up to the Fermi momentum. Now, each of the above-mentioned energy state can accommodate two nucleons of the same kind, one with spin $(1/2\hbar)$ up and the other with spin pointing down. As such, the number of nucleons of each kind, that is number of neutrons N and number of protons Z in energy states up to p_F may be obtained by multiplying Eq. (5.36) by 2. Therefore,

$$N = \frac{V(p_F^n)^3}{3\pi^2\hbar^3} \text{ and } Z = \frac{V(p_F^p)^3}{3\pi^2\hbar^3} \tag{5.37}$$

Here, p_F^n and p_F^p are, respectively, the Fermi momenta of neutron and proton. The volume V of the nucleus may be given in terms of the unit nuclear radius $r_0 = 1.21$ fm as

$$V = \frac{4}{3}\pi R^3 = \frac{4}{3}\pi r_0^3 A \tag{5.38}$$

Substituting the values for volume V in Eq. (5.37), we get

$$p_F^n = \left(\frac{9\pi}{4}\right)^{\frac{1}{3}}\left(\frac{\hbar}{r_0}\right)\left(\frac{N}{A}\right)^{\frac{1}{3}};\ p_F^p = \left(\frac{9\pi}{4}\right)^{\frac{1}{3}}\left(\frac{\hbar}{r_0}\right)\left(\frac{Z}{A}\right)^{\frac{1}{3}} \tag{5.39}$$

To estimate the magnitude of the Fermi momentum, we put $Z = N = A/2$, which means that the nucleus has equal number of neutrons and protons.

$$p_F^n = p_F^p = p_F = \left(\frac{9\pi}{8}\right)^{\frac{1}{3}}\left(\frac{\hbar}{r_0}\right) = 249.50 \approx 250 \text{ MeV/c} \tag{5.40}$$

Equation (5.40) tells that nucleons inside the nucleus move with large momentum of about 250 MeV/c. It may be noted that the Fermi momentum is almost independent of A. The Fermi energy E_F, the energy of the highest occupied level, may be obtained from Fermi momentum p_F by dividing the square of p_F by $2M$, where M is the mass of the nucleon (= 938 MeV). Hence,

$$E_F = \frac{p_F^2}{2M} = \frac{250\times 250}{2\times 938} \approx 33\,\text{MeV} \tag{5.41}$$

The rectangular (or square) potential well for neutrons and protons in a nucleus is shown in figure 5.22. Fermi level; the highest occupied energy level is also shown in the figure. As the average binding energy of the last nucleon in all nuclei is nearly 7 – 8 MeV, the total depth of the potential well is approximately 33 MeV + 7 MeV ≈ 40 MeV. It may, however, be noted that the depth of the potential well for protons is less than that for the neutrons. It is because protons in a nucleus are less bound as compared with neutrons on account of the Coulomb repulsion between them and hence the proton potential well is shallow in comparison to the neutron well. Further, the Fermi levels for both the neutron and proton wells are co-incident, that is, they have same energy. This is essential for the stability of the nucleus. If for example, the Fermi level for neutrons is at a higher energy than the Fermi level for protons, then neutrons in the nucleus will undergo β^- decay and fall in the Fermi level of protons, which is at lower energy. In the other case, if the Fermi level of protons is at a higher energy, protons will undergo β^+ decay to fall in the Fermi level of neutrons at lower energy. Thus, for the internal stability of the nucleus it is essential that the Fermi levels for the proton well and the neutron well have same energies.

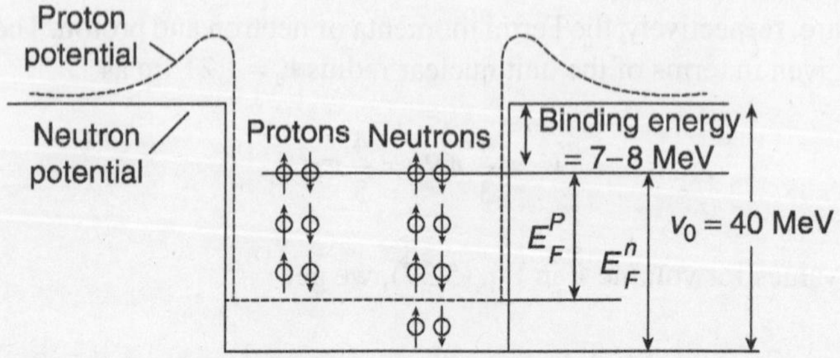

Figure 5.22 *Potential wells for neutrons and protons in a nucleus. It may be noted that the neutron well is deeper as compared with the proton well. The well depth V_0 and the average binding energy per nucleon are independent of A. Fermi level is also shown in the figure*

5.5.1 Average Kinetic Energy per Nucleon

The average kinetic energy of a nucleon $<E_{kin}>$ may be calculated using the standard method of calculating the average value of a parameter, as follows:

$$<E_{kin}>=\frac{\int_0^{E_F} E\frac{dn}{dE}dE}{\int_0^{E_F}\frac{dn}{dE}dE}=\frac{\int_0^{P_F} E\frac{dn}{dp}dp}{\int_0^{P_F}\frac{dn}{dp}dp} \tag{5.42}$$

However, from Eq. (5.35) $dn=\frac{4\pi p^2dp}{(2\pi\hbar)^3}V$ and, therefore,

$$\frac{dn}{dp}=\left(\frac{4\pi V}{(2\pi\hbar)^3}\right)p^2=\text{constant}.\,p^2 \tag{5.43}$$

Substituting Eq. (5.43) in Eq. (5.42), we obtain

$$<E_{kin}>=\frac{\int_0^{p_F} E_{kin}p^2dp}{\int_0^{p_F}p^2dp}=\frac{\int_0^{p_F}\frac{p^2}{2M}p^2dp}{\frac{1}{3}p_F^3}=\frac{3}{10}\frac{p_F^2}{M}\approx 20\,\text{MeV} \tag{5.44}$$

5.5.2 Total Kinetic Energy of a Nucleus

Let us calculate the total kinetic energy $E_{kin}(N, Z)$ of a nucleus with N neutrons and Z protons.

$$E_{kin}(N,Z)=[N<E_{kin(n)}>+Z<E_{kin(p)}>]=\frac{3}{10M}[N((p_F^n)^2)+Z(p_F^p)^2] \tag{5.45}$$

Substituting the values of p^{n}_{F} and p^{p}_{F} from Eq. (5.39) in Eq. (5.45), we may get

$$E_{\text{kin}}(N, Z) = \frac{3}{10M}\frac{\hbar^2}{r_0^2}\left(\frac{9\pi}{4}\right)^{2/3}\frac{N^{5/3}+Z^{5/3}}{A^{2/3}} \tag{5.46}$$

It follows from Eq. (5.46) that the kinetic energy is minimum when $N = Z = A/2$, and in that case the total kinetic energy is

$$\left[E_{\text{kin}}\left(N = Z = \frac{A}{2}\right)\right]_{\text{Minimum}} = \frac{3}{10M}\frac{\hbar^2}{r_0^2}\left(\frac{9\pi}{8}\right)^{2/3} A \tag{5.47}$$

In the case when N is not equal to Z, the total kinetic energy of the nucleus increases and in turn the binding energy of the nucleus decreases. This essentially means that in the absence of Coulomb repulsion when only nuclear force between the neutron and proton exists, the most strongly bound nucleus for a given value of A will be that in which the number of neutrons is equal to the number of protons. If V_0 is the depth of the potential well, then the average binding energy per nucleon will be

$$\frac{B}{A} = V_0 - \langle E_{\text{kin}}(N, Z)\rangle \tag{5.48}$$

In view of this, the total kinetic energy of a nucleus $E_{\text{kin}}(N, Z)$ may be treated as a measure of the binding energy of the nucleus. We shall now drive an expression for the total kinetic energy in terms of $(N - Z)$, the difference between the number of neutrons and protons in the nucleus.

We know that $N + Z = A$. Let $(N - Z) = x$.

Therefore, $N = \frac{(A+x)}{2}$ and $Z = \frac{(A-x)}{2}$

Substituting these values of N and Z in Eq. (5.46), we get

$$E_{\text{kin}}(N,Z) = \frac{3}{10M}\frac{\hbar^2}{r_0^2}\left(\frac{9\pi}{4}\right)^{2/3}\left[\frac{\left(\frac{A+x}{2}\right)^{5/3}+\left(\frac{A-x}{2}\right)^{5/3}}{A^{2/3}}\right]$$

$$= \frac{3}{10M}\frac{\hbar^2}{r_0^2}\left(\frac{9\pi}{4}\right)^{2/3}\left[\left(\frac{A}{2}\right)^{5/3}\left\{\frac{\left[\left(1+\frac{x}{A}\right)\right]^{5/3}+\left(1-\frac{x}{A}\right)^{5/3}}{A^{2/3}}\right\}\right]$$

$$= \frac{3}{10M}\frac{\hbar^2}{r_0^2}\left(\frac{9\pi}{4}\right)^{2/3}\left[\left(\frac{A}{2}\right)^{5/3}\left\{\frac{\left[\left(1+\frac{5}{3}\left(\frac{x}{A}\right)+\frac{5}{9}\left(\frac{x}{A}\right)^2+\ldots\right)+\left(1-\frac{5}{3}\left(\frac{x}{A}\right)+\frac{5}{9}\left(\frac{x}{A}\right)^2-\ldots\right)\right]}{A^{2/3}}\right\}\right]$$

$$= \frac{3}{10M} \frac{\hbar^2}{r_0^2} \left(\frac{9\pi}{8} \right)^{2/3} \left[A + \frac{5}{9} \left\{ \frac{(N-Z)^2}{A} \right\} \right] \quad (5.49)$$

Substituting the value of $E_{kin}(N, Z)$ from Eq. (5.49) in Eq. (5.48), we get for the average binding energy per nucleon

$$B/A = V_0 - [KA] - \left[K' \frac{(N-Z)^2}{A} \right] = K'' A - \left[K' \frac{(N-Z)^2}{A} \right] \quad (5.50)$$

where $K = \frac{3}{10M} \frac{\hbar^2}{r_0^2} \left(\frac{9\pi}{8} \right)^{2/3}$ and $K' = \frac{3}{10M} \frac{\hbar^2}{r_0^2} \left(\frac{9\pi}{8} \right)^{2/3} \left(\frac{5}{9} \right)$ and $K'' = (V_0/A - K)$ are constants. The important fact is that the Fermi gas model of the nucleus predicts that the average binding energy per nucleon is made up of two terms: the first term that depends on the mass number A and the second term that depends the square of the neutron excess divided by A, that is on $(N - Z)^2/A$. If one does not care for the coefficients, these terms may be identified as the volume energy term and the symmetry (or also called asymmetry) energy term of Weizsacker's binding energy formula, derived semi-empirically using the liquid drop model. The symmetry energy grows with the neutron or proton surplus, thereby reducing the binding. Regarding the coefficients K'' and K' it may be remarked that the depth of the potential well also changes with the difference $(N - Z)$ and the correction for this is almost as important as the change in the kinetic energy.

Solved example S-5.3

Calculate the number of energy states of a particle within the momentum interval dp from p to $(p + dp)$, confined in a space V.

Solution. Let us, for simplicity, assume that the volume V in which our system is confined is a cube of length L. We take the origin of the Cartesian coordinate system at one corner of the cube. A set of eigen functions for a particle in a cubical box may be written as

$$\psi = A \sin k_x x \ \sin k_y y \ \sin k_z z \quad \text{(S-5.3-1)}$$

Here, k_x, k_y, and k_z are, respectively, the x, y, and z-components of the linear momentum of the particle. The values of the k_x, k_y, and k_z must satisfy the boundary conditions that at $x = L$, $y = L$, and $z = L$ there are infinite high potentials and the wave function should vanish. This gives

$$k_x L = n_x \pi;\ k_y L = n_y \pi; \text{ and } k_z L = n_z \pi \text{ where } n_x, n_y, n_z = 1, 2, 3 \cdots. \quad \text{(S-5.3-2)}$$

The square of the total linear momentum p^2 is related to the three components as

$$p^2 = (k_z^2 + k_y^2 + k_z^2)\hbar^2 = \left(\frac{\pi^2 \hbar^2}{L^2} \right) [n_x^2 + n_y^2 + n_z^2] \quad \text{(S-5.3-3)}$$

Each set of integral values of n_x, n_y, and n_z gives one state for the particle with an associated value of p^2. To find out how many different sets of n_x, n_y, and n_z give a numerical value of the momentum below a certain value p, we plot the values of n_x, n_y, and n_z along the three axes as shown in the figure S-5.3.

Each unit volume in the first quadrant (in which n_x, n_y, and n_y have only positive values) contains one point, corresponding to one energy state of the particle. The positive numbers n_x, n_y, and n_z are components of a vector n, the maximum length of which is limited by the momentum p. Eq. (S-5.3-3) states

$$p_{max}^2 = \left(\frac{\pi^2\hbar^2}{L^2}\right)n_{max}^2 \quad \text{and} \quad n_{max} = \frac{p_{max}L}{\pi\pi\hbar} \qquad \text{(S-5.3-4)}$$

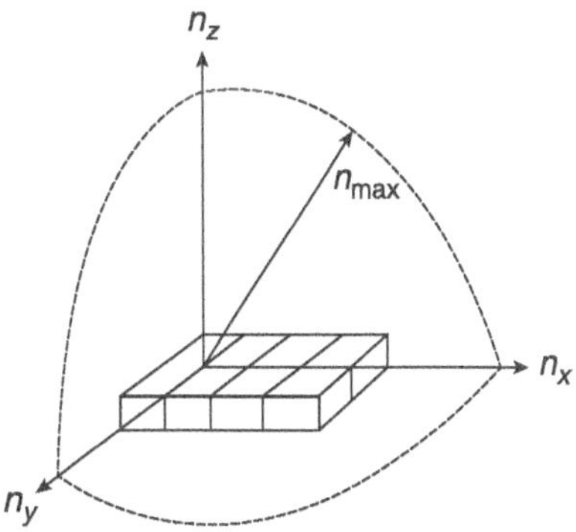

Figure S-5.3

The total number of points, each of which represents one energy state of the particle, may be found by counting the number of unit volume cells in the first quadrant with in the distance n_{max} from the origin. This is equal to 1/8 of the volume of a sphere of radius n_{max}. The number of states with linear momentum less than p_{max} is, therefore,

$$N = \frac{1}{8}\left(\frac{4}{3}\pi n_{max}^3\right) = \frac{1}{6}\left(\pi\left(\frac{L\,p_{max}}{\pi\hbar}\right)^3\right) = \frac{Vp_{max}^3}{6\hbar^3\pi^2}, \text{ here } V = L^3$$

The number of states within the momentum interval dp between p_{max} and $(p_{max} + dp)$ may be found by differentiating the expression for N.

$$dN = \frac{V\,p^2\,dp}{2\pi^2\hbar^3}$$

Thus, the required number of energy states is $dN = \dfrac{V\,p^2\,dp}{2\pi^2\hbar^3}$

Exercise p.5.15: What is a Fermi gas? Show that a nucleus in its ground state may be treated as a Fermi gas.

Exercise p.5.16: Assuming Fermi gas model of the nucleus derive expression for the average kinetic energy of nucleons in a nucleus which has N neutrons and Z protons.

Multiple choice questions

Note: In some multiple choice questions more than one alternative may be correct. All correct alternatives need to be picked or ticked.

Exercise M-5.1: The minimum average kinetic energy in MeV of nucleons in a nucleus with $A = 40$ is approximately

(a) 50.0 (b) 40.0 (c) 30.0 (d) 20.0

Exercise M-5.2: According to the vibrational model of the nucleus the maximum and the minimum excitation energies are, respectively, for the

(a) breathing mode, dipole mode
(b) monopole mode, quadrupole mode
(c) dipole mode, mixed mode
(d) dipole mode, octupole mode

Exercise M-5.3: Strong $\ell - s$ coupling in shell model is essential to explain the
(a) occurrence of islands of isomerism
(b) higher magic numbers
(c) constant density of nuclear matter
(d) excited states of the nucleus

Exercise M-5.4: The independent particle model alone could not explain
(a) Schmidt lines
(b) higher magic numbers
(c) lower magic numbers
(d) islands of isomerism

Exercise M-5.5: If the mass excess $(M - A)$ for $^{235}_{92}$U is 43915×10^{-6} u, the approximate total energy of a single $^{235}_{92}$U nucleus is
(a) 220×10^{-5} MeV (b) 22.0×10^{5} MeV (c) 2.2×10^{5} MeV (d) 0.22×10^{-5} MeV

Exercise M-5.6: The approximate percentage of the total energy of $^{235}{}_{92}$U released in the fission of a single ^{235}U nucleus is
(a) $> 5.0\%$ (b) $> 0.1\%$ (c) $< 0.1\%$ (d) $< 0.05\%$

Exercise M-5.7: According to the shell model, the quadrupole moment of a nucleus with spin j is given by

(a) $-\frac{2j+1}{2(j+1)}\left(\frac{3}{5}R^2\right)$ (b) $-\frac{2(j+1)}{2j+1}\left(\frac{3}{5}R^2\right)$ (c) $-\frac{2j+1}{2(j+1)}\left(\frac{3}{5}R^{-2}\right)$ (d) $-\frac{2j+1}{2(j+1)}\left(\frac{5}{3}R^2\right)$

Exercise M-5.8: Tick all the correct statements.
(a) No nucleus has a spin 11/2.
(b) $1h_{11/2}$ level robs an nucleon from the nearby levels to keep the number of nucleons even.
(c) Pairing energy is more in a state of lower orbital angular momentum.
(d) In light nuclei, surface energy is less than the Coulomb energy.

Exercise M-5.9: According to the shell model, the magnetic moment for odd proton jack-knife case is

(a) $\frac{1}{2}\frac{j}{j+1}(2j+3-g_s^p)$
(b) $(2j-3+g_s^p)$
(c) $\frac{1}{2}(2j+3-g_s^p)$
(d) $\frac{1}{2}\frac{j+1}{j}(2j+3-g_s^p)$

where symbols have their usual meanings.

Exercise M-5.10: According to the shell model, the magnetic moment is −1.91 nm for which of the following case(s):
(a) odd proton jack-knife
(b) for odd proton stretched
(c) odd neutron stretched
(d) odd neutron jack-knife

6

Radioactive Decay

6.1 NUCLEAR STABILITY

Atomic nuclei are made up of neutrons and protons. A neutron is a carrier of mass and strong nuclear force. A proton carries less mass than a neutron and in addition to the strong nuclear force also carries a positive charge and hence electric field. In a neutron number (N) verses proton number (Z) plot of nuclides (figure 6.1), stable nuclides fall in a narrow band called the valley of stability. Nuclei lying at the edges of the valley of stability define the upper and lower limits of the charge-to-mass ratio for stable nuclei. In the case of light nuclei, the stability prefers equal number of neutrons and protons. However, as the number of protons increases more neutrons are required to make the nucleus stable and, therefore, N increases beyond Z. For a fixed Z if the number of neutrons is increased beyond a limit, the system becomes unstable as it contains more mass than the charge Z can sustain. Thus, a neutron-rich nucleus tends to shake-off the extra mass and increase the charge by undergoing β^- decay. In β^- decay, a neutron inside the nucleus gets converted into a proton (thus increases the charge and reducing the mass) and a β^- particle and antineutrino are expelled out. Any further increase in the number of neutrons makes the system so unstable that the additional neutron shoots out spontaneously. The edge of the stability against instantaneous neutron emission defines the neutron drip line. In addition, if the number of protons is increased in a nucleus keeping the number of neutrons fixed, a limit is reached when the given mass of the system could not sustain the positive charge. The system becomes unstable again and tries to shake-off the extra charge and put on more mass to fall back into the valley of stability. This is achieved by β^+ decay.

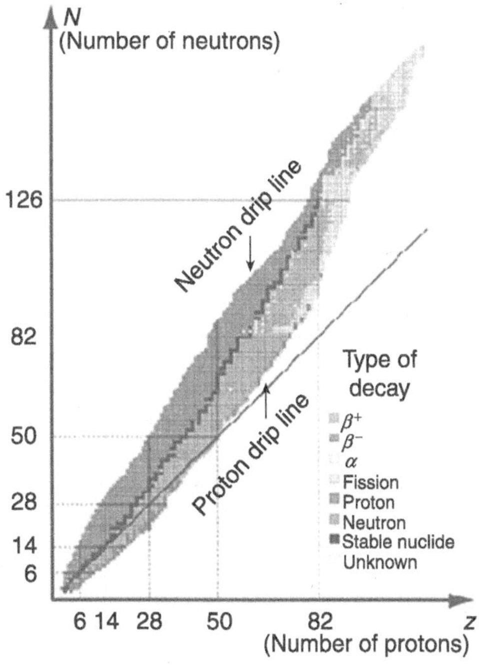

Figure 6.1 *Valley of stability and drip lines*

In proton rich nuclei, a proton inside the nucleus converts into a neutron; thus, increasing the mass and a β^+ particle along with a neutrino are emitted to reduce the charge. Another possible way for proton rich nuclei to reach the valley of stability is by the process called electron capture (EC). In this process (EC), the heavy proton-rich nucleus captures an atomic electron from the inner most electron shell of the atom, once inside the nucleus the captured electron combines with a proton to create a neutron and a neutrino, which is ejected from the nucleus. Any further increase of the number of protons results in spontaneous emission of proton leading to the proton drip line. Thus, on both side of the valley of stability, there are regions of β^- and β^+ instabilities, which end at respective drip lines. Heavy nuclei that have both mass and charge beyond the limits of stability attain stability by emitting α particles. Most of the heavy nuclei ($A > 140$) undergo α decay in which two units of charge and four units of mass are reduced in a single step of decay. In still heavier nuclei, fission takes place and the heavy nucleus divides into two nearly equal parts, few neutrons, and releases a large amount of energy. For any nucleus lying in the valley of stability, a particular configuration of nucleons has minimum energy. This minimum energy configuration is called the ground state of the nucleus. If a nucleus for which the charge-to-mass ratio is within the limits of stability possesses energy in addition to the ground-state energy, the excess energy called the excitation energy is removed from the system by γ emission. Gamma rays are quanta of electromagnetic energy.

We thus observe that to achieve stability, unstable nuclei, depending on their charge-to-mass ratio, may decay into a more stable residue via (a) β^- decay, (b) β^+ decay and or by electron capture, (c) α decay, (d) fission. As a result of these decays, the parent nucleus changes into a new nucleus, generally called daughter nucleus that has different values of N and Z. The changes in N and Z values are summarized in figure 6.2. Nuclides undergoing self-decay to attain a more stable configuration are called radioactive nuclide and the process radioactivity.

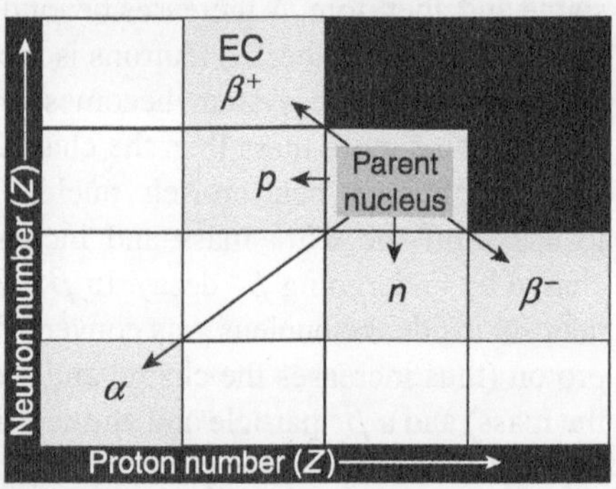

Figure 6.2 *Position of decay product on the N verses Z plot*

All these radioactive decays follow some rules. These rules are based on two experimental observations: (1) The number of nuclei that decay at a given instant t are proportional to the number of nuclei N_t present at instant t. (2) The probability of decay per unit time λ is same for each nucleus, is independent of the age of the nucleus, and remains same throughout the decay process. From the above-mentioned observations, it follows that the number of nuclei N_t at time t is related to the original number of nuclei N_0 (at time $t = 0$) through the relation

$$N_t = N_0 \, e^{-\lambda t} \tag{6.1}$$

Eq. (6.1) expresses the exponential decay of a radioactive sample. The *half-life* denoted by $T_{1/2}$ is the time in which the number of radioactive nuclei reduces to half of their initial value, and the *mean life* τ is the time in which the number of radioactive nuclei falls to $1/e$ of their original value N_0. They are related to the decay constant in the following way

$$T_{1/2} = \frac{0.693}{\lambda}, \ \tau = \frac{1}{\lambda} = 1.44\ T_{1/2} \tag{6.2}$$

The *activity* that is the rate of depletion of a sample at time t is given by

$$\left|\frac{dN_t}{dt}\right| = N_t\,\lambda = N_0\,\lambda\, e^{-\lambda t} \tag{6.3}$$

Some time a given radioactive nucleus may decay independently through two or more different modes with decay constants λ_1, λ_2, λ_3, … λ_n. As decay constant represents decay probability per unit time, the total decay probability per unit time λ is the sum of all partial decay probabilities.

$$\lambda = \lambda_1 + \lambda_2 + \lambda_3 + \cdots \lambda_n \tag{6.4}$$

The partial activity for a particular decay mode characterized by λ_i is then given by

$$\frac{dN_i}{dt} = \lambda_i N_t = \lambda_i N_0\, e^{-\lambda t} \tag{6.5}$$

And the total activity is

$$\frac{dN_t}{dt} = \sum_{i=1}^{n} \frac{dN_i}{dt} = N_t \sum_{i=1}^{n} \lambda_i = \lambda N_0\, e^{-\lambda t} \tag{6.6}$$

It may be noted that the partial activities are proportional to total activity at all times. Each partial activity falls off as $e^{-\lambda t}$ and not as $e^{-\lambda_i t}$.

The strength of a radioactive sample is measured in terms of its activity. A sample in which a single disintegration per second occurs has strength of 1 Bequerel or 1 Bq. In a sample of strength 1 Curie (or 1 Ci), 3×10^{10} disintegrations occur per second.

6.2 RADIOACTIVE DECAY CHAIN

A simple decay chain is a sequence of radioactive nuclei in which each species is populated by the decay of the preceding one.

$$N_1 \xrightarrow{\lambda_1} N_2 \xrightarrow{\lambda_2} N_3 \xrightarrow{\lambda_3} N_4 \cdots N_k \tag{6.7}$$

Eq. (6.7) describes a simple decay chain in which radio nuclides N_1 decays with a decay constant λ_1 into nuclides N_2 that decays with decay constant λ_2 into N_3 and so on until the chain reaches a stable nucleus N_k.

With
$$\frac{dN_i}{dt} = \lambda_{(i-1)} N_{(i-1)} - \lambda_i N_i \tag{6.8}$$

The first term in Eq. (6.8) describes the formation of the nuclide N_i from the decay of the preceding nuclide N_{i-1} and the second term gives the depletion of N_i due to its decay in nuclide N_{i+1}. There will be a group of k equations such as (6.8) one for each nuclide species. This group of differential equations can be solved to get

$$\begin{aligned} N_1 &= C_{11} \exp(-\lambda_1 t) \\ N_2 &= C_{21} \exp(-\lambda_1 t) + C_{22} \exp(-\lambda_2 t) \\ &\cdots \\ N_k &= C_{k1} \exp(-\lambda_1 t) + \cdots + C_{kk} \exp(-\lambda_k t). \end{aligned} \tag{6.9}$$

Here, N_k is the number of nuclide of type k at time t. The coefficients are found to be

$$C_{ij} = C_{i-1,j} \frac{\lambda_{i-1}}{\lambda_i - \lambda_j},\ i \neq j \tag{6.10}$$

And the coefficients C_{ii} are found from the boundary condition at time $t = 0$.

$$N_i(0) = C_{i1} + C_{i2} + C_{i3} + \ldots\ C_{ii} \tag{6.11}$$

Eqs. (6.9)–(6.11) forms the so-called Bateman equations as they were derived for the first time by him.

6.3 TWO-STEP DECAY

Let us consider a two-step decay $N_1 \xrightarrow{\lambda_1} N_2 \xrightarrow{\lambda_2} N_3$ (stable).

Let us further assume that at time $t = 0$ only the nuclei of type N_1 are present with their number $N_1(0)$. The initial number of nuclei of all other species is zero.

Using Eq. (6.9) and substituting $\lambda_3 = 0$ (as N_3 is stable), we get

$$N_1 = C_{11} \exp(-\lambda_1 t)$$

$$N_2 = C_{21} \exp(-\lambda_1 t) + C_{22} \exp(-\lambda_2 t)$$

$$N_3 = C_{31} \exp(-\lambda_1 t) + C_{32} \exp(-\lambda_2 t) + C_{33} \tag{6.12}$$

And using Eqs. (6.10) and (6.11) we may solve for only N_1 and N_2 to get

$$C_{11} = N_1(0);\ C_{21} = N_1(0) \frac{\lambda_1}{\lambda_2 - \lambda_2};\ C_{22} = -C_{21}.$$

Finally, substituting the values of coefficients in the set of Eqs. (6.12), we get

$$N_1 = N_1(0)e^{-\lambda_1 t} \tag{6.13}$$

$$N_2 = N_1(0)\frac{\lambda_1}{\lambda_2 - \lambda_1}\left[e^{-\lambda_1 t} - e^{-\lambda_2 t}\right] \tag{6.14}$$

An interesting situation develops when the daughter (species N_2) is short lived as compared to the parent species N_1. This means that $\lambda_2 > \lambda_1$. Eq. (6.14) may be written as

$$\lambda_2 N_2 = N_1(0)\ \lambda_1\ e^{-\lambda_1 t} \frac{\lambda_2}{\lambda_2 - \lambda_1} [1 - e^{-(\lambda_2 - \lambda_1)t}] \tag{6.15}$$

$$= N_1\ \lambda_1 \frac{\lambda_2}{\lambda_2 - \lambda_1} [1 - e^{-(\lambda_2 - \lambda_1)t}] \tag{6.16}$$

However, $N_1\lambda_1$ and $N_2\lambda_2$ are, respectively, the activities of the parent and the daughter at time t. Also, $(\lambda_2 - \lambda_1)$ is positive, and for time $t \gg \frac{1}{(\lambda_2 - \lambda_1)}$ the exponential term $e^{-(\lambda_2 - \lambda_1)t}$ becomes negligibly small and can be neglected. Therefore, after a long time

$$\text{Activity of the daughter} = \text{Activity of the parent}\left(\frac{\lambda_2}{\lambda_2 - \lambda_1}\right) \quad (6.17)$$

Or
$$\frac{\text{Activity of the daughter at time } t}{\text{Activity of the parent at time } t} = \frac{\lambda_2}{\lambda_2 - \lambda_1} > 1 \quad (6.18)$$

Eq. (6.18) states that after a time $t >> \frac{1}{(\lambda_2 - \lambda_1)}$ the ratio of the activities of the daughter to the parent becomes constant for all subsequent times and the activity of the daughter is larger than the activity of the parent.

Also from Eq. (6.17) the activity of the daughter decays with time at the same rate as the rate of decay of the parent activity. Notice that after the time $t >> \frac{1}{(\lambda_2 - \lambda_1)}$, the decay rate of the daughter does not depend on λ_2 but on λ_1.

When the ratio of the activities of the daughter to parent becomes constant and the daughter decays with the decay rate of the parent, the daughter and parent are said to be in *transient equilibrium*. Figure 6.3 shows the variation of the daughter and parent activities in transient equilibrium. Before attaining transient equilibrium, the activity of the daughter reaches a maximum. The time of the maximum activity of the daughter t_{max} is given by

$$t_{max} = \frac{\log(\lambda_2/\lambda_1)}{\lambda_2 - \lambda_1} \quad (6.19)$$

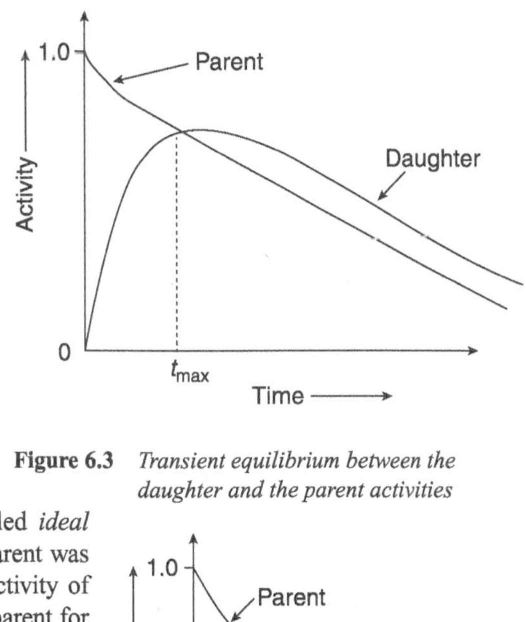

Figure 6.3 *Transient equilibrium between the daughter and the parent activities*

At t_{max}, the activities of the parent and daughter are equal for a moment and that is called *ideal equilibrium*. Before t_{max}, the activity of the parent was higher than the daughter and after t_{max} the activity of the daughter remains higher than that of the parent for all times.

In a special case of the transient equilibrium when the parent is much longer lived than the daughter, that is $\lambda_2 >> \lambda_1$, in the denominator of Eq. (6.18), λ_1 may be neglected in comparison to the much bigger λ_2. The ratio of the daughter to parent activity becomes 1. That is after a time much larger than the mean life of the daughter the activities of the daughter and the parent becomes equal and the daughter decays with the decay rate of the parent. This is called *secular equilibrium* and is represented in figure 6.4.

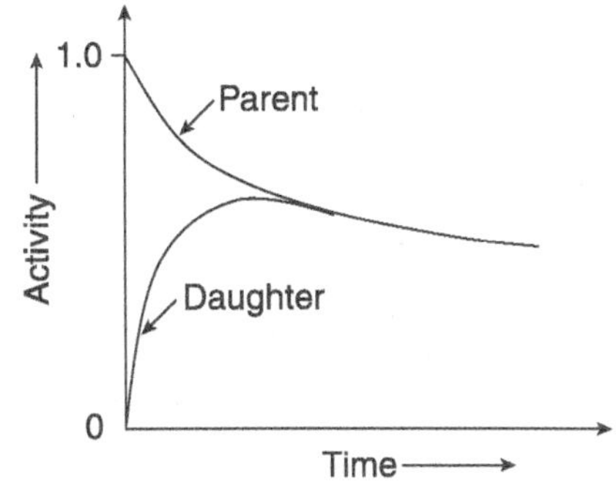

Figure 6.4 *Secular equilibrium between the daughter and the parent activities*

6.4 ACTIVATION ANALYSIS

In experiments with particle accelerators, thin samples of material of interest are irradiated with the accelerated beams of desired incident ions that have desired energy. These irradiations initiate nuclear reactions in the nuclei contained in the target material. The end product or the residual nuclei of some of these reactions are radioactive. The number of the radioactive residues on one hand increases with the time of irradiation as more and more target nuclei undergoes the specific reaction to produce the same residues. On the other hand, the radio nuclei so produced also decay simultaneously with their characteristic decay constant and thus deplete their number. The difference between the production rate through irradiation and the decay rate through radioactive decay decides the yield of the residual nuclei. Situation is comparable to the case of successive radioactive decay. In radioactive series decay, the daughter is fed by the decay of the parent while in case of irradiation the nuclei are produced by the interaction of the incident ion with the nucleus of the target material. The subsequent decay is identical in both the cases. Let us consider some example. Let a nucleus X in the target material on being hit by the accelerated incident ion 'a' get converted into a radioactive residual nucleus Y with the emission of a particle 'b'. This nuclear reaction in short may be written as X(a,b)Y. The production rate R of the nucleus Y depends on (1) the number N of nuclei X in the target, (2) the flux φ of the incident beam of ions 'a', (3) cross-section σ for the reaction X(a,b)Y, and is given by

$$R = N\varphi\sigma \tag{6.20}$$

In Eq. (6.20), R is essentially the number of nuclei Y produced per second as a result of the irradiation. As nuclei Y are radioactive, they will decay with their characteristic decay constant λ exponentially with time. Their decay rate will be equal to the product of their number R multiplied by $e^{-\lambda t}$. As such, the net accumulation rate of the nuclei Y will be given by

$$\begin{aligned}\text{Number of Y accumulated per unit time} &= \text{Production rate} - \text{depletion rate}\\ &= R - Re^{-\lambda t} = N\varphi\sigma(1 - e^{-\lambda t})\end{aligned} \tag{6.21a}$$

Here, t is the time of irradiation. It may be observed that the accumulation rate of Y increases with time t with the decrease of the exponential term. For an irradiation time $t >> 1/\lambda$, the exponential term in the bracket in Eq. (6.21a) becomes negligible and the net accumulation rate of Y becomes constant. This is called the saturation of the accumulation rate. The exponential term $e^{-\lambda t}$ becomes negligible for $t \geq 5\ (1/\lambda)$ or $5\ \tau$, where τ is the mean life of nuclide Y. Therefore, after an irradiation of nearly $5\ \tau$, the accumulation rate of Y becomes constant and does not increase. It is because after the time $5\ \tau$ as many new Y nuclei are produced per unit time by irradiation the same number of nuclei gets depleted through its radioactive decay. As such, there is no further advantage of an irradiation longer than say five or six mean (or half) lives so far as the yield of a radioactive nuclide is concerned.

Suppose the irradiation by the accelerated ion beam is stopped after an irradiation of duration t_1 and the irradiated sample with induced activity is taken to a detector. Let t_2 be the time that lapses between the stop of irradiation and the beginning of the counting of the irradiated sample by a detector, which has absolute detection efficiency ε for the radiations emitted by the irradiated sample. Let the detector counting rate be N_2 at time t_2. It follows from Eq. (6.21a)

$$[N\varphi\sigma(1 - e^{-\lambda t_1})]e^{-\lambda t_2} = N_2\varepsilon$$

Or $$\sigma = \frac{N_2 \varepsilon e^{\lambda t_2}}{N.\varphi.(1-e^{-\lambda t_1})} \tag{6.21b}$$

In an actual experiment, all quantities on the right hand side of the Eq. (6.21b) are known and, therefore, the cross-section for the reaction responsible for induced activity in the sample can be determined. This forms the basis of the activation technique, which is very much used for the measurement of reaction cross-sections.

6.4.1 Carbon-14 Dating

Several examples of the production of radioactive nuclides by irradiation may be found in nature. A very interesting example is the generation of radio nuclide $^{14}_{6}$C in atmosphere. As we know, cosmic rays are constantly bombarding the top of the earth's atmosphere. The interactions of the cosmic ray radiations with the molecules of atmospheric gases produce many new particles including neutrons. Neutrons produced by cosmic rays in the atmosphere interact with $^{14}_{7}$N nuclei producing ^{14}C from the following reaction:

$$^{1}_{0}n + ^{14}_{7}\text{N} \rightarrow ^{14}_{6}\text{C} + ^{1}_{1}\text{H}$$

^{14}C is radioactive and decays by β^- emission back into ^{14}N with a half-life of about 18.1×10^{10} s.

$$^{14}_{6}\text{C} \rightarrow ^{14}_{7}\text{N} + \beta^- + \bar{\nu} + 0.156 \text{ MeV}$$

Thus, the production and decay of ^{14}C nuclides in atmosphere has been going on for a very long time that the production and decay rates of the nuclide have reached equilibrium. As a result, the amount of ^{14}C in atmosphere has become constant. Actual experiments conducted to determine the concentration of ^{14}C nuclides indicated that it is nearly constant at all altitudes. Living organism inhales ^{14}C in the form of $^{14}CO_2$ and thus the ratio of ^{12}C to ^{14}C in living organism has also become constant. However, after the death of the organism, the percentage of ^{14}C in the dead body of the organism decreases as ^{14}C is not replenished by inhaling. Prof. W.F. Libby of the University of Chicago suggested that the measurement of the relative activity of ^{14}C in a dead body may be used to determine the time when the organism went dead, that is the age of the dead body. This method of determining the age of the dead organisms is called 'carbon-14 dating'. Prof. Libby was awarded Nobel Prize in 1960 for this pioneering work.

Solved example S-6.1

Define the mean life of a radioactive material and derive an expression for it.

Solution. In radioactive decay, each nucleus (or atom) of the radioactive sample has the same decay probability per unit time but some nuclei live longer than the other. For example, half of the initial nuclei that have survived after one half-life, have lived longer than those half that have decayed. This is because of the statistical nature of the decay process that some nuclei decay earlier than the other. As each nucleus of the radioactive material lives for a different duration of time, it is reasonable to find the average life time or the mean life time of the radioactive atom. As usual, the mean or average life time τ of a radioactive sample is defined as

$$\tau = \frac{\text{Sum of the life time of all radioactive nuclei}}{\text{Total number of radioactive nuclei}} \tag{A}$$

Suppose we start with N_0 initial nuclei that have a decay constant λ. The number of nuclei N_t present at time t is given by

$$N_t = N_0 e^{-\lambda t} \tag{B}$$

On differentiating Eq. (B) with time, we get dN_t, the number of nuclei that have lived for time 't' and will decay in the next 'dt' time. As 'dt' may be very small approaching zero in the limit, the life time of dN_t nuclei may be taken equal to 't' without any appreciable error. Therefore, the sum of the life time of all atoms may be found by integrating ($t\, dN_t\, dt$) from $t = 0$ to $t = \infty$. The total number of nuclei in the sample are N_0. Hence,

$$\tau = \frac{\text{Sum of the life time of all radioactive nuclei}}{\text{Total number of radioactive nuclei}} = \frac{\int_0^\infty t\, dN_t\, dt}{N_0}$$

Or

$$\tau = \lambda \int_0^\infty t e^{-\lambda t}\, dt = \frac{1}{\lambda} \tag{C}$$

It is easy to see that after a time equal to one mean life, the number of radioactive nuclei drops to $1/e$ of its initial value. As in one half-life, $T_{1/2}$, the number of nuclei drops to 1/2 of the initial value, τ and $T_{1/2}$ are related by the relation

$$T_{1/2} = 0.693\ \tau$$

A different physical interpretation of the mean life may be given in terms of the linear decay of the radioactive sample. In actual radioactive decay, the decay rate changes with time as it is proportional to the number of radioactive nuclides present at that instant. Initially, the decay rate is large and it decreases with time. This gives rise to the exponential decay. However, if for a minute it is assumed that the decay rate is constant, that is it does not change with time then the total amount of the sample will decay out in one mean life as shown in the following figure.

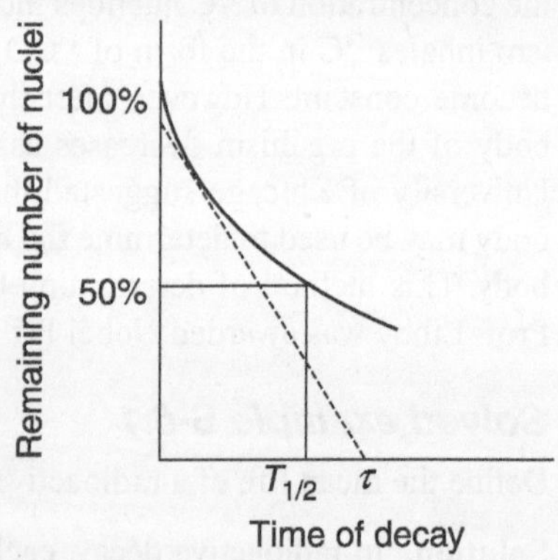

Thus, the mean life of a radioactive material may be defined as the time in which whole of the sample may have decayed out, if the decay rate has remained constant, equal to the initial decay rate.

Solved example S-6.2

Charcoal samples taken from an area surrounding the pyramids in Egypt has an activity of 0.215 s^{-1} due to the ^{14}C atoms per gram of carbon. If the percentage of ^{14}C compared to ^{12}C in living trees is 1.35×10^{-10}%, calculate the age of the area. Given that the decay constant of ^{14}C is 3.92×10^{-10} s^{-1} and its atomic weight is 12.

Solution. Charcoal is produced from the remains of living organism after their death. Long back when the organism was alive the concentration of ^{14}C atoms in it would have been same as it is in the trees and other living organism of today. Therefore, the first thing is to calculate the number of ^{14}C atoms in living organism (trees) of today. It is given that in trees the percentage of ^{14}C is $1.35 \times 10^{-10}\%$ of ^{12}C. Let us first calculate the number of carbon-12 atoms, $N(^{12}C)$, in 1 g of tree material and then the number, $N(^{14}C)$, of ^{14}C atoms in 1 g of tree material. We make use of the Avogadro's number = 6.02×10^{23} per gram mole.

$$N(^{12}C) = [\{1 \times 6.02 \times 10^{23}\}/12] \text{ and } N(^{14}C) = N(^{12}C) \times 1.35 \times 10^{-10} \times 100$$

Hence,

$$N(^{14}C) = [\{1 \times 6.02 \times 10^{23}\}/12] \times 1.35 \times 10^{-10} \times 100 \qquad \text{(A)}$$

Now, the present decay rate of the old charcoal due to ^{14}C atoms = $|N(^{14}C) \times \lambda \times e^{-\lambda t}|$

$$0.215 = ([\{1 \times 6.02 \times 10^{23}\}/12] \times 1.35 \times 10^{-10} \times 100)\,(3.92 \times 10^{-10})\; e^{-3.9\times 10^{-10}t}$$

Or

$$e^{-3.9\times 10^{-10}t} = 0.0081$$

which gives $t = 1.228 \times 10^{10}$ s = 390 years.

Ans: The age of the region is about 390 years or 1.228×10^{10} s.

Solved example S-6.3

Consider the two-step decay chain shown in the following figure.

$$\text{A} \xrightarrow{\beta^-,\ 5 \text{ day half life}} \text{B} \xrightarrow{\beta^-,\ 138 \text{ day half life}} \text{C (Stable)}$$

A freshly purified sample of A weighs 5×10^{-10} g at time $t = 0$. If the sample is left undisturbed, calculate the time at which the number of atoms of B will be a maximum. Also, find this number. It is given that the atomic weight of A is 210.

Solution. Let us denote by λ_x and N_x^t, respectively, the decay constant and the number of atoms of type x at time t; x may be A, B, or C.

Now,

$$N_A^t = N_A^0\, e^{-\lambda_A t} \text{ and } \frac{dN_A^t}{dt} = -\lambda_A N_A^t$$

Also,

$$\frac{dN_B^t}{dt} = \lambda_A N_A^t - \lambda_B N_B^t \qquad \text{(A)}$$

Eq. (A) states that the decay rate of B is equal to its formation rate through the decay of A minus its depletion rate through its own decay.

A solution of (A) is

$$N_B^t = C_1 e^{-\lambda_A t} + C_2 e^{-\lambda_B t} \qquad \text{(B)}$$

Substituting the boundary condition that at $t = 0$, $N_B{}^0 = 0$, we get $C_2 = -C_1$ and therefore,

$$N_B^t = C_1(e^{-\lambda_A t} - e^{-\lambda_B t}) \qquad \text{(C)}$$

Differentiating Eq. (C) and substituting $t = 0$, we may get

$$\frac{dN_B^0}{dt} = \lambda_A N_A^0 = C_1(-\lambda_A + \lambda_B) \quad \text{or} \quad C_1 = \frac{\lambda_A N_A^0}{\lambda_B - \lambda_A}$$

Substituting this value of C_1 in Eq. (C), we get

$$N_B^t = \frac{\lambda_A N_A^0}{\lambda_B - \lambda_A}(e^{-\lambda_A t} - e^{-\lambda_B t}) \tag{D}$$

Differentiating expression (D) for N_B^t with time and equating it to zero give the desired value of the time when the number of B is maximum.

$$t_{max} = \frac{1}{(\lambda_B - \lambda_A)}\left[\ln\frac{\lambda_B}{\lambda_A}\right]$$

Now, let us calculate N^0_A, λ_A and λ_B

$$N_A^0 = \frac{6.02 \times 10^{23} \times 5 \times 10^{-10}}{210} = 1.43 \times 10^{12};\ \lambda_A = \frac{0.693}{5} = 0.138\,\text{d}^{-1},$$

And $$\lambda_B = \frac{0.693}{138} = 0.0052\ \text{d}^{-1}$$

When these values are substituted in Eqs. (D) and (E), we get

$$t_{max} = 24.6\ \text{d and } N_B^{max} = 1.37 \times 10^{10}$$

Ans: The maximum number of B atoms will be 1.37×10^{10} and that will happen after 24.6 days.

Exercise p-6.1: For two-step series decay, derive an expression for the yield of the stable end product. Find the half-life of a radioactive element if its activity decreases by 10% in 1 month.

Exercise p-6.2: There are two stable isotopes, ^{235}U and ^{238}U, in the natural uranium. The ratio of ^{235}U to ^{238}U is 0.7. Assuming that this ratio was 1 when the earth was formed, calculate the age of earth. Given that the mean lives of ^{235}U and ^{238}U are, respectively, 6.52×10^9 and 1.02×10^9 years.

Exercise p-6.3: Calculate the activity of a radioactive sample after one half-life if the initial number of atoms was N. What will be the activity of 4.0 microgram of an radioactive element of half-life 3.64 d.

Exercise p-6.4: Discuss transient and secular equilibriums.
$^{90}_{38}$Sr decays into $^{90}_{39}$Y with a half-life of 28 years, which in turn decays into $^{90}_{40}$Zr with a half-life of 64 hours. Identify the mode of decay of the two nuclides. If the end product is stable and a pure sample of $^{90}_{38}$Sr is allowed to decay, determine the composition of the sample after 1 hour and after 10 years.

6.5 THE α DECAY

Most of the heavy nuclei ($A > 140$) decay by α emission. It sounds strange that a nucleus which is stable against neutron or proton emission may spontaneously decay by emitting an α particle that contains two neutrons and two protons. The answer lies in the separation energy of the last nucleon, which though fluctuates from one nucleus to the other but has an average value of about 7 MeV for heavy nuclei. In addition, the separation energy of α particle is high, of the order of 10 MeV for light nuclei but decreases with the size of the nucleus and around $A \approx 140$ becomes zero. The emission of α particle for heavier nuclei ($A > 140$) becomes energetically possible. Another way of looking to the problem is to consider the formation of α particle in the nucleus. Assuming that the average binding energy of both neutrons and protons in the nucleus is 7 MeV, the total energy required to pull out two neutrons and two protons will be $4 \times 7 = 28$ MeV. In addition, in the nucleus when two neutrons and two protons join together and form an α particle 28.3 MeV, which is the binding energy of the α particle, is released. As such, there is a net gain of energy in the whole process and the emission of α particle becomes energetically possible.

6.5.1 The α Decay Energy or Q-value

In nuclear transformations, the total energy of a system is conserved. The total energy consists of the rest mass energy and the kinetic energy. If M_i, M_f, E_i, and E_f, respectively, represent the initial and final masses and initial and final kinetic energies then

$$M_i + E_i = M_f + E_f \tag{6.22a}$$

It may be kept in mind that all quantities in Eq. (6.22a) are in same units; often in nuclear physics nuclear masses are given in MeV. From Eq. (6.22a) one may now define the decay energy E_d or the Q-value as

$$E_d = (M_i - M_f) = (E_f - E_i) \tag{6.22b}$$

If the initial mass energy is more than the final mass energy, that is E_d is positive the initial system will be energetically allowed to transform to the final system. The decay energy E_d will be released in the transformation. In addition, if E_d is negative, that is, the final mass energy is more than the initial mass energy, the transformation will not be able to proceed on its own unless energy equivalent to the magnitude of E_d is supplied to the system by some external means.

Suppose a nucleus ${}^{(A+4)}_{(Z+2)}\text{Y}$ at rest transforms into the nucleus ${}^{A}_{Z}\text{X}$ by emitting an α particle of kinetic energy E_α then

$$\begin{aligned} {}^{(A+4)}{}_{(Z+2)}\text{Y} &= {}^{A}{}_{Z}\text{X} + {}^{4}{}_{2}\text{He} + E_f \\ E_d &= [M_Y - M_X - M_\alpha] = E_f \end{aligned} \tag{6.22c}$$

Here, M_Y, M_X, and M_α are, respectively, the rest mass energies of nuclei Y, X, and the α particle. The decay energy E_d is equal to the total kinetic energy in the final system. As it is assumed that Y decays at rest, the residual nucleus X and the α particle will move out in opposite directions to conserve the linear momentum. We denote the recoil velocity of the nucleus X and the α particle, respectively, by V_x and v_α and the mass of α particle by m_α. Then, for the conservation of linear momentum

$$M_x V_x = m_\alpha v_\alpha$$

or

$$V_x = \frac{m_\alpha}{M_x} v_\alpha$$

The total kinetic energy in the final system E_f is equal to the sum of the kinetic energies of the α particle and the recoiling nucleus X.

$$E_d = E_f = \frac{1}{2} M_x V_x^2 + \frac{1}{2} m_\alpha v_\alpha^2 = \frac{1}{2} M_x \left[\frac{m_\alpha}{M_x} v_\alpha \right]^2 + \frac{1}{2} m_\alpha v_\alpha^2 = \frac{1}{2} m_\alpha v_\alpha^2 \left(1 + \frac{m_\alpha}{M_x} \right) \quad (6.22d)$$

We observe that if the kinetic energy of the α particle $E_\alpha \left(= \frac{1}{2} m_\alpha v_\alpha^2 \right)$ is measured experimentally than the decay energy E_d can be calculated using Eq. (6.22d). Further simplification may be done by taking the mass of the nucleus X and of α particle proportional to their mass numbers, which reduces Eq. (6.22d) to

$$E_d = E_\alpha (1 + \frac{4}{A}) \quad (6.22e)$$

If the residual nucleus is heavy ($A >> 4$), the decay energy is nearly equal to the kinetic energy of the emitted α particle $E_d \approx E_\alpha$.

6.5.2 Characteristics of a Particle Spectrum

With the availability of high-resolution charged particle detectors and magnetic spectrometers, it is now possible to measure very accurately the energy of the α particles emitted from radioactive sources. A typical α particle energy spectrum contain groups of α particles of different discrete energies E_1, E_2, ... The intensities of the energy peaks are all different. The α energy spectrum can be co-related with the excited states of the residual nucleus. For example, ^{241}Am decays by α emission into ^{237}Np. The α spectrum has six groups of αs of energies 5.314, 5.379, 5.433, 5.476, 5.503, and 5.535 MeV, which feed different excited states of the nucleus ^{237}Np, as shown in figure 6.5. The α group of highest energy 5.535 MeV feeds the ground state of ^{237}Np. The higher excited states of the residual nucleus eventually decay to the

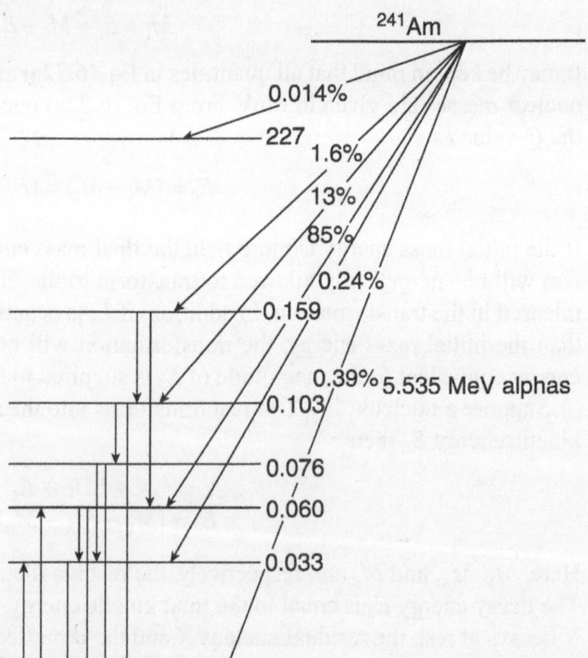

Figure 6.5 *Decay scheme of ^{241}Am*

ground state via γ rays. As such, discrete energy α particles are accompanied with corresponding γ rays in α spectrum. As the energy difference between different groups of αs is generally very small and the peaks due to these groups lie close to each other in α energy spectrum, this is often referred as the α fine structure.

The fine structure of α particle spectrum of ^{241}Am is shown in figure 6.6.

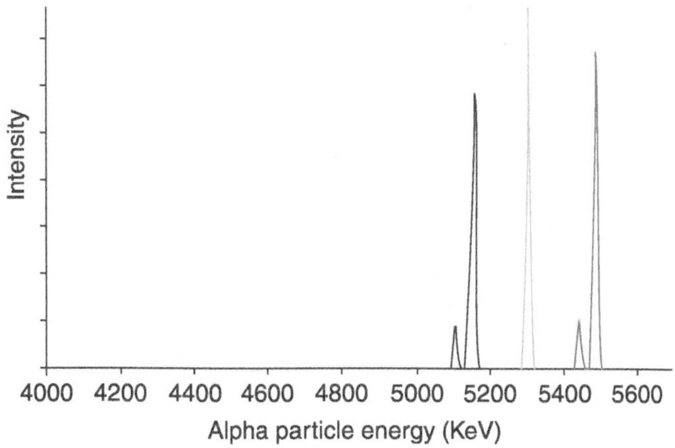

Figure 6.6 *Fine structure of α spectrum of ^{241}Am*

The mean life of an α emitter depends very strongly on the energy of the emitted α particle. Larger the energy of the α particle smaller is the mean life. Geiger and Nuttal had shown that the relationship between the decay constant λ and the energy E of the α particle (in MeV) may be given as

$$\log_{10}\lambda = C - \frac{D}{\sqrt{E}} \quad (6.22f)$$

where C and D are parameters that vary slowly with the atomic number Z of the α emitting nucleus, but they do not depend on its neutron number. Large spreads in the half-lives of α emitters from a fraction of a micro second to 10^{17} years have been experimentally observed. There might be other α emitters with even smaller and larger half-lives but to some extent the limits are set by the experimental measurement of very large and very small half-lives. Another equivalent expression for Geiger–Nuttal law is

$$\log\lambda = k\ \log x + C \quad (6.22g)$$

where k and C are constants and x is the range of the emitted α particle.

A good theory of α decay must explain all the above-mentioned characteristics of the α decay.

6.5.3 Theory of α Decay

As has already been indicated, the separation energy of α particle for nuclei with $A > 140$ is zero or negative, which means that if a nucleus with $A > 140$ emits an α particle the total energy of the residual nucleus plus the α particle will be less than the energy of the parent nucleus. Hence, from the perspective of stability it will be advantageous for the nucleus to immediately emit an α particle and become more stable. However, many heavy nuclides with $A > 140$ either do not emit α particles or emit them with a long half-life.

Being energetically possible is an essential requirement for α decay but is not the only requirement. An α particle for which it is energetically allowed to go out of the nucleus finds a potential barrier when it attempts to break through the nucleus. To understand the concept of the potential barrier, we consider the reverse process. Let an α particle $^4_2$He be shot towards a stationary target nucleus ^A_ZX as shown in figure 6.7. At large distance of separation r between the α particle and the nucleus X, the α particle will experience the repulsive Coulomb potential $U(r) = \frac{2Ze^2}{4\pi\varepsilon_0 r}$, which will increase as r decreases. When the α particle will reach within the range of the nuclear field of the nucleus X, the strongly attractive nuclear potential will somewhat reduce the repulsive Coulomb potential. As the α particle will enter into the target nucleus, the strong nuclear force will further reduce the potential barrier and finally the incident α will break into two protons and two neutrons, which will occupy their respective single particle orbits. The incident α particle loses its identity once it is well inside the target nucleus and a composite nucleus ${}^{A+2}_{Z+2}\mathrm{Y}$ is formed. Now, we consider the α decay of the nucleus Y and assume that inside nucleus Y two neutrons and two protons join together to form an α particle, which tries to come out. This α particle coming out from nucleus Y will also encounter the same potential barrier as is shown in figure 6.7. Whether the α particle will be able to come out of the nucleus Y will depend on two factors: (1) Is it energetically allowed or not? and (2) whether it will be able to cross the potential barrier or not. Classically, the α particle will be able to cross the potential barrier if its kinetic energy inside the nucleus is equal to or more than the maximum barrier height. However, for all α emitter nuclei in the valley of stability, the kinetic energy of the preformed α is less than the barrier height and classically their emission is not possible, though it is allowed energetically. In quantum mechanical treatment, an α particle with kinetic energy lower than the height of the barrier can still penetrate through the barrier by tunnelling and may go out of the parent nucleus.

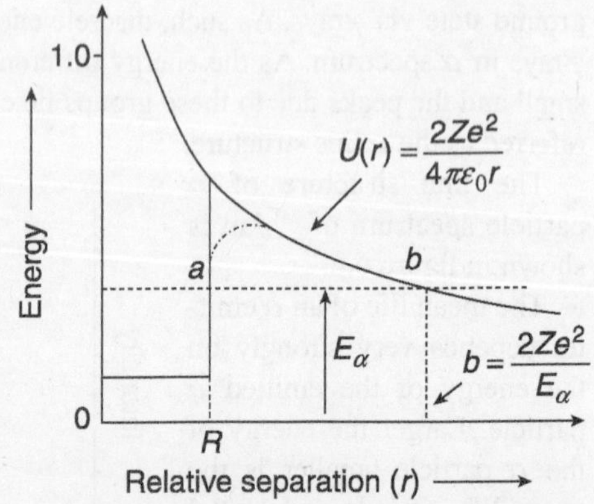

Figure 6.7 *Potential barrier faced by an α particle*

G. Gamow for the first time treated α decay in quantum mechanical framework and applied the concept of tunnelling of the potential barrier.

For simplicity, one may treat α decay as a three-step process. In the first step, in the parent nucleus, two neutrons and two protons join together to make an α particle. In case the decay energy E_d is more than the maximum barrier height, the α particle formed in step one will move out of the parent nucleus. In case E_d is less than the barrier height, the α particle moves across in the parent nucleus, reaches the nuclear surface again and again and attempts to tunnel out. Attempts made by the α particle to tunnel out constitute the second step. In one of the several thousand attempts, the α particle succeeds and moves out of the nucleus, in the third step. One may estimate the probabilities for each of the above-mentioned three steps.

6.5.3.1 Step I: Probability for the formation of α particle

The rough estimate for the probability of the formation of an α particle inside the parent nucleus may be done in the following way:

The probability, P_a, that two nucleons inside the nucleus are within a distance X from each other may be taken equal to the ratio of the volumes of the sphere of radius X and the total nuclear volume $V = 4/3\ \pi R^3$, where R is the radius of the nucleus. As such,

$$P_a = \left(\frac{X}{R}\right)^3 \tag{6.23}$$

Now, to form an α particle, the distance X between the centres of the two nucleons must be equal to the $2r_\alpha$, where r_α is the radius of the α particle. Hence,

$$P_a = \left(\frac{2r_\alpha}{R}\right)^3 \tag{6.24}$$

The probability, P_b, that the third nucleon is also within the volume of the first two nucleons will be

$$P_b = (r_\alpha/R)^3$$

The probability that the fourth nucleon is also within the volume of the three nucleons will also be equal to P_b. Thus, the lower limit to the formation of a cluster of four nucleons with in a spherical volume of radius r_α is equal to the multiplication of $P_a \times P_b \times P_b$. Hence,

$$P_{min} = 8(r_\alpha/R)^9 \tag{6.25}$$

Other, more rigorous estimates give a larger value for $P_{min} = 64\ (r_\alpha/R)^9$. A compromise may, however, be made by taking the probability $P(1)$ for the first step

$$P(1) = 25\ (r_\alpha/R)^9 = 25\left[\frac{r_0\ (4^{1/3})}{r_0\, A^{1/3}}\right]^9 = 25\left(\frac{4}{A}\right)^3 \tag{6.26}$$

Equation (6.26) gives a rough estimate of the probability that an α particle is formed within the parent nucleus.

6.5.3.2 Step II: The frequency with which the α particle hits the nuclear surface

Let the velocity of the nucleons inside the parent nucleus be V, then the number of times a nucleon will reach the nuclear surface per unit time N will be

$$N = \left(\frac{V}{2R}\right)$$

However, the probability $P(2)$ for step-two is equal to N; hence,

$$P(2) = N = \left(\frac{V}{2R}\right) \tag{6.27}$$

On multiplying $P(2)$ with $P(1)$, we get the probability of the first two steps, that is, the probability of the frequency of appearance of α particle at nuclear surface. In the absence of any potential barrier (which essentially means that the kinetic energy of the α particle is larger than the barrier height), the α particle would have gone out of the nucleus each time when it reached the surface of the nucleus. Hence, in the absence of the potential barrier, the decay constant for α emission λ_0 would have been

$$\lambda_0 = P(2)\cdot P(1) = \left(\frac{V}{2R}\right)\left\{25\left(\frac{4}{A}\right)^3\right\}$$

λ_0 is called the *reduced decay constant* and refers to the number of α decays per second in the absence of the potential barrier. The reciprocal of λ_0 is the reduced mean life τ_0. The reduced width ΔE_0 is defined as $\Delta E_0 = \hbar/\tau_0$ and gives the natural width of the level of parent nucleus that has undergone α decay. If the value of $\hbar = 6.58 \times 10^{-16}$ eV-sec is used the reduced width may be given as

$$\Delta E_0 = 6.58 \times 10^{-16} \times \frac{1}{\tau_0}\ \text{eV} = 6.58 \times 10^{-16} \times \lambda_0$$

In a heavy nucleus, the average kinetic energy of the most energetic nucleons are of the order of 30 MeV and the radius is about 8 F, the value of $P(2)$ comes out to be $\approx 5 \times 10^{21}\ \text{s}^{-1}$ and reduced decay constant λ_0 becomes

$$\lambda_0 = 6.5 \times 10^{17} \text{ per second} \tag{6.28}$$

Equation (6.28) states that α particles are formed and reach nuclear surface 6.5×10^{17} times in 1 second. The value of the reduced decay constant obtained is much smaller than the values obtained from the analysis of α decay energies. However, these estimates are very rough and further refinements in calculations of probabilities are expected to improve the estimates and bring them closure to the experimental values.

6.5.3.3 Step III: Barrier penetration and transmission coefficient

An α particle trying to come out from the nuclear surface encounters a potential barrier as shown in figure 6.5. The horizontal line in the figure shows the decay energy E_d released in the breaking of the parent nucleus into the residual nucleus ${}^{A}_{Z}\text{X}$ and the α particle ${}^{4}_{2}\text{He}$. In a heavy nucleus, E_d is nearly equal to the kinetic energy of the α particle E_α.

The radial part of Schrödinger equation for α particle in spherical polar coordinates and assuming a spherically symmetric nuclear potential may be written as

$$\frac{d^2u_\ell}{dr^2} + k^2(r)u_\ell = 0 \tag{6.29}$$

Here, u_ℓ is the reduced radial wave function related to the radial wave function $\chi(r)$ by

$$\chi(r) = \frac{u_\ell}{r}\, Y_{\ell m}(\theta,\phi) \tag{6.30}$$

And k is given by

$$k^2 = \frac{2M}{\hbar^2}\left[W - V(r) - \frac{\ell(\ell+1)\hbar^2}{2Mr^2}\right] \tag{6.31}$$

Here, M is the reduced mass of the α particle and the residual nucleus and is given by

$$M = \left(\frac{M_x M_\alpha}{M_x + M_\alpha}\right) \approx \left(\frac{4}{1+\frac{4}{A}}\right) \tag{6.32}$$

W is the kinetic energy of the α particle inside the nucleus and is nearly equal to E_d. $V(r)$ is the potential shown in figure 6.5 and is equal to the Coulomb potential $U(r)$ for large values of r and to the sum of the nuclear and Coulomb potentials for r of the order of the radius R of the residual nucleus X. ℓ is the orbital angular momentum of α particle with respect to the residual nucleus X.

Gamow showed that the probability of tunnelling $T(\ell)$ through the potential barrier may be written as

$$T(\ell) = e^{\left(-2\int_a^b k\, dr\right)} \tag{6.33}$$

For known $V(r)$ and E_d, k can be calculated from Eq. (6.31) for each value of ℓ. Using the calculated value of k, the integral in Eq. (6.33) may be evaluated by numerical integration and the value of $T(\ell)$ may be obtained. The limits of integration, a and b, are called the classical turning points and their values depend on the kinetic energy E_α available to the α particle from the breaking of the parent nucleus. If E_α is large, the magnitude of the integral in Eq. (6.33) is small, which means that the negative power of the exponent is small and therefore the transmission coefficient T is large. As is obvious from Eq. (6.33), T depends very strongly on E_α because of the exponential dependence. Transmission coefficient T also depends on the angular momentum ℓ. A larger value of ℓ means a higher barrier, as the total barrier is the sum of the nuclear potential $V(r)$ and the centrifugal potential. Therefore, for a fixed value of E_α, transmission coefficient for higher ℓ will be smaller as compared to that for the lower ℓ. Theoretical calculations of the transmission coefficient have shown that they vary slowly with the mass number A and strongly with the atomic number Z. A typical plot of the barrier transmission coefficient $T(0)$ for $\ell = 0$ αs as a function of E_α for different Z values is shown in figure 6.8. It may be observed from the graph that: (1) the absolute values of $T(0)$ are very small, (ii) for a fixed value of E_α, $T(0)$ decreases rapidly with the increase in Z, (3) for a fixed Z, $T(0)$ increases rapidly with the increase in E_α.

Finally, the α decay constant λ may be written as the multiplication of the probabilities for the above-mentioned three steps.

$$\lambda = P(1) \times P(2) \times T(\ell) = \lambda_0 \times T(\ell) \tag{6.34}$$

In the simple theory of α decay, the energy (E_α) and Z dependence of the decay constant λ (or the mean life) are contained largely in $T(\ell)$. For example, at $Z = 80$, $T(0)$ values for $E_\alpha = 4$ MeV and 6 MeV are $\approx 10^{-31}$ and 10^{-21}, respectively. This shows that for a change in energy by a factor of 1.5 the transmission coefficient changes by a factor of 10^{10}. This is in agreement with the experimental observation of α energies and half-lives. The biggest success of the simple theory of α decay based on quantum mechanical tunnelling of the potential barrier is that the theory could explain the large variations in observed mean lives of α emitters and their dependence on α energy.

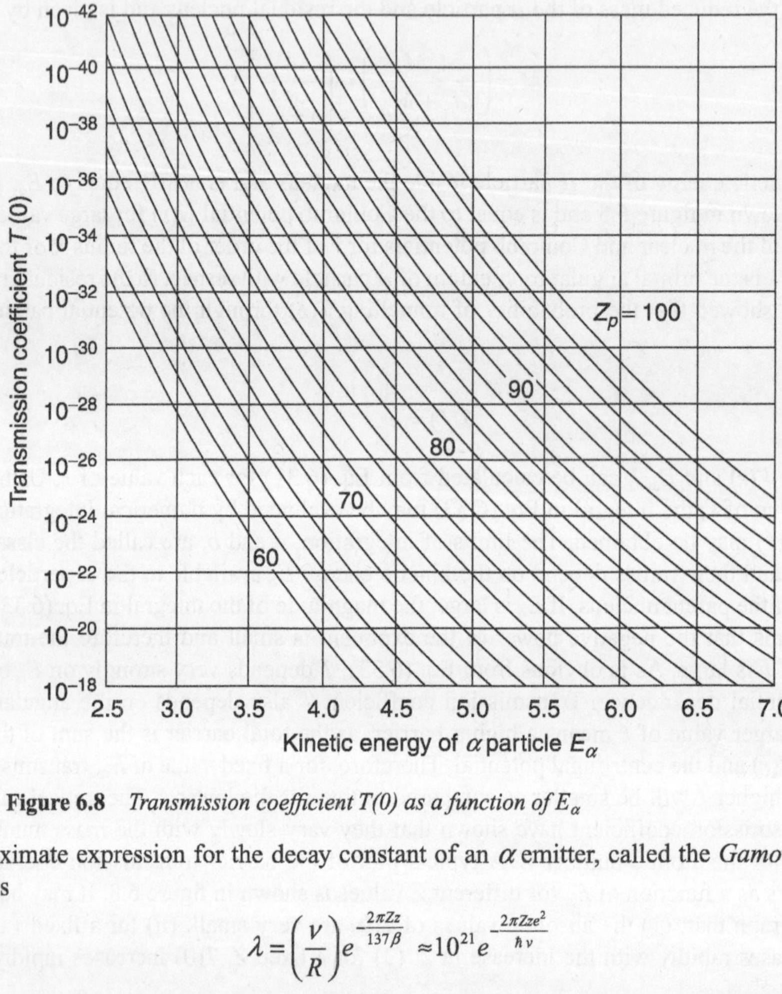

Figure 6.8 *Transmission coefficient $T(0)$ as a function of E_α*

An approximate expression for the decay constant of an α emitter, called the *Gamow formula*, is given as

$$\lambda = \left(\frac{v}{R}\right) e^{-\frac{2\pi Zz}{137\beta}} \approx 10^{21} e^{-\frac{2\pi Zze^2}{\hbar v}}$$

where λ is the decay constant, Z and z are, respectively, the atomic numbers of the emitting nucleus and of the emitted particle, v the velocity of the emitted particle and R the radius of the emitting nucleus. $\beta = \left(\frac{2E}{Mc^2}\right)^{1/2}$, E being the kinetic energy and M the mass of the emitted particle.

Solved example S-6.4

Show that the Geiger–Nuttall law follows from Gamow theory of α decay.

Solution. Gamow theory of α decay is based on the quantum mechanical tunnelling of the potential barrier faced by an α particle going out from the nucleus undergoing α decay. The decay constant λ_α for the α decay is related to the reduced decay constant λ_0 through the relation

$$\lambda_\alpha = \lambda_0 T = \lambda_0\, e^{\left(-2\int_a^b k\,dr\right)}.$$

To find the value of λ_α, it is required to evaluate the transmission coefficient T, as the approximate value of λ_0 has already been obtained to be of the order of $(v/2R)$. We will now evaluate the term $2\int_a^b k\,dr$.

$2\int_a^b k\,dr = \frac{2}{\hbar}\int_R^b \left[2\mu\left(\frac{Z_\alpha Z_D e^2}{r} - E_\alpha\right)\right]^{1/2} dr$, where μ is the reduced mass of the system, Z_α, Z_D are, respectively, the atomic numbers of the α particle and the residual nucleus. The α particle energy E_α is related to its distance of the closest approach b from the residual nucleus as

$$E_\alpha = \frac{Z_\alpha Z_D e^2}{b}.$$

Hence,
$$2\int_a^b k\,dr = \frac{2}{\hbar}\sqrt{2\mu E_\alpha}\int_R^b \left(\frac{b}{r}\right)^{1/2} dr = \frac{2b}{\hbar}\sqrt{2\mu E_\alpha}\left[\cos^{-1}\sqrt{\frac{R}{b}} - \sqrt{\frac{R}{b}}\sqrt{1-\frac{R}{b}}\right]$$

$$= 2\sqrt{\frac{2\mu}{\hbar^2 E_\alpha}}\,(Z_\alpha Z_D e^2)\left[\cos^{-1}\sqrt{\frac{R}{b}} - \sqrt{\frac{R}{b}}\sqrt{1-\frac{R}{b}}\right]$$

For thick barrier, $\frac{R}{b} \ll 1$ as $\left(\frac{E_\alpha}{V_c} \ll 1\right)$, where Coulomb barrier $V_C = \frac{Z_D Z_\alpha e^2}{(R_D + R_\alpha)}$

And, therefore, one may approximate, $\cos^{-1}\sqrt{\frac{r}{B}} \approx \frac{\pi}{2} - \sqrt{\frac{R}{b}}$

This makes $2\int_a^b k\,dr = 2\sqrt{\frac{2\mu}{\hbar^2 E_\alpha}}\,(Z_\alpha Z_D e^2)\left(\frac{\pi}{2}\right)$

Therefore, the transmission coefficient $T = e^{\left(-2\sqrt{\frac{2\mu}{\hbar^2 E_\alpha}}(Z_\alpha Z_D e^2)\left(\frac{\pi}{2}\right)\right)}$

Or $\lambda_\alpha = \lambda_0 T = \lambda_0\, e^{\left(-2\sqrt{\frac{2\mu}{\hbar^2 E_\alpha}}(Z_\alpha Z_D e^2)\left(\frac{\pi}{2}\right)\right)}$

$$\log\lambda_\alpha = \log\lambda_0 - \frac{\sqrt{\frac{2\mu\pi}{\hbar^2}}(Z_\alpha Z_D e^2)}{\sqrt{E_\alpha}}$$

This equation represents Gieger–Nuttall law.

Exercise p-6.5: Explain why heavy nuclides decay by α emission but not by the emission of nucleons. $^{239}_{94}$Pu decays by α emission with decay constant 2×10^{-13} s^{-1}. Calculate the number of α particles emitted per second by a source of 1 gram.

Exercise p-6.6: What is reduced decay constant? Obtain an expression for it.

Exercise p-6.7: On what factors does the half-life of an α emitter essentially depends? Calculate the energy that must be imparted to an α particle to force it inside a $^{238}_{92}U$ nucleus.

Exercise p-6.8: Two α emitter nuclides $^{A1}_{Z1}X$ and $^{A2}_{Z2}Y$ ($A1 > A2$; $Z1 > Z2$) have same value of decay energy. Which of the two nuclides will have a longer mean life? Explain your answer.

Exercise p-6.9: Two isobars of elements of atomic numbers 88 and 90 emit α particles, respectively, of energies 3.0 and 5.0 MeV. Calculate the ratio of the decay constants of the two isobars.

[**Hint:** We use Gamow formula $\lambda = \left(\frac{V}{R}\right) e^{\frac{2\pi Zz}{137\beta}}$ to get, $\log \frac{\lambda_1}{\lambda_2} = \frac{Z_1\beta_2}{Z_2\beta_1} \log \sqrt{\frac{E_1}{E_2}}$,

Also, $\frac{\beta_1}{\beta_2} = \frac{\sqrt{2E_1}}{\sqrt{2E_2}}$, since both are isobars their radii and masses may be taken to be equal.]

6.6 THE β DECAY

Neutron rich nuclei attempt to acquire stability by β^- transformation in which a neutron inside the nucleus converts into a proton, a β^- particle (electron), and an antineutrino. The reaction may be written as

$$n \rightarrow p + \beta^- + \bar{\nu} \tag{6.35}$$

Beta negative (β^-) is an electron that is created in a nuclear process. The antineutrino is an elementary particle the presence of which is essential for the conservation of angular momentum in the decay process. The residual nucleus of β^- decay has one unit of additional charge compared to the parent nucleus and the same mass number A.

$$^A_ZX \rightarrow {}_{Z+1}^{A}Y + \beta^- + \bar{\nu} \tag{6.36}$$

Proton rich nuclei decay either by β^+ emission or by electron capture. In the former, a proton inside the nucleus transforms into a neutron a β^+ (or positron) and a neutrino

$$p \rightarrow n + \beta^+ + \nu \tag{6.37}$$

Beta positive or positron is the antiparticle of electron. Similarly, neutrino and antineutrino are antiparticles of each other. Although a free neutron can decay into a proton (Eq. (6.35)) but a free proton cannot decay into a heavier neutron. However, decay represented by Eq. (6.37) may take place inside the nucleus. The residual nucleus of β^+ decay has a charge one unit less than the charge of the parent nucleus and the same A.

$$^A_ZX \rightarrow {}_{Z-1}^{A}Y + \beta^+ + \nu \tag{6.38}$$

In the process of electron capture (EC), an orbital electron of the atom is captured by the nucleus. This is possible mostly in heavy atoms for which the K-electron orbit passes through the nucleus.

The captured atomic electron interacts with the proton in the nucleus and converts it into a neutron. A neutrino is also created in the process, which moves out from the nucleus. The newly formed neutron remains in the nucleus.

$$p + e^- \rightarrow n + \nu \tag{6.39}$$

$$^{A}_{Z}\mathrm{X} + e^- \rightarrow {}^{A}_{Z-1}\mathrm{Y} + \nu \tag{6.40}$$

Electron capture competes with positron decay in heavy nuclei.

6.6.1 Decay Energy or *Q*-value

6.6.1.1 Beta negative (β^-) decay

In real experiments, generally we deal with atoms and not with nuclei as it is not possible to remove all the electrons of all the atoms of a given sample. In β decay also it is the parent atom the nucleus of which emits a β particle. In case of β^- decay, the parent atom is initially electrically neutral; however, after emitting a β^- particle the charge of the nucleus increases by one unit. As a result, the atom, which was initially neutral, becomes a positive ion and a β^- particle plus an antineutrino are emitted. The positive ion is not stable and captures an electron from the surroundings to become a neutral atom. When the positive ion captures an electron from the surroundings, energy equal to the binding energy B_e of the captured electron in the residual atom is liberated. As the electron captured by the ion goes to the outermost orbit of the atom, B_e is of the order of few eV. To calculate the decay energy or the energy that will be released in β^- decay, we need to calculate the difference in the initial and final mass energies.

Decay energy for β^- = Initial mass energy – final mass energy
= (mass of the atom $^{A}_{Z}\mathrm{X}$)c^2 – [mass of the ion $^{A}_{Z+2}\mathrm{Y}$ + mass of β^- + mass of neutrino]c^2

If masses are measured in units of energy, the factor c^2 can be dropped. Further, when positive ion Y captures an electron from the surroundings and becomes a neutral atom, its mass increases by the amount of the mass of an electron m_e and energy B_e is released. As has already been mentioned, B_e is the binding energy of the outermost electron in the atom. Also, the rest mass of neutrino is zero. Considering these, we get

Energy released in β^- decay or the Q-value of $\beta^- = Q^{\beta^-}$

$$= [M_x - \{(M_y - m_e - B_e) + m_\beta\}] \tag{6.41}$$

In Eq. (6.41), M_x and M_y denotes the atomic mass energies of neutral atoms X and Y, respectively, and m_β the mass energy of the β^- particle which is same as the mass energy of an electron and therefore $m_e = m_\beta$.

$$Q^{\beta^-} = M_x - M_y + B_e \tag{6.42}$$

6.6.1.2 Beta positive (β^+) decay

In β^+ decay, a neutral atom $^{A}_{Z}\mathrm{X}$ emits a positron, a neutrino, and becomes the negative ion $^{A}_{Z-1}\mathrm{Y}^-$. The negative ion being unstable ejects one electron from the outermost shell to become a neutral atom and releases the binding energy of the electron B_e. Therefore, in this case, we have

Decay energy of $\beta^+ = Q^{\beta^+} = M_x - [(M_y + m_e - B_e) + m_\beta]$

Or $$Q^{\beta^+} = M_x - M_y - 2m_e + B_e \qquad (6.43)$$

6.6.1.3 Electron capture (EC)

The probability of electron capture is largest from the innermost K-shell of the atom. The binding energy of electron in the K-shell B_k is few hundred keV while B_e is only few eV. When a K-shell electron is captured in the nucleus an electron vacancy is created in the K-shell, which is filled by electrons from higher shells. As a result, electrons from higher shells fall in to the inner shells and the electron vacancy created in the innermost shell reaches the outermost shell of the atom. Ultimately, energy equal to the difference of B_k and B_e is released. In the case of EC, no electron or β particle is emitted but a neutrino of zero rest mass energy comes out of the nucleus. Therefore,

$$Q^{EC} = M_x - [M_y + B_k - B_e] = M_x - M_y - B_k + B_e \qquad (6.44)$$

It may be kept in mind that the masses M_x, M_y, and m_e appearing in the above equations are the rest mass energies of neutral atoms and electron (in unit of energy, i.e. in MeV or so). If masses in atomic mass unit or unified mass unit are used then they must be converted into energy by multiplying with c^2. Further, as B_k and B_e are much smaller than the atomic mass energies of X and Y, they may be neglected without any appreciable errors.

The β or EC decay will be energetically allowed if the corresponding Q-value is zero or positive. It is clear from Eqs. (6.43) and (6.44) that $Q^{EC} > Q^{\beta^+}$ and hence, whenever electron capture is energetically allowed β^+ decay will also be allowed. However, the converse is not true. Rarely, however, β^+, β^-, and EC occur in the same nucleus. One such example is the decay of ${}^{40}_{19}K$. The decay scheme of ${}^{40}_{19}K$ nucleus is shown in figure 6.9.

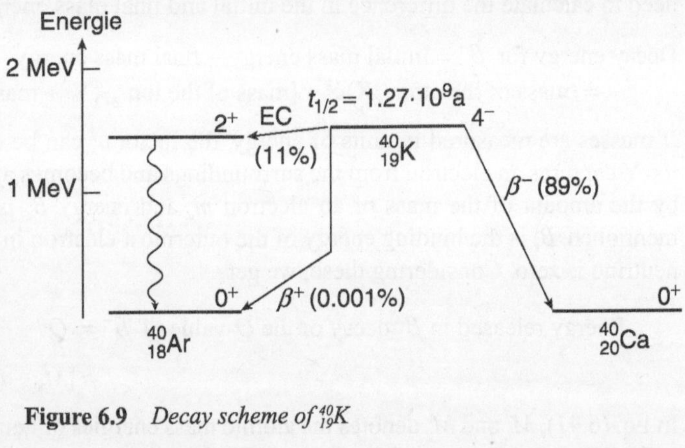

Figure 6.9 *Decay scheme of* ${}^{40}_{19}K$

6.6.2 Characteristics of β Particle Spectrum

Both in β^- and β^+ decays, the decay energy is essentially shared between two particles, β^- and antineutrino in case of β^- decay and between β^+ and neutrino in the case of β^+ decay. It is because the recoil energy of the residual atom is negligible. In a decay where there are two particles in the final state, the exact amount of energy carried by each particle depends on the angle of emission. As a result, the energy spectrum of emitted β particles is continuous both for β^+ and β^- decays as shown in figure 6.12. However, there is a little difference between the two spectra. β^- spectrum has relatively more β^- particles with smaller energies as compared with the spectrum of β^+, which has got a relatively large component of higher energy β^+ particles. The difference in the two spectra comes from the interaction of the Coulomb field of the nucleus with the Coulomb field of the outgoing β particle. The maximum energy E_{max} also called the end point energy is the energy of the most energetic β particle in the spectrum and corresponds to the situation when neutrino (or antineutrino) is emitted with zero kinetic energy. As such, the end point energy is equal to the decay energy if the recoil energy of the residual atom is neglected. Accurate measurement of end point energy is, therefore, very important to determine the decay energy.

6.6.3 Simple Theory of β Decay

Simple theory of β decay, also called Fermi theory, essentially predicts the shape of β spectrum. The starting point of the theory is the so-called 'Golden rule of Fermi' according to which the decay constant λ for any transformation is given by

$$\lambda = \frac{2\pi}{\hbar}\left(|\mathcal{H}'_{\text{if}}|\right)^2 \frac{dN}{dE_0} \tag{6.45}$$

The decay constant λ in Eq. (6.45) gives the number of transformations per unit time, $\mathcal{H}'_{\text{if}}$ the matrix element for the transformation from an initial state i to the final state f, and $\frac{dN}{dE_0}$ is the number of states of the final system per unit interval of total energy E_0. The matrix element can be written as

$$\mathcal{H}'_{\text{if}} = \int \psi_{\text{f}}^* H' \psi_{\text{i}}\, dv \tag{6.46}$$

where ψ_{f} and ψ_{i} are the wave functions for the final and initial states of the system and dv is the volume element.

If one wants to apply Eq. (6.45) to the problem of β decay then the values of $\frac{dN}{dE_0}$ and $\mathcal{H}'_{\text{if}}$ appropriate to β decay, need to be substituted in Eq. (6.45).

6.6.4 The Density of States $\frac{dN}{dE_0}$

The decay constant λ in Eq. (6.45) is the total decay constant that gives the total number of transformations in unit time and dN/dE_0 is the corresponding total density of final states. However, if for a given transformation one wants to calculate partial decay constant for those events in which the final system is left in some specific states N' out of the total states N, one may do so by using dN'/dE_0 instead of dN/dE_0 in Eq. (6.45). As the purpose of this simple theory is to see how far the theory can predict the shape of the energy spectra of β particles, we will derive an expression for the probability $P(p_\beta)dp_\beta$ that corresponds to those decays in which β particles are emitted with their momentum lying between p_β and $(p_\beta + dp_\beta)$. The relative momentum distribution obtained from $P(p_\beta)dp_\beta$ may then be compared with the momentum distribution obtained experimentally. $P(p_\beta)dp_\beta$, therefore, corresponds to the partial decay constant that transforms the initial system to final states in which the β particles are emitted with their momentum between p_β and $p_\beta + dp_\beta$.

In β decay, the initial state is a nucleus that is unstable and undergoes β decay. This decaying nucleus has some mean life say τ. From the uncertainty principle, the energy of this state is not well-defined and the uncertainty in energy ΔE may be given as

$$\Delta E \tau = \hbar \tag{6.47}$$

As such, there is a finite spread in the energy of the initial state that is shown as ΔE on the left hand side of figure 6.10. Some discrete energy levels of the initial (decaying nucleus) system lie within the energy spread ΔE. Now, let us look at the final state, that is the system after decay.

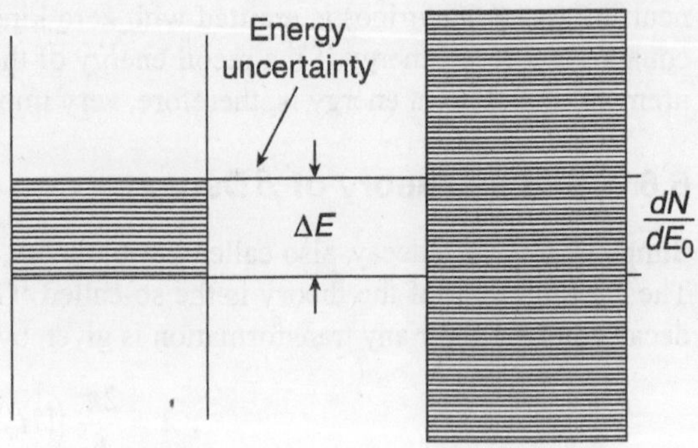

Figure 6.10 *The widths of the initial and the final states in the overlapping region*

The final state of β decay consists of the residual nucleus, β particle, and neutrino (or antineutrino). Both β particle and neutrino are jointly classified as leptons. The decay energy is essentially shared between the two leptons, if the recoil energy of the residual nucleus is neglected.

Total energy available from transformation = decay energy $E_d = E_\beta + E_\nu$

Therefore,

$$E_\nu = E_d - E_\beta \tag{6.48}$$

And for a fixed value of E_β

$$dE_\nu = dE_d \tag{6.49}$$

Also, both neutrino and β particles are relativistic; therefore,

$$E_\nu^2 = (p_\nu c)^2 + m_\nu^2 c^4 \text{ and } E_\beta{}^2 = (p_\beta c)^2 + m_e^2 c^4 \tag{6.50}$$

where m_ν and m_e are the rest masses of the neutrino and the β particle or electron.

The amount of energy carried by the β particle and the neutrino depends on the angle of their emission with respect to the nucleus. There are several allowed emission angles and, therefore, several possible ways in which the decay energy is shared between the leptons. Each allowed way of lepton emission constitutes a quantum mechanical final state. As such, there is large number of allowed quantum mechanical states of the final system. Transitions from initial states lying in the energy spread ΔE to the energy states of the final system overlapping within energy ΔE are represented by the transition probability given by Eq. (6.45). The density of states $\frac{dN}{dE_0}$, in Eq. (6.45), for the case of β decay, corresponds to the density of final states in the overlapping energy region as shown in figure 6.10.

To calculate $P(p_\beta)dp_\beta (= \lambda_{(p,\, p+dp)})$, we assume that the β particle is emitted with a fixed energy E_β. As the energy of the β particle is now fixed, different final states are due to different possible values of neutrino energy E_v. The $\frac{dN'}{dE_0}$ in this case may be obtained from the multiplication of $\frac{dN_v}{dE_d}$ and dN_β, where $\frac{dN_v}{dE_d}$ is the density of the neutrino states per unit total decay energy and dN_β is the number of final states in which β particles are emitted with momentum in the range p_β and $p_\beta + dp_\beta$.

$$\frac{dN'}{dE_0} = \frac{dN_v}{dE_d} \times dN_\beta \tag{6.51}$$

To calculate the number of states of a particle with the momentum in the range p and $p + dp$, we make use of the phase space and calculate the volume of the phase space occupied by the particle of above momentum range. Dividing the occupied volume by $(2\pi\hbar)^3$ gives the desired numbers of states as each state occupies a volume $(2\pi\hbar)^3$ of phase space. As shown in figure 6.11, the states with momentum p and $p + dp$ are contained in an annular spherical volume with $p + dp$ outer and p inner radii. The volume of this annular spherical shell is $4\pi p^2\, dp$ (neglecting terms of dp^2 and higher order).

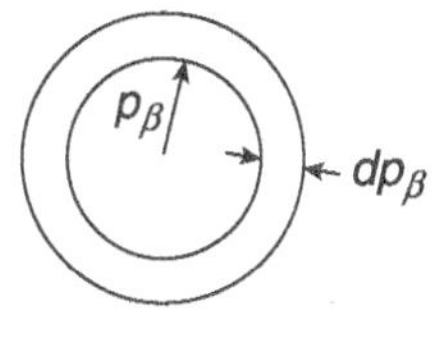

Figure 6.11 *Volume of the phase space occupied by the states of momentum in the range p and p + dp*

The number of states dN_p per unit spatial volume

$$= \{4\pi p^2\, dp/(2\pi\hbar)^3 \tag{6.52}$$

Using Eq. (6.52), we get

$$dN_\beta = \frac{4\pi p_\beta^2\, dp_\beta}{(2\pi\hbar)^3} \tag{6.53a}$$

And

$$dN_v = \frac{4\pi\, p_v^2\, dp_v}{(2\pi\hbar)^3} \tag{6.53b}$$

Next, we calculate $\left(\frac{dN_v}{dE_d}\right)$

From Eq. (6.50), we have $E_v^2 = (p_v c)^2 + m_v{}^2 c^4$, which on differentiation gives

$$dp_v = \frac{E_v}{p_v c^2}\, dE_v \tag{6.53c}$$

Multiplying both sides of Eq. (6.53c) by p_v^2 and substituting $dE_v = dE_d$ from Eq. (6.49), we get

$$p_v^2\, dp_v = \frac{p_v E_v}{c^2}\, dE_d \tag{6.54}$$

Again, from Eq. (6.50) on substituting $E_v^2 = (E_d - E_\beta)^2$, we may write

$$p_v = \frac{1}{c}\left[(E_d - E_\beta)^2 - m_v^2 c^4\right]^{1/2} \tag{6.55}$$

However, from Eq. (6.54) and (6.55)

$$dN_{\mathrm{v}} = \frac{4\pi\, p_{\mathrm{v}}^2\, dp_{\mathrm{v}}}{(2\pi\hbar)^3} = \frac{4\pi}{(2\pi\hbar)^3}\frac{p_{\mathrm{v}} E_{\mathrm{v}}}{c^2} dE_{\mathrm{d}} = \frac{4\pi}{(2\pi\hbar)^3} \frac{\frac{1}{c}\left[(E_{\mathrm{d}} - E_\beta)^2 - m_{\mathrm{v}}^2 c^4\right]^{1/2}(E_{\mathrm{d}} - E_\beta)}{c^2} dE_{\mathrm{d}}$$

Or

$$\frac{dN_{\mathrm{v}}}{dE_{\mathrm{d}}} = \frac{1}{2\pi^2\hbar^3 c^3}\left[(E_{\mathrm{d}} - E_\beta)^2 - m_{\mathrm{v}}^2 c^4\right]^{1/2}(E_{\mathrm{d}} - E_\beta) \tag{6.56}$$

Therefore,

$$\frac{dN_{\mathrm{v}}}{dE_{\mathrm{d}}} \times dN_\beta = \frac{1}{2\pi^2\hbar^3 c^3}\left[(E_{\mathrm{d}} - E_\beta)^2 - m_{\mathrm{v}}^2 c^4\right]^{1/2}(E_{\mathrm{d}} - E_\beta) \times \frac{4\pi\, p_\beta^2\, dp_\beta}{(2\pi\hbar)^3}$$

Or from Eq. (6.51), we get

$$\frac{dN'}{dE_0} = \frac{1}{4\pi^3\hbar^6 c^3}\left[(E_{\mathrm{d}} - E_\beta)^2 - m_{\mathrm{v}}^2 c^4\right]^{1/2}(E_{\mathrm{d}} - E_\beta)\, p_\beta^2\, dp_\beta \tag{6.57}$$

6.6.5 The Matrix Element

Matrix element defined by Eq. (6.46) is a measure of the overlap of the wave functions for the initial and the final states. Larger the overlap more is the transition probability and in turn smaller mean life. The initial state wave function ψ_{i} is the wave function for the parent nucleus and may be written as ψ_{ni}. The final state of β decay contains two leptons and the residual nucleus. The final state wave function may be written as the multiplication of the three wave functions one each for the β particle, the neutrino, and the residual nucleus, respectively, ψ_β, ψ_{v} and ψ_{nf}. As such, we may write

$$\psi_{\mathrm{f}} = \psi_{\mathrm{v}}\psi_\beta\psi_{\mathrm{nf}}$$

And

$$\mathcal{H}'_{\mathrm{if}} = \int \psi_\beta^* \psi_\beta^* \psi_{\mathrm{nf}}^*\, H'\, \psi_{\mathrm{ni}}\, dv \tag{6.58}$$

Eq. (6.58) can be evaluated under the following simplifying assumptions suggested by Fermi:

1. The operator H' that couples the initial and final states via weak nuclear interaction is not exactly known and may be taken as a constant G, being the simplest operator.
2. In view of the fact that at energies under consideration, the de Broglie wave lengths for both the β particle and the neutrino are much larger than the size of the nucleus, the wave functions for both the leptons may be taken as free particle plane waves $\psi_\beta = e^{-ik_\beta \cdot r}$ and $\psi_{\mathrm{v}} = e^{-ik_{\mathrm{v}} \cdot r}$, which have a constant value over the dimensions of the nucleus. For the purpose of normalization, it may be assumed that the system is contained in space of unit volume.
3. The assumed plane wave function for β particle will suffer some distortion because of the electrostatic interaction between the nuclear charge of the nucleus and that of the β particle. The effect of this distortion on the spectrum of β particles with momentum p_β and

$p_\beta + dp_\beta$ is considered by multiplying $(|\psi_\beta|)^2$ by a factor $F(Z, p_\beta)$, which depends on the charge Z of the residual nucleus and for non-relativistic case is given as

$$F(Z, p_\beta) \cong \frac{2\pi y}{1-e^{-2\pi y}}. \tag{6.59}$$

With $y = \pm Z\alpha \frac{E_\beta}{p_\beta}$, where positive and negative signs are, respectively, for β^- and β^+ decays, α is the fine structure constant $\left(\approx\frac{1}{137}\right)$ and $|Z|$ is the atomic number of the residual nucleus. For relativistic case, a more detailed value for F, based on Dirac's electron theory is used. The factor F is called the Fermi factor or simply Coulomb factor. The net effect of F on the spectrum of the emitted β particles is to decelerate β^- due to the attraction by the positively charged residual nucleus and to accelerate the β^+ because of repulsion. It may be observed in Eq. (6.59), for β^- emission the exponential term in the denominator is negligibly small and, therefore, F varies almost linearly with E_β. However, for β^+ emission the exponential term rises sharply and drastically reduces the value of $F(Z, p_\beta)$ for low momentum positrons. As such, the relative components of low and high energy β particles in the two spectrums are different. The positrons coming out of the nucleus encounter a potential barrier (just like α particle) and only higher energy positrons can tunnel through the barrier. Thus, the experimental fact that positron spectrum has relatively larger component of high-energy positrons may be understood as an effect of $F(Z, p_\beta)$, which in turn is due to the potential barrier that a positron encounters while coming out of the nucleus. No such potential barrier is faced by the negatron, but the energy of each out coming β is reduced due to its attraction by the nucleus.

With the above-mentioned substitutions the matrix element becomes

$$|\mathcal{H}'_{\text{if}}|^2 = G^2\, F(Z, p_\beta)|(\int \psi^*_{\text{nf}} \psi_{\text{ni}}\, dv|)^2$$

Or

$$|\mathcal{H}'_{\text{if}}|^2 = G^2\, F(Z, p_\beta)\, |M_{\text{if}}|^2 \tag{6.60}$$

where

$$|M_{\text{if}}| = \int \psi^*_{\text{nf}} \psi_{\text{ni}}\, dv \tag{6.61}$$

Matrix element M_{if} determines the extent of overlap of the initial and final state nuclear wave functions.

Substituting the values of $\mathcal{H}'_{\text{if}}$ from Eq. (6.60) and of $\frac{dN'}{dE_{\text{d}}}$ from Eq. (6.57) in Eq. (6.45) we get for the partial decay constant

$$P(p_\beta)dp_\beta = \frac{2\pi}{\hbar} G^2 F(Z, p_\beta)\, |M_{\text{if}}|^2 \times \frac{1}{4\pi^3 \hbar^6 c^3}\, [(E_{\text{d}} - E_\beta)^2 - m_\nu^2 c^4]^{1/2} (E_{\text{d}} - E_\beta) p_\beta^2\, dp_\beta$$

$$= \frac{G^2 F(Z, p_\beta)}{2\pi^2 \hbar^7 c^3} |M_{\text{if}}|^2\, [(E_{\text{d}} - E_\beta)^2 - m_\nu^2 c^4]^{1/2}\, (E_{\text{d}} - E_\beta)\, p_\beta^2\, dp_\beta \tag{6.62}$$

Equation (6.62) gives the dependence of β particle momentum distribution on the energy E_β. The term containing the mass of the neutrino is included only to show how an upper limit to the neutrino mass m_ν can be obtained from the comparison of the predicted and the experimental momentum distributions. If the mass of neutrino is taken as zero then Eq. (6.62) may be written as

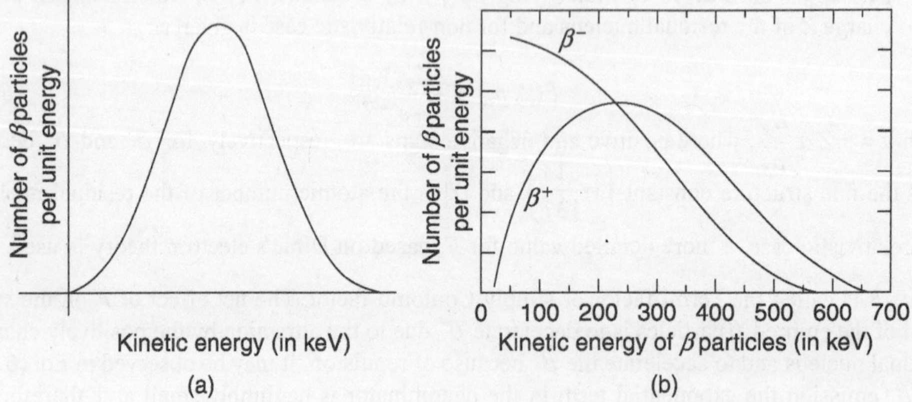

Figure 6.12 *Energy distributions for Z = 0 and Z = 29 nuclei for β^+ and β^- decay*

$$p(p_\beta)dp_\beta = \frac{G^2F(Z, p_\beta)}{2\pi^2\hbar^7c^3}|M_{if}|^2 (E_d - E_\beta)^2 p_\beta^2 dp_\beta \qquad (6.63)$$

According to Eq. (6.63), the terms $(E_d - E_\beta)^2$ and p_β^2 determine the shape of the β particle momentum distribution. At lower values of the momentum p_β, the momentum distribution is governed by p_β^2. It is because momentum is also lower when value of E_β is low and $(E_d - E_\beta)^2$ is essentially equal to E_d^2. Towards the higher momentum end where E_β is also large, the momentum distribution varies as $(E_d - E_\beta)^2$. The maximum energy of the emitted β particles $E_{\beta_{max}}$ is equal to E_d. To see the effect of $F(Z, p_\beta)$ on the momentum distribution, momentum distribution for $Z = 0$ is plotted in figure 6.12 and is marked (a). Curve (a) shows the effect of the p_β^2 and $(E_d - E_\beta)^2$ terms only. The other two curves show the effect of $F(Z, p_\beta)$ for electron and positron emissions from the decay of ^{64}Cu ($Z = 29$).

In general, expression (6.64) derived using a very simple approach out lined by Fermi predicts the shape of momentum distributions remarkably well. The comparison of experimental momentum distributions, particularly for the cases in which β particles of low energy are emitted with the predictions of the Fermi theory indicates that the rest mass of neutrino is either very small (<250 eV) or zero. Other experiments have conclusively proved that neutrino rest mass is zero.

6.6.5.1 Screening by atomic electrons

Fermi factor $F(Z, p_\beta)$ accounts for the electrostatic interaction between the β particle and the residual nucleus. However, the atomic electrons also affect the shape of β spectrum. It has been estimated that screening by atomic electrons reduces the probability of β^- emission by about 2% for 50 keV β^- at Z of the order of 50. The effect increases as Z increases or E_β decreases. Screening by atomic electrons enhances the probability of β^+ emission by about 37% for 50 eV positrons at Z around 50. The effect decreases with the increase of Z and is about 3% for $Z \approx 90$.

6.6.6 Fermi–Kurie Plot

From Eq. (6.62), we may write

$$\sqrt{\frac{P(p_\beta)dp_\beta}{F(Z,p_\beta)p_\beta^2}} = \frac{1}{(2\pi^2\hbar^7c^3)^{1/2}} G\, M_{\text{if}}\, (E_\text{d} - E_\beta) \tag{6.64}$$

In expression (6.64), G and the matrix element M_{if} remain constant and, therefore, essentially the term on the LHS is proportional to $(E_\text{d} - E_\beta)$. A graph between the quantities on the LHS against E_β is expected to be a straight line for allowed β transitions, if the theory is correct. Such graphs are called Fermi–Kurie or simply Kurie plots and are used to test the theory. The straight line graph cuts the energy axis at $E_\beta = E_\text{d}$ and gives the maximum value of the β ray energy, generally denoted by $E_{\beta_{\text{max}}}$. Any deviation of the Fermi–Kurie plot from the straight line is taken as an indication of either the transition is not allowed or the curve is due to the mixture of more than one allowed β decay processes. The Fermi plot for the decay of free neutron is worth mentioning. A free neutron decays into a proton, a β^-, and an antineutrino. Nuclear reactors provide neutron beams of good flux and the β decay of neutrons has been studied in detail. In case of neutron decay, the Coulomb factor $F(Z, p_\beta)$ may be taken almost constant over the energies of the emitted β particles and, therefore, may not be considered. Kurie plot for neutron decay is shown in figure 6.13a. As may be seen, there is excellent agreement between the theory and experiments at higher energies. A few points towards the lower energy do not fall on the straight line and may be due to the fact that the detectors used in the experiment were not very sensitive at low energies. Similarly, ${}^3_1$H (tritium) also undergoes β^- decay but the end point energy in this case is very low ≈18 keV. Gas-filled proportional counters, which have very high detection efficiency for low energy βs, were used in these measurements. All data points in this case fall on the straight line. Kurie plots at very low energies are particularly sensitive to the rest mass of neutrino. The ${}^3_1$H Kurie plot shown in figure 6.13b puts

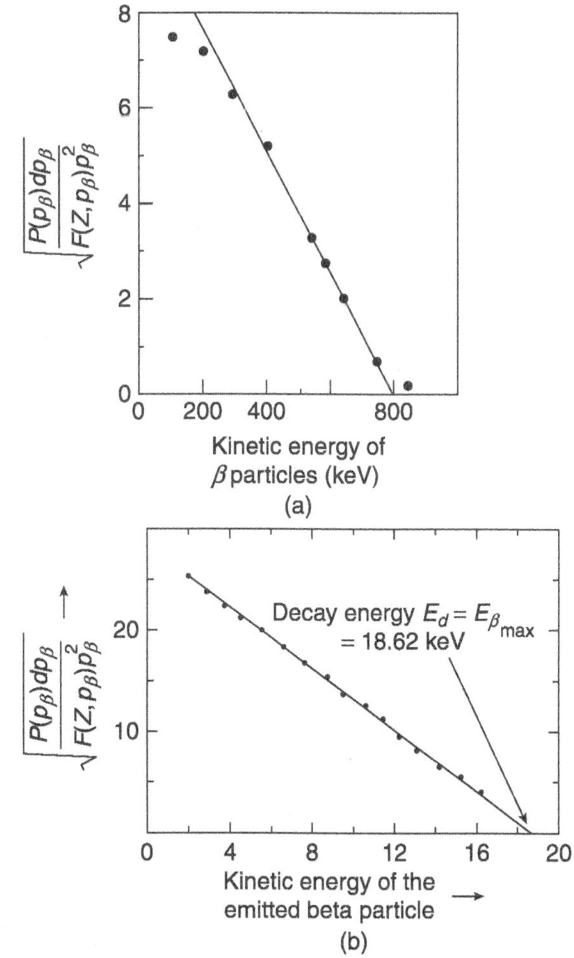

Figure 6.13 *Kurie plots for (a) decay of free neutron and (b) decay of ${}^3_1H$*

an upper limit to the rest mass of neutrino to be less than 250 eV. As expected, the Kurie plot for forbidden transitions deviate from straight line.

6.6.7 The Comparative Half-life and log$_{10}$(ft)-value

The total decay constant λ may be obtained from the partial decay constant $P(p_\beta)dp_\beta$ by integrating over momentum values.

$$\lambda = \frac{\ln 2}{t_{1/2}} = \int_0^{p_{\max}} P(p_\beta)dp_\beta$$

$$= \frac{m_e^5 c^4 G^2}{2\pi^2 \hbar^7} M_{\text{if}}^{\ 2} \int_0^{p_{\max}} F(Z, p_\beta) \frac{p_\beta^2}{m_e^{\ 2} c^2} \frac{(E_d - E_\beta)^2}{m_e^2 c^4} \frac{dp_\beta}{m_e c} \tag{6.65}$$

In Eq. (6.65), each quantity in the integral has been made dimensionless by dividing them by the appropriate power of $m_e c$. The integral on the right hand side can be evaluated by numerical integration. If the value of the integral is denoted by $f(Z, E_\beta)$ then

$$f(Z, E_\beta)\, t_{1/2} = \frac{1}{|M_{\text{if}}|^2} \left(\frac{2\pi^2 \hbar^7 \ln 2}{m_e^5 G^2 c^4} \right) \tag{6.66}$$

$f(Z, E_\beta)\, t_{1/2}$ is called the ft value or the comparative half-life as it has been made a dimensionless quantity. It is inversely proportional to the square of the nuclear matrix element, and the square of the transition operator G. The LHS of Eq. (6.66) can be found from experiments. If one of the two M_{if} or G is known, the other can be determined. For example, the square of matrix element $|M_{\text{if}}|^2$ is estimated to be ≈ 2 for β transitions between $J = 0$ states in mirror nuclei and hence from the experimental values of $t_{1/2}$ and the numerical integration of $f(Z, E_\beta)$, G for such cases has been determined to be $\approx 1 \times 10^{-4}$ MeV-F^3.

The ft value is inversely proportional to the square of the nuclear matrix element. A large value of the overlap matrix element means that the nuclear wave functions of the final and the initial states have large overlap and, therefore, the transition will proceed with a short half-life and a smaller value of ft. As the spread of ft values for allowed transitions is quite large, log$_{10}$ (ft) instead of ft is used to classify the β transitions. A histogram of log$_{10}$(ft) – values obtained from the analysis of experimental data are shown in figure 6.14.

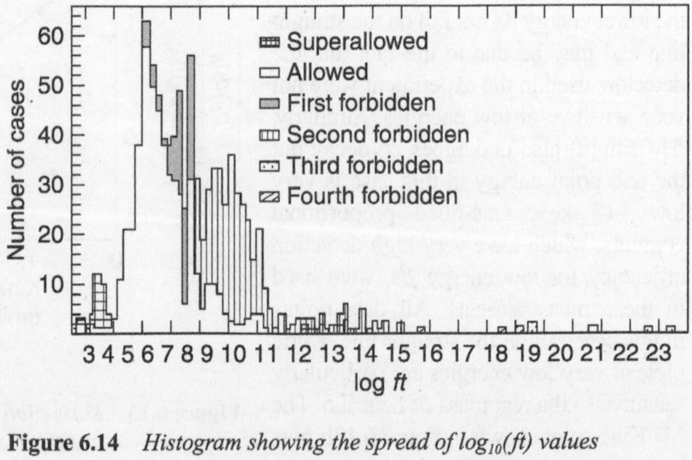

Figure 6.14 *Histogram showing the spread of log$_{10}$(ft) values*

The value of the decay constant λ given by Eq. (6.65) under the condition that the maximum energy $E_{\beta_{\max}}$ of the emitted β

particle is much larger than its rest mass energy and for the allowed transitions is approximately given by

$$\lambda = \frac{G^2 \,|\, M_{\text{if}} \,|^2 \, E_{\beta_{\max}}}{60\pi^3 (\hbar c)^6 \hbar}$$

In general, the shape of the β particle spectrum, that is the number $N(E)$ of β particles emitted with energy E may be well represented by the expression

$$N(E)dE \; = C\sqrt{E}(E_{\beta_{\max}} - E)^2 dE$$

where $E_{\beta_{\max}}$ is the end point energy or the maximum β particle energy (which may be taken equal to the decay energy without much error) and C is a constant.

6.6.8 Allowed and Forbidden β Transitions

In the derivation of the simple theory of β decay, the two leptons were represented by plane waves. The plane waves may be decomposed into partial waves each with a particular value of the orbital angular momentum l. As only $l = 0$ or S-partial waves fall within the range of the nuclear field of the nucleus, the outgoing leptons do not carry any orbital angular momentum. Now, each lepton has intrinsic spin 1/2 and if their spins are antiparallel (i.e. they are emitted in singlet state), then the spin change in the transition ΔJ will be zero and there will also be no change in the parity. Such transitions in which $\Delta J = 0$ and no change of parity takes place are called β decay by Fermi interaction. However, if the neutrino and the β particle are emitted with their spins pointing in the same direction (i.e. in triplet state), the change ΔJ in the transition will either be zero or ∓ 1. These transitions will, however, not include those cases in which $J_i = j_f = 0$. The β transitions with $\Delta J = 0, \mp 1$, (not including $J_i = J_f = 0$) and with no change in the parity are called Gamow–Teller transitions.

As has been already discussed, according to the simple theory of β decay leptons with zero orbital angular momentum are most likely to be emitted. It is because only S-partial waves of leptons fall within the nuclear dimensions. Beta transitions in which leptons with higher ℓ-values are emitted are, therefore, called forbidden transitions. The values of ft or $\log_{10}$ft for forbidden transitions are larger as the overlap integral is small. Thus, the extent to which a β transition is allowed or forbidden is determined by the ft or $\log_{10}$ft value. Classification of β transition, their selection rules and the spread of the ft-value are given in table 6.1.

Table 6.1 *Classification of β transitions*

Class of the transition	Change in J ΔJ	Change in parity	Range of $\log_{10}$ft value
Super-allowed	$0, \mp 1$	No	3–4
Allowed	$0, \mp 1$	No	4–6
First forbidden	$0, \mp 1$	Yes	6–8
Unique first forbidden	∓ 2	Yes	8–9

(*Continued*)

Table 6.1 *(Continued)*

Class of the transition	Change in J ΔJ	Change in parity	Range of $\log_{10}$ft value
Second forbidden	∓ 2	No	10–13
Unique second forbidden	∓ 3	No	12
Third forbidden	∓ 3	Yes	18
Unique third forbidden	∓ 4	Yes	15
Fourth forbidden	∓ 4	No	23
Unique fourth forbidden	∓ 5	No	19

The transitions for which $\log_{10}$ft values lie in the range of 3–4 are called super allowed transitions. Such small $\log_{10}$ft values indicate that the initial and the final state wave functions have large overlap. Large overlap may happen when the initial and final state wave functions are nearly identical. In the case of β transition between mirror nuclei, the wave function of the initial nucleus and the final nucleus are almost identical and, therefore, have large overlap and high transition probability. The β transitions between mirror isobars are, therefore, super allowed transitions.

6.7 THE ELECTRON CAPTURE DECAY (EC)

The simple theory of electron capture is just like the Fermi theory of β decay. The mathematical expressions obtained in case of β decay may be used for the EC decay as well but with the following modifications:

1. The decay energy E_d is the Q-value of EC.
2. As no β particle is emitted $\frac{dN'}{dE_d}$ is equal to $\frac{dN_v}{E_d}$ and should not be multiplied by dN_β. Thus, $\frac{dN'}{dE_d} = \frac{dN_v}{E_d} = \frac{E_d^2}{2\pi^2\hbar^3c^3}$ that may be obtained from Eq. (6.56) by substituting $E_\beta = 0$ and $m_v = 0$.
3. In case of β decay, the wave functions for both the electron and the neutrino were assumed to be the free particle plane waves. In EC decay, the wave function ψ_v is still treated as a plane wave but ψ_e is taken as the wave function for the electron in the atomic K-orbit. The amplitude of the Kth electron wave function at the origin is approximately given by $\left[\frac{1}{\sqrt{\pi}}\left\{\frac{Z\,m_e e^2}{4\pi\varepsilon_0\hbar^2}\right\}^{3/2}\right]$.

With above-mentioned modifications, the decay constant for electron capture decay may be written as

$$\lambda_{EC} = 2\frac{2\pi}{\hbar}\,G^2\,\frac{E_d^2}{2\pi^2\hbar^3c^3}\,|M_{if}|^2\left[\frac{1}{\sqrt{\pi}}\left\{\frac{Z\,m_e e^2}{4\pi\varepsilon_0\hbar^2}\right\}^{3/2}\right]^2 \tag{6.67}$$

The first factor 2 in Eq. (6.67) considers the fact that there are two electrons in the K-orbital of the atom. It may be noted that the matrix element $|M_{\text{if}}|^2$ is same for both β decay and EC decay. From Eqs. (6.65) and (6.67), the ratio of EC to β^+ decay may be obtained as

$$\frac{\lambda_{\text{EC}}}{\lambda_{\beta^+}}=\frac{Z^3e^6E_{\text{d}}^2}{16f(Z,E_{\text{d}})\hbar^3\pi^2c^7\varepsilon_0^3m_{\text{e}}^2} \tag{6.68}$$

The ratio $\frac{\lambda_{\text{EC}}}{\lambda_{\beta^+}}$, also called the branching ratio, is infinite only if EC is energetically allowed but β^+ decay is energetically not possible. The branching ratio depends both on the decay energy and also on the atomic number Z of the nucleus. For a fixed value of the decay energy, the ratio $\frac{\lambda_{\text{EC}}}{\lambda_{\beta^+}}$ rises sharply with the increase of Z and for a fixed Z it increases almost exponentially with the decrease of the energy.

The EC leaves an excited neutral atom with an electron vacancy in the K-shell. The excited atom may emit characteristic K-X rays or the electron vacancy may shift to some other higher shell such as L-shell. In such a case, Auger electrons may also be emitted. Auger electrons are the electrons emitted, generally from the L-shell, when one electron from the shell reverts to the K-shell to fill the vacancy there and the difference in the energy of the K and L shell is directly given to the other electron of the shell which go out as Auger electron. Thus, X-rays and Auger electrons emitted in EC are used to study the electron capture process experimentally.

6.8 NON-CONSERVATION OF PARITY IN β DECAY

Parity is a property that is associated with the object and its mirror image. If the mirror image of an object is identical to the object then the object is said to have a definite parity. In case the mirror image is not identical to the object, then the object does not have a definite parity. When an object has a definite parity, which means that the mirror image of the object is identical to the object then there are two possibilities: the wave function for the mirror image is same as the wave function of the object or it is equal to the wave function of the object with a minus sign. In the former case, the object is said to have a positive or even parity and in the later an odd or negative parity. The parity operator amounts to take the mirror image of the system.

There are some conservation laws that hold good also in microscopic systems. For example conservation of total energy, linear momentum, and angular momentum are found to hold in case of atomic and nuclear systems. Parity of a system is found to be conserved in case of nuclear strong interactions and electromagnetic interactions. However, Lee and Yang way back in 1956 pointed out that there was no proof that parity is conserved in electro-weak interactions. Processes such as β decay involve nuclear weak interactions and, therefore, efforts were made to test these processes for the parity conservation. Now, what are the criteria for the conservation of parity in a process? As systems of definite parity have identical mirror images, *if a process and the mirror image of the process are both doable experiments then parity is conserved in the process.* In addition, if only one of them, that is either the original process or its mirror image is doable experiment, then parity is not conserved in the process. The fact that parity is conserved in processes governed by strong nuclear interactions is due to the fact that mirror symmetric potentials are used to describe nuclear strong interactions.

Before describing the actual experiment carried out by Wu to test the parity conservation in the β decay of ^{60}Co nucleus, let us see how the object (^{60}Co nucleus undergoing β^- decay)) and its mirror image should look if the parity is conserved and in the case parity is not conserved. $^{60}{}_{27}$Co has the spin or angular momentum $J = 5^+$ and it decays into $^{60}{}_{28}$Ni* ($J = 4^+$) by β^- emission.

$$^{60}{}_{27}\text{Co} \rightarrow {}^{60}{}_{28}\text{Ni}^* + \beta^- + \overline{\nu}$$

The Ni nucleus is left in an excited state and it comes to ground state by emitting γ rays.

An isolated nucleus of ^{60}Co is shown in figure 6.15. As may be observed, the spin J (=5$^+$) of the object nucleus is pointing upwards. The image of the ^{60}Co nucleus is shown in the mirror. It may be observed that in the mirror image J points downwards. The nucleus ^{60}Co emits β^- particles. Now, there are two possibilities: (1) the same number of β^- particles are emitted per unit time both in the direction of the spin J and in the direction opposite to it, as shown in figure 6.15a. (2) Unequal number of β^- particles are emitted, say, more β particles in the direction opposite to the direction of J and less along the direction of J (figure 6.15b). It may now be noted that in figure 6.15a both the object and its mirror image are identical while in figure 6.15b the object and the mirror image are not identical. Therefore, if β emission from ^{60}Co is like the one represented in figure 6.15a then parity is conserved. However, if the β emission is like the one represented in figure 6.15b then parity will not be conserved in β decay.

From the discussion, it is clear that to test ^{60}Co nucleus for parity conservation the following settings are required: (1) All the ^{60}Co nuclei in the sample must have their spins pointing in one direction, that is they should be polarized. (2) It should be possible to monitor the degree of polarization of ^{60}Co nuclei. (3) It should be possible to flip the direction of J so that the number of β particles emitted along the direction of J and in the direction opposite to J may be counted by the same β particle detector.

Wu in the actual experiment took a small sample of ^{60}Co in some nitrate compound. The atoms of the nitrate could be aligned by a small external magnetic field. Once the atoms of nitrate

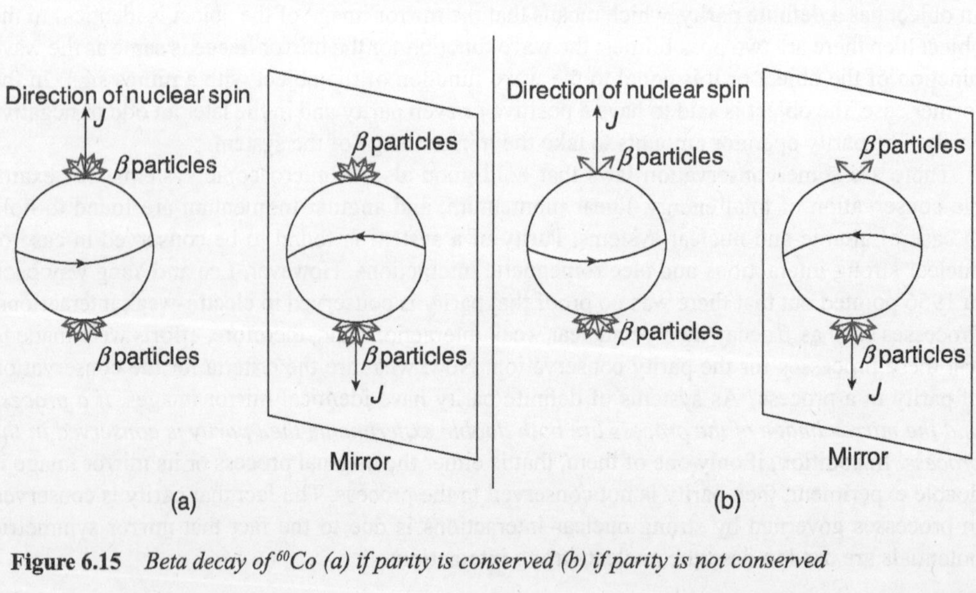

Figure 6.15 *Beta decay of ^{60}Co (a) if parity is conserved (b) if parity is not conserved*

are aligned, these aligned atoms produce a strong internal magnetic field which aligned the spin of ^{60}Co nuclei. In this way, a small external magnetic field was used to polarize the cobalt sample. The direction of polarization of cobalt nuclei can be easily flipped by reversing the direction of the external magnetic field. A thin anthracene scintillation crystal coupled to a photomultiplier tube via light pipe was used to detect β particles emitted from the decay of cobalt.

As the magnetic moment of cobalt nuclei is very small they easily lose their spin alignment due to thermal motion. Therefore, to reduce thermal motion the cobalt sample was initially cooled to 0.01 K and the temperature was further reduced by adiabatic demagnetization. The adiabatic demagnetization cycles were repeated after a fixed time interval. On completion of the cycle of adiabatic demagnetization, the sample starts getting heated and slowly the cobalt nuclei start losing the alignment of spins. Two γ detectors (NaI detectors) one in the polar direction and the other in the equatorial direction were employed to record γ rays emitted by the excited residual nucleus ^{60}Ni. The alignment of spins of ^{60}Co is transferred to ^{60}Ni. Polarized ^{60}Ni nuclei emit more γ rays in the equatorial plane and less in the polar plane. Therefore, when cobalt and nickel nuclei are polarized the γ counting rate in the equatorial counter is more than the polar counter. As the sample heats up more and more nuclei of Ni lose polarization and become randomly oriented and, therefore, the count rate of the two γ counters approach a mean value. Ultimately, when polarization is totally lost the count rate in the two γ counters becomes equal. Thus, the degree of alignment or polarization can be monitored by the difference of the γ count rates in the polar and equatorial counters. Further, the γ count rates depend only on the degree of polarization and the temperature but not on the direction of the external magnetic field.

The experimental lay out is sketched in figure 6.16. As the cobalt sample has to be kept at low temperature, it was kept in an evacuated cryostat. The β particle detector, a thin anthracene crystal, was also placed just above the cobalt sample. The scintillations produced in the thin anthracene crystal were transported to the photomultiplier tube at the top using a light pipe. The thickness of anthracene crystal was kept small to reduce the background. The NaI-γ detectors were kept around the top and on the side of the cryostat. Range of γ rays is large and they may be easily detected at a distance.

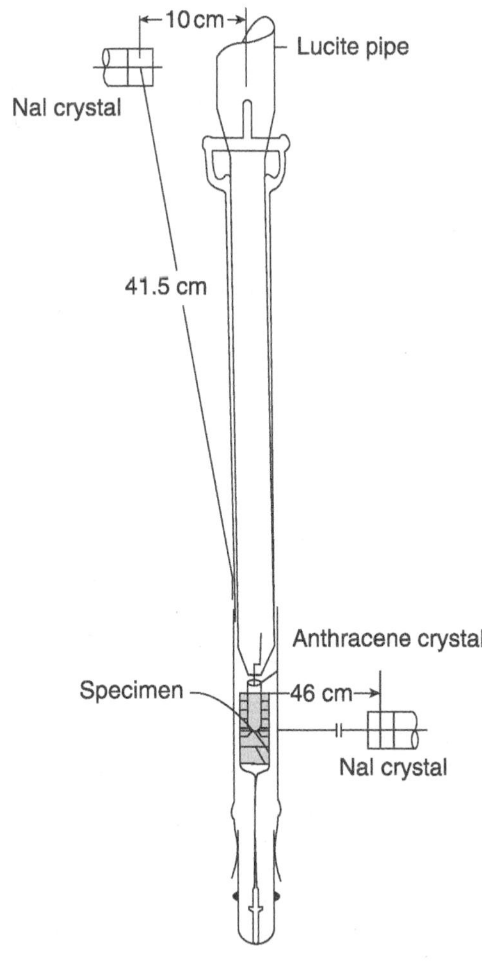

Figure 6.16 *Experimental layout*

The anisotropy of the γ detectors and the β counter were recorded for two directions of the magnetic field. These anisotropies are shown in the graphs. It may be seen that the count rate for β particles is more in the direction opposite to the direction of the spin of ^{60}Co, as indicated by the direction of the external magnetic field. The anisotropies of the γ and β counters decrease with time as the temperature rises. The situation is like the one shown in figure 6.15b that confirms that parity is not conserved in β decay.

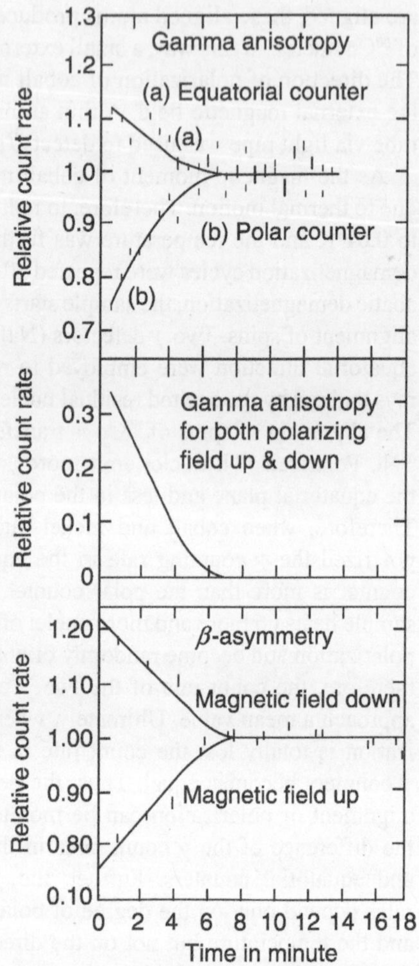

Figure 6.17 *Observed variation in anisotropies of β and γ detectors with time*

6.8.1 Why more β particles are emitted in opposite direction to spin J of ^{60}Co nucleus?

^{60}Co nucleus (spin $J = 5$) decays to ^{60}Ni ($J = 4$) via Gamow–Teller interaction. In this interaction, the outgoing β particle and the antineutrino are in triplet states and, therefore, the spins of neutrino and of β particle are parallel. To conserve the total angular momentum, the spin vectors of both neutrino and the β particle should also be parallel to the spin of the nucleus. As antineutrino is right handed, that is its linear momentum and spin directions are same; it is emitted in the direction of the spin J of the nucleus. However, to conserve linear momentum, β particles are emitted opposite to the direction of emission of neutrino and hence opposite to the direction of J of the nucleus. Let us denote the spins and linear momentum of neutrino and the β particle, respectively, by S_ν, P_ν, S_β, and P_β, then vectors S_ν, P_ν, S_β, will be parallel to J but P_β will be opposite to P_ν and hence to J. Emission of β particle in a direction opposite to J is due to the right-handedness of antineutrino. To conserve the total momentum, the β^- particle behaves as a left-handed particle (spin S_β opposite to P_β).

6.9 THE NEUTRINO AND ITS REST MASS

Initially, the presence of neutrino in β decay was assumed to explain the energy and momentum conservation. However, excellent agreement between the predictions of Fermi theory of β decay and the experimentally observed momentum distributions of emitted β particles confirmed that neutrinos are definitely emitted in β decay. The Kurie plot for the low energy β decay of $^3_1$H put an upper limit on the rest mass of neutrino to be less than 250 eV. Later detailed experiments, such as the one carried out by Kofoed–Hansen conclusively proved that the rest mass of neutrino is zero. Kofoed et al. studied the EC decay of $^{37}{}_{18}Ar$.

$$^{37}_{18}Ar + e^- = ^{37}_{17}Cl + \nu^- \qquad (6.69)$$

As the electron capture occurs in a stationary atom of argon, the ^{37}Cl atom and the neutrino move in opposite directions to conserve the linear momentum. Further, the total decay energy E_{dEC} is shared between the recoiling chlorine atom and the neutrino. Kofoed et al. carefully measured the recoil energy E_r of ^{37}Cl atom and from E_r calculated the momentum P_{Cl} of chlorine atom. Now, the invisible neutrino should also be emitted with same momentum as that of the recoiling Cl atom. That is, $P_v = P_{Cl}$. Assuming that the rest mass of neutrino is zero and that it travels with the velocity of light, the energy $E_v = P_v c$. Therefore, the sum of the energies of neutrino and the recoiling atom of Cl is given by $E_{total} = \frac{P_{Cl}^2}{2M_{Cl}} + P_{Cl}.c$. If the two assumptions made are true then E_{total} should be equal to the decay energy E_{dEC}, which can be calculated using Eq. (6.44). In actual experiment, the value of E_{total} was determined to be 812 ±8 keV. The value of E_{dEC} calculated from the mass difference between Ar and Cl atoms and the difference in the binding energies of the K-shell electron and the valence electron was 813± 2 keV. Excellent agreement in the two values of the E_{total} and E_{dEC} proved beyond any doubt that the rest mass of neutrino is zero and that it travels with the velocity of light.

It is very difficult to directly detect neutrinos as they have very weak interaction with matter. As a result, their mean free path is of the order of light years. It is estimated that neutrinos generated just after the Big Bang are still roaming in the universe.

6.9.1 Neutrino Helicity

An elementary particle in motion has a linear momentum vector p and may have an intrinsic angular momentum vector s, also called spin. When the vectors s and p are parallel, the particle is classified as right handed. It is because its motion is like the motion of a right-handed screw, which moves forward when it is rotated clockwise. In addition, if vector s points opposite to vector p, the particle is classified as left handed. Right or left handedness of a particle is also related to the helicity $\mathcal{H}$ defined as

$$\mathcal{H} \equiv \frac{s.p}{|p|} \qquad (6.70)$$

To determine the helicity of a particle, it is required to know: (1) the direction of its motion (direction of vector p) and (2) the direction of spin vector s. In case of neutrino, which moves with the velocity of light and has almost negligible interaction with matter, it is very difficult to determine the directions of p and s by directly tagging the neutrino. Goldhaber, Grodzins, and Sunyar devised an ingenious method to determine the direction of motion and the direction of spin of neutrino without tagging neutrino. Before describing their experiment, it will be better to understand nuclear resonance fluorescence as their method is based on it.

Suppose there is a nucleus at rest and an excited level of the nucleus at energy E_0 emits a γ ray. Normally, the energy of the emitted γ ray is taken equal to E_0. However, actually the energy of the γ ray is less than E_0 because the emitted γ ray carries a momentum $P_\gamma = (E_\gamma/c)$ and to conserve the linear momentum the emitting nucleus recoils back. The recoil energy is provided by the excitation

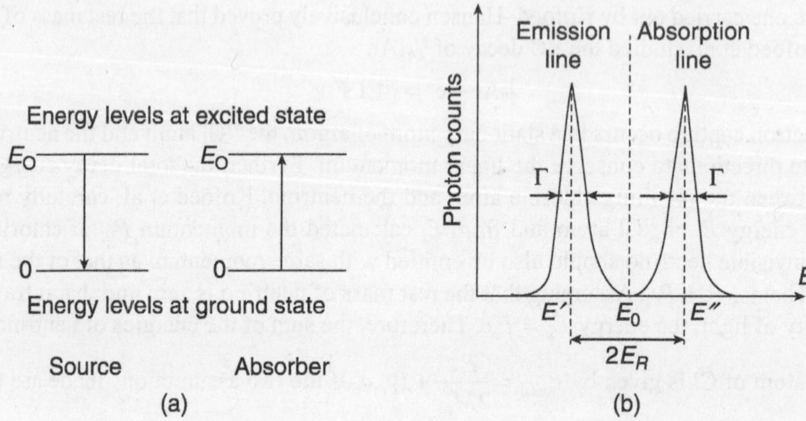

Figure 6.18 *Energy difference between the emission and absorption lines*

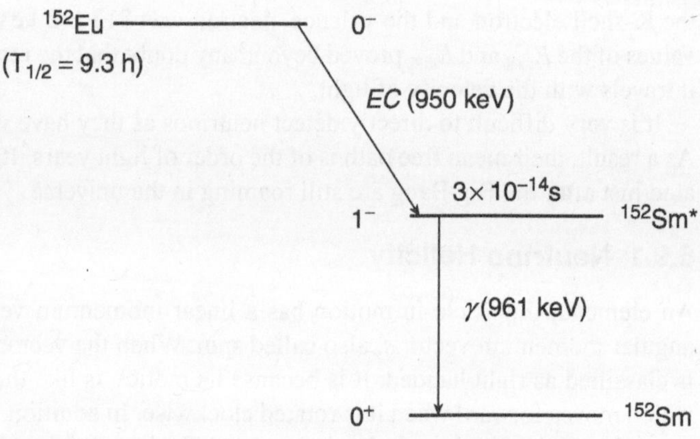

Figure 6.19 *Relevant portion of the decay scheme of ${}^{152}Eu$*

energy E_0. As such, the actual energy of the emitted γ ray is less than E_0 by the amount of the recoil energy. Now, suppose this γ is made to hit an identical nucleus in ground state from which it was emitted. The nucleus in ground state should absorb the incident γ and rise to the excited level. However, this does not happen again because the absorbing nucleus recoils to conserve the linear momentum. Therefore, the energy of the emission line and the absorption lines differ by a factor of twice the recoil energy. This is shown in figure 6.18. For re-absorption of the γ ray, the energy difference between the two lines should be reduced so that the emission and absorption lines overlap. This is called resonance fluorescence. As a result of resonance between the absorption and emission lines, the γ ray is emitted and absorbed simultaneously. One way to achieve resonance fluorescence is to give a push to the emitter nucleus in a direction opposite to the direction of recoil (i.e. in the direction of emission of γ ray) to compensate for the recoil energy loss. Such a push may be delivered by a preceding emission of some other radiation from the nucleus. Goldhaber et al. made use of this technique in their experiment on the determination of the helicity of neutrino. They used the radioactive isotope ${}^{152}Eu$ that decays by electron capture to the excited level at 961 keV of ${}^{152}Sm^*$. The decay scheme of ${}^{152}Eu$ is shown in figure 6.19 and the decay equation is as follows:

$$ {}^{152}_{63}\text{Eu} + e^- \rightarrow {}^{152}_{62}\text{Sm}^* + \nu \tag{6.71} $$

As is clear from Eq. (6.71), after EC the excited nucleus $^{152}Sm^*$ and the neutrino will move in opposite directions. The moving $^{152}Sm^*$ will emit γ rays of 961 keV. Now, in those cases where the 961 keV γ is emitted from $^{152}Sm^*$, which is moving in the direction of the emitted γ ray, the recoil energy will be compensated by the motion of $^{152}Sm^*$. As such, there will be no loss of energy and γ ray of full 961 keV will be emitted. If this full energy γ falls on another nucleus of ^{152}Sm in ground state, it will be absorbed and will be re-emitted. It is important to note that: (1) resonance fluorescence will occur only for those γ rays that are emitted opposite to the direction of motion of neutrino and (2) in such cases the nucleus $^{152}Sm^*$ was moving in the direction of the emission of γ rays, that is opposite to the direction of motion of neutrino. Thus, by observing resonance fluorescence of 961 keV γs the direction of the motion of the neutrino or of its linear momentum p_ν may be determined. The experimental setup used by Goldhaber is shown in figure 6.20. The radioactive sample of ^{152}Eu is placed in a cavity in an electromagnet. The magnetic field is used to align the spins of the nuclei ^{152}Eu of the source. The 961 keV γs emitted on EC travel up to the scatterer Sm_2O_3, which contains atoms of ^{152}Sm in ground state. The scatterer is spread in an annular ring around the lead shielding that protects the photomultiplier tube and the scintillating detector crystal from direct γs of ^{152}Eu. The γ spectrum recorded by the detector is shown in figure 6.21. When some scatterer other than Sm_2O_3 was used no γs of 961 keV were recorded as is indicated by the dotted curve. However, when ^{152}Sm was used as scatterer the resonance fluorescence peak of 961 keV in the spectrum is clearly visible. Therefore, from this part of the experiment it has been identified that the direction of motion of neutrino is upwards.

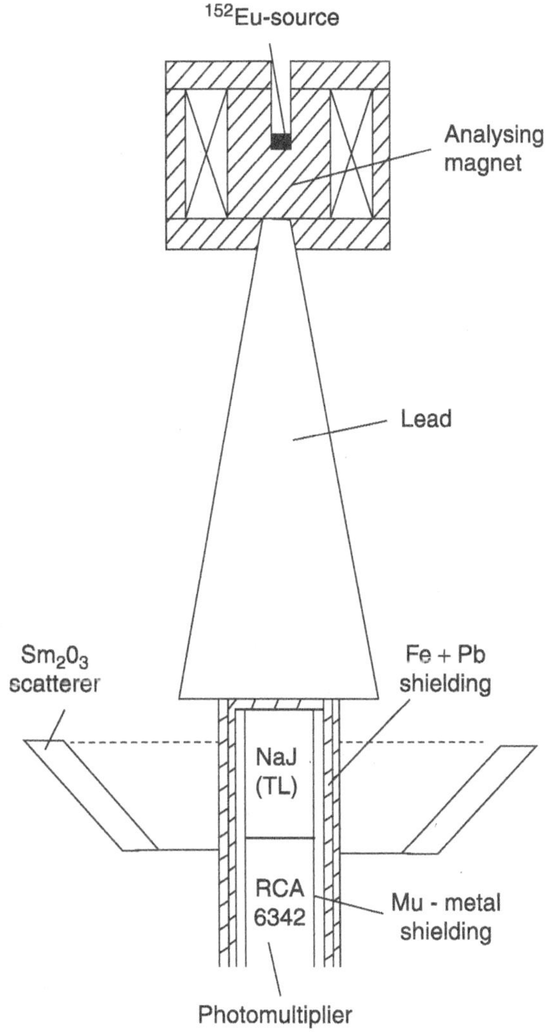

Figure 6.20 *Experimental setup of Goldhaber et al*

The next task was to determine the direction of spin of the neutrino. The spin of the ^{152}Eu is 0^-, when it captures an electron the spin of the composite system ($^{152}Eu + K^-$ electron) becomes $\frac{1}{2}^-$. From this composite system, a neutrino is emitted. If it is assumed that the neutrino is left-handed and that it is emitted in the upward direction (as concluded from the resonance fluorescence experiment) then the spin of the excited level of residual nucleus ^{152}Sm should be 1^-. This excited level emits a γ ray of 961 keV, which carries the angular momentum of 1 unit and leaves the ground level with spin 0^+. This is possible when the spin direction and the direction of motion of the γ ray are opposite to each other. In other words, it may be said that the handedness of neutrino and the 961 keV γ rays will be same. To determine the handedness of γ rays, they were made to hit an iron scatterer. The scattering cross-section for left-handed γs with iron is more than for the right-handed γs. The iron-scattering part of the experiment confirmed that 961 keV γ rays are actually left handed. This in turn proved that neutrinos emitted in EC are also left handed.

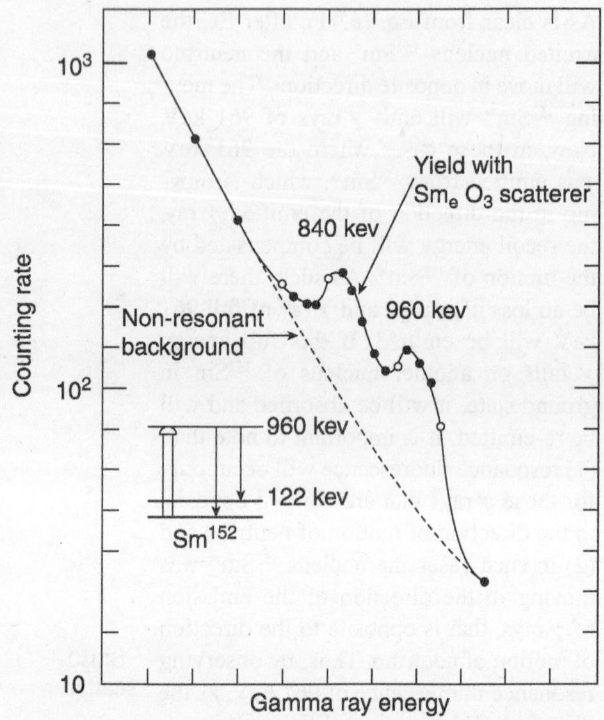

Figure 6.21 *γ spectrum showing resonance fluorescence of 961 keV γ*

6.9.2 Two Types of Neutrinos

Pi-masons (π^0 and $\pi^\mp$) are produced when materials are bombarded by accelerated particles of kinetic energy greater than the threshold for their production. The charged pions decay into muons and neutrino

$$\pi^\mp \rightarrow \mu^\mp + \nu_\mu \tag{6.72}$$

The neutrinos associated with the reactions involving muons are called muonic neutrinos while those which are associated with reactions involving β particles or electrons are called electronic neutrinos and are denoted by ν_e. To verify whether ν_μ and ν_e is one and the same particle or different, Danby et al. carried out an experiment at Brookhaven, USA in 1962. Using 30 MeV proton beam, they produced a large number of charged pions. The pions were allowed to pass through a 13.5-m thick steel shielding. A large spark counter capable of distinguishing electrons from muons was placed on the other side of the shielding. Pions decayed in flight and produced muonic neutrinos. Muonic neutrinos on interaction with nucleons present in the shielding were expected to produce almost equal number of muons and electrons if they were identical to the electron-associated neutrinos. However, in the experiment it was found that very few electrons and predominantly muons were produced. This suggested that the muon-associated neutrinos are different from the electron-associated neutrinos in some property.

Solved example S-6.5

Assuming that the energy spectrum of β particles emitted in natural radioactive decay may be represented by the expression

$$N(E)\,dx = C\,\sqrt{E}\,(E_{\beta_{max}} - E)^2\,dE,$$

where $N(E)$ is the number of β particles with energy E and $E + dE$, C is a constant and $E_{\beta_{max}}$ is the maximum energy of the emitted β particles. Calculate (i) the energy with which largest numbers of β particles are emitted (ii) the average energy of emitted β particles.

Solution. (i) To calculate the energy with which maximum number of particles is emitted, we differentiate $N(E)$ with respect to E and equate it to zero, which gives

$$\frac{dN(E)}{dE} = \frac{d}{dE}\left[C\sqrt{E}\,(E_{\beta_{max}} - E)^2\right] = 0$$

Or

$$2\sqrt{E}\,(E_{\beta_{max}} - E)(-1) + (E_{\beta_{max}} - E)^2 \cdot \frac{1}{2}E^{-1/2} = 0$$

As the energy of the particles that have maximum number cannot be $E_{\beta_{max}}$, we get

$$4E = (E_{\beta_{max}} - E),\ \text{or}\ E = E_{\beta_{max}}/5$$

Ans: Maximum number of β particles will be emitted with energy equal to one-fifth of the end point energy $E_{\beta_{max}}$.

(ii) Suppose there is a function $f(x)$ that varies from $x = a$ to $x = b$ and we wish to find the average value of x, then by definition

The average value of $x = \bar{x} = \dfrac{\int_a^b x f(x)dx}{\int_a^b f(x)dx}$. Using this method we may write

$$\bar{E} = \frac{\int_0^{E_{\beta_{max}}} E\,N(E)dE}{\int_0^{E_{\beta_{max}}} N(E)dE} = \frac{\int_0^{E_{\beta_{max}}} E\sqrt{E}\,(E_{\beta_{max}} - E)^2 dE}{\int_0^{E_{\beta_{max}}} \sqrt{E}\,(E_{\beta_{max}} - E)^2 dE} = \frac{E_{\beta_{max}}}{3}.$$

Ans: The average energy of emitted β particles is one-third of the end point energy.

Exercise p-6.9: Draw rough sketches for the energy spectrum of α particles and β particles emitted in natural radioactive decay. In what respect do the two spectrums differ? Give reasons for the difference.

A radioactive source that emits 3.7×10^{13} β^- particles per second with the end point energy of 1.87 MeV is used to power a fuel cell. Calculate the rate at which heat energy is deposited in the cell by the radioactive source.

[**Hint:** Energy deposited per second = No. of emitted βs per second × average energy of emitted β particle.]

Exercise p-6.10: Using golden rule, derive an expression for the density of levels at energy E_d for the emission of β particles with momentum p and $p + dp$.

Exercise p-6.11: What is the significance of log ft values?

In the following β transitions, the spin, and parity of the parent and daughter nucleus is given in bracket adjacent to their symbol. Identify the type of the transition.

(a) ${}^{17}_{9}F(5/2)^+ \rightarrow {}^{17}_{8}O\left(\frac{5}{2}\right)^+ + \beta^+ + \nu$; (b) ${}^{18}_{9}F(1)^+ \rightarrow {}^{18}_{10}Ne(0)^+ + \beta^- + \bar{\nu}$

(c) ${}^{20}_{9}F(2)^+ \rightarrow {}^{20}_{10}Ne(0)^+ + \beta^- + \bar{\nu}$; (d) ${}^{14}_{8}O(0)^+ \rightarrow {}^{14}_{7}N(0)^+ + \beta^+ + \nu$

(e) ${}^{6}_{2}He(0)^+ \rightarrow {}^{6}_{3}Li(1)^+ + \beta^- + \bar{\nu}$

Exercise p-6.12: It is possible to have electron capture without positron emission but the reverse is not. Why?

In an experiment, β particles from a radioactive source were passed through a magnetic field and from the curvature of a particular β particle its momentum was found to be 2.0 MeV/c. If the decay energy of the source is 2.55 MeV, calculate the energy of the accompanying neutrino.

[**Hint:** As electrons become relativistic at very low energies, we have to use the relativistic expression for energy of the β particle or the electron. If E is the total energy of the β particle that has a momentum 2.0 MeV/c then $E^2 = (p^2c^2 + m_e^2c^4)$; substituting the values of $m_ec^2 =$ 0.511 MeV and $pc = 2.0$ MeV, $E = 2.064$ MeV. Hence, the kinetic energy of the β particle $E_{kin} =$ $E - 0.511 = 1.55$ MeV. The energy of the accompanying neutron is = decay energy – kinetic energy of β particle = 2.55 – 1.55 = 1.0 MeV.]

Exercise p-6.13: What is Fermi–Kurie plot? What information does it provide about the neutrino and the nature of β decay?

Decay of neutrons from nuclear reactors has been thoroughly studied. These experiments established that 0.790 MeV of energy is liberated when a neutron decays into a proton, β^-, and a neutrino. From these data, calculate the minimum kinetic energy of the neutrino that may initiate the inverse reaction

$$P + \nu \rightarrow N + \beta^+.$$

[**Hint:** Neutron decay may be written as $N \rightarrow P + \beta^- + \bar{\nu} + 0.790$ MeV. In terms of energy balance it means $M_N = M_P + M_e + M_\nu + 0.790$ MeV. Here, masses are in MeV. We add the mass of a β particle (i.e. an electron) on both sides of the equation to get $M_N + M_e = M_P + 2M_e + 0.790\,MeV = M_P + 2 \times 0.511 + 0.790 = M_P + 1.812$ MeV (the rest mass of neutrino is zero). To find the minimum energy of the neutrino for the inverse reaction, we work in the CM frame where the total momentum is zero. If the CM energy of neutrino is taken as E_0 then the total energy in the CM for the inverse reaction is $(M_P + E_\nu)^2 =$ $(M_N + M_e)^2$ but we have derived that $(M_N + M_e) = (M_P + 1.812)$, comparing the two values, we get $E_\nu = 1.812$ MeV.]

Exercise p-6.14: Discuss with necessary details Wu's experiment on parity non-conservation in β decay.

Exercise p-6.15: What is helicity of a particle? How was the helicity of neutrino experimentally determined?

Exercise p-6.16: Write a note on neutrino.

6.10 GAMMA (γ) DECAY

An excited nucleus within the two drip lines, first tries to emit heavy particle such as α particle to get rid of the excess energy. If the excitation energy is less than the separation energy of the particle, it de-excites by the emission of electromagnetic radiations, generally called γ rays. The γ emission from an excited nucleus is comparable to the emission of visible and other electromagnetic radiations from an excited atom. However, the energy of γ rays is much larger ≈MeV, as compared to the energy of photons ≈eV and keV emitted in atomic de-excitation. γ rays have much shorter wavelength as compared to the wavelength of visible light.

The atomic number Z and the mass number A of the nucleus do not change in γ decay, only the configuration of nucleons inside the nucleus changes to a more stable state. As a result of the change of nucleonic configuration, oscillations of electric charge and current distributions occur and electromagnetic radiations are generated. γ rays of energy E_γ are produced when an excited nucleus shifts from a level at excitation energy E_2 to a level at lower energy E_1 such that $(E_2 - E_1) = (E_\gamma + E_r)$. Here, E_r is the recoil energy of the emitting nucleus, which moves back to conserve the linear momentum. As the nucleus is much heavy, the recoil energy is small and for all practical purposes the energy of the emitted γ ray is considered to be $(E_2 - E_1)$. However, sometimes the recoil energy E_r plays important role and cannot be neglected, as in the case of resonance fluorescence of γ rays.

Gamma (γ) ray of energy E_γ carries a linear momentum $\frac{E_\gamma}{c}$ and also carry angular momentum in integer multiple of $\hbar$, the minimum value being $1\hbar$. The value of the angular momentum quantum number l depends on the spin and parity of the initial and final states.

6.10.1 Multipolarity of γ rays

Classically, a distribution of static charges in space can be represented as a superposition of electric multipoles, that is of a monopole, dipole, quadrupole, and so on. Similarly, a distribution of current elements may be represented by a superposition of magnetic dipole, quadrupole, octupole, and so on (there can be no magnetic monopole). The elements of both electric and magnetic multipoles may be defined in terms of the spherical harmonics of various orders. In a nucleus, both electric charges and current elements are present because of the charged protons and the motion of both protons and neutrons. Quantum mechanical theory of radiation also uses the classical concept of oscillating electric and magnetic multipoles as the source of radiations. γ ray emitted from an excited nucleus may, therefore, be assigned either electric or magnetic multipolarity depending on the change of the angular momentum and the parity brought about by it. The multipole order of γ transition is 2^ℓ. Electric dipole, electric quadrupole, electric octupole, and so on are denoted, respectively, by $E1$, $E2$, $E3$... while the corresponding magnetic transitions by $M1$, $M2$, $M3$,.... The angular momentum of a γ ray with respect to an emitting or absorbing nucleus is determined by the same type of quantum numbers ℓ and m, as for a material particle. The absolute value of the angular momentum ℓ is $\hbar\sqrt{\ell(\ell+1)}$ with projection on any arbitrary axis $m_\ell\hbar$. As angular momentum is conserved in γ transitions,

$$\ell = |J_i - J_f| \tag{6.73}$$

Here, J_i and J_f are, respectively, the spins of the initial and final states of the nucleus, reached by the emission of γ rays of multipolarity ℓ. Conversely, between states of spins J_i and J_f, the possible values of the angular momentum ℓ carried by the γ ray may have any non-zero integer value given by

$$\Delta J \equiv |J_i - J_f| \leq \ell \leq J_i + J_f \tag{6.74}$$

The magnetic quantum number m of the γ ray is the difference of the magnetic quantum numbers of the initial and final nuclear levels, or

$$m = m_f - m_i \tag{6.75}$$

Although from Eq. (6.74), γ rays of different ℓ values within the limits defined by the equation can be emitted in a transition but from the point of transition probability, ΔJ is the most likely angular momentum of the emitted γ ray. As a thumb rule, γ transitions favour lower multipole order and electric transitions. A transition of type $E(\ell + 1)$ is suppressed by few orders of magnitudes as compared to $E(\ell)$ and a transition of type $M(\ell)$ is nearly as probable as transition $E(\ell + 1)$. Because of the transverse nature of γ rays, monopole transition of $\ell = 0$ are not possible, as a consequence γ transitions between $J_i = 0$ to $J_f = 0$ are not possible. The change of parity $\Delta\pi$ brought about by the γ ray of multipole order ℓ is given by

$$\Delta\pi = (-1)^{\ell} \text{ for electric multipole transitions}$$

$$(-1)^{\ell+1} \text{ for magnetic multipole transitions} \tag{6.76}$$

The change of parity in any transformation is derived from the transition matrix element $\int \psi_i O_{if} \psi_f^* d\tau$ where ψ_i and ψ_f are, respectively, the initial and final state wave functions and O_{if} the transition operator. $d\tau$ is the volume element. In case of electric dipole transition, the transition operator is the electric dipole moment, which may be of the form $(e{\cdot}r)$ and r-changes sign under parity operation $(r \to -r)$. For a non-zero value of the transition matrix, the parities of the initial and final states must be opposite to each other. In the case of $E2$, the transition operator for electric quadrupole moment has the form $(e{\cdot}r^2)$ and does not change sign under parity operation, hence in $E2$ transitions parity does not change.

The transition probabilities for both electric and magnetic multipoles have been calculated by Weisskopf and Moszkowski using single particle wave functions. Weisskopf estimates are

$$\lambda_E(\ell) = \frac{4.4(\ell+1)10^{21}}{\ell[(2\ell+1)\,!!]^2}\left(\frac{3}{\ell+3}\right)^2\left(\frac{E_\gamma}{197}\right)^{2\ell+1} R^{2\ell}\ s^{-1} \tag{6.77}$$

And

$$\lambda_M(\ell) = \frac{1.9(\ell+1)10^{21}}{\ell[(2\ell+1)\,!!]^2}\left(\frac{3}{\ell+3}\right)^2\left(\frac{E_\gamma}{197}\right)^{2\ell+1} R^{2\ell-2}\ s^{-1} \tag{6.78}$$

In expressions (6.77) and (6.78), E_γ is the γ ray energy in MeV and R the nuclear radius in Fermi. The theoretical estimates of the decay rates involve many drastic assumptions, and, therefore, even a satisfactory agreement between theory and experimental values is not expected.

The selection rules for γ rays may be summarized as in table 6.2.

Table 6.2 *Selection rules for γ transitions*

Type of transition	Symbol	Angular momentum ℓ	Parity change
Electric dipole	E1	1	Yes
Magnetic dipole	M1	1	No
Electric quadrupole	E2	2	No
Magnetic quadrupole	M2	2	Yes
Electric octupole	E3	3	Yes
Magnetic octupole	M3	3	No
Electric-16 pole	E4	4	No
Magnetic-16 pole	M4	4	Yes

As the angular momentum of the emitted γ ray is correlated to the spins of the initial and final levels, for an isolated nucleus the angular distribution of γ rays is not isotropic and depends on the multipolarity of the transition. However, in an unpolarized sample where nuclei are randomly oriented the angular distribution appears isotropic. To study the angular distribution of γ rays it is, therefore, required to have a polarized sample. It can be shown that for both the electric and magnetic transitions the angular distribution is proportional to $A_{\ell m}$, which is given by

$$A_{\ell m} = \frac{1}{2\ell(\ell+1)} [\{\ell(\ell+1) - m(m+1)\}] \, |Y_{\ell,m+1}|^2$$

$$+ [\{\ell(\ell+1) - m(m-1)\}] \, |Y_{\ell,m-1}|^2 + 2m^2 \, |Y_{\ell,m}|^2 \tag{6.79}$$

The angular distribution for any combination of ℓ and m may be obtained from Eq. (6.79). As for example for $\ell = 1$ and $m = 1$

$$A_{11} = \frac{3}{16\pi} (1 + \cos^2 \theta)$$

In the special case, when γ_1 emitted from a level of spin J_i feeds a short-lived level of spin j, which in turn emits γ_2 leading to a final stable level of spin J_f, the angular correlation between γ_1 and γ_2 gives information about J, even when the sample is not aligned. Angular correlation studies of γ rays have been extensively used to extract information about the nature of the transition.

6.10.2 Isomeric Transitions

The typical mean life time of γ rays varies from 10^{-15} to 10^{-9} seconds. However, sometimes excited nuclear levels have considerably longer mean lives. Such excited states are called isomeric states or meta-stable states and in the decay scheme meta-stable states are indicated by the superscript m. Generally, the decay of an excited level to a level at lower excitation is very much hindered if the difference in the spins of the two levels is large and the energy difference between the levels is small. The so called islands of isomerism occur for odd-A nuclei just before the magic numbers 50 and 82. Goldhaber et al. studied in details the properties of a large number of isomers of half-lives in the range of 1 second to 8 months. They found that about 50% of them were cases of M4 transitions in which the spin difference between the initial and final levels was $4\hbar$ and the remaining were E3, M3, and E4 transitions. Some isomers with mean lives in the range of 10^{-5} to 10^{-9} s were found to be the cases of E2, M2, and M1 transitions.

6.10.3 Internal (or Inner) conversion (IC)

Oscillations in the electric and magnetic multipole fields of the nucleus give rise to the emission of γ rays. However, it is also possible that these nuclear multipole fields directly interact with the orbital electrons of the atom and transfer the excess energy to the atomic electrons. As a result of this direct interaction between nuclear multipole fields and the atomic electrons, atomic electrons from the inner atomic shells are ejected instead of the emission of γ rays. This is called internal or inner conversion. Internal conversion competes with γ emission. If E_d is the decay energy for the transition then the kinetic energy E_c of the emitted conversion electrons is given by

$$E_c = E_d - B_e \tag{6.80}$$

Being closest to the nucleus, the maximum probability of conversion is for K-electrons. However, electrons from L and M shells are also emitted but with lower probabilities. As the binding energy for K-electrons is largest as compared to the L- and M-electrons, the kinetic energy of K-conversion electrons is least. The energy of conversion electrons is generally measured by magnetic spectrometers where sector magnets are used as momentum selectors and electrons are sorted out according to their momentum. In the energy spectrum of electrons taken by magnetic β spectrometers monoenergetic conversion electron lines are found superimposed over the continuous β spectrum. It is because β decay often leaves the nucleus in excited state, which decays to the ground or to a state of lower excitation by γ emission and internal conversion.

The total decay constant λ for an excited nuclear level is the sum of the decay constants for γ emission λ_γ and the decay constant for internal conversion λ_c.

$$\lambda = \lambda_\gamma + \lambda_c = \lambda_\gamma(1 + \lambda_c/\lambda_\gamma) = \lambda_\gamma(1 + \alpha)$$

where $\alpha = \lambda_c/\lambda_\gamma$ is called the conversion coefficient. The total conversion coefficient α may itself be written as the sum of partial conversion coefficients for the K, L, and M shells.

$$\alpha = \alpha_K + \alpha_L + \alpha_M + \cdots$$

Theoretical estimate of α_K made by Weisskopf may be given as

$$\alpha_K = Z^3\left(\frac{1}{137}\right)^4 \frac{\ell}{\ell+1}\left[\frac{2m_e c^2}{\hbar\nu}\right]^{\ell+5/2} \tag{6.81}$$

As may be seen from Eq. (6.81), the K-conversion coefficient increases with the cube of the charge of the nucleus (more for heavy nuclei) and with the decrease of the energy $h\nu$ of the γ ray. As such, internal conversion is more likely in heavy nuclei for low-energy γ transitions.

Because of the transverse nature of γ ray no γ emission can take place from $J_i = 0$ to $J_f = 0$ level. The only mode of decay for such cases is through internal conversion. There is no competition with γ emission in such cases. Further, in $0 \to 0$ transition, the orbital angular momentum $\ell = \Delta J = 0$, hence energy transfer to the electron takes place in $\ell = 0$ or S-state. As S-state falls within the nucleus, only K- or L-electrons which have finite probability of being inside the nuclear field can get energy from the nuclear multipoles. As a result, mostly K-conversion electrons with very weak L and M conversion lines are observed in these transitions. The complete absence of γ rays is the most characteristic property of $0 \rightarrow 0$ transitions.

In those cases where $\Delta J = \ell > 0$, the nuclear multipole fields extend beyond nuclear dimension. Therefore, contributions to the internal conversion come from both, the region within the nuclear dimensions and the region beyond the nuclear dimensions.

6.11 RESONANCE FLUORESCENCE OF γ RAYS

Resonance fluorescence is a common phenomenon in atomic spectra. A well-known example is the Fraunhoffer lines observed in the spectrum of sunlight. The spectra of sunlight contain a large number of dark lines. These dark lines are assigned to the frequencies of light absorbed by the vapours in the atmosphere of sun. If a particular atom in an excited state emits a line of frequency ν, the same atom in its ground state will absorb the radiation of frequency ν and will thus go to the excited state and will re-emit the radiation of the same frequency. This simultaneous absorption and re-emission of radiations of a specific frequency is called resonance fluorescence. Elements in the core of the sun emit continuous spectrum; atoms in the outer atmosphere of the sun at a lower temperature absorb their characteristic frequencies producing dark lines. As a matter of fact, these dark lines provide valuable information about the elements present in the outer atmosphere of the sun.

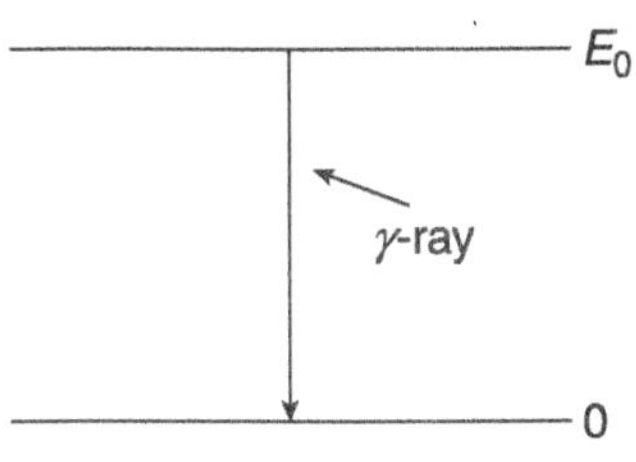

Figure 6.22 *An excited nuclear level at energy E_0 emits a γ ray*

Let us consider an excited level of the nucleus at energy E_0 that emits a γ ray as shown in figure 6.22. If the nucleus was initially at rest it will recoil back to conserve the linear momentum when the γ ray is emitted. If the linear momentum of the nucleus is denoted by p_N and that of the γ ray by p_γ, then

$$|p_N| = |p_\gamma|$$

And the recoil energy E_{rN} of the recoiling nucleus of mass M_N will be

$$E_{rN} = \frac{p_N^2}{2M_N} = \frac{p_v^2}{2M_N} \tag{6.82}$$

However, $p_v = \dfrac{E_v}{c}$, where E_v is the energy of the emitted γ ray. Therefore, the recoil energy of the nucleus is given by

$$E_{rN} = \frac{E_v^2}{2M_N c^2} \tag{6.83}$$

The recoil energy is provided by E_0, the excitation energy of the nuclear level. As such, the actual energy of the emitted γ ray is not E_0 but less than this by the amount of recoil energy E_{rN}. Up to this point the description of the recoil process is equally applicable to the atomic case. In atomic case, also the actual energy of the emitted photon is less than the excitation energy of the emitting level. However, the difference enters through the recoil energy. In atomic case, the energy E_γ is of the order of few keV and, therefore, the recoil energy is relatively much small. In addition, in nuclear case the γ ray energies are in MeV and, therefore, the recoil energy is considerably larger. Now, similar reduction in the energy of the γ ray occurs when it is absorbed by a stationary nucleus of the same type in ground state. As such, the emission line energy and the energy of the absorption line lie $2E_{rN}$ apart. This is shown in figures 6.23 and 6.24. Both the absorption and the emission lines are broad due to their natural line widths and also because of the thermal and Doppler broadening effects. As may be seen in figure 6.24, in atomic case because of the thermal and Doppler broadening the absorption and emission lines have considerable overlap and in the overlap region absorption and re-emission is quite possible. In nuclear case, however, because of the large E_{rN} there is none or very small overlap between the emission and absorption lines and hence no detectable absorption and simultaneous re-emission occurs.

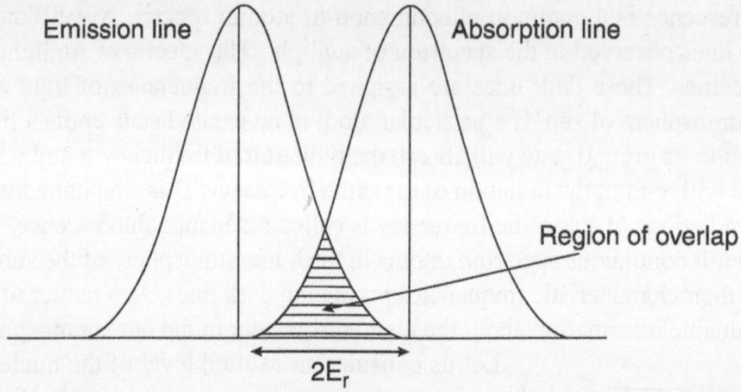

Figure 6.23 *In the atomic case, the emission and absorption lines have considerable overlap*

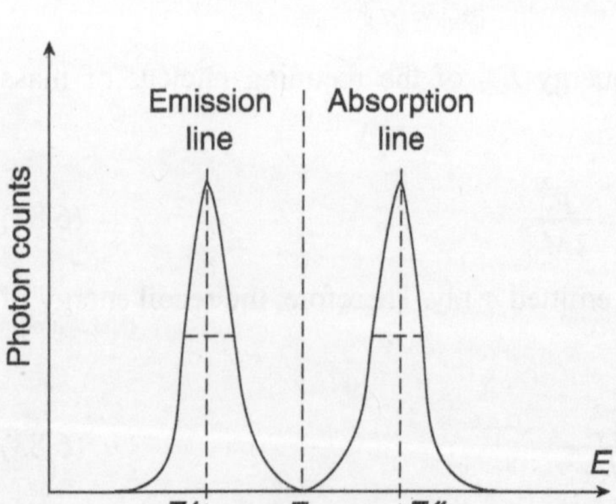

Figure 6.24 *In the nuclear case, the emission and absorption lines either have none or very little overlap*

To observe resonance fluorescence in nuclear case, it is essential that the emission and the absorption lines have sufficient overlap. This requires reduction in the recoil energy. As recoil energy depends on the square of the γ ray energy, very low energy γ transitions stand a better chance of resonance fluorescence.

It is because of this reason that 0.0144 MeV level of ^{57}Fe fed by the EC of ^{57}Co is often used in such studies.

Earlier it was thought that overlap between the emission and absorption lines may be achieved by raising the temperature so that both the lines become broader and overlap. However, it was soon realized that it is counter-productive because, by broadening the lines their intensities fall sharply without any appreciable overlap. On the contrary, it was realized that broadening of the lines should be avoided and, therefore, both the source of emission line and the absorber are rather cooled to low temperatures to reduce the line broadening by thermal motion.

The other two methods by which overlap in the emission and absorption lines may be achieved are: (a) mechanically compensate for the recoil energy (b) to have recoil-less emission and absorption.

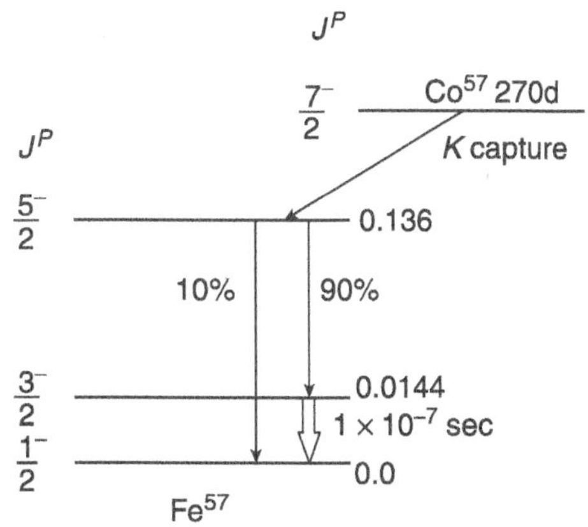

Figure 6.25 *Decay scheme of ^{57}Co*

Mossbauer used the setup shown in figure 6.26 for measuring the natural line width of the γ ray by mechanically compensating for the recoil. S is a low temperature rotating cryostat in which the source of γ rays is mounted. The cryostat is shielded by lead from all sides except a small hole for the γ rays to go out towards a thin absorber foil having same nuclei in ground state. The absorber is also cooled by a low temperature cryostat. A γ detector tuned to the energy of the emitted γ rays is placed behind the absorber foil. The source of the γ rays can be rotated at any desired speed and the readings in the detector are recorded for various speeds of the source when moving towards and away from the absorber.

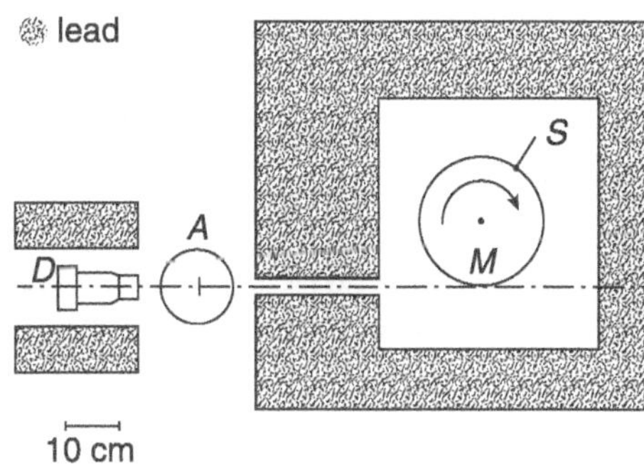

Figure 6.26 *The original set-up used by Mossbauer*

Figure 6.27 shows the fall in the intensity of the γ ray with the speed of the source. Initially, there is no resonance absorption and therefore ΔI, the change in the counting rate or intensity is zero. However, as the speed of the cryostat is increased some absorption starts taking place and it becomes maximum when the speed of the cryostat just compensate for the recoil energy. The structure of the absorption curve gives the natural shape of the emission line.

The recoil-less emission and absorption may be produced if the source and the absorber are embedded in a crystalline lattice. According to Einstein model, the crystalline structure may be considered as an ensemble of harmonic oscillators of frequency ω and the lattices vibrations have discrete energy levels at intervals of $(\hbar\omega)$. If the recoil energy E_{rN} is less than the energy $\hbar\omega$, no recoil will be possible.

The Debye model provides a more realistic description of the spectrum of lattice vibrations. In this model, the relevant quantity is Debye temperature Θ. At temperature $T > 0$ K, modes of lattice vibrations are excited and the probability of energy transfer from γ ray to the lattice increases with T. Quantitatively, the probability P (recoil-less) of recoil less emission is given by

$$P\,(\text{recoil-less}) \approx \left[-\left\{\frac{3}{2}\frac{E_{rN}}{k\Theta}\left(1+\frac{2}{3}\left\{\frac{\pi T}{\Theta}\right\}^2\right)\right\}\right] \quad \text{for } T << \Theta \tag{6.84}$$

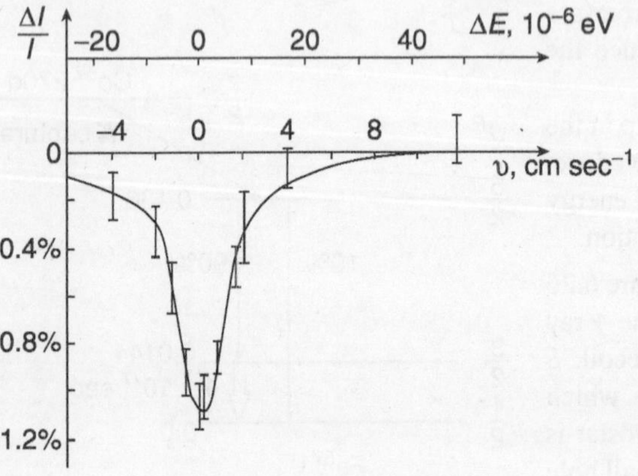

Figure 6.27 *Relative intensity of the γ ray*

As such, P (recoil-less) will be appreciable only when the recoil energy $E_{rN} << k\Theta$.

In modern Mossbauer setup, a γ ray source that is embedded in a suitable crystalline lattice and emits recoil-less γ rays is used. Let the frequency of the γ ray be ν so that its energy is $E_0 = h\nu$. The source is mechanically provided a velocity, say, υ. As a result of this mechanical motion, the frequency and the energy of the emitted γ ray undergo Doppler shift so that the frequency of the γ ray emitted in the direction of the motion becomes $\nu' = \nu\left(1+\frac{\upsilon}{c}\right)$ and the energy $E' = E_0\left(1+\frac{\upsilon}{c}\right)$. The relative change in the energy or the frequency $\frac{\Delta\nu}{\nu} = \frac{\Delta E_0}{E_0} = \frac{\upsilon}{c}$. Assuming that the velocity υ is 1 mm s^{-1}, $\frac{\upsilon}{c} \approx 3.3 \times 10^{-10}$ and, therefore, very small variations in the relative change in energy may be achieved. If the velocity υ is changed continuously the γ rays emitted by the moving source will have a distribution of energies that may compensate for the loss of their energy in recoil when absorbed by n absorber. The absorption peak of the γ ray will now be shifted in energy by an amount $\frac{\upsilon}{c}$.

It is known that the natural shape of all radiation lines is Lorentzian with the width at half maximum denoted by Γ. If the mean life of the level is τ, then from the uncertainty principle, $\Gamma\ \tau = \hbar$. For the most common Mossbauer isotope, ^{57}Fe, the line width Γ is 5×10^{-9} eV. Compared to the γ ray energy of 14.4 keV, this gives a resolution of 1 part in 10^{12} which is like one sheet of paper in the distance between the sun and the earth. This extremely fine resolution makes Mossbauer spectroscopy a very sensitive tool for probing fine and hyperfine splitting of energy levels.

6.11.1 The Red Shift

Although the rest mass of a photon is zero, a moving photon of frequency ν possesses an inertial or gravitational mass equivalent to $\frac{h\nu}{c^2}$. Therefore, when a photon drops through a height H in the gravitational field of earth its energy decreases by the amount $\left[\left(\frac{h\nu}{c^2}\right)gH\right]$. This loss of energy reduces the frequency and increases the wavelength of the photon, which is referred as red shift. The relative change in energy or frequency of the photon due to the red shift, $\frac{\Delta\nu}{\nu}$ is equal to $\frac{g \cdot H}{c^2}$. It is obvious that the magnitude of the red shift is very small; however, it has been successfully measured using the Mossbauer technique.

6.11.2 Applications of Resonance Fluorescence

Resonance fluorescence provides a tool to determine very precisely the energy of γ rays. The energy levels of a nucleus are very much influenced by the environment around the nucleus, particularly in lattice of a solid. The physical and chemical environment around a nucleus in a crystalline lattice may shift and/or split the energy levels of the nucleus.

6.11.2.1 Isomer shift (also called chemical shift)

The electrostatic energy of the nucleus in the potential caused by the atomic electrons may be different for the ground state and the γ-active excited state, due to slightly different nuclear radii of these two states. These shifts are different for different chemical bonds and can thus be observed if the source and the absorber have different chemical composition.

6.11.2.2 Magnetic dipole splitting

A magnetic field at the nuclear site lifts the degeneracy of the magnetic nuclear levels, leading to a γ ray spectrum with several lines.

6.11.2.3 Electric quadrupole splitting

An electric field gradient interacts with the nuclear electric quadrupole moment leading to level splitting for the different nuclear orientations and corresponding splitting in the γ ray spectrum.

Mossbauer technique has been extensively used in biology, for example in determining the chemical state of iron in haemoglobin. It has also been used in astronomy to determine the chemical state of iron in rocks of heavenly bodies such as Mars. Accurate determination of gravitational red shift has also been achieved using this technique.

Exercise p-6.17: Discuss how γ rays are produced and what is meant by their electric and magnetic multipolarity.

An excited state of an even–even nucleus decays to the ground state in two steps: first by emitting a electric dipole radiation of 800 keV followed by a 600 keV M2 transition. What may be the possible values of the spin and parity of the excited state?

Exercise p-6.18: In an experiment red shift for the 14 keV, γ rays from ^{57}Fe source was determined using the Mossbauer method. The γ source was placed at the top of a 30 m high tower at the bottom of which the absorber iron foil was kept on a platform that may be moved up or down with the desired speed. Determine the speed and the direction of motion of the absorber to compensate for the red shift.

Exercise p-6.19: What is electron capture? When EC has no competition with γ emission?

The nucleus of a stationary atom of mass M emits γ ray and reverts to a level that is lower in energy by the amount ΔE. Calculate the frequency of the emitted γ ray.

[**Hint:** If ν is the frequency of the emitted γ ray than its momentum P will be $P = (h\nu/c)$. Same will be the magnitude of the linear momentum of the recoiling nucleus. Therefore, the recoil energy of the nucleus of mass M will be $E_{recoil} = (p^2/2M = (h\nu)^2/2Mc^2$.

Therefore, $h\nu = \Delta E - \frac{(\nu h)^2}{2Mc^2}$. This is a quadratic equation in ν and may be solved to get the roots. Only one root will be physically possible. The square root term may be expanded using Binomial theorem to get $= \left(\frac{\Delta E}{h}\right)\left[1 - \frac{\Delta E}{Mc^2}\right]$

Exercise p-6.20: Resonance fluorescence is common in atomic transitions but not in the case of the nuclear transitions. Give reasons and discuss methods to achieve it in case of γ rays.

In a Mossbauer experiment, the absorption peak of 114 keV γ ray from ^{57}Fe source was found to get shifted by the amount of its natural width Γ at the source speed of 1 cm s^{-1}. Determine the value of Γ.

Multiple choice questions

Note: Some of the following questions have more than one correct alternative. Select all possible correct alternatives in such cases.

Exercise M-6.1: β^+ emission always competes with

(a) β^- emission (b) α particle emission
(c) emission of γ rays (d) electron capture

Exercise M-6.2: K-conversion is the only mode of decay for the transition

(a) $J_i = 0$ to $J_f = 0$ (b) $J_i = 0$ to $J_f = 1/2$ (c) $J_i = 0$ to $J_f = 1$ (d) $J_i = 1/2$ to $J_f = 3/2$

Exercise M-6.3: A nuclide lying beyond the neutron drip line is formed in a nuclear reaction. It will instantly decay by emitting

(a) β^- (b) β^+ (c) a proton (d) neutron

Exercise M-6.4: At some instant '*t*' the ratio of the parent activity to the daughter activity is 1/2. The decay constants of the parent and the daughter are in the ratio

(a) 1:2 (b) 2:1 (c) 1:1 (d) 4:1

Exercise M-6.5: In a two-step series decay, the decay constant of the daughter is twice that of the parent and the time of the maximum activity is 0.1 log2. The mean life of the daughter in seconds is

(a) 20 (b) 10 (c) 5 (d) 2.5

Exercise M-6.6: The fine structure in α particle energy spectrum is due to the

(a) groups of α particles that feed different excited levels of the residual nucleus
(b) difference in the separation energies of neutrons and protons
(c) non-conservation of parity in the decay
(d) γ rays that are emitted by the nucleus that emit α particles

Exercise M-6.7: A sample containing 10^{18} nuclei of type X is hit by an accelerated proton beam of flux 10^7 protons per metre^2 per second (m^{-2} s^{-1}). The X(p,n) reaction produces the residual nucleus Y of half-life 1 second. If the cross-section for the X(p,n)Y reaction is 100 barn, the approximate number of nuclide Y accumulated in an hour of irradiation will be

(a) 120 (b) 60 (c) 30 (d) 15

Exercise M-6.8: According to Gamow theory of α decay, the half-life of an α emitter is very sensitive to the kinetic energy of the emitted α particle because

(a) the height of the potential barrier changes rapidly with the energy of the emitted α particle.
(b) the magnitude of the transmission coefficient changes rapidly with the kinetic energy of the emitted α particle
(c) the total energy available for decay changes rapidly with the kinetic energy of the emitted α.
(d) the width of the potential barrier changes with the kinetic energy of the emitted α.

Exercise M-6.9: The nature of the energy spectrum of β particles emitted in natural radioactive decay is

(a) continuous
(b) discrete mono energetic
(c) discrete with fine structure
(d) continuous with fine structure

Exercise M-6.10: A non-linear Fermi–Kurie plot indicates that

(a) the transition is forbidden.
(b) the curve is due to the mixture of more than one allowed β transitions.
(c) the transition is super allowed.
(d) the rest mass of neutrino is not zero.

Exercise M-6.11: Parity is not conserved in

(a) strong nuclear interactions.
(b) weak nuclear interactions.
(c) electromagnetic interactions.
(d) gravitational interactions.

Exercise M-6.12: An excited level of a nucleus has spin 3/2 and parity positive. It decays to a level of spin 1/2 and parity negative by γ emission. The possible types of γ transitions are

(a) E1, M1 (b) E1, M2 (c) E1, E2 (d) M1, E2

Exercise M-6.13: The activity of a radioactive sample drops by a factor of 32 in 1 hour and 15 minutes. The half-life of the material is

(a) 25 seconds (b) 25 minutes (c) 15 seconds (d) 15 minutes

Exercise M-6.14: In an experiment using Mossbauer technique, the relative change in the frequency due to red shift for two γ ray sources of energies E_1 and E_2 was found to be respectively X_1 and X_2. If E_1/E_2 is 1.5 then the ratio X_1/X_2 will be

(a) 0.44 (b) 0.67 (c) 1.0 (d) 1.5

Exercise M-6.15: At approximately what number of neutrons or protons does the ratio 1:1 of neutrons and protons start to produce unstable nuclei?

(a) 10 (b) 20 (c) 30 (d) 50

Exercise M-6.16: The basis of carbon-14 dating method is

(a) ^{14}C is very unstable and is readily lost from the atmosphere.
(b) Living tissues will absorb ^{12}C and not ^{14}C.
(c) The ratio of ^{12}C to ^{14}C in atmosphere is constant.
(d) The amount of ^{14}C in all objects is same.

Exercise M-6.17: Which of the following can be done to shorten the half-life of radioactive decay of ^{252}Cf?

(a) Convert to CfF_6 (b) Oxidize to the +2 oxidation state
(c) Bombard it with α particles (d) None of these

Exercise M-6.18: The half-life of the radioactive ^{137}Cs is 30 years. How many years must pass to reduce a 25 mg sample to 8.7 mg?

(a) 3.6 (b) 36 (c) 46 (d) 56

Exercise M-6.19: An excited nuclear level of spin 0 and parity even decays by the emission of γ rays of type E1 to an intermediate level that in turn undergoes M1 transition to the ground state. The spin and parity of the ground state may be

(a) 0^- (b) 1^+ (c) 2^+ (d) 3^-

Exercise M-6.20: The gravitational mass of a photon of frequency ν is

(a) $\left(\frac{h\nu}{c^2}\right)$ (b) $\left(\frac{h\nu}{c}\right)$ (c) $\left(\frac{\nu^2}{hc}\right)$ (d) $\left(\frac{c^2}{h\nu}\right)$

7

Nuclear Radiations and Detectors

Nuclear radiations may be of two types: charged particles such as protons, α particles, and heavier ions and charge-less radiations such as γ rays, neutrons, and neutrinos. Charged particles may be further classified as heavy charged particles, such as protons, α particles, fission fragments, and heavy ions, and light charged particle, such as electrons and charged muons. Electron, positron, or β particles are rather special because of their very small mass. All radiations, irrespective of the nature of their charge, lose energy while they pass through matter. As the amount of energy lost by the radiation is specific to the radiation, it is used in designing radiation detectors. Knowledge of energy loss processes is also important for radiation therapy, designing radiation shielding, and for radiation dosimetery. It is, therefore, important to know the processes by which different radiations lose energy in passing through different materials.

When a beam of radiation of any kind penetrates matter, some of the radiation may be absorbed completely, some may be scattered, and some may pass straight through without any interaction at all. The processes of absorption and scattering can be described and explained in terms of interactions between incident radiations and the electrons, atoms, and molecules of the target material. There are two broad kinds of process by which a particle travelling through matter can lose energy. In the first kind, the energy loss is gradual, and the particle loses energy nearly continuously through many interactions with the surrounding material. In the second kind, the energy loss is catastrophic, and the incident radiation moves without any interaction at all through the material until, in a single collision, it loses all its energy. The motion of charged particles through matter is characterized by gradual energy loss whereas photon interactions are of the 'all-or-nothing' type. All-or-nothing type of interactions are also called 'one-shot' processes.

7.1 ATTENUATION COEFFICIENTS

If the interactions are of the 'all-or nothing' type, then the *attenuation* of a beam of particles with identical energies, all travelling in the same direction, is described by an exponential law. If at some distance into the material N_0 particles are moving through a slab of material, then after penetrating an extra distance x it is found that the number of particles in the beam is reduced to: $N(x)$, then

$$N(x) = N_0\, e^{-\mu_t x} \tag{7.1}$$

This exponential attenuation law follows from the fact that, over any short distance, the probability of losing a particle from the beam is proportional to the number of particles left. Where there are many particles many will be lost, but as the number of particles that have left decreases so does the rate of loss. Gamma ray photons lose their energy by such 'one shot' process while passing through matter (figure 7.1).

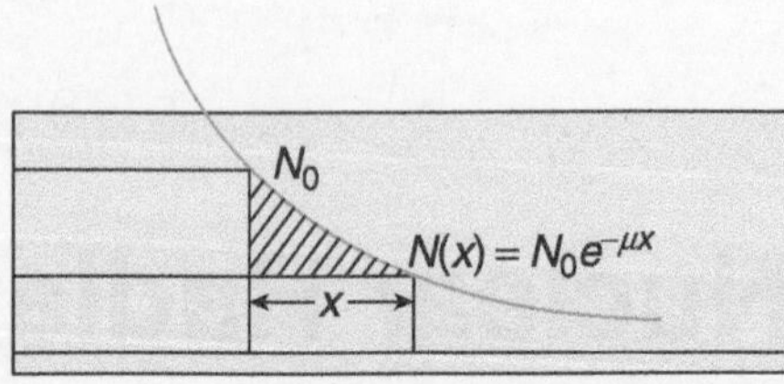

Figure 7.1 *Reduction in the number of particles after travelling additional distance x*

An alternative way of expressing the exponential attenuation law is to replace the linear attenuation coefficient, $\mu\ell$ by its reciprocal:

$$\lambda = \frac{1}{\mu_\ell}$$

so that

$$N(x) = N_0\, e^{-\frac{x}{\lambda}} \tag{7.2}$$

Here, λ is known as the ***attenuation length*** or ***mean free path***. It is the average distance travelled by a photon before it is absorbed. The distance over which one half of the initial beam is absorbed is called the *half thickness* $x_{1/2}$. It is related to the linear attenuation coefficient and the mean free path by

$$X_{1/2} = \frac{\ln 2}{\mu_\ell} = (\ln 2)\lambda = 0.693\ \lambda \tag{7.3}$$

As photons interact with individual atoms, the probability that a photon will interact somewhere within a slab of matter depends on the total number of atoms ahead of it along its path. Therefore, the attenuation of radiation depends on the amount of material in the beam's path and not on how it is distributed. It is useful, therefore, to describe the attenuation process in a way that does not depend on the density of material, only on what kind of stuff it is. We can achieve that by defining the *mass attenuation coefficient* μ_m, which is related to the linear attenuation coefficient by

$$\mu_\ell = \mu_m\,\rho \tag{7.4}$$

Here ρ is the density of the material. This means that the mass attenuation coefficient is the same for ice, liquid water and steam whereas the linear attenuation coefficients will differ greatly. The total attenuating effect of a slab of given type of material can be described by quoting the mass attenuation coefficient, which is characteristic of the material's chemical composition and the photon energy, together with the material's density and its thickness. We now have yet another way of writing the attenuation law:

$$N(x) = N_0 e^{-\mu_\ell x} = N_0\, e^{-\mu_m \rho x} \tag{7.5}$$

The product ρx, the mass per unit area of a thickness x of the attenuating material, is also called the *density-thickness*. It is often quoted instead of the linear thickness x. Although the SI unit of density-thickness is kg.m^{-2}, the obsolete unit g.cm^{-2} is still used in practice. If an absorber is made up of a composite of materials the mass attenuation coefficient is readily calculated by adding together the products of the mass attenuation coefficient and the relative percentage of the mass due to each element, for all the elements present in the material, i.e.

$$\mu_m^{total} = \sum p^i \mu_m^i \tag{7.6}$$

Where p^i and μ_m^i are respectively the percentage of the mass and the mass attenuation coefficient of the i-th component material. The exponential law will always describe the attenuation of the ***original*** radiation by matter. If the radiation has changed, degraded in energy (and not totally absorbed) or if secondary particles are produced then the effective attenuation is less so the radiation will penetrate more deeply into matter than is predicted by the exponential law alone. Indeed it is possible to get an increase in the number of particles with depth in the material. The process is called *build-up* and has to be taken into account when evaluating the effect of radiation shielding.

7.2 ENERGY LOSS BY ELECTROMAGNETIC RADIATION

Electromagnetic radiations are named according to the mode of their origin. γ rays are produced in nuclear interactions. X-rays and characteristic X-rays originate in electronic transitions in atoms while bremsstrahlung is produced when a charged particle accelerates or decelerates.

In principle there may be twelve different ways in which a quanta of electromagnetic energy, called photon, may interact with atoms, atomic electrons, and the nucleus of the atom of the material through which it passes. However, some of the processes have measurable but very small cross sections while some others have not been observed because of their negligible cross sections. For example, **Rayleigh scattering** in which the incident electromagnetic radiations interact coherently and elastically with the tightly bound electrons of the atom is important only for very low energy photons when they interact with high Z atoms. Essentially there is no transfer or loss of energy by the incident photon in Rayleigh scattering and the probability of the process decreases sharply with the increase in the incident photon energy. Processes like the **Delbruck scattering,** where the incident photons of energy greater than 1.02 MeV are expected to produce virtual electron-positron pairs through elastic scattering, has not been observed.

The three interactions, namely, **photo electric effect, Compton scattering** and **pair production** are the dominant modes of energy loss by photons. Each of the above interaction peaks and dominates in a different energy range as is shown in figure 7.2.

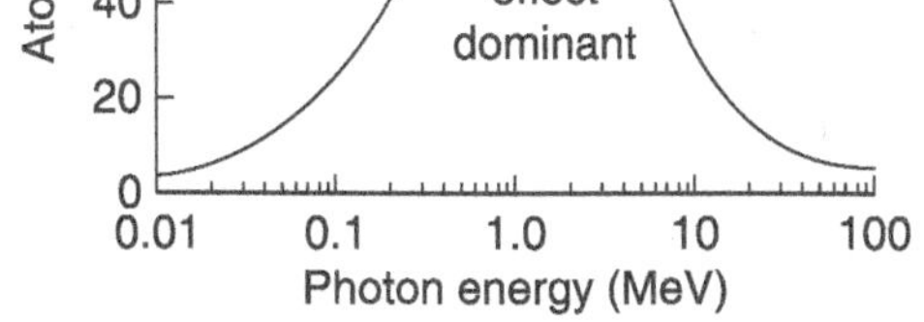

Figure 7.2 *Regions of dominance of energy loss processes as a function of photon energy*

7.2.1 Photo Electric Effect

In photo electric effect the incoming photon interacts with the bound electron of the atom, ejects the electron out, loses all its energy and disappears in a 'one-shot' process. Obviously the energy

$h\nu$ of the photon must be more than the binding energy B_e of the electron with the atom. The incoming photon brings in a linear momentum $\frac{h\nu}{c}$. The target atom recoils to conserve the linear momentum, however, the recoil energy is much too small because of the larger mass of the atom, it is neglected. The ejected electron called photoelectron, moves with a kinetic energy E_e given by

$$E_e = h\nu - B_e \tag{7.7}$$

As the target atom loses an electron and becomes an ion, the photoelectric effect is also called photo-ionization. Although in photoelectric effect the recoil energy of the atom is neglected, the momentum and the energy conservations must hold for the photoelectric effect to take place. The probability for the recoil of the atom depends on how tightly the photoelectron is bound to the atom. As K-shell electron in an atom is most tightly bound, the probability of photoelectric effect is largest with the K-shell electrons of the atom.

In physics, relative values of quantities are often more important than their absolute values. For example, in photoelectric effect the recoil energy may be neglected in comparison to the kinetic energy of the photo electron. In a similar way, the binding energy of the electron to the atom may become insignificant if the energy of the incoming photon is very large. In that case, to a high energy photon each electron of the atom will appear as if it is free, not bound to the atom. Consequently, the probability of photo electric effect, which requires a bound electron for momentum conservation, will fall with the increasing photon energy. As such with the increase of the photon energy, the photo electric effect gives way to another process called *Compton scattering*. Linear momentum is still conserved in Compton scattering as well, but between the electron and the photon, and not between the photon and the atom.

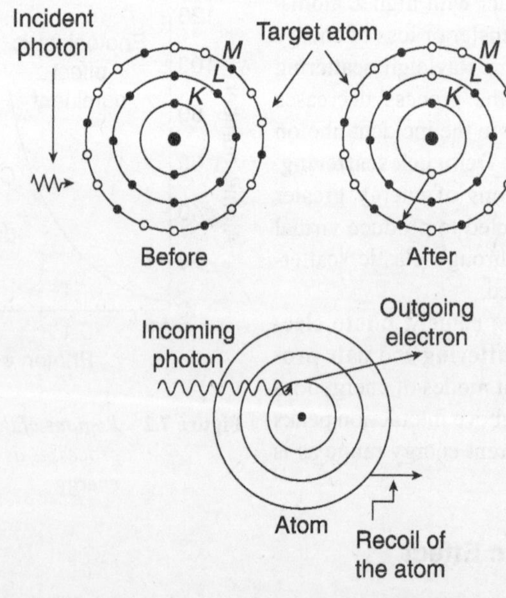

Figure 7.3 *Schematic representation of photo electric effect*

Exact quantum mechanical treatment of photoelectric effect is quite involved as it requires the use of Dirac's relativistic equations for bound electrons. Most of the information about photoelectric effect is, therefore, empirical, and approximate theoretical estimates are used to extrapolate or supplement the experimental data. Following conclusions may be drawn from the analysis of the experimental as well as theoretical data:

1. For photons of very low energy ($h\nu \ll$ rest mass of the electron but more than B_e) for K-shell electron) the photoelectric cross-section is proportional ($\propto$)$Z^5 E_v^{-3.5}$, where E_v is the energy of the photon.
2. At higher photon energies, the Z-dependence remains almost as the 5th power of Z, but the energy dependence changes from (−3.5) to (−1) at very high energy.
3. The photo electrons are emitted mostly in the direction of the incident photon at high photon energies. However, the peak of the photon emission shifts to higher emission angles as the energy of the incident photon decreases. For example, for 0.02 MeV photon the most probable angle of emission is around 70° with respect to the direction of the incident photon.
4. The photoelectric effect leaves the target atom with an electron vacancy in inner shell. Electrons from higher shells fill the vacancy emitting characteristic X-rays or Auger electrons. The ion, before capturing an electron from the surrounding, may produce ionization/excitation of the nearby atoms of the material.

The mass attenuation coefficient for photoelectric absorption decreases with increasing photon energy, that is as general rule, high energy photons are more penetrating than low energy radiations. For a fixed value of the energy, the attenuation coefficient increases with the atomic number Z of the substance.

On a graph (figure 7.4) showing how attenuation varies with photon energy, the general decrease of μ_m with increasing energy is interrupted by a series of jumps, called *absorption edges*, which occur at a unique set of energies for each element. An absorption edge occurs when the photon is energetic enough to eject an electron from one of the deep energy levels in the atom. If a photon's energy is even a little less than that required for raising the tightly bound electron to a vacant energy level, then that photon cannot interact with the inner electron whereas, if the photon's energy were to be made equal to or greater than the required energy, the interaction suddenly becomes possible. Therefore, there are sudden changes in interaction probabilities and attenuation coefficients at photon energies corresponding to quantum jumps between filled and unfilled energy levels of atoms in the absorber. The edge with the highest photon energy

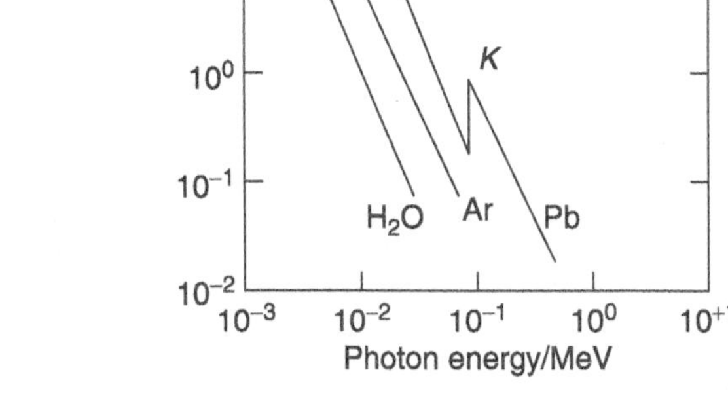

Figure 7.4 *Mass absorption coefficient for photoelectric effect in water, argon, and lead*

corresponds to an interaction with an electron in the K-shell. The next set of edges corresponds to the group of closely spaced energy levels in the L-shell. The energies of the absorption edges are characteristic of each element and, such as the atomic spectra, they can be used to identify elements.

In metals, the valence electrons are not attached to individual atoms but they are relatively free to hop from one atom to the other within the metallic lattice. These electrons are called free electrons and their average binding energy is smaller as compared to the binding energy of inner electrons strongly bound to the atom. Incident photon may eject a free electron from the metallic surface. This process is also called photoelectric effect.

In case of semiconductors, valence electrons are shared between neighbouring atoms giving rise to the covalent bonding. The electron energy levels in semi-conductors are grouped in two distinct bands called the conduction and the valence bands with a forbidden energy gap ($\approx$1 eV) between them. Electrons in the conduction band contribute to the flow of current in semi-conductors. Incident photon in case of semi-conductors may lift electron from valence band to the conduction band as shown in figure 7.5.

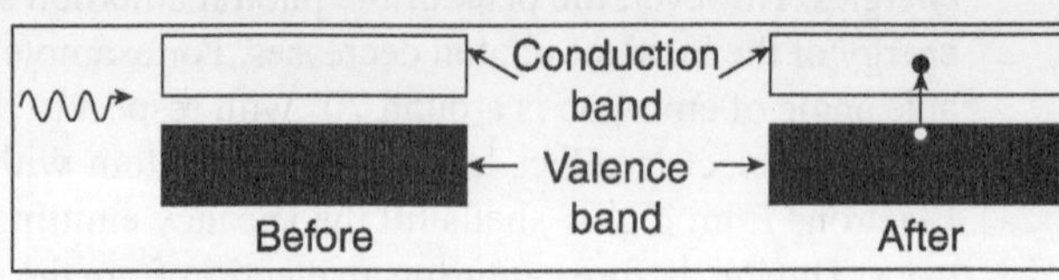

Figure 7.5 *An incident photon in semi-conductor may lift an electron from valence band to conduction band*

Most of the detectors used for γ ray energy measurements utilize the photoelectric effect. It is because in photoelectric effect almost all the energy of the γ ray is transferred to the photoelectron. Trapping the energy of photoelectron and converting it into electrical signal makes it possible to know the energy of incident photon. Materials with high Z are more suitable for such detectors. Further, the large thickness of detecting material makes the detector more efficient for photoelectric effect and in turn for γ detection.

7.2.2 The Compton Scattering

If the energy of the incoming photon (or γ ray) is much larger than the binding energy of the electron in the atom, the photon is scattered by the atomic electron treating the electron unbound and free. In Compton scattering, the incident photon transfers a part of its energy to the atomic electron and itself gets scattered with reduced energy. Compton scattering is also a 'one shot process' as far as the original photon is concerned. After scattering, there is a scattered photon that has less energy than the original photon and is different from it. Hence, in Compton scattering, the original photon is removed from the beam in a single collision with the electron. The probability of Compton scattering is more with the electron that is most weakly bound to the atom, that is, with the valence electron. The energy and momentum conservation in Compton scattering takes place between the incident photon, the scattered electron and the scattered photon (figure 7.6).

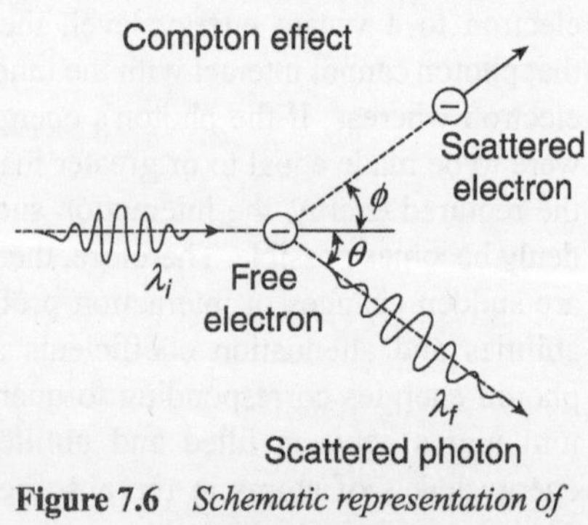

Figure 7.6 *Schematic representation of Compton scattering*

If p denotes the linear momentum of the scattered electron, E_e its kinetic energy, and ν and ν', respectively, the frequencies of the original and scattered photons, the conservation of momentum and of energy lead to the following relations:

$$\frac{h\nu'}{c}\sin\theta - p\sin\phi = 0 \tag{7.8}$$

$$\frac{h\nu'}{c}\cos\theta + p\cos\phi = \frac{h\nu}{c} \tag{7.9}$$

and

$$pc = \sqrt{E_e(E_e + 2m_e c^2)} \tag{7.10}$$

It is easy to obtain the following relations from Eqs. (7.8)–(7.10):

The Compton shift: $$\Delta\lambda = \frac{c}{\nu'} - \frac{c}{\nu} = \lambda' - \lambda = \frac{h}{m_e c}(1 - \cos\theta) \tag{7.11}$$

Energy of the scattered γ ray: $$h\nu' = \frac{m_e c^2}{1 - \cos\theta + \left(\frac{1}{\alpha}\right)} \tag{7.12}$$

and

$$\frac{\nu'}{\nu} = \frac{1}{1 + \alpha(1 - \cos\theta)} \tag{7.13}$$

where

$$\alpha = \frac{h\nu}{m_e c^2} \tag{7.14}$$

The energy of the struck electron $$E_e = h\nu\frac{\alpha(1 - \cos\theta)}{1 + \alpha\,(1 - \cos\theta)} \tag{7.15}$$

Maximum energy transferred to electron $$E_e^{max} = \frac{h\nu}{1 + \left(\frac{1}{\alpha}\right)} \tag{7.16}$$

Following conclusions may be drawn from the set of equations:

1. The Compton shift in wavelength in any direction is independent of the energy of the incident photon.
2. On the contrary, the Compton shift in energy is very strongly dependent on the energy of incident photon.
3. For a fixed angle, the high-energy photon suffers a very large change in energy while the low-energy photon suffers only a moderate change in energy.
4. The length $h/m_e c = 2.426 \times 10^{-10}$ cm is called the Compton wavelength λ_c and it is equal to the wavelength of a photon of energy equal to the rest mass energy of electron.

The energy distribution of scattered electrons for two energies of incident γ rays is shown in figure 7.7. It may be observed that the number of scattered electrons remains nearly same for all energies below the energy of the incident γ ray, increases a little bit towards the higher energy end, and drops down to zero abruptly just below the energy of the incident photon. The rise of the distribution curve at the higher energy end is called the Compton edge.

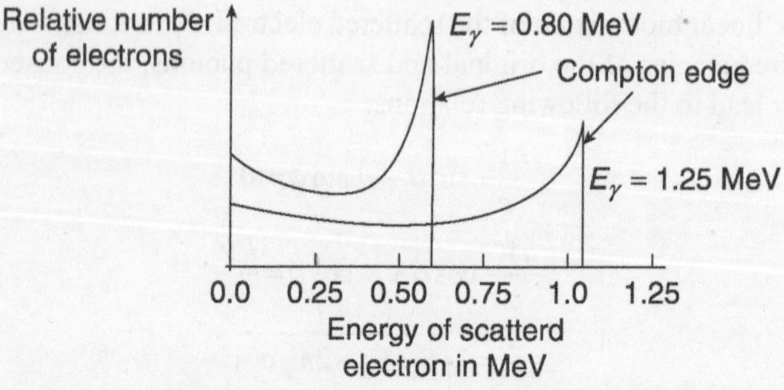

Figure 7.7 *Energy distribution of scattered electrons*

Klein and Nishina calculated Compton scattering cross-section per electron σ^{e}_{comp} using Dirac's relativistic electron equation as follows:

$$\sigma^{e}_{comp} = \left(\frac{\pi^2 e^2}{4\pi\varepsilon_0 m_e c^2}\right)^2 \left[\left(1 - \frac{2\{\gamma+1\}}{\gamma^2}\right)\ln(2\gamma+1) + \frac{1}{2} + \frac{4}{\gamma} - \frac{1}{2(2\gamma+1)^2}\right] \tag{7.17}$$

where $\gamma = \dfrac{h\nu}{m_e c^2}$. The total cross section per atom $\sigma^{atom}_{comp} = Z\,\sigma^{e}_{comp}$. Here, Z is the number of electrons per atom. It may be noted that Compton scattering cross-section for a material is proportional to the atomic number Z of the material and is inversely proportional to the energy of the incident γ ray. It dominates between 0.5 MeV and 5.0 MeV γs for most elements.

We have seen that in photoelectric effect each incident γ produces one photoelectron and one recoiling ion. In Compton scattering, each incident γ produces one electron, one scattered γ of lower energy, and an ion of the target atom. Therefore, the total number of radiations after the interaction increases both in photoelectric effect and in Compton scattering. This increase in the number of radiations plays an important role in designing radiation shielding (figure 7.8).

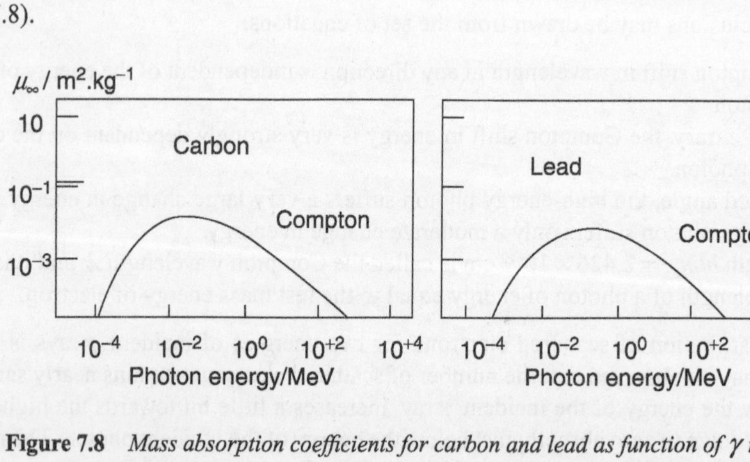

Figure 7.8 *Mass absorption coefficients for carbon and lead as function of γ ray energy*

7.2.3 Pair Production

Gamma rays of energy $E_\gamma > 1.02$ MeV may also produce a pair of electron and positron. The exact theory of pair production is derived from the field theory. However, it may also be understood in terms of the Dirac's theory of the sea of negative energy electrons. In this theory, vacuum is considered to be filled with electrons of negative energy, which could not be detected. There is a forbidden energy gap of $2m_ec^2$ between the top of the negative energy electron states and the bottom of the positive energy states. A photon that has energy in excess of $2m_ec^2$ may lift an electron from the negative energy state to a state of positive energy. The positive energy electron can be detected as a free electron while the absence of an electron from the sea of negative energy electrons behaves like a positron (figure 7.9).

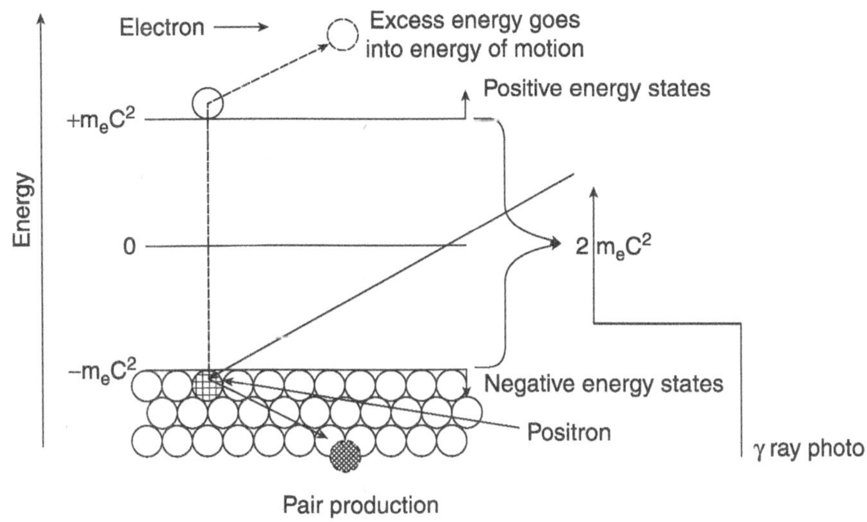

Figure 7.9 *Representation of pair production*

The pair production is possible only:

1. If the energy of incident photon is larger than the sum of the rest mass energies of the positron and the electron ($=2m_ec^2$).
2. Another essential condition for pair production is the presence of a field like that of a nucleus. It is required for simultaneous conservation of the energy and momentum. If the total energy E_γ of the incident photon is conserved between the rest mass and kinetic energies of the positron–electron pair, then it can be shown that the γ ray momentum E_γ/c is always more than the sum of the linear momentums $(p_{e+} + p_{e-})$ of the electron positron pair, irrespective of their directions of emission. Therefore, with the conservation of energy, momentum conservation will not hold unless the difference of momentum $\{E_\gamma/c - (p_{e+} + p_{e-})\}$ is taken by some other body. The balance momentum is transferred to the nucleus via its electric field. Thus, to conserve both the momentum and the energy in pair production it is necessary that it occurs in the field of some nucleus that carries the balance momentum (figure 7.10).

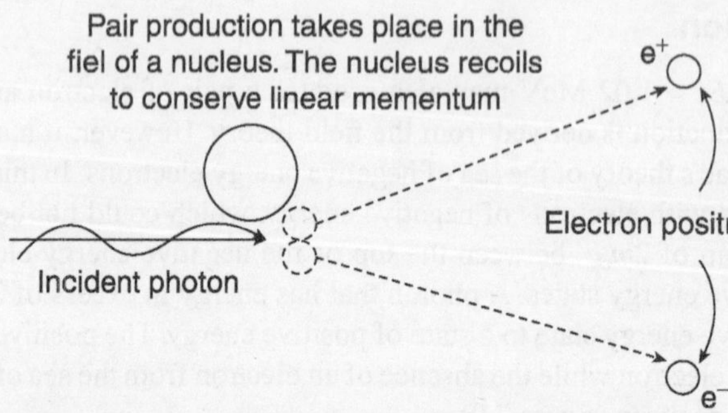

Figure 7.10 *Schematic representation of pair production*

The cross-section σ_{pair} for the nucleus of atomic number Z is given as

$$\sigma_{\text{pair}} = \frac{Z^2}{137}\left(\frac{e^2}{4\pi\varepsilon_0 m_e c^2}\right)^2 \left\{\frac{28}{9}\ln\frac{2h\nu}{m_e c^2} - \frac{218}{27}\right\} \tag{7.18}$$

Equation (7.18) tells that the probability of pair production (and hence the mass absorption coefficient) is proportional to the square of the atomic number Z^2 and to ($\ln E_\gamma$) for γ rays of higher energies. Variation of mass absorption coefficient with energy for pair production in lead is shown in figure 7.11.

The end products of pair production are two charged particles, e^+ and e^- and a recoiling nucleus. Again the number of radiations after pair production is more than before.

The variation of the photoelectric, Compton scattering, and pair production mass absorption coefficients along with the total mass absorption coefficient in aluminium and lead with the energy of the incident γ ray is shown in figure 7.12.

The reduction of the intensity of the incident γ rays after traversing a thickness 'd' of some material, because of the above-mentioned processes is accounted for by the exponential relation $I_d = I_0 e^{-\mu_m d}$.

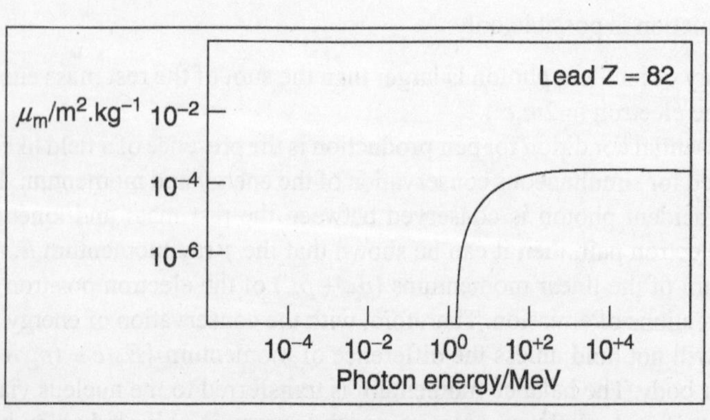

Figure 7.11 *Mass absorption coefficient for pair production in lead as a function of γ ray energy*

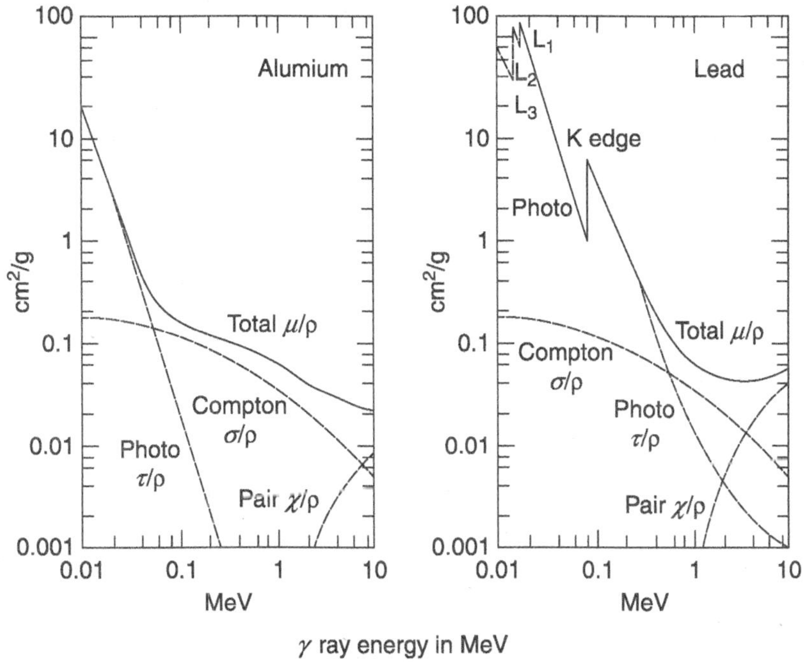

Figure 7.12 *Mass absorption coefficients for photoelectric, Compton, and pair production with total mass absorption coefficient as a function of energy of γ ray*

However, the intensity of the radiations that includes the remaining incident γ rays along with all other secondary radiations produced in the three interaction processes is more than I_d because of the radiation build-up. The ultimate radiation intensity I_f may be written as

$$I_f = I_d B_r$$

where B_r is called the dose build-up factor and is essentially determined experimentally in actual cases. In some idealized situations, B_r may also be calculated theoretically. Important information regarding the three dominant modes of energy loss by γ rays is summarized in table 7.1.

Table 7.1 *Summary of dominant γ interactions*

Type of interaction	Interaction probability	Comments
Photoelectric effect	$\propto Z^5 E_\gamma^{-3.5}$	Dominates for low-energy photons with heavy materials. All most all photon energy is converted to the kinetic energy of photoelectron
Compton scattering	$\propto Z E_\gamma^{-1}$	Dominates for photons of 0.5–5.0 MeV. A part of photon energy is converted into kinetic energy of electron. A new photon of lesser energy is produced.
Pair production	$\propto Z^2 (E_\gamma - 1.022\,\text{MeV})$ $\propto Z^2 \text{ in } (E_\gamma)$	For $E_\gamma < 2.5$ MeV For $E_\gamma > 2.5$ MeV (E_γ −1.022 MeV) energy is shared between e^+ and e^- as kinetic energy

7.2.4 γ Ray Energy Loss by Nuclear Reactions

Electromagnetic radiations may also initiate nuclear reactions. For example, photons of energy larger than 2.22 MeV may break deuteron into its constituent neutron and proton. Such reactions are called photo-disintegration reactions. If the energy of the incident photon is more than the binding energy of the neutron or proton in a nucleus, it may reject it from the nucleus. In resonance fluorescence, γ rays are absorbed by the nucleus to reach an excited state. Photo fission has also been observed in case of some heavy nuclei. However, *the probability of initiating nuclear reactions by a γ ray is much smaller than the probability for other processes such as Compton scattering and photoelectric effect*. Therefore, from the perspective of energy loss by γ rays, nuclear interactions are not important.

Solved example S-7.1

Show that photoelectric effect cannot take place with a free electron.

Solution. Let us for a moment assume that photoelectric effect may take place with a free electron. Let the (kinetic) energy of the photon be E_ν. If the struck electron is free and in the process photon disappears, then the total energy of the photon must be transferred to the electron and to conserve the momentum the electron must move in the forward direction with the momentum $p_e = \frac{h\nu}{c}$, the momentum of the photon. Thus, the kinetic energy of the electron will be $E_e = h\nu$. As electron becomes relativistic at low kinetic energies, we use relativistic expression for its total energy E_T = [kinetic energy of electron E_e + rest mass energy (m_ec^2)] accordingly,

$$E_T^2 = (E_e + m_ec^2)^2 = p_e^2c^2 + m_e^2c^4$$

Or
$$(h\nu + m_ec^2)^2 = \left(\frac{h\nu}{c^2}\right)^2 c^2 + m_e^2c^4$$

Or
$$(h\nu)^2 + m_e^2c^4 + 2m_eh\nu c^2 = (h\nu)^2 + m_e^2c^4$$

This gives $2m_eh\nu c^2 = 0$, which is an absurd result as neither m_e nor any other quantity on the LHS is zero. It means that our initial assumption that photoelectric effect may take place with a free electron is not correct as it leads to an absurd result.

Solved example S-7.2

A 10 GeV photon hits a stationary and free electron head-on and is back scattered. Calculate the energy and the momentum of the electron and the photon after the collision.

Solution. We use Eq. (7.11)

$$\Delta\lambda = \frac{c}{\nu'} - \frac{c}{\nu} = \lambda' - \lambda = \frac{h}{m_e c}(1-\cos\theta) \qquad (1)$$

In the present case, we denote the momentum of the incident and the scattered photon, respectively, by p_i and p_f. As the photon is back scattered, the scattering angle is 180°. Therefore, $\lambda' = \frac{h}{p_f}$; $\lambda = \frac{h}{p_i}$ and $\cos 180 = -1$. Substituting the values in Eq. (1), we get

$$\frac{h}{p_f c}-\frac{h}{p_i c}=\frac{2h}{m_e c^2} \text{ or } \frac{h}{p_f c}=\frac{2h}{m_e c^2}+\frac{h}{p_i c}=\frac{h\left[2p_i c+m_e c^2\right]}{m_e c^2 p_i c}$$

Or
$$p_f c=\frac{m_e c^2 p_i c}{2p_i c+m_e c^2} \tag{2}$$

However, $p_i c$, the energy of the incident photon is 10 GeV and is much larger than the rest mass energy $m_e c^2$ of the electron; hence, it may be neglected in comparison of $2p_i c$ in Eq. (2). Therefore,

$$p_f c=\frac{m_e c^2 p_i c}{2p_i c+m_e c^2} \approx \frac{m_e c^2}{2} = \frac{0.511}{2}\text{MeV} = 0.256\,\text{MeV}$$

As such, the energy of the scattered photon is only 0.256 MeV. From the law of the conservation of energy, it follows that the energy of the scattered electron is 10 GeV − 0.256 MeV = 9.999744 GeV.

We thus see that almost all the energy of the incident photon is transferred to the electron. The momentum of the scattered photon p_f = 0.256 MeV/c and that of the electron is 9.999744 GeV/c.

Exercise p-7.1: How does a 10 MeV cosmic ray photon lose energy while passing through atmosphere? The absorption cross-section σ for ^{27}Al for 1.0 MeV γ rays is 10 barn per atom. What is the linear attenuation coefficient of ^{27}Al for these γ rays? Consider the density of aluminium as 2.3 g cm^{-3}.

[**Hint:** linear attenuation coefficient $\mu = \frac{\sigma \times \text{Avogadro number} \times \text{density}}{\text{Atomic mass number}}$]

Exercise p-7.2: Calculate the difference in the energies of photons scattered at 90° and 180° by a photon of 5 MeV.

Exercise p-7.3: What are the units of mass and linear absorption coefficients? The linear absorption coefficient for 0.2 MeV γ rays in lead is 3.0 cm^{-1}. What fraction of γ rays will be left after traversing a thickness of 2-mm thick lead sheet? What is the mass absorption coefficient of lead for these γ s?

Exercise p-7.4: The mass absorption coefficients, of different processes for a 10 MeV γ ray in m^2 kg^{-1}, are (A) 1.2×10^{-3}, (B) 5×10^{-3}, and (C) 0.006×10^{-3}. Identify the processes to which each of them belongs.

Exercise p-7.5: Describe pair production and discuss the conditions necessary for it to occur.

UV light of wavelengths 7000 nm and 5000 nm when allowed to fall on hydrogen atom in ground state librated photo electrons of 8.100 eV and 7.098 eV kinetic energy. Calculate the value of the Planck's constant from these data.

Exercise p-7.6: A photo cell has anode made of a metal of work function 4.7 eV. Calculate the maximum wavelength of light that may produce any current through the cell.

Exercise p-7.7: The atomic absorption cross-sections for 660 keV γ rays in aluminium for photoelectric effect and Compton scattering are respectively, 4.0 b and 8.1 b. Calculate how much intensity of the beam will be reduced by 3.7 g cm^{-2} thickness of the aluminium.

Exercise p-7.8: What is the wavelength of the annihilation γ rays?

7.3 ENERGY LOSS BY HEAVY CHARGED PARTICLES

In principle, charged particles passing through some absorbing material may lose their energy by three processes: (1) Ionization and/or excitation of the atoms of the material through which they pass, (2) Bremsstrahlung, the process in which electromagnetic radiations, that is photons are emitted when charged particles are accelerated or decelerated in the nuclear field of the target atoms, and (3) Cherenkov radiations that are emitted when a high-energy charged particle travels in a medium with a velocity greater than the velocity of light in that medium. The amount of energy lost in each case depends on the charge and energy of the particle as well as on the atomic number and the density of the absorbing material.

Proton, α particle, deuteron, other ions, fission fragments, and so on come under the category of heavy charged particles. These particles lose their kinetic energy predominantly by ionizing or exciting the atoms of the material through which they pass. These interactions are characterized by small amount of energy loss per collision, large number of collisions, and almost straight line motion of the incident heavy charged particle. Ionization and excitation of the target material atoms takes place through the Coulomb interaction between the incident charged particle and the electrons of the atoms of target material. As the range of Coulomb force is infinite, many electrons may simultaneously interact with the incident charged particle.

We follow a classical treatment of the energy loss process under the following assumptions:

1. The electrons in the target material are assumed to be free and at rest.
2. The energy transferred to the electron is more than its binding energy to the target atom.
3. The incident charged particle moves rapidly as compared to the electrons of the atom.

Under these assumptions the interaction between the incident charged particle and the atomic electron becomes elastic.

7.3.1 Maximum Energy Transfer to Electron

The maximum energy is lost by the particle in a head-on collision with electron. To calculate the maximum energy loss per collision, let us consider a particle of rest mass M moving with initial velocity V_p and colliding head-on with a stationary electron of mass m_e. As electron becomes relativistic at quite low energies, it is necessary to do relativistic calculations.

Let $\gamma \equiv \dfrac{1}{\sqrt{1-\left(\dfrac{V_p^2}{c^2}\right)}}$, where Vp is the velocity of the particle.

$$E_p^i = \text{Initial (total) energy of the particle} = \gamma Mc^2 \tag{7.19a}$$

Initial total energy of the particle + electron $E_i^{total} = E_p^i + E_e^i = \gamma Mc^2 + m_e c^2$ (7.19b)

Further, the relativistic energy of a particle is related to its momentum by

$$E^2 = p^2c^2 + m^2c^4 \text{ and so } p^2 = \frac{E^2}{c^2} - m^2c^2 \tag{7.19c}$$

$$\text{The initial momentum of the particle } p_p^i = \left[\gamma^2M^2c^2 - M^2c^2\right]^{1/2} \tag{7.19d}$$

We shall denote quantities before collision and after collision by superscript i and f.

Let us assume that an amount of energy ΔE is transferred from the particle to the electron, and therefore, the energies of the particle and the electron after head-on collision have respectively become

$$E_{\mathrm{p}}^{\mathrm{f}} = E_{\mathrm{p}}^{\mathrm{i}} - \Delta E = \gamma Mc^2 - \Delta E \quad E_{\mathrm{e}}^{\mathrm{f}} = E_{\mathrm{e}}^{\mathrm{i}} + \Delta E = m_{\mathrm{e}}c^2 + \Delta E \tag{7.19e}$$

The momentum of the particle and the electron after collision using Eq. (7.19(c)) may be written as

$$p_{\mathrm{p}}^{\mathrm{f}} = \left[\frac{\{\gamma Mc^2 - \Delta E\}^2}{c^2} - M^2c^2\right]^{1/2}, \quad p_{\mathrm{e}}^{\mathrm{f}} = \left[\frac{\{m_{\mathrm{e}}c^2\}^2}{c^2} - m_{\mathrm{e}}^2c^2\right]^{1/2} \tag{7.19f}$$

Conservation of momentum requires

$$p_{\mathrm{p}}^{\mathrm{i}} = p_{\mathrm{e}}^{\mathrm{f}} + p_{\mathrm{p}}^{\mathrm{f}} \tag{7.19g}$$

Now, it must be realized that for maximum energy transfer the incident particle after collision must go in the direction opposite to the incident direction, that is the sign of $p_{\mathrm{p}}^{\mathrm{f}}$ will be negative. Substituting the values of $p_{\mathrm{p}}^{\mathrm{i}}$, $p_{\mathrm{p}}^{\mathrm{f}}$, and $p_{\mathrm{e}}^{\mathrm{f}}$ from Eqs. (7.19d) and (7.19f) and using the negative sign in Eq. (7.19g), we get

$$\left[\gamma^2M^2c^2 - M^2c^2\right]^{1/2} = \left[\frac{\{m_{\mathrm{e}}c^2 + \Delta E\}^2}{c^2} - m_{\mathrm{e}}^2c^2\right]^{1/2} - \left[\frac{\{\gamma Mc^2 - \Delta E\}^2}{c^2} - M^2c^2\right]^{1/2} \tag{7.19h}$$

Squaring both the sides of Eq. (7.19h) and solving, we get

$$\left[\frac{\{m_{\mathrm{e}}c^2 + \Delta E\}^2}{c^2} - m_{\mathrm{e}}^2c^2\right]^{1/2}\left[\frac{\{\gamma Mc^2 - \Delta E\}^2}{c^2} - M^2c^2\right]^{1/2} = \left(\frac{\Delta E}{c}\right)^2 + (m_{\mathrm{e}} - \gamma M)(\Delta E) \tag{7.19i}$$

Squaring both the sides again and solving, we get

$$\left[M^2 + 2\gamma Mm_{\mathrm{e}} + m_{\mathrm{e}}^2\right](\Delta E)^2 - 2m_{\mathrm{e}}M^2c^2(\gamma^2 - 1)(\Delta E) = 0 \tag{7.19j}$$

As ΔE is not zero, we divide Eq. (7.19j) by ΔE to get

$$\Delta E = \frac{2m_{\mathrm{e}}M^2c^2(\gamma^2 - 1)}{\left[M^2 + 2\gamma Mm_{\mathrm{e}} + m_{\mathrm{e}}^2\right]} \tag{7.19k}$$

Now, $(\gamma^2 - 1) = \gamma^2\frac{V_{\mathrm{p}}^2}{c^2}$ and for a non-relativistic particle $\gamma = 1$. Substituting these values in Eq. (7.19k), we get

$$\Delta E = \frac{4m_{\mathrm{e}}M\left(\frac{1}{2}MV_{\mathrm{p}}^2\right)}{(M + m_{\mathrm{e}})^2} = \frac{4m_{\mathrm{e}}ME_{\mathrm{p}}^{\mathrm{i}}}{(M + m_{\mathrm{e}})^2} \tag{7.19l}$$

Further, if $M >> m_e$, m_e may be neglected in the denominator of Eq. (7.19l) to get

$$\Delta E = \frac{4m_e E_p^i}{M} \tag{7.20}$$

Equation (7.20) may be used to calculate the maximum possible energy loss in a single collision by any particle. For example, a 10 MeV proton may lose a maximum of 0.022 MeV (0.22%) of energy in collision with a free electron. Thus, a very small fraction of the energy is lost by the particle per collision even in the case of maximum energy transfer. The particle, therefore, makes several hundred to several thousand collisions depending on its energy before coming to a stop.

7.3.2 The Stopping Power

Now, we consider a collision between an incident heavy charged particle that passes at a distance from a free electron as shown in figure 7.13. Let M and m_e represent the masses of the incident charged particle and the electron. Let ze (say positive) be the charge on the incident particle and its velocity be V. The particle is assumed to be moving in the positive X-direction in a straight line at perpendicular distance b from the electron; b is called the impact parameter. When the particle is at the point A it attracts the electron with the force F which is given by

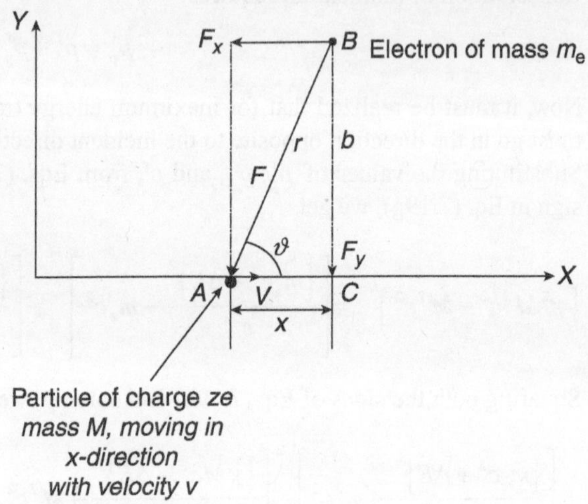

Figure 7.13 *A charged particle passes in front of a stationary electron*

$$F = \frac{ze^2}{4\pi\varepsilon_0 r^2} = \frac{ze^2}{4\pi\varepsilon_0 \left(x^2 + b^2\right)} \tag{7.20a}$$

Force F can be resolved into two components: F_x and F_y. As it is assumed that the particle passes swiftly across the electron say in time dt, the electron will experience a push because of the sudden passage of the charged particle. We now calculate the impulse of the force on the electron in time dt and integrate it over $x = -\infty$ to $+\infty$ to calculate the total change in the momentum of the particle.

$$\text{Impulse} = \int_{x=-\infty}^{+\infty} F\,dt = \int_{x=-\infty}^{+\infty} \left(F_x + F_y\right) dt = \int_{x=-\infty}^{+\infty} F_y \frac{dx}{V} \tag{7.20b}$$

As is clear from the figure, the $\int_{x=-\infty}^{+\infty} F_x\,dt = 0$, because of the equal and opposite contributions to the integral from $-\infty$ to origin and from origin to $+\infty$. Thus, the total impulse received by the electron is given by the last integral in Eq. (7.20b). However, total impulse is equal to the change in momentum. As such, the momentum p_e received by the electron is

$$p_e = \int_{x=-\infty}^{+\infty} F_y \frac{dx}{V}$$

Here, we have used the relation $V = \frac{dx}{dt}$ and $dt = \frac{dx}{V}$

Or

$$p_e = \int_{-\infty}^{+\infty} F \sin\theta \frac{dx}{V} = \int_{-\infty}^{+\infty} \frac{ze^2}{4\pi\varepsilon_0 (x^2+b^2)} \left(\frac{b}{(b^2+x^2)^{1/2}} \right) \frac{dx}{V} = \frac{ze^2 b}{4\pi\varepsilon_0 V} \int_{-\infty}^{+\infty} \frac{dx}{(x^2+b^2)^{3/2}}$$

$$= \frac{2ze^2}{4\pi\varepsilon_0 Vb} \tag{7.20c}$$

The kinetic energy gained by the electron $E_e = p_e^2/2m_e = \frac{4z^2e^4}{32\pi^2\varepsilon_0^2 m_e V^2 b^2}$ (7.20d)

The energy gained by the electron is equal to the energy lost by the heavy charged particle. Hence, the incident charged particle loses energy E_e given by Eq. (7.20d) to each electron that has impact parameter b and $(b + db)$. To calculate the number N_e^b of the electrons with impact parameter between b and $(b + db)$, we consider figure 7.14.

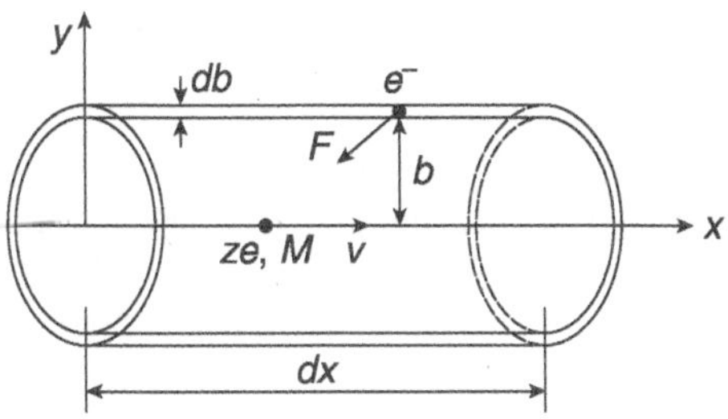

Figure 7.14 *Electrons contained in the cylindrical shell of thickness db*

All those electrons that are present in the annular cylindrical shell of inner radius b and outer radius $(b + db)$ and length dx have their impact parameters between b and $(b + db)$. Hence,

N_e^b = the number of electron per unit volume of the absorbing material

×volume the annular cylinderical shell of outer radius $(b+db)$ and inner radius b

$$= Zn\left[\pi(b+db)^2 - \pi b^2\right]dx = Zn\,(2\pi b\,db)dx \tag{7.20e}$$

where Z is the number of electrons per atom (=atomic number of the absorbing material) and n is the number of electrons per unit volume of the absorber. The number density n may also be written as

$$n = (N_A \rho)/A$$

where N_A is Avogadro's number, ρ the density of the absorbing material, and A its atomic weight.

Thus, the energy loss $(-dE)_b^{b+db}$ suffered by the incident particle in travelling a distance dx of the absorbing material in giving energy to electrons with impact parameter b and $b + db$ is

$$(-dE)_b^{b+db} = N_e^b \times E_e = \frac{4z^2e^4}{32\pi^2\varepsilon_0^2 m_e V^2 b^2} Zn\,(2\pi b\,db)\,dx$$

Hence, the energy loss per unit path length may be found by integrating this expression over b.

$$-\frac{dE}{dx} = \int_{b_{\min}}^{b_{\max}} \frac{z^2 e^4}{8\pi^2 \varepsilon_0^2 m_e V^2 b^2} Zn\ (2\pi b db) \tag{7.20f}$$

The two limiting values of b are to be determined from the physical conditions of the problem. The minimum value of b, that is $b_{\min}$, is determined from the uncertainty principle. The relative linear momentum of the electron is $(m_e V)$, where V is the relative velocity of the particle with respect to the electron. From uncertainty principle, the multiplication of the linear momentum of electron and the uncertainty in its position Δx should be of the order of h (Planck's constant).

Therefore,

$$m_e V \cdot \Delta x \approx \text{h}$$

Or

$$\Delta x \approx (h/m_e V)$$

$$\text{Considering } b_{\min} = \Delta x/2 = (h/2m_e V) \tag{7.20g}$$

$b_{\min}$ is essentially the effective size of the electron as seen by the moving particle.

Classically, the electrons in the absorber atom are in oscillatory motion in different orbits with different frequencies ν_1, ν_2, ν_3 and so on. If the average of these frequencies is taken as ν, then the average time period of the oscillatory motion of the electron is given by $T = 1/\nu$. It may be argued that if the interaction time dt of the particle with the electron is not smaller than the time period T of the oscillations of the electron, the particle will not be able to change the state of the electron. Now, the maximum value of dt may be taken to be of the order of $b_{\max}/V$ and this should in the limiting case be equal to T. Therefore,

$$b_{\max} = V/\nu \tag{7.20h}$$

When the values of $b_{\min}$ and $b_{\max}$ are inserted in Eq. (7.20f) and integration is done, we get

$$S = -\frac{dE}{dx} = \frac{z^2 e^4 Z n}{4\pi\varepsilon_0^2 m_e V^2} \ln\left[\frac{2m_0 V^2}{h\nu}\right] = \frac{z^2 e^4 Z n}{4\pi\varepsilon_0^2 m_e V^2} \ln\left[\frac{2m_0 V^2}{I}\right] \tag{7.20i}$$

The quantity $h\nu$ may be put equal to the average ionization potential I. The empirical estimates of average ionization energy gives $I = kZ$, where $k = 19$ for hydrogen and 10 for lead. Generally, I is taken as a free parameter and its value is determined from the comparison of experimental data with Eq. (7.20i). Eq. (7.20i) is also called *Bethe formula* for stopping power.

$S = -\frac{dE}{dx}$, called the linear energy loss, the linear or simply stopping power, or linear energy transfer (LET) is of great importance from the point of the biological effects of charged particle radiation loss. The mass stopping power $S_m = S/\rho$ also called specific stopping power, may be obtained by dividing the linear stopping power by the density of the absorbing material.

A more detailed quantum mechanical derivation that includes relativistic effects, called *Bothe–Bloch formula*, which gives

$$S = -\frac{dE}{dx} = \frac{z^2 e^4 Z n}{4\pi\varepsilon_0^2 m_e V^2}\left[\ln\frac{2m_e V^2}{\text{I}} - \ln\left\{1 - \frac{V^2}{c^2}\right\} - \frac{V^2}{c^2}\right] \tag{7.21}$$

The following important observations may be made from the simple formula for stopping power given by Eq. (7.20i)

1. **Nuclear and electronic collisions:** The incident particle may well interact and lose energy with the nucleus of the atoms of the absorbing material in the same way as it interacts with the electron of the atom. Yes, that is true but the energy loss by nuclear interaction may be obtained by multiplying Eq. (7.20i) by the charge Z of the nucleus and dividing by M_n/m_e where M_n is the mass of the nucleus. Now, the nuclear charge Z cannot be more than 100 (no stable nucleus has atomic no. more than 100) but M_n/m_e is at least of the order of 10^3; hence, the mass factor wins over the charge factor and the energy loss by collision with nucleus is negligible.
2. **Effect of the mass of the incident particle:** The mass of the incident particle does not enter in the expression for the energy loss. It means that two non-relativistic particles of same charge z but of different masses having same velocity V will have same stopping power in a given medium.
3. **Effect of the energy of the incident particle;** Stopping power in a given medium decreases with the increase of the kinetic energy of the incident particle

$$-\frac{dE}{dx} \propto \frac{1}{V^2} \propto \frac{1}{E}$$

4. **Effect of different absorbing materials:** The stopping power of a given incident particle in different absorbing materials is proportional to the number density n and the atomic number Z of the material.

$$-\frac{dE}{dx} \propto nZ \propto \frac{\rho Z}{A}$$

5. **Effect of the charge of the incident particle:** The stopping power for same velocity particles of different charge z is proportional to the square of the charge.

$$-\frac{dE}{dx} \propto z^2 \quad \text{(for constant } V\text{)}$$

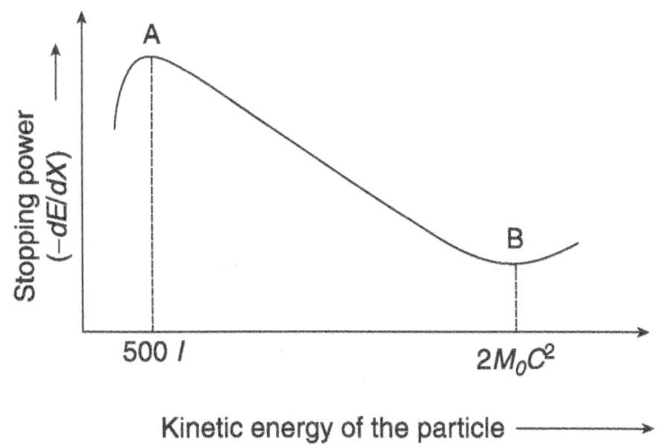

Figure 7.15 *Variation of the stopping power as a function of the kinetic energy of the particle*

Figure 7.15 shows a sketch of the dependence of stopping power on the kinetic energy of the incident particle obtained experimentally. The special features of the curve is a slow decrease in stopping power after a little peak at very low energy (at around 500 I, where I is the ionization energy), a broad minimum at relatively high energies and then a very slow increase. The part of the curve between the points A and B is well explained by the simple formula (7.20i). A peak around point A and decrease of stopping power at still lower energies is the result of electron capture by the incident particle. This reduces the effective charge z of the particle that in turn decreases stopping power. Slow rise after point B at higher energies (~3 M_0c^2) may be understood in terms of the relativistic effect.

7.3.3 Bragg's Curve

The energy and the velocity of an incident charged particle gradually decrease as it moves deeper inside the absorbing material. As a result, the energy loss per unit path length slowly increases. Ultimately, the stopping power attains a maximum value and within a very short distance the particle loses all its energy. The particle then picks up electrons from the surroundings and becomes charge neutral and comes to a stop in the stopping medium. The average distance travelled by the particle in the medium before coming to stop is called the range of the particle and is denoted by R.

Figure 7.16 shows the variation of the stopping power with the depth of the absorbing medium. The curve is called *Bragg's absorption curve*. It shows that a heavy charged particle deposits most of its energy towards the end of its path or range and comparatively small amount of energy is lost in the part of the medium lying between the points of entry to the point just before the stop of the particle. This curve is very important from the point of radiation therapy. For example, if there is a cancerous growth say x-cm deep inside the body and it is desired to kill the cancerous cells by charged particle radiations without damaging the healthy cells in the length x, heavy charged particles of energy such that their range corresponds to a thickness x may be used. The energy lost by such particles in the distance x will be small and will not harm the healthy cells but the cancerous cells at the depth x will receive a blast of energy and will be killed. Heavy charged particles such as charged pi-mesons and protons are now routinely used for radiation therapy and medical accelerators that produce charged particles of desired energy are in use in various hospitals worldwide.

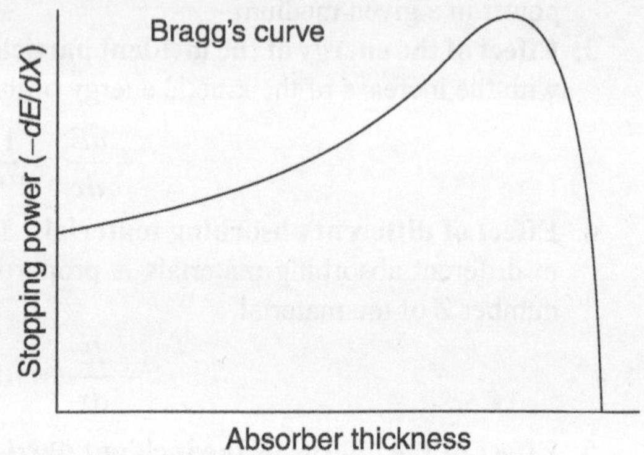

Figure 7.16 *Variation of the stopping power with the depth in absorbing medium*

7.3.4 Formation of Ion-pairs

Heavy charged particles passing through an absorbing material lose energy by ionising the atoms of the material and produce primary ion-pairs. The electrons and positive ions produced in primary ionization are often enough energetic to initiate further ionization, called secondary ionization. The total number of ion pairs produced by the passage of the incident particle is a measure of the energy lost by it. The number of ion-pairs per unit path length may be calculated from the linear stopping power S by dividing it by the energy required to create one ion-pair.

Another quantity used to describe particle energy loss is specific ionization. The *specific ionization* is equal the number of ion pairs (i.p.) formed per unit distance travelled by the charged particle.

$$\text{Specific ionization} = \frac{dE/dx}{w}$$

where w is the average energy spent to create an electron–ion pair or electron–hole pair. The energy required to create an electron–hole pair depends on the irradiated material. Experimental results have shown that the actual energy necessary to create an ion pair is about 2–3 times greater than the ionization potential. For instance, the ionization potentials of oxygen and nitrogen are 13.6 and 14.5 eV, respectively, whereas the mean energy to create an ion pair in air is about 34 eV. For other materials, such as air, silicon (Si), germanium (Ge), gallium-arsenide (GaAs), and silicon dioxide (SiO_2), the average energy for ion-pair/electron–hole generation is respectively, 34 eV, 3.6 eV, 2.8 eV, 4.8 eV, and 17 eV. The total negative and positive charges liberated in the formation of ion pairs may be collected by applying an electric field and electrical signal proportional to the collected charge may be produced using appropriate electronic circuits. Ionization chambers, gas filled counters, and solid-state detectors utilize this for the detection and energy analysis of heavy charged particles.

7.3.5 Range of Heavy Charged Particles

Because of their large mass heavy charged particles do not deviate from their almost straight line path when traversing an absorbing medium, though they lose small amount of energy in each successive interaction with the electrons of the atoms of target material. After large number of collisions, depending on the energy, the incident particle comes to a stop in the stopping medium. The distance travelled in the stopping medium before coming to stop is called range and is denoted by R. In principle, range can be calculated from the stopping power $S = -\frac{dE}{dx}$ using the relation

$$R = \int_0^R dx = \int_{E_{\text{inc}}}^{0} \frac{dx}{dE} dE = \int_0^{E_{\text{inc}}} \left(-\frac{1}{\left(\frac{dE}{dx}\right)} \right) dE \tag{7.22a}$$

Here, E_{inc} is the incident energy of the heavy charged particle. The stopping power values of Bethe or for more detailed calculations, values from Bethe–Bloch formula can be used. However, both the expressions for stopping power are not valid for low energies and as such Eq. (7.22a) cannot be integrated for all values of energies. However, if the range R_1 of the particle up to some energy E_1 is measured than the range R_2 for energy E_2 may be obtained as

$$R_2 = R_1 + \int_{E_1}^{E_2} \left(-\frac{1}{\left(\frac{dE}{dx}\right)} \right) dE \tag{7.22b}$$

Experimental determination of the range of a given particle in a medium is possible by using foils of the absorbing medium of different thicknesses and counting the number of particles stopped in foils as a function of the foil thickness. In such experiments, a typical absorption curve shown in figure 7.17 is obtained. Some important features of the transmission curve are as follows:

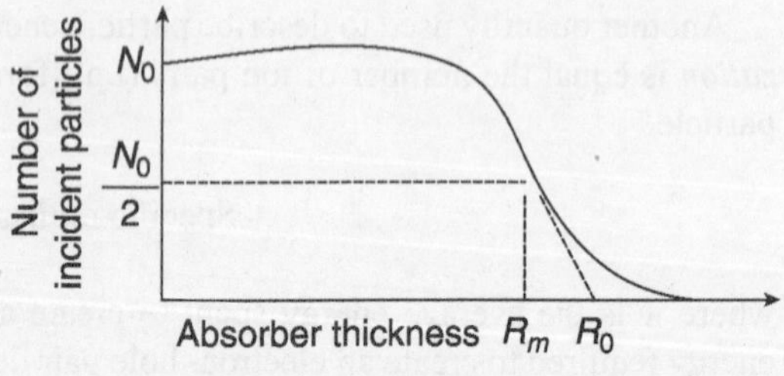

Figure 7.17 *A typical transmission curve for heavy charged particle*

1. The particle intensity remains almost constant for a large value of absorber thickness.
2. The intensity does not fall to zero sharply at a fixed absorber thickness, instead decreases slowly with thickness.

It is clear from the nature of the transmission curve that all incident particles though initially of same energy do not lose their energy in exactly the same thickness of the absorber. Some particles travel smaller absorber thicknesses while some others a little more. This is because of the statistical nature of the energy loss process. There are two aspects of this. First, suppose there are two incident particles of exactly same energy and both interact with two identical but different atoms of the absorber. The energy loss by the two particles may not be exactly same. Second, the total number of collisions suffered by two different incident particles before coming to stop may not be exactly same. As such, the absorber thicknesses for total loss of the energy by different incident particles differ a little bit from each other. Another cause for the difference in the total absorption thicknesses is the change in the energy profile of the incident beam as it traverses the thickness of the material. The energy profile of a mono energetic incident beam becomes broader and skewed as it moves deeper in the absorbing medium. This is shown in figure 7.18. As the beam moves towards the range thickness its profile has more particles of lower energy than that of higher energy. This also adds to the difference in the stopping thicknesses for different particles of a monoenergetic incident beam. The fact that the total stopping thickness (or range) for particles of same energy differ slightly (ΔR) for different particles and is distributed around the mean range $\bar{R}$ is called *range straggling*. If σ is the standard deviation of the range straggling, then the ratio $\left(\frac{\sigma}{\bar{R}}\right)_A$ for a particle of atomic mass number A is related to the same quantity for α particles through the relation

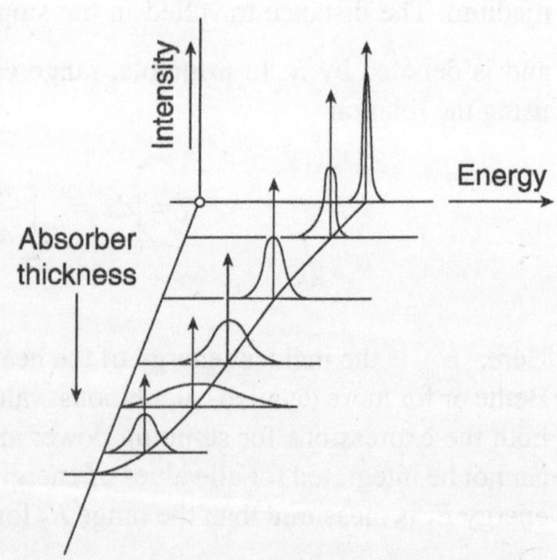

Figure 7.18 *Change in the energy profile of the incident beam with absorber thickness*

$$\frac{\left(\frac{\sigma}{\bar{R}}\right)_A}{\left(\frac{\sigma}{\bar{R}}\right)_\alpha} = \left(\frac{4}{A}\right)^{1/2}$$

In addition, if a beam of monoenergetic incident particles traverses a fixed thickness Δt of the absorber, there will be fluctuations in the energy of the emerging particles of the beam about a mean value. This is called *energy straggling*. The variance of the energy straggling is given by

$$\sigma^2 = 4\pi n z^2 e^4 \Delta t$$

Here, n is the number of electrons per cm^3 of the absorbing material and ze the charge of the incident particle.

The mean range $\bar{R}$ or R_{mean} is defined as the absorber thickness at which the intensity (or the initial number of particles) falls to half of its initial value (see figure 7.17). The extrapolated range R_0 is obtained by linear exploration at the point of inflection of the transmission curve. This variation in the absorber thickness for the total energy loss, that is range straggling, is often represented by the straggling parameter α. Some statistical calculations have shown that

$$R_{mean} = R_0 - \frac{\sqrt{\pi}}{2}\alpha \text{ and } \alpha \approx 0.015 R_{mean} \tag{7.22c}$$

In practice, empirical range energy relations are often used. An empirical relation between the energy E and range R of a particle of charge Z may be given as

$$E = k Z^{2n} M^{(1-n)} R^n$$

Here k and n are empirical constants that depend on the nature of the absorber and M is the mass of the particle in terms of the proton mass. If two particles of mass M_1 and M_2 and charges Z_1 and Z_2 enter a medium with same velocity, then the ratio of their ranges R_1/R_2 is given as

$$\frac{R_1}{R_2} = \left(\frac{M_1}{M_2}\right)\left(\frac{Z_2}{Z_1}\right)^2$$

For the rough estimate, we can use the Bragg–Kleeman rule

$$\frac{R}{R_1} = \frac{\rho_1\sqrt{A}}{\rho\sqrt{A_1}} \tag{7.22d}$$

where subscript 1 denotes the quantities for the reference medium, which for example may be taken as air at 15°C and one atmospheric pressure for which $\sqrt{A_1} = 3.81$ and $\rho_1 = 1.226 \times 10^{-3}$ g/cm^3. As such,

$$R = 3.2\times10^{-4}\,\frac{\sqrt{A}}{\rho} \times R_{air} \tag{7.22e}$$

Another empirical relation gives the range of α particles in terms of its energy as

$$R_{air}^{\alpha} = \begin{Bmatrix} 0.56E_\alpha & \text{for} \quad E_\alpha < 4\,\text{MeV} \\ 1.24E_\alpha - 2.62 & \text{for} \quad 4\,\text{MeV} < E_\alpha < 8\,\text{MeV} \end{Bmatrix} \tag{7.22f}$$

Here, the range is in cm and the α particle energy in MeV.

Geiger's Rule: According to this rule the range in air is proportional to the cube of the velocity of the particle, R = constant v^3

For α particles of 4–10 MeV, the range in air may be given by $R = 0.32\ E^{3/2}$, where R is in cm and the energy E is in MeV.

Similarly, for protons the range in air may be given by the empirical relation

$$R_{\text{air}}^{\text{p}}\ (\text{in metres}) = \left(\frac{E_{\text{P}}\ \text{in MeV}}{9.3}\right)^{1.8} \tag{7.22g}$$

This relation holds good for proton energies of few MeV to about 200 MeV.

7.3.6 Bremsstrahlung and Cerenkov Radiation Losses For Heavy Charged Particles

Charged particles also lose energy by emitting electromagnetic radiations if they are accelerated or decelerated. As these electromagnetic radiations are produced as a result of sudden braking/stopping of charged particles they are called breaking radiations (in German language bremsen means brake and strahlung means radiations). When a particle of mass M and charge ze fly passes close to a nucleus of charge Ze, it is decelerated. The deceleration is proportional to (Zze^2/M). The probability of bremsstrahlung is proportional to the square of the acceleration or deceleration. Therefore, the chance of energy loss by bremsstrahlung is proportional to the square of the charge z^2 of the particle and inversely to the square of the mass M of the particle. Thus, the probability of energy loss by bremsstrahlung is negligible for heavy charged particles such as protons, αs and other heavy ions. However, it is quite significant for electrons and other light charged particles.

When a charged particle passes through a medium and if its velocity V is larger than the velocity of light v in that medium ($v = c/\mu$, μ being the refractive index of the medium and c the speed of light in vacuum), the particle loses energy in the form of visible light called Cerenkov radiations. Only light charged particles of very high relativistic energies can fulfil the condition $V > v$, and, therefore, for heavy charged particles that are generally involved in nuclear structure and reaction studies energy loss by Cerenkov radiations is unlikely.

7.4 ENERGY LOSS BY LIGHT CHARGED PARTICLES

Electron, positron, and muons come under the category of light charged particles. These particles lose their energy by three processes: (1) by ionizing and exciting the atoms of the material through which they pass. This is the same process by which heavy charged particles also lose their energy. However, the main difference is due to the very small mass that is equal to or comparable to the mass of the electrons of the target atoms. (2) By bremsstrahlung, when light charged particles come near to the nucleus of the absorbing atom they are accelerated or decelerated depending on their charge. As a result braking radiations in the form of X-rays or lower frequency electromagnetic waves are emitted resulting in the energy loss of the incident particle. (3) Light charged particles of large kinetic energies also emit Cerenkov radiations in transparent media such as water. We shall discuss these processes taking the example of electron as the representative of light charged particles.

7.4.1 Energy Loss of Electron by Ionization and Excitation of Target Atoms

Like heavy charged particles, electrons also lose energy by ionizing and exciting the atoms of the target material through which they pass. Energy loss by this mode is also called collision loss. The main differences in the collision loss by heavy charged particles and the electron are as follows:

1. Larger fraction of incident energy is lost per collision. As the mass of the incident electron and the struck electron of the target atom is equal, up to half of the kinetic energy of incident electron may be lost in a single collision.
2. Large deviation in the direction of motion. Again because of the equal mass the incident and the struck electrons may scatter in any direction. Further, it is not possible to distinguish between the incident and the struck electron and hence the path of the incident electron in the absorbing material is not a straight line, it is zigzag. Also, the projected range, that is the projection of the travel path on the incident direction, is much smaller than the actual distance travelled by the incident electron.
3. High energy of scattered electrons. The kinetic energy of both the incident and the scattered electron are generally large as more energy is lost per collision. The scattered electrons also called delta-rays (δ-rays) further ionize other atoms of the material.

The linear stopping power of electron due to collision S_e^C is given by

$$S_e^C = -\frac{dE_e}{dx} = \left(\frac{e}{4\pi\varepsilon_0}\right)^2 \frac{2\pi e^2 N Z}{m_e c^2 \beta^2} \left[\begin{array}{c} \ln \frac{m_e V^2 E_e}{2I^2 (1-\beta^2)} - (\ln 2)\left(2\sqrt{1-\beta^2} - 1 + \beta^2\right) + \\ (1-\beta^2) + \frac{1}{8}\left(1-\sqrt{1-\beta^2}\right)^2 \end{array} \right] \tag{7.23a}$$

Here, E_e is the kinetic energy and V is the velocity of the electron. Z is the atomic number of the target atom, I its average ionization potential, and N the number density of the atoms in the absorbing material. $\beta = \frac{V}{c}$ is the ratio of the electron velocity to the velocity of light.

Like the case of the heavy charged particles, the linear stopping power for electron is inversely proportion to the square of the electron velocity or to the energy of the electron. The collision energy loss decreases with the increase of energy and ultimately gives way to the radiation energy loss.

7.4.2 Radiative Energy Loss

High energy electrons are accelerated when they come close to the nucleus of the absorber atom. Accelerated charge emits electromagnetic radiations and loses energy. This is called radiative loss. The probability of radiative loss is proportional to the square of the acceleration. High-energy electrons may reach close to the nucleus of the target atom and, therefore, suffer large acceleration losing substantial amount of energy. For electrons of kinetic energy $E_e > m_e c^2$, Bethe and Heitler gave the following expression for the radiation loss per unit path length:

$$S_e^r = -\left(\frac{dE_e}{dx}\right)_r = \left(\frac{e}{4\pi\varepsilon_0}\right)^2 \frac{NZ\,(Z+1)E_e e^2}{137 m_e^2 c^4}\left[4\ln\frac{2E_e}{m_e c^2} - \frac{4}{3}\right] \tag{7.23b}$$

The radiation stopping power is proportion to the kinetic energy of the incident electron. The mass stopping power obtained by putting $X' = \rho x$ may be written as

$$-\frac{dE_e}{dX'} = \frac{E_e}{X_0} \tag{7.23c}$$

where X_0 is a constant for a given material.

Equation (7.23c) on integration gives

$$E_e = (E_e)_0 \, e^{-\frac{x'}{x_0}} \tag{7.23d}$$

Here, $(E_e)_0$ is the initial value of the kinetic energy.

Equation (7.23d) tells that the energy loss by radiation becomes exponential as the particle moves into the absorbing material. The thickness X_0 is called the radiation length and is equal to the thickness of the material (in g/cm^2) that reduces the energy loss to $1/e$ of its initial value.

The ratio of the collision loss to the radiation loss may be roughly given as

$$\frac{S_e^C}{S_e^r} \approx \frac{E_e \text{ (in MeV) } Z}{700} \tag{7.23e}$$

Figure 7.19 shows electron stopping power as a function of its kinetic energy both for collision and radiation losses. It may be observed that for higher energies radiation loss increases almost exponentially. At lower electron energies, collision loss dominates over the radiation loss.

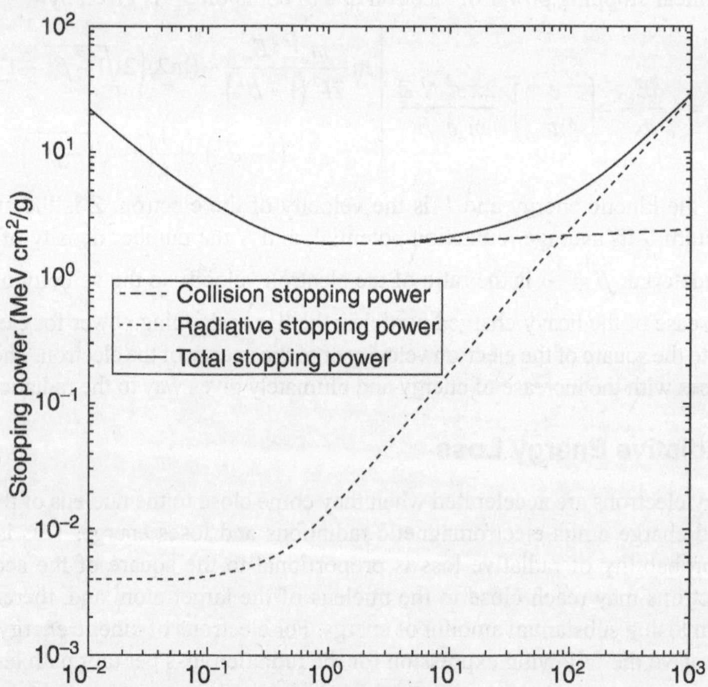

Figure 7.19 *Stopping power of electron in liquid water as a function of the electron energy*

7.4.3 Energy Loss by Cerenkov Radiations

Light charged particles with speed greater than the speed of light in the transparent medium emit the so-called Cerenkov radiations and lose energy. The essential condition for the production of Cerenkov radiations is that the velocity of the charged particle in the transparent medium V should be greater than the velocity of light $(=\frac{c}{\mu})$ in the medium, where μ is the refractive index of the medium. It means that $\beta\mu > 1$. Threshold energy for Cerenkov radiations is given by

$$E_{\text{thre}}^{\text{Cer}} = m_e c^2 \left[-1 + \sqrt{1 + \frac{1}{\mu^2 - 1}} \right] \tag{7.24a}$$

These radiations are generated as a result of the interaction of the electromagnetic field carried by the incident charged particle with the electrons of the transparent material. Under the influence of the electromagnetic field, electrons emit radiations which cancel in all directions if $\beta\mu < 1$. However, if $\beta\mu > 1$, then there will be one direction θ in which there will be constructive interference and radiations will be emitted in that direction. The angle θ is defined as

$$\cos\theta = \frac{1}{\beta\mu}$$

The energy loss per unit path length by Cerenkov radiations is given by

$$\left(\frac{dE_e}{dx} \right)_{\text{Cer}} = \frac{4\pi^2 z^2 e^2}{c^2} \int \left(1 - \frac{1}{\beta^2 \mu^2} \right) \nu \, d\nu \tag{7.24b}$$

Integration in Eq. (7.24b) is to be carried out over all frequencies; however, the relative contribution from higher frequencies are dominant. The emission cone of Cerenkov radiations is shown in figure 7.20.

The energy loss by Cerenkov radiations is very small ≈ 1 keV per cm of path length and is really negligible as compared to the collision loss. The importance of Cerenkov radiations lies in their use in detecting relativistic charged particles by specially designed Cerenkov detectors.

Radioactive waste from nuclear reactors is often kept in water pools for initial cooling before further processing. Electrons and positrons coming out from the radioactive waste produce bluish Cerenkov radiations illuminating the pool during night (figure 7.20).

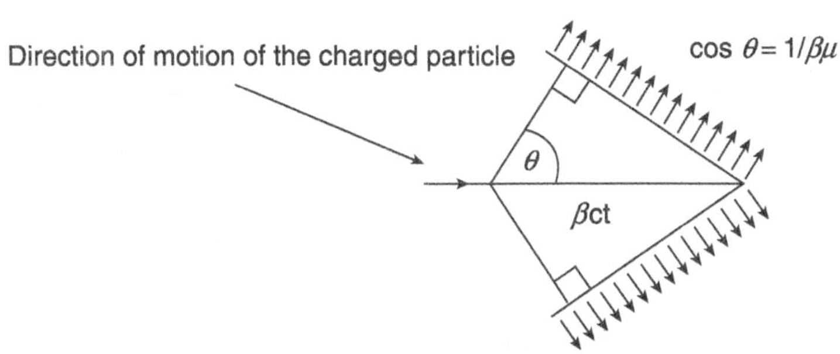

Figure 7.20 *Emission cone of Cerenkov radiations*

Exercise p-7.6: Bring out the main differences in the energy lost by a photon, a proton and an electron, each of 1 MeV energy when they pass through the same absorber material.

Exercise p-7.7: Discuss the energy dependence of the energy lost by electrons through radiative and collision processes.

7.5 ENERGY LOSS BY UN-CHARGED NUCLEAR PARTICLES

Two types of un-charged nuclear particles are the heavy particles such as neutrons and very light particles such as neutrino. We shall discuss the energy loss processes of chargeless particles with reference to the neutron and neutrino.

7.5.1 Energy Loss by Neutrons

Energy loss by neutrons depends on the energy of the neutron and on the properties of the absorbing material through which they pass. However, broadly speaking the neutron interactions can be divided into two types: (1) interaction in which the neutron interacts with the atom (or nucleus) of the absorbing material, loses a part of its energy but remains in the material. (2) Interactions in which the neutron interacts with the nucleus of the target atom initiate a nuclear reaction and disappear from the system. The first kind of interactions called scattering are similar to the energy loss processes of heavy charged particles, while the second kind of interactions are like 'one-shot' processes by which electromagnetic radiation lose their energy.

Scattering events can be subdivided into elastic and inelastic scattering. In elastic scattering, the total kinetic energy of the neutron and the nucleus remains unchanged by the interaction. During the interaction, a fraction of the neutron's kinetic energy is transferred to the nucleus. For a neutron of kinetic energy E encountering a nucleus of atomic weight A, the average energy loss is $(2EA)/(A+1)^2$. This expression shows that to reduce the speed of neutrons (that is, to moderate them) with the fewest number of elastic collisions, the target nuclei with small A should be used. By using hydrogen, with $A = 1$, the average energy loss has its largest value of $E/2$. A neutron with 2 MeV of kinetic energy will (on the average) have 1 MeV left after one elastic collision with a hydrogen nucleus, 0.5 MeV after a second such collision, and so on. To achieve a kinetic energy of only 0.025 eV would take a total of about 17 such collisions. (A neutron of energy 0.025 eV is said to be in thermal equilibrium with molecules of the surrounding medium and is considered a 'thermal neutron'. From the relation $E = kT$, where k is Boltzmann's constant, an energy E of 0.025 eV corresponds to a temperature T of about 30°C.) In general, after 'n' elastic collisions, the kinetic energy of the neutron will change to E_n from the initial value E_0, where

$$E_n = E_0\left[\frac{(A^2+1)}{(A+1)^2}\right]^n$$

In other words, to reach to the kinetic energy E_n from E_0, the required number of collisions 'n' will be

$$n = \frac{\log\left(\frac{E_n}{E_0}\right)}{\log\left\{\frac{\left(A^2+1\right)}{\left(A+1\right)^2}\right\}} \text{ collisions, on the average}$$

Inelastic scattering is similar to elastic scattering except that the nucleus undergoes an internal rearrangement into an excited state from which it eventually releases radiation. The total sum of the kinetic energies of the outgoing neutron and nucleus is less than the kinetic energy of the incoming neutron; part of the original kinetic energy is used to put the nucleus into the excited state. It is no longer easy to write an expression for the average energy loss per in-elastic collision, because it depends on the energy of the excited levels in the nucleus. However, the net effect on the neutron is again to reduce its speed and change its direction. If all the excited states of the nucleus are too high in energy to be reached with the energy available from the incoming neutron, inelastic scattering is impossible. In particular, the hydrogen nucleus does not have excited states so only elastic scattering can occur in that case. In general, scattering moderates or reduces the energy of neutrons and provides the basis for some neutron detectors (for example, proton recoil detectors).

Instead of being scattered, a neutron may be absorbed by the nucleus of the target atom initiating various types of reactions, such as (n, γ), (n, p), (n, d), (n, α), ... (n, f). The probability of each type of reaction depends on the energy of the incident neutron and the target nucleus. Some of these reactions are used to detect neutrons. Boron-10, helium-3, and uranium-235 have large cross-sections for the absorption of low energy neutrons and producing charged particles. Another important reaction is $^{113}Cd\ (n, \gamma)\ ^{114}Cd$, which has a very high cross-section $\approx 57 \times 10^3$ barns for thermal neutrons. In natural cadmium, the relative % of 113 isotope is about 12%; hence, even natural cadmium is a powerful poison for thermal neutrons. Cadmium rods are, therefore, used in reactors for removing/absorbing thermal neutrons and controlling the fission rate. Cadmium is also used as shielding material for thermal neutrons. However, absorption of thermal neutrons by cadmium results in the production of γ rays.

The intensity of a collimated neutron beam falls off exponentially with the thickness of the absorbing material. It is because of the fact that both the absorption of the incident neutrons by nuclear reactions as well as the neutrons scattering by nuclei remove the incident neutrons from the original beam. The scattered neutrons suffer large deviations in their direction of motion. Therefore, both scattering and nuclear reactions are like 'one-shot' processes leading to exponential decay in intensity with thickness of the absorbing material

$$I = I_0 e^{-\mu_m d_m} \tag{7.25a}$$

Here, I_0 and I are, respectively, the intensities of the neutron beam before and after traversing the thickness d_m (in g/cm^2) of the absorbing material of mass absorption coefficient μ_m for the material. The mass absorption coefficient can be calculated if the cross-sections for scattering and for absorption by reactions are known.

7.5.2 Energy Loss by Neutrino

Neutrinos have no charge, zero (or almost zero) mass, and they interact with matter only through weak interactions, which have very small cross-sections. As a result the energy loss by neutrinos while passing through matter is extremely small. To understand the point, let us consider the reaction in which an antineutrino is captured by a proton and a neutron and positrons are formed.

$$\bar{\nu}_e + P \rightarrow N + e^+ \tag{7.25b}$$

The cross-section for this reaction is $\approx 10^{-43}$ b. Assuming that 1 cm^3 of matter contains $\approx 10^{24}$ protons, the interaction probability = 10^{24} (cm^{-3}) × 10^{-43} (cm^2) = 10^{-19} (cm^{-1}). It means that an antineutrino on average will travel almost 10^{19} cm of distance in the matter before it is captured. This is more than 10 light years.

There are three different kinds (flavours) of neutrinos, the electron neutrino, ν_e, *the muon nutrino* ν_μ and *tau nutrino* ν_τ. All the three different kinds of neutrinos have different properties and interact with matter with different probabilities. Detection of neutrino is done through their reaction products.

Exercise p-7.9: The mass stopping power for 100 MeV protons in lead is 0.35 MeV m^2 kg^{-1}. How much energy will be lost by the proton in a 1-mm thick lead sheet? The density of lead is 11.4×10^3 kg m^{-3}.

Exercise p-7.10: A deuteron and a proton are both moving with the same velocity. What will be the ratio of their ranges in air?

Exercise p-7.11: Calculate the number of collisions that a 1 MeV neutron will suffer in paraffin wax and in carbon pile before attaining thermal energy (0.025 eV).

Exercise p-7.12: Discuss how the stopping power of a heavy charged particle changes with (i) the energy of the particle and (ii) with the thickness of the absorbing medium.

Calculate the energy of α particle that will have approximately the same specific ionization as a 30 MeV proton.

Exercise p-7.13: A β negative, an α particle and a proton all of the same energy are incident on a thick aluminium plate. Discuss their energy loss and indicate with reason which will have the largest range.

Exercise p-7.14: Which particle out of all nuclear particles you know has the maximum mean free path? What is the source of the production of these particles?

Exercise p-7.15: what is Bragg's cure? How can you explain it? Discuss one of its important applications.

A proton and a deuteron lose the same amount of energy in passing through a tin sheet of lead. Show that their energies are in the ratio 1:2.

7.6 NUCLEAR DETECTORS

In nuclear physics experiments, detection, identification, determination of energy, tracking the path, and measuring the intensity of radiations are required. The radiations themself either be charged or uncharged particles such as electron, positron, protons, αs, heavy ions, neutrons, or energy quanta of γ rays. Depending on the requirements, single or multiple detectors are used in experiments. Generally, detectors are based on the principle of converting the energy lost by the incident radiations in the material of the detector into some measurable form, such as variations in current and/or voltage. The current or voltage pulses obtained from the detector are pre-amplified/ amplified, shaped, and digitalized to improve the detection efficiency and the signal-to-noise ratio. The electronically processed pulses may be directly counted by a counter (also called scalar) to check the presence and to determine the intensity of the radiations. Further analysis of the pulses may give information about the nature and energy of the radiations. Detector set-up that is used just to count the pulses is called a radiation counter while set-ups in which energy distribution of the incident radiations is also determined are called spectrometers. The basic layout of a detector system is shown in figure 7.21.

Detector set-ups may be classified on the basis of their functionality/detection mode as:

1. Counters: Detect the number of interactions produced by the radiations in a given time.
2. Spectrometers: Detect and give the energy distribution of the incident radiations.
3. Dosimeters: Gives the total energy deposited by the radiations.
4. Imaging systems: Photographic plates, cloud chamber, CT-scan, NMR-scans, solid-state track detectors, and so on.

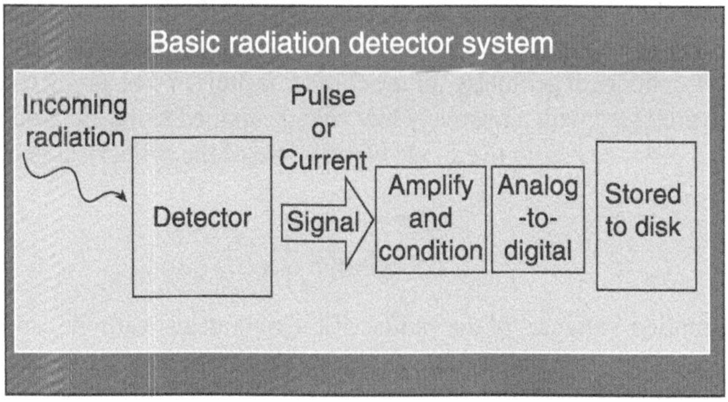

Figure 7.21 *Block diagram of a radiation detector system*

Another way of classifying detector systems is on the basis of their physical composition as follows:

1. Gas-filled detectors
2. Scintillation detectors (liquid and solids)
3. Solid-state detectors

7.6.1 Gas-filled Detectors

When some nuclear radiation enters the active volume of a detector it either directly (such as charged particles) or indirectly (such as γ rays that may produce electrons by photoelectric and Compton effects) produces ionization and excitation of the atoms of the material of the detector.

If the excited/ionized atoms come to ground state by emitting photons of visible light, the photons may be recorded by a suitable device such as a photomultiplier tube and the event may be registered. This is the principle of operation of scintillation and Cerenkov detectors. In case of solid-state detectors, the incident radiation produces electron–hole pairs and constitutes a current. The current pulse formed as a result of passage of the radiation through the detector may be processed and recorded. In case of a gas-filled counter, the incident radiations lose energy and produce electron–positive ion pairs. The charge liberated by the radiation is collected by the electric field, produced by applying a potential difference between two electrodes.

7.6.1.1 Construction

A gas-filled counter is generally made of a cylindrical glass or metallic container with a thin central tungsten wire as anode. In case of a glass container, a hollow metallic cylinder insulated from the central wire envelopes the anode wire, which is stretched along the axis of the cylinder. The metallic hollow cylinder works as cathode. In the case of metallic container, the container itself works as cathode. In either case, the thickness of the metallic and/or glass casing does not allow low energy charged particles such as βs and αs and soft X-rays to enter the volume of the counter, though high energy X and γ rays may reach inside the volume. To allow low energy charged particles and electromagnetic radiations, a thin window, generally of mica or mylar, is fixed at one end of the cylindrical counter. Some inert gas, such as argon is filled in the counter at low pressure $\approx 1/8$ of atmospheric pressure. Some quenching gas, such as alcohol or any other long-chain hydrocarbon gas is also mixed with the inert gas in a ratio of 1:10 by volume, for self-quenching. The need of quenching will be discussed later. A cylindrical geometry for the counter is preferred as strong electric field near the anode wire may be produced with a relatively low voltage applied to the electrodes. The strength of the electric field at a distance r from the anode wire in case of the cylindrical geometry is given by

$$E = \frac{V_0}{r\log\left(\frac{b}{a}\right)} \tag{7.26a}$$

where 'V_0' is the applied voltage, 'b' the radius of the cylindrical cathode, and 'a' the radius of the anode wire. In addition, in parallel plate geometry the strength of the electric field is uniform between the plates and is given by

$$E_{\text{parallel}} = \frac{V_0}{d} \tag{7.26b}$$

where d is the gap between the parallel plates. For d = 1 cm, to produce an electric field of 11.92×10^6 V/m, the required value of V_0 is 119.2 kV, which is an extremely large voltage. However, the same field (11.92×10^6 V/m) at the surface of the anode wire of radius 80×10^{-4} cm and cathode radius b = 1 cm can be obtained in a cylindrical geometry by only V_0 = 2 kV. Thus, in proportional and Geiger–Muller counters where large electric field near the anode is required for gas multiplication, cylindrical geometry for the counters is preferred. Another advantage of using cylindrical geometry for the counter is that the electric field profile is symmetrical around the anode wire, which means the electric field strength is same for all points that are situated at a given perpendicular distance from the anode. Parallel plate geometry is used in ionization chambers, where gas multiplication is not desired. The volume of the counter between the anode and the cathode is called the active volume. The electrical pulses formed in the gas-filled counters are fed to a preamplifier, which is basically an impedance matching device. In proportional and Geiger–Muller

(GM) counters, the pulse size also depends on the physical size of the counter. The pulses obtained from proportional and GM counters are generally big enough to be registered by a counter (scalar). However, analogue to digital (A to D) converter and amplifier may be used optionally.

7.6.1.2 Working of a gas-filled counter

Passage of some ionizing radiation through the active volume of the counter creates electron–(positive) ion pairs. The number of primary ion pairs depends on the amount of the energy ΔE lost by the radiation. If w denotes the amount of energy required to create one ion pair, then the number of primary ion-pairs N_i formed is

$$N_i = \frac{\Delta E}{w} \tag{7.26c}$$

The value of w depends on the gas and its pressure in the counter. However, the average value of w is of the order of 32 eV.

If the applied potential difference V_0 is zero the ion pairs created in the initial ionization will recombine and no charge will be collected at the electrodes. If V_0 is very small that the electric field is very low, only those ion pairs that were formed near the electrodes will be collected by the electrodes. On further increasing V_0, the strength of the electric field will increase and more and more electrons and positive ions will be collected at the respective electrodes. Ultimately, at a particular value of the applied potential V_i the electric field will be strong enough to collect almost all the ions and electrons created in the initial process. For a certain voltage interval V_i to $V_i + \Delta V_i$, the strength of the electric field remains such that most of the primary charges are collected by electrodes. The amount of charge ΔQ transported to the electrodes may be given by

$$\Delta Q = 1.6 \times 10^{-19} N_i\, \eta \ \text{Coulomb} \tag{7.26d}$$

Here, η is the charge collection efficiency of the gas-filled counter. The charge collection time Δt is generally quite small (10^{-9} t0 10^{-4} s) and, therefore, the small current Δi that flows through the resistance R (figure 7.22) produces a voltage pulse Δv given by

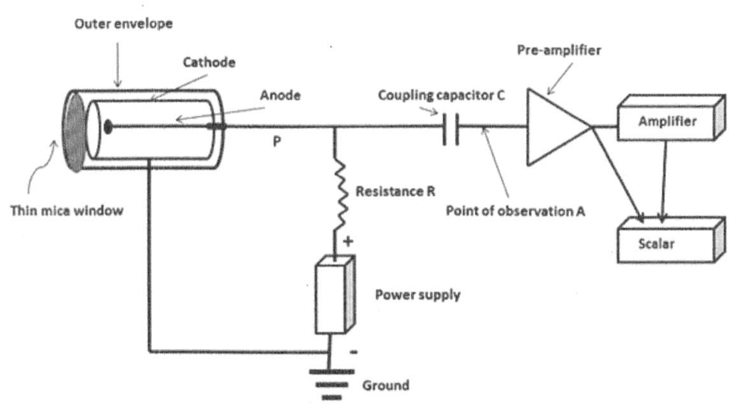

Figure 7.22 *Layout of a gas-filled end-window counter and associated electronic circuits*

$$\Delta v = R\Delta i = R\frac{\Delta Q}{\Delta t} \tag{7.26e}$$

The voltage pulse Δv or current pulse Δi are called signals. In solid-state detectors, the current signal is directly fed to a pre-amplifier. In ionization chamber, the current pulse is measured by a very sensitive current device. If a constant flux of ionizing radiations falls on the counter, a current of constant value flows through the resistance R. The incident flux is proportional to the magnitude of the current. When a gas-filled counter is operated in the ionization plateau region from V_i to $V_i + \Delta V_i$ it is called ionization chamber. Ionization chambers are used to determine the flux of the incident radiations. They may also be used as charged particle spectrometers. One problem that is encountered in ionization chambers is the drift time of the ion to their respective electrodes. As a matter of fact electrons, because of their small mass, have larger drift velocity as compared to the positive ions and reach the central wire (anode) in a much smaller time than the positive ions. The ion collection time is, therefore, decided largely by the collection time of the positive ion, which in turn depends on the location where the positive ion was formed in the active volume of the chamber. However, this difficulty may be overcome by putting a mesh grid called Frisch grid between the anode and the cathode. A potential almost half that at the anode is applied to the grid and the ionizing radiations are directed in the region beyond the grid. As a result, only the fast moving ions pass through the grid and the voltage pulse formed between the grid and anode has a fast rise time.

7.6.1.3 Gas multiplication

When the applied voltage is increased beyond the ionization region, the electric field near the anode becomes quite large. The electrons and the positive ions in this region of the field get accelerated. Accelerated electrons in particular collide with other molecules of the gas and produce secondary ionization. Electrons suffer collisions with other gas molecules on an average after travelling a distance of the order of their mean free path. As a result, the number of electrons increases with the distance as one move towards the anode. The rate of increase in the number of electron with distance may be given as

$$\frac{dn}{dx} = \alpha(x)n \text{ and so } n(x) = \mathrm{n}(0)\, e^{\alpha x} \qquad (7.26\text{f})$$

Equation (7.26f) is called *Townsend equation* and 'α' the Townsend's coefficient. Thus, the number of electrons increases exponentially with the distance. This is called gas multiplication and results in the collection of much larger charge at anode and cathode than was produced in the initial ionization. However, in the region of proportionality, the charge collected at the electrodes is proportional to the initial ionization. The ratio of the charge collected to the charge produced in the initial ionization K is called the gas multiplication factor, which depends on the value of the applied voltage. At a given value of the applied voltage K has a fixed value that does not depend on the initial ionization. The magnitude of K increases with the magnitude of the potential applied to the counter.

7.6.1.4 Pulse formation

Passage of an ionizing radiation through the active volume of the counter creates initial ionization. Electrons and positive ions formed in the initial event move, respectively, towards the anode and the cathode. Electrons being faster reach quickly near anode where there is large electric field and produce secondary ionization giving rise to gas multiplication. Collection of electrons at anode constitutes a current through resistance R and the potential at point P sharply falls from the initial value V_0 by the amount Δv in electron collection time t_e. If C is the capacitance of the counter, then

$$\Delta v = \frac{\Delta Q}{C} = \frac{KN_i\eta}{c} \qquad (7.26g)$$

Slowly positive ions also start reaching the cathode, take electrons from the cathode, and become neutral molecules of the gas. As ions, particularly those created in secondary ionization near the anode, take longer time to reach the cathode and become neutral molecules, the condition of the counter slowly recovers back to the one existed before the passage of the ionizing radiation. Thus, a negative pulse of short rise time and long tail superimposed over a large DC potential V_0 is formed whenever an ionizing radiation passes through the active volume of the counter. The coupling capacitor C blocks the DC potential and only the transient negative pulse appears at point A in figure 7.23. This pulse after passing through pre-amplification stage is often passed through a pulse discriminator to reduce noise. Electronic noise in the form of small random pulses is always generated in electronic circuits. To filter out the spurious noise pulses, discriminator circuit allow only those pulses that are larger in size (voltage) than the selected discriminator level to pass through it.

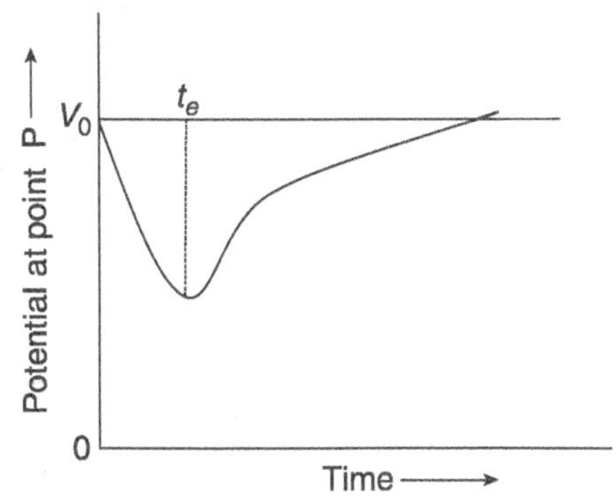

Figure 7.23 *Variation in the potential at point P with time*

Let us study the condition of the counter just at the moment when all the electrons have reached the anode and the slow moving positive ions are moving towards the cathode. These positive ions make a sheath around the anode, which in a way isolates the anode from the cathode. As a result, the potential difference between the anode and the cathode drops to a very low value. If some ionizing radiations now enter the active volume, they may create ion pairs but these ion pairs will not be collected at their respective electrodes because of the very small electric field. Hence, these ion-pairs will recombine and no net charge due to the passage of the radiation will be released. As a result, the radiations will not be counted. The effect of the ion-sheath decreases as more and more ions reach cathode and get neutralized. The electric field between the anode and the cathode recovers with the depletion of the ion-sheath. Ultimately, at time t_d the electric field has recovered to a value just sufficient to collect the ion-pairs. If an ionizing radiation passes through the active volume of the counter after time t_d it will make a pulse but of smaller size because of the smaller value of the multiplication factor K. The time t_d for which the counter will remain dead for the counting of radiation is called the dead time of the counter. The dead time depends on many factors including the operating voltage, physical size of the counter, gas pressure and on the type of the gas used. As the ion-sheath neutralizes, the electric field and the magnitude of K increases and pulses bigger in size are formed. However, they may not be counted if the size is less than the discriminator level. Ultimately the electric field attains a value such that pulses of size larger than discriminator level are produced. The time lapse between the dead time and the time when pulses above discriminator level are formed is called *recovery time*. The time-lapse between two successive detectable pulses is called the *resolving time* of the system. As is obvious, the resolving time also depends on the discriminator level and the pulse detecting electronics (scalars) (figure 7.24).

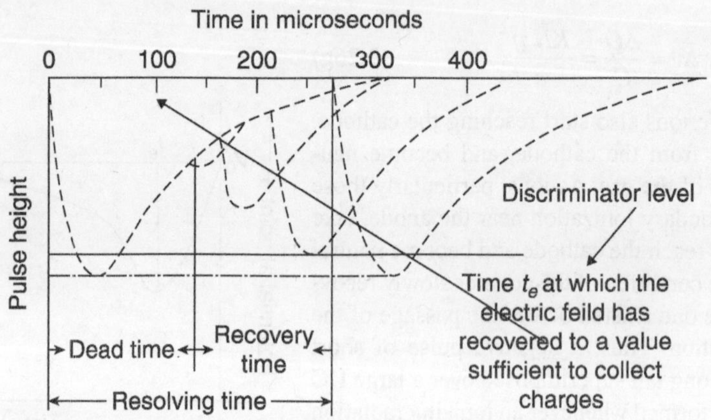

Figure 7.24 *Dead time, recovery time, and resolving times of a counter*

7.6.1.5 Variation of the pulse size with the applied potential V_0

Figure 7.25 shows how the pulse size increases with the potential V_0 applied to the counter. There are two curves in this figure: the lower one corresponds to the incident radiation that loses 1 MeV of energy in the active volume of the counter while the upper one is for a more ionising radiation that loses 2 MeV of energy in the counter. The initial ion pairs formed by the more ionizing radiation (upper curve) will be roughly twice the ion pairs formed by the other. For very low value of V_0, the electric field is not enough to collect all ion pairs, most of them re-combine and only a very small amount of charge reaches electrodes forming a small pulse. This region, where recombination of electron–ion pairs is the dominant process is called '*Recombination region*' and is of no practical application. On further increasing the applied potential, a stage comes when the electric field between the electrodes is enough to pull all electrons and ions produced in the initial interaction to the respective electrodes, before their recombination. However, the field is not enough to initiate secondary ionization and gas multiplication. As such, the pulse size is due to the charge released in primary ionization as shown in figure 7.25. This condition prevails for the voltage interval V_1 to V_2 and a counter operated in this region works as ionization chamber.

When V_0 is increased beyond V_2 gas multiplication takes place and the final charge collected at the electrodes is proportional to the initial ionization. Hence, the power of distinguishing between more ionizing and less ionizing radiations by the counter is retained. The gas multiplication factor K provides an internal amplification that is linear, that is at fixed value of V_0 the multiplication factor K has a fixed value and the initial number of ion-pairs N_i, which may have different values for different radiations, are multiplied by the same K irrespective of the magnitude of N_i. The value of K may be as high as 10^4 initially but may increase to as high as 10^9 if V_0 is increased from V_2 to V_3. The operating region between V_2 to V_3, called *proportional region*, is very much used in making proportional counters. Proportional counters can distinguish between a more ionizing

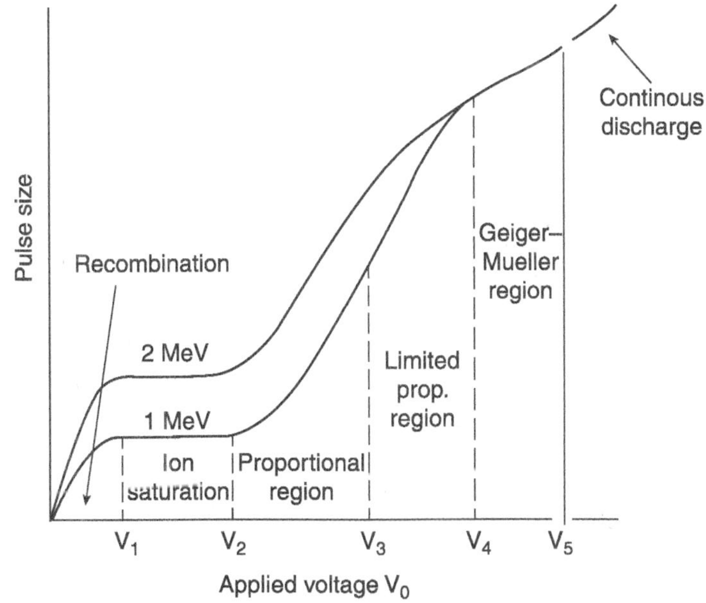

Figure 7.25 *Variation of the pulse size with the applied potential V_0*

and a less ionizing radiation. If the incident radiation is stopped in the counter losing all its energy in the active volume, the energy of the radiation may also be determined from the pulse size.

When potential V_0 is increased beyond V_3 the electric field becomes so high that the positive ions while getting neutralized at cathode emit photons that produce photo-electrons from the metallic cathode. These photo-electrons released near the cathode rush towards the anode producing secondary ionization. As a result, ionization avalanches one after the other are generated following the passage of a single ionizing radiation. The ionization avalanche created in this way is called *Townsend avalanche*. The net result of this self-propagating avalanche mechanism is that after the original pulse formed by the passage of an ionizing radiation, spurious pulses one after the other are generated without the passage of any radiations. As a consequence, linearity between the pulse size and initial ionization is lost and the counter is said to be in the region of limited proportionality. Further, the counter loses the property of generating a pulse only when some radiation passes through it.

7.6.1.6 Quenching of avalanche

The ionization avalanches need to be stopped if the counter is used as a detector. The mechanism by which the generation of secondary avalanches is stopped is called *quenching*. Ionization avalanches are produced because of the two things: (1) Very high electric field between the electrodes and (2) emission of photons near cathode when ions neutralize. Quenching can be achieved in two different ways. In the first method, called *external quenching*, the resistance R is kept large so that the large potential drop across the resistance reduces the electric field between the electrodes to a value so low that no secondary ionization is produced by photo electrons emitted near the cathode. The big drawback of external quenching is that a large value of resistance R also increases the recovery time of the counter.

In the other method called *internal quenching* photons that are produced when ions neutralize are not allowed to be produced. This is done by mixing some long-chain hydrocarbon such as alcohol vapours with the counter gas. As the ion sheath moves towards the cathode it takes electrons from the organic vapours and transfers their charge to them. As a result, the charged molecules of the long-chain hydrocarbon reach cathode, get neutralized, and instead of emitting photons, undergo cracking. In this way, the emission of photons and in turn of photoelectrons from the cathode is avoided. Later, the cracked quenching gas molecules recombine and are ready to quench again.

7.6.1.7 Geiger–Muller region

If the potential difference across a quenching incorporated counter is increased beyond V_4 the electric field in the counter becomes so large that even a single ion pair ionizes whole of the counter gas. *The pulse size does not depend on the initial ionization. All radiations with small or large initial ionization produce pulses of same size.* This is called the Geiger–Muller (or simply GM) region. For the operating potential between V_4 and V_5 the pulse size is nearly constant and independent of the nature of the radiation. The near flat portion of the curve between V_4 and V_5 is called the *GM Plateau*. A counter operating in the GM region cannot distinguish between an α particle, a β particle, or a γ ray. However, the big advantage is that the pulse size is very large. Basically, the size of the pulse depends on the amount of the counter gas, which itself depends on the physical dimensions of the counter. For example, in case of the ionization chamber the pulse size is of the order of few tens of microvolts, in proportional counter of millivolts and in GM counter of volts.

If the applied voltage is increased beyond V_5 the electric field becomes so strong that gas molecules tear-apart by the electric field and electrons are emitted. This is also called field emission of electrons. Thus, electrons are produced in the counter without the passage of any radiation. A continuous discharge occurs through the counter.

Solved example S-7.3

Drive an expression for the dead time of a GM counter using two-source method.

Solution. In the two-source method of determining the dead time of a counter two-radiation sources S_1 and S_2 are taken and one of the source, say S_1, is placed at some place near the counter and the count rate N_1 (number of count in one second) is recorded. Now, without disturbing the source S_1, the other source S_2 is also placed near the counter and the counting rate N_{12} due to the two sources together is recorded. Finally, the first source S_1 is taken away without disturbing the position of source S_2 and the count rate N_2 due to the source S_2 is recorded. If B is the count rate of the background and T the dead time of the counter then the true count rates for the three cases will be given by the following equations:

$$N_1^{\text{true}} = \frac{N_1 - B}{1 - TN_1}; \quad N_{12}^{\text{true}} = \frac{N_{12} - B}{1 - TN_{12}}; \quad N_2^{\text{true}} = \frac{N_2 - B}{1 - TN_2}$$

In the three expressions, the denominator in each case gives the actual time for which the counter was active while TN is the time for which the counter was inactive. Now, the sum of the true count rates of the individual sources should be equal to the true count rate when both the sources were put together. Hence,

$$\frac{N_{12}-B}{1-TN_{12}}=\frac{N_1-B}{1-TN_1}+\frac{N_2-B}{1-TN_2}$$

Using the known values of N_1, N_2, N_{12}, and B, the dead time T can be calculated from the above-mentioned expression. An approximate expression for the dead time that follows from the expression if terms of higher order in T are neglected is

$$T=\frac{N_1+N_2-N_{12}+B}{2(N_1-B)(N_2-B)}$$

7.6.2 Scintillation Detector and Spectrometer

Scintillation detectors are another class of detectors that are very frequently used in nuclear physics experiments and in medical applications. Although liquid and gaseous scintillators are used for special applications, it is the solid scintillators detectors that are used more frequently. The main advantage of solid scintillators is their high atomic number (Z) and density both of which are advantageous from the perspective of the energy loss by the radiations. There are other advantages of scintillators detectors that will be discussed later.

Scintillation is a general term referring to the process of giving off light; it is used both literally and figuratively. More specifically in science, scintillator is a material that can release a photon in the ultraviolet (UV) or visible region when an excited electron of the material returns to the ground state. These scintillation photons are detected by a photomultiplier tube (PMT) and are converted into an electrical signal. Many different kinds of materials have the ability to scintillate. Many organic conjugate ring compounds scintillate when their orbital electrons de-excite. Anthracene has the highest detection efficiency amongst the organic scintillators such as plastic. Some gases such as xenon and helium do scintillate when charged particles pass through them. Glasses mixed with enriched ^{6}Li are used for the detection of neutrons. However, all these scintillators have low atomic number and are not very efficient for the detection of γ rays.

The scintillator that is most frequently used for γ ray detection is the thallium activated sodium iodide crystal, NaI(Tl) that was developed in 1948 by Robert Hofstadter. This detector is best suited for low-energy γ rays. Other scintillators such as bismuth germinate (BGO), lutetium oxyorthosilicate, and gadolinium oxyorthosilicate, because of their high Z and density are better suited for the detection of high energy γ rays.

Most desirable property of sodium iodide crystal is the excellent scintillation light yield. Roughly 38,000 photons/MeV of energy loss by the radiations are emitted by NaI(Tl) crystal. Large yield of scintillation photons make measurements more accurate and precise.

7.6.2.1 Working of NaI(Tl) scintillator

Pure sodium iodide (NaI) crystal has delocalized molecular bonding with a forbidden energy gap of about 5.5 eV between the valence and conduction bands. If any radiation such as γ ray lose energy in the crystal and excite an electron from the valence to the conduction band, the excited electron will have higher probability of de-excitation by emitting photon of frequency not falling in the range of ultraviolet or visible light. As such, the pure NaI crystal cannot be used as an efficient scintillator. However, if a small amount of impurity of thallium (Tl) is added to the crystal it becomes a highly efficient scintillator.

The energy band diagram of the thallium activated NaI(Tl) crystal is shown in figure 7.26. The energy levels of the thallium impurity lie in the forbidden energy gap of pure NaI as is shown in the figure. When an incoming γ ray lifts an electron from the valence band to the conduction band, the free electron and the hole created in the process move, respectively, in the conduction and valence bands and ultimately migrate, respectively, to the excited levels and the ground state of the impurity thallium. The important aspect of the process is that the probability for the excited electron to de-excite directly to the valance band of NaI is much smaller than the probability of its falling into an excited level of the thallium impurity. Electron from the excited levels of thallium de-excite to its ground state by emitting scintillations of visible light. In this way, scintillations of visible light are emitted from electronic transitions in thallium atoms making thallium activated NaI(Tl) crystal as one of the most efficient scintillating material.

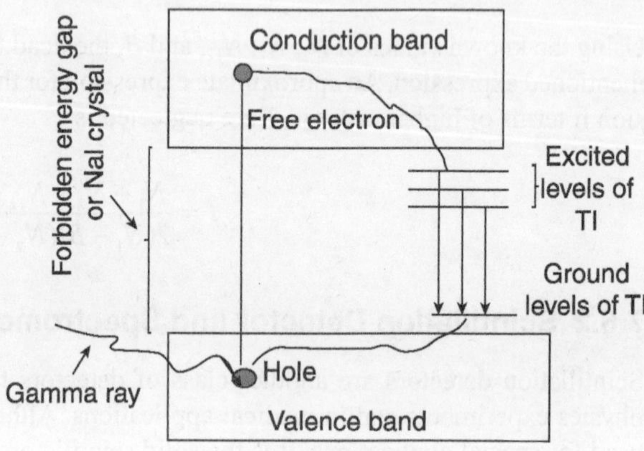

Figure 7.26 *Energy band diagram of NaI(Tl) crystal*

There are three very important aspects of scintillation process: First, the number of visible light photons produced is proportional to the energy lost by the radiation in the crystal. For example, if a γ ray of 250 keV is totally stopped in the crystal and produces X number of scintillation photons, then a 500 keV g ray will emit nearly 2× number of photons and a 750 keV γ ray 3 × photons, provided the γ rays are stopped in the crystal and their energy is totally absorbed.

Second, the total processes of energy absorption to the emission of photons take a very small time. A γ ray entering the crystal deposits its energy within a few nanoseconds. The emission of scintillation photon reaches its maximum within 30 nanoseconds and then decays with a half period of about 150 nanoseconds. As such almost 68% of the light is emitted within 230 nanoseconds of the entry of the γ ray. This is called the decay time or decay constant of the scintillator. Small decay time makes the system very fast. Moreover, if the source of the γ ray is not very strong, each γ ray will be detected before the entry of the next, reducing the chance of pulse pile-up.

Third, the wavelength of the emitted visible light peaks at λ_{max} = 420 nanometre (nm) for which the detection efficiency of photomultiplier tubes (PMT) is maximum.

The NaI(Tl) crystal is quite fragile and highly hydroscopic. It readily absorbs moisture from the surroundings if left in contact with air. Moisture in the crystal makes it opaque to its own radiations, that is the visible light emitted during scintillation process, could not pass through the crystal. As such, the crystal is first polished in vacuum to remove any cracks and surface irregularities so that the visible light photons are not scattered and may reach to the photocathode of the PMT. The polished crystal is then mounted in an aluminium container covering the polished surface with transparent glass cover. A very thin layer, free of any bubbles, of DC-200 silicon oil is sandwiched between the glass covering and the crystal surface to match the refractive

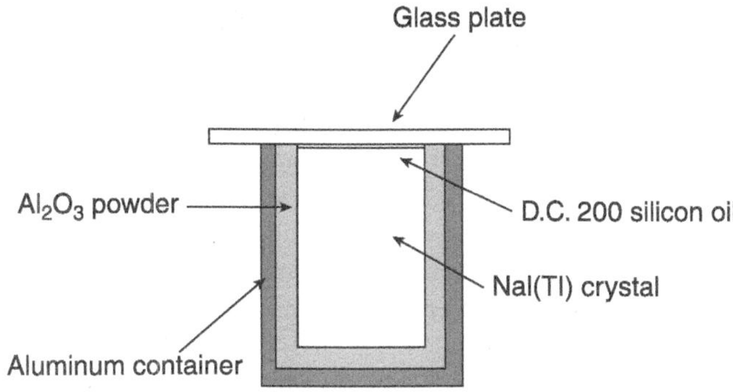

Figure 7.27 *A NaI(Tl) crystal mounted in an aluminium container*

index of the glass with that of the crystal. This matching of the refractive index is required for smooth passing of scintillation light from the crystal to the photomultiplier tube. A thin layer of Al_2O_3 powder, that has a refractive index slightly less than that of the NaI crystal, is filled in gaps between the crystal and the aluminium container on the sides of the crystal as is shown in figure 7.27. The photons of visible light scintillations moving out of the crystal surface are turned back by total internal reflection in aluminium oxide and are not lost. Finally, the crystal is hermetically sealed.

7.6.2.2 The photomultiplier tube (PMT)

Production of scintillations by the detector crystal is not enough. These scintillations must be converted into an electronic pulse. A photomultiplier tube converts visible or UV light photons into electronic pulses of measurable size. The basic structure of the photomultiplier tube is shown in figure 7.28. Scintillation-photons from the scintillator crystal falls on the photocathode of the PMT. The photocathode is made of a photo emissive material that emits photoelectrons when flashes of visible or UV light from the scintillator crystal hit it. There are a number of

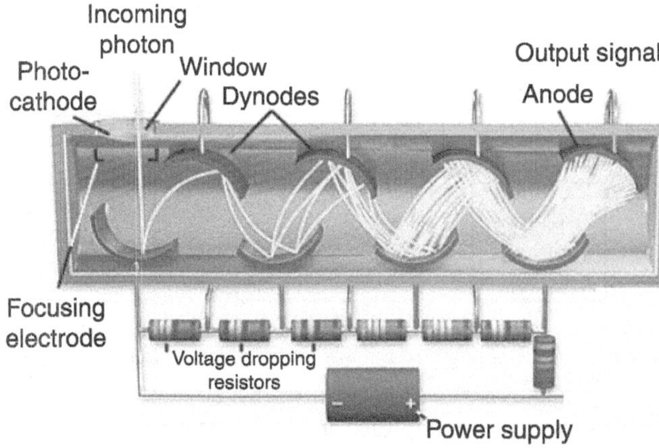

Figure 7.28 *Photomultiplier tube*

electrodes called dynodes in the PMT, which are kept at successively higher potentials by a chain of resistances connected in series to a power supply. Photoelectrons from the cathode (at ground potential) are attracted to the first dynode where they produce secondary electrons. Secondary electrons from the first dynode are attracted to the second then to the third dynode and so on. At each step of attraction, electrons acquire more and more kinetic energy and produce more secondary electrons. As such, the number of electrons goes on increasing as they move from one dynode to the next. In general, there are 9–19 dynodes in a PMT. The dynodes are arranged such that the secondary electrons emitted from one are focused on the next. Ultimately, a very large number of electrons proportional to the number of visible or UV photons that hit the cathode reach the last dynode, called anode. A small voltage pulse of the size proportional to the amount of energy deposited by the radiation in the crystal may be developed by passing the electron current from the anode through a resistance. Although the number of electrons from the first dynode to the last gets multiplied by a factor of 10^4 or 10^5, but the pulse formed at anode is still quite small, of the order of few hundred micro volts. The anode signal is then passed to a pre-amplifier that is generally integrated with the photomultiplier tube.

7.6.2.3 Scintillation spectrometer

Scintillation detectors, unlike gas filled GM counter, are used not only for the detection of radiations but also to determine their energy. The associated electronic modules used for the purpose are a linear amplifier, a discriminator which is generally in-built with the amplifier, a single channel pulse height analyser and a scalar. The block diagram of a scintillation spectrometer is shown in figure 7.29.

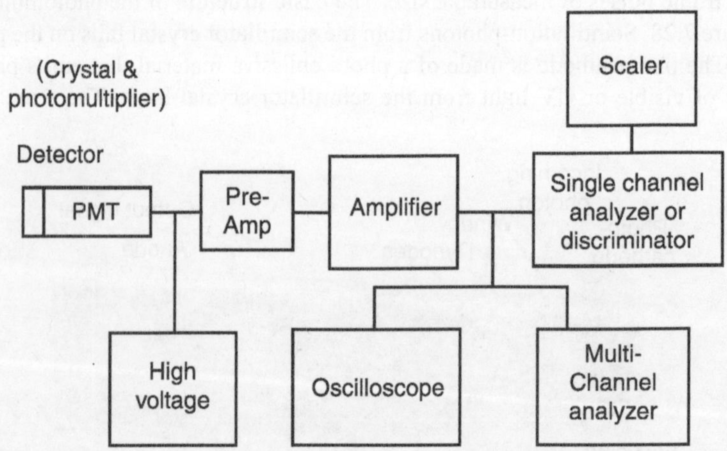

Figure 7.29 *Block diagram of a scintillation spectrometer*

Small electronic pulses from the pre-amplifier are coupled to the input of a linear amplifier of adjustable gain. The amplifier first shapes the input pulses to improve the signal-to-noise ratio using RC network. The pulses are then fed to the input of a system of amplifiers cascaded to each other. The special property of amplifiers is their linearity. Linearity means that the gain of the amplifier is same for pulses of all sizes irrespective of their size. The overall gain of the amplifier can be adjusted to any desired value by the control switches on the front panel. A discriminator circuit is also provided with in the amplifier. The discrimination level may be set to the desired value and pulses larger in size than the discriminator level are allowed to be passed. It is also possible to get a uni-polar or bipolar pulse in the output of the amplifier. Amplifier output contains pulses of different sizes corresponding to the amount of energy lost by the radiations in the scintillation crystal. As the linearity is maintained from the crystal to the PMT and to the amplifier, the pulse height (size) is directly proportional to the amount of energy deposited in the crystal. The intensity (number of pulses) of a pulse of a given height as well as the pulse height spectrum may be recorded by using a single channel analyser (SCA). In principle, an SCA consists of two discriminators, one with discriminator level V_1 called the LLD (lower level discriminator) and the second with $(V_1 + \Delta V)$ called the upper level discriminator (ULD). Level V_1 called the baseline may be set to any desired value by a front panel control. Potential difference ΔV is called the window-width and like the baseline can be set to any desired value. Now, suppose V_1 is set to 1 V and ΔV also to 1 V. The output of the SCA will contain pulses equal in number to the pulses whose pulse height lies between 1 V and 2 V. If now the baseline is changed to 2 V, the SCA output will have pulses equal in number to the pulses whose height lies between 2 V and 3 V. In this way by changing the base line voltage, the pulse height spectrum can be recorded. The output pulse of the SCA is a logic pulse of fixed shape and size that may derive a scalar for recording the number of pulses in a given time. In this way, the number of pulses in a given time having heights between V_1 and $V_1 + \Delta V$ may be recorded by the scalar. The pulse height spectrum of the amplifier output may thus be recorded by recording scalar count rate for each value of baseline voltage V_1. Nowadays, it is also possible to record the pulse height spectrum in one go by using a multichannel analyser (MCA). MCA module also has an in-built oscilloscope. The pulse height spectrum, which corresponds to the distribution of energies lost by the radiation in the scintillation crystal, can be projected on the oscilloscope screen. MCA also provide facilities and tools for analysing the pulse height spectrum.

A typical pulse height spectrum of a ^{137}Cs source taken by a NaI(Tl) spectrometer is shown in figure 7.30. ^{137}Cs source emit mono energetic γ rays of 662 keV. These γ rays enter the scintillation crystal and interact with the atoms of NaI. Some of the γs lose their entire energy (almost 662 keV) by photoelectric effect. This is the largest amount of energy loss that the γ ray can suffer. The photoelectrons of kinetic energy almost equal to the energy of the incident γ rays lose their energy by ionization and excitation of target atoms. Ultimately, photoelectrons lose all their energy and come to a stop. The biggest peak towards the right end of the spectrum (figure 7.30) is due to those events in which γ ray lost energy by photoelectric effect. Many of the incident 662 keV γs interacted with the crystal atoms by Compton scattering. The energy distribution of Compton electrons shows an edge that is also present in the spectrum in figure 7.30.

Another small peak, levelled as back-scattered peak, is due to those 662 keV γ rays that collided with heavy materials lying around the detector and got scattered back into the detector but with smaller energy. The position of the back-scattered peak depends on the amount

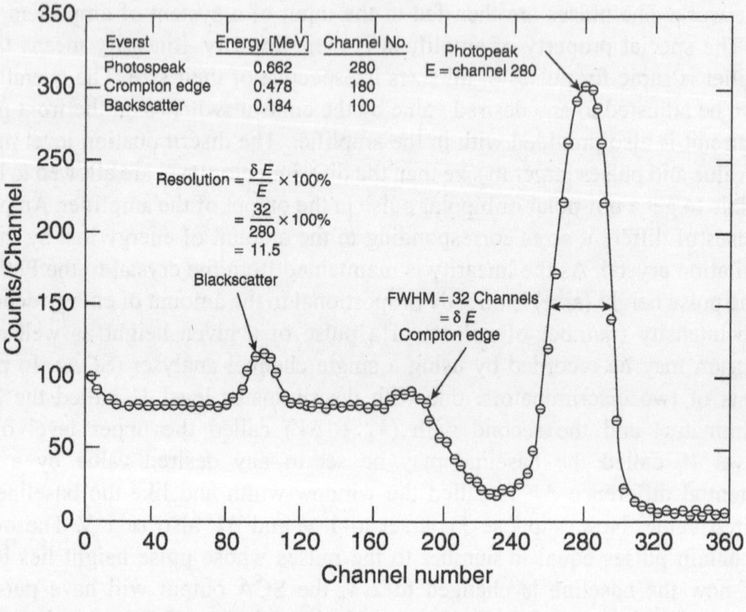

Figure 7.30 *A typical γ ray spectrum of ^{137}Cs source taken by a NaI(Tl) crystal spectrometer.*

and distribution of heavy material surrounding the detector and, therefore, shifts if the detector location is changed.

7.6.2.4 Photo peak resolution

Although the ^{137}Cs source emits γs of a fixed energy (662 keV) but in the spectrum we see that the photo peak is not a line but has a finite width. This is due to the statistical nature of various processes involved from energy loss to pulse formation. First, the energy of photo electrons emitted from the sodium atom and iodine atoms are slightly different because of their different binding energies in the two atoms. Second, the photo electrons when lose energy by excitation and ionization of sodium and iodine atoms deposit different amounts of energy in each interaction. Electron multiplication in PMT is yet another statistical process responsible for the broadening of the peak. Some broadening is also brought by the electronic processing of the pulse in various electronic modules. If ΔE is the width of the peak at half maxima and the maximum value of the peak occurs at $E = E_0$, then the percentage resolution of the peak is defined as $\left(\frac{\Delta E}{E_0}\right)\times 100$. Percentage resolution tells about the quality of the spectrometer. A small value of the percentage resolution means a better system as it will be able to resolve two different γ ray peaks, which differ by a very small amount of energy. The percentage resolution of a NaI(Tl) scintillation spectrometer is generally in the range of 5–10%. Gamma ray spectrometers using solid-state detectors have much better resolution ≈ 1–2% and are, therefore, much better than NaI(Tl) scintillators.

The x-axis or the pulse height (also called channel number) of the spectrum can be calibrated in terms of γ ray energy by taking spectra for several γ sources of different energies with the same spectrometer. A graph between the pulse height and the corresponding γ energy is generally a straight line. This straight line curve, called the calibration curve, may be used to find out the energy of an unknown γ ray that occurs at a certain pulse height in the spectrum.

7.6.2.5 Important characteristics of a scintillator

Flashes of UV or visible light are emitted when energy is deposited by an incident radiation in a scintillator. This flash of visible (or UV) light is then made to fall on the cathode of a photo multiplier tube that ejects photoelectrons, which are multiplied by dynodes and ultimately a current pulse is formed. Further processing of the current pulse is done by electronic modules. For successful generation and recording of pulses when radiations pass through the detector, the scintillator crystal must have some properties that are characterized by the following parameters:

1. Effective atomic number Z: Scintillators that are used for γ detection should have high Z and large value of density. It is because the γ ray energy is determined through the photo peak, which is formed as a result of photo electric effect. The probability of photoelectric effect increases as Z^5 and, therefore, scintillators of dense materials and high Z value are better for γ detection. Often compounds containing more than one element are used in for scintillation. The effective atomic number Z of the scintillating material is the weighted average of the Z-values of different elements.
2. Light or photon yield per unit energy: The energy ΔE lost by the radiation in the scintillator is converted in the photons of UV or visible light. The number of photons per unit energy loss, called light or photon yield, is an important parameter. A large photon yield means a more intense light flash and a more easily detectable signal. It is given in units of number of photons per MeV.
3. Wavelength of maximum emission λ_m: Flash of light or photons emitted by the scintillator has a distribution of wavelengths. However, the distribution peaks at a certain wavelength λ_m. The wavelength λ_m, corresponding to the maximum number of emitted photons, is called the wavelength of maximum emission. This is an important parameter because the photocathode of the PMT must be sensitive to the wavelength of maximum emission.
4. Decay time constant: The scintillation flash lasts for a certain time. It starts, its intensity increases reaching a maximum, and then it decays out. The time required for scintillation emission to decrease to $\frac{1}{e}$ of its peak value is called the decay constant or decay time of the scintillator. In one decay time, almost 68% of the photons are emitted by the scintillator. A good scintillator should have small decay time so that flashes from successive radiations do not overlap.
5. Detection efficiency: If a system detects N radiations out of the N_0 falling on it, the ratio $f = \frac{N}{N_0}$ is called the detection efficiency of the system. Often detection efficiency is given in percentage ($f \times 100$). The efficiency of a system depends on several factors, such as the efficiency of the detector, which in turn depends on the size and other qualities of the detector, and the efficiency of other electronic modules.

Apart from these special characteristics, a scintillator must be transparent to its own radiations, should not be damaged by the radiations and should be stable.

6. Percentage resolution: A good detector system should be able to differentiate between two radiations of same kind and of nearly same energy. This can be possible if the signal peaks are sharp and of small widths so that the overlapping between the peaks is least. This property of resolution is quantified in terms of 'percentage resolution'. If a signal peak corresponding to energy E has a width ΔE at its half maxima then:

$$\text{Percent resolution} = \frac{\Delta E}{E} \times 100$$

A small value of percent resolution means a good detecting system. For a given detector setup, percentage resolution decreases (becomes better) with energy. As a matter of fact, the resolution of a total detector set up, including the detector and the following electronics is important. The resolution of the detector alone depends on the square root of the energy $\sqrt{w}$, required to create one pair of charge carriers. w, in turn depends on the ionization potential of the detector material and the forbidden energy gap if the detector is made of a solid material. Another factor called Fano factor also plays important role. It arises because of the additional possibility of lattice excitation in case of solid-state detectors.

7.6.2.6 Some important scintillators

Anthracene ($C_{14}H_{10}$): The pure organic scintillator anthracene is a very efficient detector for β particles. Spectrometer using anthracene crystals are often employed to study the energy spectrum and Fermi–Kurie plots for β particles emitted in radioactive decay. Some important scintillation parameters for anthracene are molecular weight = 178; density = 1.25 g/cm^3; wavelength of maximum emission λ_m = 445 nm; decay time = 30 ns; light output: photons = 20,000/MeV.

Stilbene ($C_{14}H_{12}$): This organic scintillator is used in neutron detection and spectroscopy as it is good for pulse-shape discrimination between neutron and γ rays. Some of its specifications are molecular weight = 180; density = 1.22 g/cm^3; wavelength for maximum emission = 390 nm; decay time = 3.5 ns; light output: photons = 14,000/MeV.

Plastic: Various types of plastic scintillators are available with decay times ranging from 0.5 ns to 0.9 ns. The average light output of plastic detectors is about 45% of anthracene (≈8000 photon/MeV). The mean wavelength for maximum emission is around 400 nm. These scintillators are used for fast counting and in pulse-shape discrimination between neutron and γ pulses.

Bismuth germanate (BGO; $Bi_4\ Ge_3\ O_{12}$): This scintillator is specially used for the recording of high-energy γ rays in the energy range of 0.5–20 MeV. It is because of its high effective atomic number $Z = 75$. The probability of producing photoelectric effect decreases with the energy of the γ ray. Hence, in low Z scintillators (NaI(Tl) etc.) the chance that a high energy γ will lose energy by photoelectric effect is very low. However, in a high Z material such as BGO the probability of producing photoelectric effect by high-energy γs is high as the photoelectric probability increases as Z^5. BGO is one of the best materials for detecting and carrying out spectroscopy of high-energy γ rays. The important scintillation characteristics of BGO are effective $Z = 75$; density= 7.13 g/cm^3; wavelength of maximum emission = 480 nm; decay time = 300 ns; relative intensity of light emission = 3800 photons/MeV (about 10% of NaI(Tl)).

Thallium activated caesium iodide CsI(Tl): Like NaI(Tl) scintillator, caesium iodide crystal activated with thallium is a good scintillator for charged particles and low-energy γ rays. It is less hygroscopic than sodium iodide and, therefore, has a longer life than it. The drawback is its decay time that is much larger. Its light yield is about 45% of NaI(Tl).

Density = 4.51 g/cm$_3$; slightly hygroscopic; wavelength of maximum emission = 550 nm, light yield = 45% of NaI; decay time = 1000 ns.

7.6.2.7 Pulse-shape discrimination

A radiation detector provides an electronic pulse for each particle detected by it. There are two timing components in every pulse: the fast component that decides the rise time of the pulse and the slow component that governs the tail of the pulse. Both the fast and the slow components depends on two factors: (1) the detector properties and (2) the energy loss process or (dE/dx) of the radiation. As a result, the pulse shapes of different radiations in the same detector are different. Further, the pulse-shape difference is more visible in the slow component (the decay or tail part) of the pulse. The pulse output from the photomultiplier tube of a typical scintillator is shown in figure 7.31.

As may be observed in the figure, the shapes of the pulse for α particle, fast neutron and the γ ray are different and the difference is pronounced in the slowly decreasing tail portion of each pulse. It may be noted that the height of the pulse in each case will be proportional to the energy deposited by the radiation in the detector but figure 7.31 shows the output of the PMT where the pulses are not much amplified, the difference in heights is, therefore, masked. It is possible by

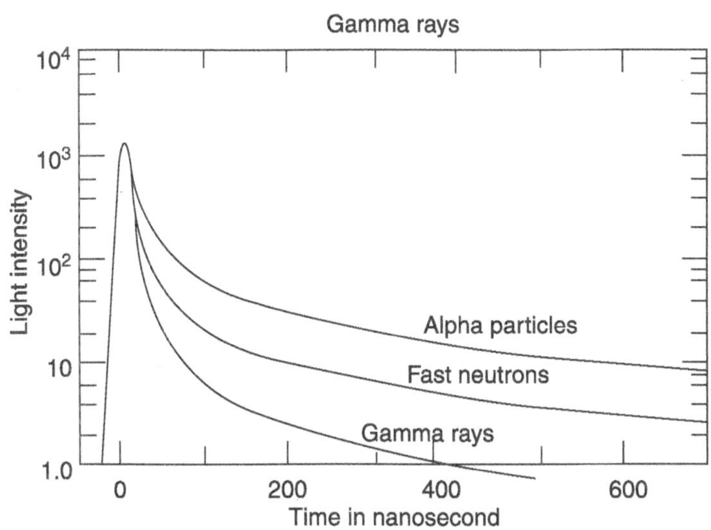

Figure 7.31 *Photomultiplier tube out put for a typical scintillation detector*

means of electronic circuits to sense each pulse and identify the radiation responsible for the pulse. This is called pulse-shape discrimination. Pulse-shape discrimination is often required in in-beam experiments where measurements are done in high background of neutrons and/or γ rays.

7.6.3 Semi-conductor Detectors

A p–n junction has a depletion layer that behaves as an insulator. Any charges created in the depletion layer are readily swept by the potential barrier across the depletion layer. A reverse biased p–n junction has a very small reverse saturation current that strongly depends on temperature. If a radiation passing through the depletion layer of the junction, ionizes the atoms of the depletion layer releasing electron–hole pairs the reverse saturation current will suddenly increase. The current pulse formed by the passage of the radiation through the depletion region may be used to identify the particle and to determine its energy. As such, a p–n junction in principle may be used as a radiation detector. However, the following problems make the use of a normal p–n junction as a detector difficult.

1. p- and n-type materials are made by controlled doping of intrinsic material with impurities. The impurity sites in the crystal acts as trap-centres for the charges released by the radiation. This reduces the strength of the current pulse and also adversely affects the linearity between the size of the current pulse and the amount of energy deposited by the radiation.
2. The depletion layer is generally in the central part of the crystal where low-energy radiations having small range may not reach.
3. The thickness of the depletion layer is not much.
4. A slight variation in the ambient temperature changes the reverse saturation current by a large amount.

7.6.3.1 Lithium-drifted germanium (Ge(Li)) detector

Lithium is an n-type material. It is diffused in the p-type germanium. To achieve the diffusion, one piece of p-type germanium is kept in contact with piece of n-type lithium and a reverse bias is applied between the two sides while the temperature of the system is raised. As a result of high temperature and reverse bias, some lithium atoms diffuse in the p-type germanium, creating a depletion layer and a region where the concentration of n- and p-type impurities neutralize each other. This region behaves like an intrinsic material and supplements the width of the depletion region. When the width of the intrinsic region reaches the desired value, the system is quickly cooled to liquid nitrogen temperature and the bias is removed.

The incident radiation may release charges in the total width of the intrinsic and depletion layers, which increases the detection efficiency of the system. Further, at liquid nitrogen temperature the reverse current is small and does not change. However, the system has to be kept at liquid nitrogen temperature always, even when it is not in use. If the temperature of the crystal is allowed to rise, the lithium ions that have drifted move back, rendering the crystal unfit for the detection of radiations. If due to some unavoidable circumstances, such as prolonged power failure, the crystal got heated, it may be sent to the manufacturer for repairs. However, it is costly and a repaired crystal has a poor resolution.

A typical Ge(Li) detector system consists of a vacuum-enclosed Ge(Li) crystal, which is coaxial in shape and attached to a copper cold finger through an insulator. The crystal is kept under vacuum to prevent frost forming on the crystal, and damage caused by impurities in the air. The cold finger is immersed in liquid nitrogen in a Dewar. The crystal is subject to failure, should its temperature be raised to room temperature, due to lithium ion drift and increased electronic noise.

7.6.3.2 Advantages of Ge(Li) detectors

Ge(Li) detectors offer an advantage of high resolution (i.e. the ability to differentiate between γ photo peaks, which differ in energy by a very small amount). It is mainly because of the fact that much less energy ≈3 eV is required to create one electron–hole pair in Ge(Li) detector. As a result, for the same loss of energy in the detector much more current carriers are generated in this detector as compared to the gas-filled detectors in which ≈33 eV energy is required to create an electron–ion pair. It can be shown that the percentage resolution $(\Delta E/E_0) \times 100$ of a peak is proportional to the square root of the energy required to create a pair of charge carriers. Another important factor regarding the resolution of a detector is the Fano factor F. Detailed analysis of statistical variations have indicated that the percentage resolution needs to be multiplied by the Fano factor which has a value 1 for gas detectors and less than 1 for solid-state detectors. For semi-conductor detectors, F has the smallest value. Fano factor arises because in detectors made of solid materials the energy deposited by the incident radiation may also excite lattice vibrations that do not happen in gaseous detectors.

The percent resolution of semi-conductor detectors is best (smallest) because of the very small energy required to create an electron–hole pair and the smallest value of Fano factor. Amongst the semi-conductor detectors, the peak resolution is best for germanium-based detectors.

A detector setup with small numerical value of percentage resolution has sharp peaks, with less overlap of nearby peaks. As such, the system has the power of resolving two peaks which differ in their energy by a small amount. A system with 10% resolution will be able to separate two peaks of energy 100 and 110 keV. However, it may not be able to resolve peaks of 100 and 105 keV as two different peaks. The two peaks will merge into one broad peak.

7.6.3.3 High Purity Germanium (HPGe) Detector

The charges (electron–hole pairs) created by the radiation in the depletion region of a semi-conductor detector are collected by the reverse bias and a current pulse is formed. The pulse is used for the detection and further energy analysis of the radiations. In normal semi-conductor detectors, the width of the depletion layer is only about a millimetre thick. Soft X-rays and charged particles may be totally absorbed in this thickness of the depletion layer and may be detected. However, such detectors are not efficient for the detection of high-energy γ rays, which require depletion thickness at least of the order of 1 cm. The depletion width d is related to the reverse bias voltage V_0 and the impurity concentration N in the semiconductor material through the relation

$$d \cong \left(\frac{2eV_0}{N}\right)^{1/2} \tag{7.27a}$$

In normal grade semiconductor material, the impurity concentration in the parent material is so large that even for a very high reverse bias potential reaching the break-down value the depletion depth d is only a few millimetre. To get $d \approx 1$ cm, the impurity concentration N must be less than 10^{10} atoms/cm^3. At this concentration with $V_0 = 1$ kV depletion depths of 1 cm may be achieved. Impurity concentration of 10^{10} atoms/cm^3 means almost 1 part in 10^{12}, which is very difficult to achieve. However, with innovative refining techniques hyper pure germanium crystals with impurity concentration of better than 1 part in 10^{12} have been achieved. Hyper pure germanium crystals still have impurities, which are generally of p-type. A thin layer of several hundred micrometre thickness of n-type lithium is diffused at one end of the crystal, which neutralizes the p-type residual impurity, forms a thin n–p junction, and also serves as the electrode for biasing. On the opposite side a much thinner layer of less than 1 μm thickness of boron: a p-type material is diffused to make the other electrode. It may be noted that no p–n junction is formed at the boron end as both the crystal and the boron are p-type materials. A thin beryllium window is mounted on the front face of the crystal to allow the radiations (γ rays) to enter the detector. The layout of a planer HPGe detector is shown in figure 7.32. As may be noted almost whole of the crystal width works as the depletion layer. HPGe detectors are very good detectors for

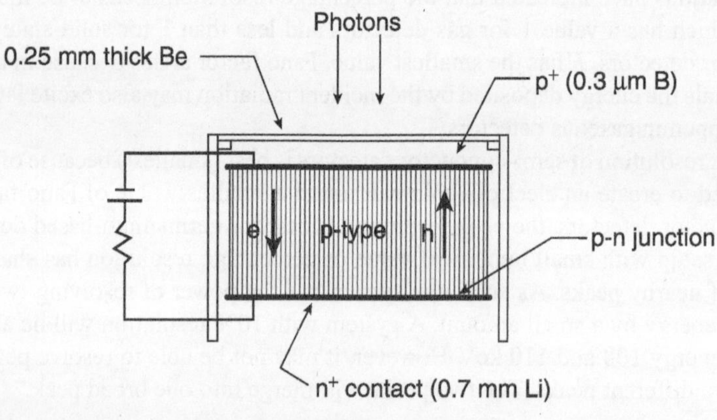

Figure 7.32 *A planer high purity germanium detector*

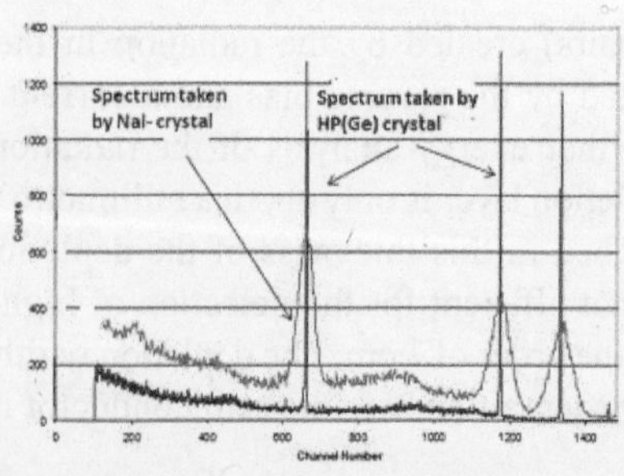

Figure 7.33 *Gamma spectrum of the same source taken by NaI(Tl) and HPGe detectors*

γ spectroscopy because they have the best resolution. Figure 7.33 shows the spectra of the same source the same source recorded by NaI(Tl) and HPGe spectrometers. The effect of the resolution of the two detectors is clearly visible in the spectrums. The resolution of semi-conductor detectors is very good (small numerical value) because they work at very low (liquid nitrogen) temperature that reduces thermal noise and dark current, the energy required to create one electron–hole pair is very low and the Fano factor for these detectors is also small ≈0.1.

7.7 SOME SPECIAL DETECTORS

7.7.1 Anti-Compton Shield

The scintillation spectrum of mono-energetic γ rays consists of a photo peak and peaks due to the Compton scattering of the γ rays in the detector material as well as in the material surrounding the spectrometer. If a source of γ s have group of mono-energetic γs, say of E1, E2, E3 … energies, each γ will give its own photo peak as well as a huge background of Compton scattered γ s. The Compton contributions of several γs make the observed spectrum very complicated. It sometime becomes impossible to separate the photo peak from the Compton background of γs of higher energies. One general method of reducing Compton γ background is to use anti-Compton shield. Anti-Compton shield consists of a number of γ detectors surrounding the main detector. The γ rays that undergo Compton scattering in the main detector and in the surrounding material, pass through the surrounding detectors. As such, both the main detector and the surrounding detectors give pulses almost simultaneously, corresponding to the events in which γ rays have undergone Compton scattering. The pulses from the outer surrounding detectors (Compton shield) are put in anti-coincidence with the pulses from the main detector. As a result, whenever there are pulses simultaneously both in the main detector and the Compton shield detectors, the pulses are rejected. Simultaneous pulses in both detector setups occur only for those events in which the incident γs undergo Compton scattering and these events are rejected using the anti-coincidence circuit. It may be noted that those events in which the incident γs undergo photoelectric effect, no signal is generated in the Compton shield and the event is registered. In this way, Anti-Compton shield arrangement cleans the γ spectrum by removing pulses generated by Compton scattered γs.

7.7.2 Phoswich Detector

In the experimental study of nuclear reactions, it is sometime required to record radiations of low intensity in high radiation background. Such a situation often arises when either neutrons or γ rays of low energy are required to be recorded in the presence of the other.

Phoswich is the short form for 'phosphor sandwich'. Combination of detectors with dissimilar pulse-shape characteristics optically coupled to each other and to a single or multiple photomultiplier tubes makes phoswich detector. They are generally used to detect and record the low-intensity and low-energy radiations including charged particles and γ rays in a high-energy ambient background. This is made possible by the pulse-shape analysis of the recorded signals, rejecting those which are present in all detectors and finding out from the pulse shape in which detector the event has originated. Successful mapping of low concentration xenon radio isotopes in reactor background could be done using a sandwich of plastic and CsI(Tl) scintillators with BGO anti-Compton shield.

7.7.3 Position Sensitive Single and Multiwire Proportional Counters (MWPC)

An ordinary single-wire proportional counter is used to detect the type and the energy of the radiation passing through the counter. Initial ion pairs in a proportional counter get multiplied by the process of secondary ionization in a region very near to the anode wire where the electric field intensity is high. The positive ion sheath receding from the anode induces a charge ΔQ on the anode wire and a pulse of voltage $\Delta V = \frac{\Delta Q}{C}$ is formed at the anode. In an ordinary proportional counter, anode pulse is taken from one end of the anode wire and the anode wire has a very low electric resistance.

A position sensitive single wire proportional counter differs from the normal proportional counter in two respects: (1) pulses can be drawn from both ends of the anode wire and (2) the anode wire is a resistive wire, that is it has a finite resistance per unit length. Now, suppose that the initial ion pairs by a radiation are produced at some point P, very near to the anode inside the detector as shown in figure 7.34. The point P divides the anode wire of total length ℓ into two parts: of length X, from P to A and of length $(\ell - X)$ from P to B. The charge ΔQ collected at point P of the anode wire will find two paths to flow, one from P to A and the other from P to B.

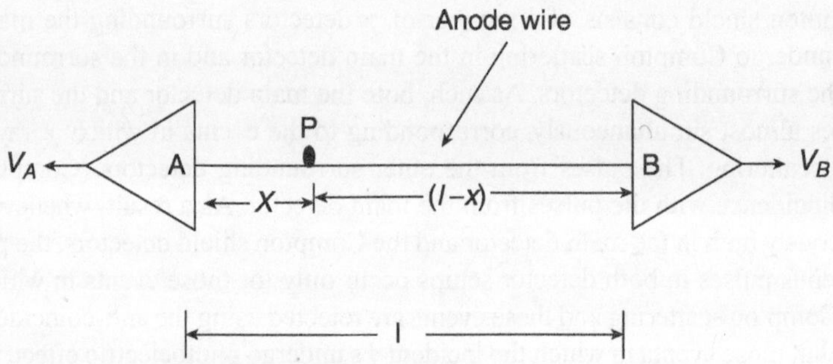

Figure 7.34 *Method of charge division in position sensitive single wire proportional counter*

Let ΔQ_a and ΔQ_b be the parts of the total charge ΔQ $(=\Delta Q_a + \Delta Q_b)$ that flow through the two paths. As the amount of the charge flow is inversely proportional to the resistance of the path,

$$\Delta Q_a \propto \frac{1}{X} \quad \text{and} \quad \Delta Q_b \propto \frac{1}{(\ell - X)}$$

Therefore,

$$\frac{\Delta Q_a}{\Delta Q_b} = \frac{(\ell - X)}{X} \quad \text{Or} \quad X = \frac{Q_b}{(Q_a + Q_b)}$$

The electronic pulses recorded at A and B will have their heights V_A and V_B proportional, respectively, to ΔQ_a and ΔQ_b. Thus, the position of the initial ionization (point P at distance X from end A) may be determined from the ratio $\frac{V_B}{(V_A + V_B)}$ and the total energy of the radiation (which is proportional to ΔQ) may be determined from the sum of the pulse heights ($V_A + V_B$).

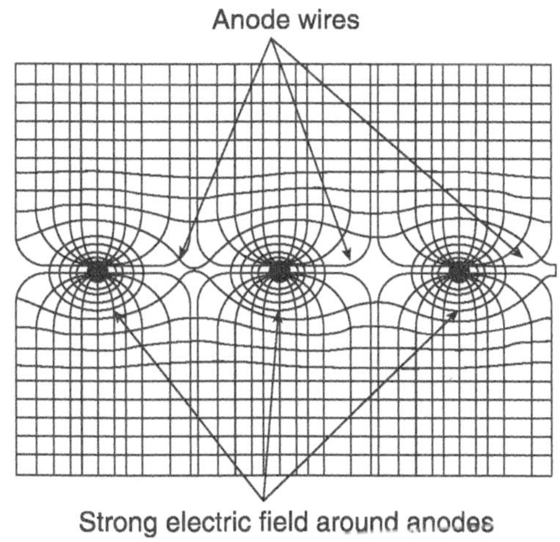

Figure 7.35 *Electric field around anodes in a multiwire proportional counter*

Multiple-wire proportional counters are used to determine the position and the energy of the ionizing radiations. In their simplest form, an MWPC consists of a number of thin parallel anode wires separated from each other by small distances (≈few mm) and sandwiched between two cathode planes. Cathode planes may be flat conducting plates or metallic wires like anodes. The separating distance between anodes and cathode is only few mm so that a small DC potential of about 1 kV produces strong electric field around anodes, which decreases rapidly as distance from anode. The lines of electric field around anode are shown in figure 7.35. The system is enclosed in an envelope with a thin mylar window to allow the passage of radiations. The airtight outer envelope has ports for feeding counter gas and the high voltage.

A noble gas such as argon mixed with some quenching gas is filled in the counter at atmospheric pressure or slightly less than that. As charged particles or ionizing radiations enter the counter they initiate initial ionization. The electrons produced in initial ionization process move towards the anode nearest to them. Very near to the anode, in the region of high electric field, the electrons produce secondary ionization giving rise to gas multiplication. The positive ion sheath encircling anode moves out towards the cathode. However, the positive ion sheath receding towards cathode induces negative charge on anodes. The quantity of the induced charge is largest on the anode nearest to the point where initial ionization took place. A negative pulse of potential $\Delta V = \frac{\Delta Q}{C}$, where ΔQ and C are respectively the magnitudes of the induced charge and the capacitance of the detector. Pulses formed at the anode carry information both about the energy of the radiation, which is proportional to the pulse height and the location of the initial ionization that is near to the anode, which has the biggest pulse. To have such a system it is necessary to have individual pulse recording systems (a pre-amplifier, an amplifier, MCA etc.) associated with each wire. This is both costly and unmanageable if the number of wires is large. The alternate method is to use the method of charge division. In this method, one end of alternate anode wires is connected to each other and thus there are two outputs one from each side of the anode wires. The difference in the size of the signal potential from the two set of outputs gives information about the position of the event while the sum of the pulse heights from the two outputs give the energy of the radiation.

MWPCs are operated under gas flow mode and care is to be taken that the applied potential does not drive the system to the GM or to the region of limited proportionality.

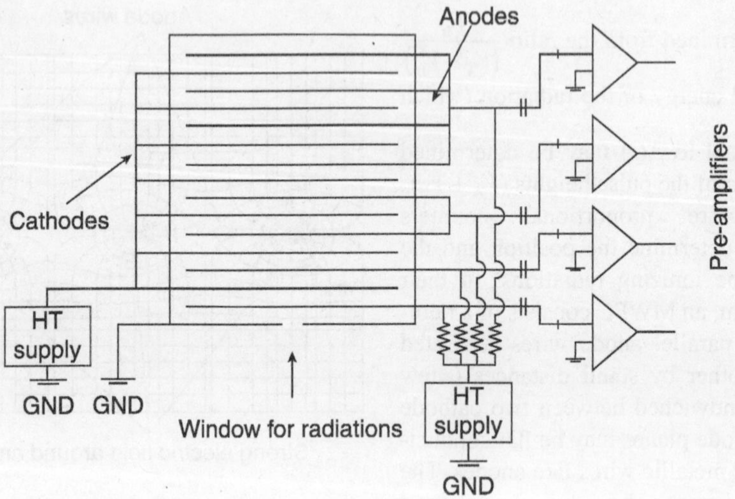

Figure 7.36 *A multiwire proportional counter setup*

7.8 DETECTION OF NEUTRON

Neutrons being neutral do not interact with atomic electrons of the absorbing medium. Neutron interaction with electron through the magnetic moment is negligible. As such, detection of neutron is not direct. Neutrons scatter nuclei that recoil with sufficient energy if they are light. The maximum energy E_{max} that may be given by a neutron of kinetic energy E to a nucleus of atomic mass number A is

$$E_{max} = \frac{4AE}{(A+1)^2}$$

Light nuclei such as protons (nucleus of hydrogen atom) recoil with large energies if scattered by a fast neutron. Detecting of the recoiling nucleus is one way of detecting neutrons. Charged recoiling nuclei can be easily detected either by a gas-filled counter or a scintillation spectrometer. As the energy distribution of recoiling nucleus is intimately related to the energy of the neutron, with proper bookkeeping it is possible to determine the energy of the incident neutron.

The other possible method of detecting neutron is through nuclear reactions initiated by neutron. For example, thermal neutrons have large cross-sections (probability) for the following reactions:

$$^{10}_{5}\mathrm{B} + {}^{1}_{0}\mathrm{n} = {}^{4}_{2}\mathrm{He} + {}^{7}_{3}\mathrm{Li}^{*} + 2.3\,\mathrm{MeV}$$

$${}^{7}_{3}\mathrm{Li}^{*} = {}^{7}_{3}\mathrm{Li} + {}^{0}_{0}\gamma + 0.48\,\mathrm{MeV}$$

and

$${}^{235}_{92}\mathrm{U} + {}^{1}_{0}n = \text{ Two fission fragments} + 2\text{ or }3\text{ fast neutrons} + 200\,\mathrm{MeV}$$

In the first reaction, a boron nucleus on absorption of a thermal neutron emits an α particle and leaves the excited nucleus ${}^{7}_{3}\mathrm{Li}^{*}$, which decay to the ground state by emitting γ rays. The emitted

α particle can be easily detected either by a gas-filled counter operated in the ionization region, proportional region or GM region. Gas-filled counters having boron trifluoride gas as counter gas are often used to detect the flux of thermal neutrons. Fast neutrons may also be detected if they are converted into thermal neutrons by passing them through a block of paraffin wax. Paraffin wax is a very good neutron moderator (material that quickly reduce the energy of neutrons without much loss in their numbers.) as it contains light nuclei (hydrocarbon). Alternately, a thin coating of boron on the cathode of a normal gas counter makes it a thermal neutron detector.

In the second reaction, thermal neutrons has a large probability to initiate fission in ^{235}U that results in the production of two heavy fission fragments each having high charge. Fission fragments being highly ionizing can be easily detected in an ionization chamber. Ionization chambers with a thin coating of enriched ^{235}U on cathode (called fission chambers) serve as good thermal neutron detector.

Counters filled with CH_4 or ^{4}He gas are sensitive to high-energy neutrons as the scattered protons or α particles may be recorded. If the counter is operated in proportional region information about the energy of the incident neutron may also be obtained.

Charged particles produced as a result of neutron scattering or in nuclear reactions initiated by neutrons, may also be detected by a scintillation detector. Lithium-loaded glass scintillators are used for neutron detection. Plastic scintillators, which have high content of hydrogenous material, may themselves serve as scatterer. Protons emitted as a result of neutron scattering are registered by the detector. Neutron detection, particularly with scintillation detectors, the presence of large γ ray background often creates problem. Neutron interaction with any material, in general, produces almost 10 times more flux of γ rays than charged particles. However, proper shielding and designing of the experiment may solve the problem. For example, γ rays of about 1 MeV emitted in nuclear fission are absorbed up to 90% by a 5 cm thick lead shielding, while only 0.1% fission neutrons are absorbed by the same thickness of lead.

The shape of the electronic pulse produced in any detector depends on how the energy is deposited in the detector by the radiation. As different types of particles deposit energy by different processes, particularly the decay time of the pulse is quite different for different particles. Electronic analysis of detector pulses may be used to sort out the pulses produced by desired radiation. This is called pulse-shape discrimination and is used to remove γ background in neutron detection by plastic scintillators.

Determination of the kinetic energy of neutrons is more involved than that of other radiations. A very common method is to use *the time of flight technique*. In this method, as the name suggests, the time that neutron takes in covering a known distance is measured which gives the velocity of the neutron. The kinetic energy may be determined from the speed and the mass of the neutron. In actual experiment, a thin detector (say a plastic detector) and a total absorbing thick detector (liquid scintillator) are kept a few meters apart in the path of the neutron. The time difference between the neutron pulse in thin detector and the thick detector is very accurately determined. Fast electronic circuits can very accurately determine time differences of the order of 10^{-11} seconds or less. Knowing the distance between the detectors and the time-of-flight of neutron the energy of neutron may be determined. A check on the energy value determined from the time of flight may be made from the thick detector pulse that has correlation with neutron energy. Time of flight method may also be used for determining the energy distribution of neutrons. The distribution of flight time is a replica of the energy distribution. In actual experiments, flight time is converted into electronic pulses and from the distribution of pulse heights neutron energy distribution may be found.

7.9 DETECTION OF NEUTRINO

Neutrino's feeble interaction with matter makes them uniquely valuable as astronomical messenger. Neutrinos move out undiverted through a huge mass of matter and do not suffer any deflection in interstellar magnetic field. Most of the neutrinos floating around were born some 15 billion years back soon after the birth of universe. New neutrinos are constantly generated at nuclear power stations, accelerator centres, explosion of supernova, in collision, and death of stars. There are three flavours of neutrinos. Neutrinos may interact with matter via neutral current (Z-boson) or through charged current (W-boson). In neutral current interaction, neutrino loses some energy and momentum but no information about the flavour is left. In charged current interaction, the neutrino is transformed into its partner lepton. Experiments based on following methods are being carried out at international level to detect and study the properties of neutrino.

7.9.1 Chemical Method

(i) Neutrino converts a chlorine atom into an argon atom via charge current interaction. Chlorine detectors contain large tanks filled with carbon tetrachloride (approx 470 metric ton). The fluid is periodically purged with helium gas that removes the argon. Helium is then cooled to separate out argon. Chlorine detectors have been used to study solar neutrino problem.

(ii) Anti-neutrino interacts with proton to give a neutron and a positron. The positron readily annihilates with an electron emitting two γ rays, 511 keV energy each in opposite directions. Reines and Cowan used this reaction to study neutrino interaction with matter. They took cadmium chloride solution in water and focused on the emission of two annihilation γs. The neutron produced in the reaction on absorption by cadmium excites it to an excited state which in a short time decays by emitting 8 MeV γ rays. Successful recording of two 511 keV γs in opposite directions followed by the emission of 8 MeV γ confirms the neutrino interaction.

$$\nu^- + p = n + \beta^+;\ \beta^+ + e^- = 2\gamma\ ;\ {}^{114}_{48}Cd + {}^{1}_{0}n = {}^{115}Cd^* = {}^{115}Cd + \gamma$$

7.9.2 Cerenkov-ring Detectors

The charged leptons created in interaction of neutrinos with matter may emit Cerenkov radiations in a transparent medium if they have sufficient energy and move in the medium with a velocity larger than the velocity of light. Cerenkov radiations, generally bluish in colour, move out in a cone. A detector, such as photomultiplier tube detects a circular ring of bluish colour. When viewed by an array of detectors the coloured ring pattern may give information about the direction of motion, the amount of energy lost and the flavour of the neutrino responsible for the registered Cerenkov pattern.

Three big projects – KAMIOKONADE, IMB, and AMANDA – use large detector arrays for tracking Cerenkov radiations.

7.9.2.1 Kamiokonade

It is a neutrino observatory under mount Kamioka in the city of Hida, Japan. The observatory is 1000 m below the mountain. Cylindrical stainless steel tank of 41.4 m length and 39.3 m diameter holds 50,000 tons of ultrapure water. Two arrays of photomultiplier tubes consisting of, respectively, 11,146 and 1885 tubes surround the tank and look for Cerenkov rings. The observatory tracked neutrinos emitted from the supernova 1987a.

7.9.2.2 Irvine–michigan–brookhaven (IMB)

This is a joint project of Irvine, Michigan, and Brookhaven universities to track neutrinos. The laboratory has been set at Marton Salt Company's Fairport mine at the shore of Lake Erie, USA. A cubical tank of 17 × 17.5 × 23 m holds 2.5 million gallons of ultrapure water. The tank is surrounded by 2048 photomultiplier tubes (figure 7.37).

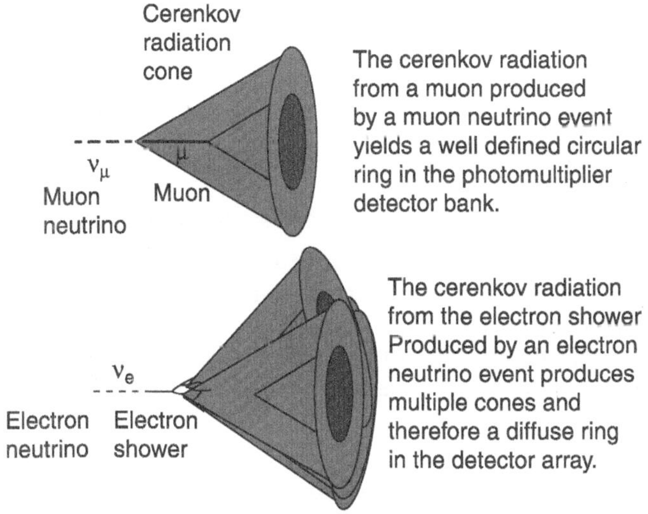

Figure 7.37 *The Cerenkov radiation cones for muon- and electron-neutrino induced events*

7.9.2.3 Antarctica muon and neutrino detector arra (AMANDA)

It is a neutrino telescope located beneath the Amundsen-Scott South Pole Station. In 2005, after 9 years of operation, AMANDA officially became part of its successor project, the IceCube Neutrino Observatory.

AMANDA consists of optical modules, each containing one photomultiplier tube, sunk in Antarctic ice cap at a depth of about 1500–1900 m. In its latest development stage, known as AMANDA-II, AMANDA is made up of an array of 677 optical modules mounted on 19 separate strings that are spread out in a rough circle with a diameter of 200 m. Each string has several dozen modules, and was put in place by 'drilling' a hole in the ice using a hot-water hose, sinking the cable with attached optical modules in, and then letting the ice freeze around it. AMANDA detects very high energy neutrinos (50+ GeV), which pass through the earth from the northern hemisphere and then react just as they are leaving upwards through the Antarctic ice. The neutrino collides with nuclei of oxygen or hydrogen atoms contained in the surrounding water ice, producing a muon and a *hadronic* shower. The optical modules detect the Cherenkov radiation from these latter particles, and by analysis of the timing of photon hits one can approximately determine the direction of the original neutrino with a spatial resolution of approximately 2°.

7.9.2.4 Sudbury neutrino observatory (SNO)

It is situated 2 km underground in Vale Inco's Creighton Mine in Sudbury, Ontario, Canada uses heavy water. Neutrino may break up deuterons in heavy water producing neutrons, which may be re-absorbed giving burst of γ rays.

7.9.2.5 MiniBooNE

It is an experiment of Fermi Lab, USA. A beam of muon neutrinos is directed at a tank filled with mineral oil which has scintillation properties. The scintillator tank is viewed by 1280 PMTs. The detector will be able to detect low-energy neutrinos that do not produce Cerenkov radiations.

7.9.2.6 MINOS (Main Injector Neutrino Oscillation search)

It is an experiment to study neutrino oscillations. Two detector setups, the near one at the Fermilab and another 735 km away at Minnesota, detected neutrinos to look for their oscillations.

Neutrino oscillation is a quantum mechanical phenomenon in which a neutrino of a specific flavour is detected with a different flavour as it moves. These flavour oscillations occur periodically. Neutrino oscillation may lead to the finite mass of the neutrino and may result in the fall of standard model.

7.9.3 Indian Initiative for Neutrino Studies

A laboratory named '*Indian Neutrino Observatory (INO)*', to study the properties and interactions of neutrinos, is going to be developed at the Bodi-West hills on the coast of southern Tamil Nadu state. The environment and forest ministries have given clearance to the project. Thick cover of rock (about 1000 m high) will provide the natural filter for removing all other radiations except the neutrinos. Moreover, a very large electromagnet will further remove the charged particle background. It is expected that the laboratory will become a leading international facility for the study of neutrinos. Facility costing \$270 m will be the fifth such facility in the world (figure 7.38).

Figure 7.38 *Proposed site for the Indian Neutrino Observatory*

7.10 SOLID-STATE NUCLEAR TRACK DETECTORS (SSNTD)

Heavy charged particles passing through insulating solids including crystals, inorganic glasses, and plastics produce sub-microscopic tracks that may be seen by a high-resolution electron microscope. The linear region of radiation-damaged material is called the track.

Though the initial tracks are of sub-microscopic size, their size may be enlarged, fixed, and stabilized by chemical treatment, called etching, with a suitable chemical such as NaOH, hydrofluoric acid, and strong alkalis depending on the material of the detector. The particle tracks after etching may be observed by an ordinary optical microscope.

7.10.1 Mechanism of Track Formation

Heavy charged particle passing through an insulating material ionizes the atoms along its trajectory forming a cylindrical region filled with positively charged ions, which violently repeal each other. This almost cylindrical space along the trajectory of the particle has strain locked in the region. Such regions because of their higher diffraction contrast and chemical reactivity may be seen by a high-resolution electron microscope. Figure 7.39 shows the photograph of a track of fission fragment in an SSNTD.

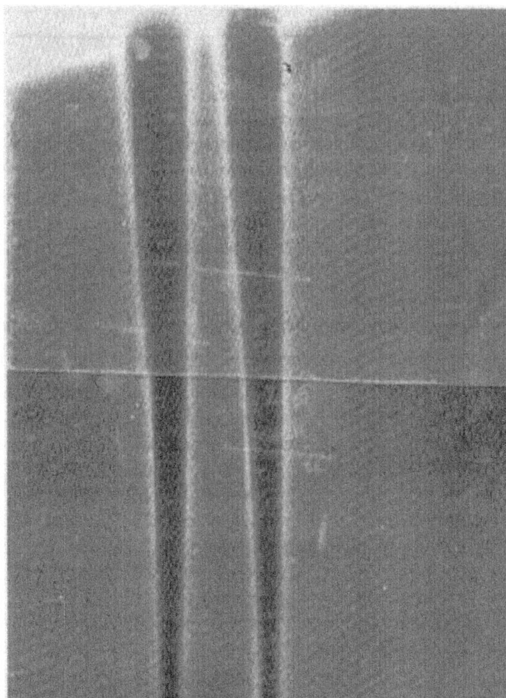

Figure 7.39 *Photograph of a fission fragment track in a SSNTD*

Different insulating materials are sensitive to different types of radiations. For example, organic polymers are good material for nuclear particles such as αs and other heavier ions. Some organic polymers even show tracks of deuterons and protons.

The selectivity of the detecting material is due to the 'minimum threshold damage density', which has a well-defined value for each material and below which no tracks may be formed.

A big advantage of SSNTD, particularly over nuclear emulsions, is its capability of detecting heavy and highly charged particles such as fission fragments, with great efficiency in presence of large background of γ rays and neutrons. However, no timing information about the recorded event may be obtained from SSNTD.

Because of its relatively low cost, ease of handling, permanent record, non-requirement of intricate electronics, and so on, SSNTD are finding applications in nuclear physics, geochronology, cosmology, biology, bird altimetry, seismology, elemental analysis, lithography, material science, and so on.

7.11 CHOOSING A DETECTOR

The choice of a detector depends on the type of the radiation to be detected, expected energy of the radiation, the intensity (the expected number of radiations per unit time) and the environment where the experiment is to be performed. Broadly speaking, there may be two types of experimental set-ups; the first where experiments are done off-beam and the second in-beam experiments. In off-beam experiments, there is no beam of accelerated particles and, therefore, the unwanted radiation background is not very much. Normal background due to cosmic rays and natural radioactivity can be shielded using lead shields of appropriate thickness. As such, in off-beam experiments there is not much problem regarding the breakdown of the detector under high radiation background. However, in the case of those experiments that require measurements to be done in-beam, when the beam of accelerated particles is hitting the target, special detectors that do not get damaged by the high background radiations need to be used.

Some experiments require only to detect the presence of the radiations and to count their number in a given time. GM counter is often used for this purpose. Although a GM counter could not distinguish between different radiations but it has almost 100% detection efficiency, once the radiation has reached the active volume of the counter. End-window GM counter further facilitates the entry of low-energy radiations in the active volume. Moreover, GM counter is rugged, stable, and requires very little electronic modules for operation.

In case the type and the energy of the radiation are to be determined, detectors such as proportional counter, scintillation detectors, and solid-state detectors may be used. In such cases, the size or thickness of the detector becomes very important. As is obvious, the thickness of the detector must be more than the range of the radiation in the detector material so that the radiation is completely stopped in the detector and deposit all its energy in the detector. A detector of larger size has higher detection efficiency, but may not be of advantage if radiations of low energy are to be recorded. For example, a 4″ × 4″ × 4″ NaI(Tl) crystal is very good for the recording of 100 keV to about 1.5 MeV γ rays, but if it is required to record only 100 keV γ s then a detector of smaller size, say 1″ × 1″ × 1″ is good. It is because a smaller detector will have very small detection efficiency for high energy γ s and large detection efficiency for lower energy γ s. As a result, the low energy γ s will be efficiently detected and recorded by a small crystal and at the same time the background activity due to the higher energy γ s present in the environment will be considerably reduced due the low detection efficiency of the detector for high-energy γ rays. It is for of this reason that thin detectors are used for the detection and recording of X-rays, β, and other charged particles.

Expected intensity (count rate) of the radiations also plays an important role in selecting the detector. If the expected intensity is high, a detector that produces fast pulses of low-pulse duration is to be used. This will avoid pile-up of pulses. Fast electronic modules that may handle high count rates are required to be used in such cases.

From the foregoing discussion, it is clear that proper planning of the experiment is very essential. As a matter of fact most of the time computer simulation of the experiment is done before actually doing the experiment to check the suitability of the experimental set-up.

Exercise p-7.16: Describe the working of a gas-filled proportional counter, explaining the process of gas multiplication and quenching. Why are cylindrical counters preferred over parallel plate type?

Exercise p-7.17: Discuss the working of a GM counter explaining the pulse formation, dead time, and recovery time of the counter. On what factors does the pulse size of a GM counter depend?

Exercise p-7.18: Describe the process of scintillation and discuss the characteristics of a good scintillator.

Exercise p-7.19: With the help of a block diagram explain the working of a NaI(Tl) scintillation spectrometer. What is the purpose of thallium activation of the crystal?

Exercise p-7.20: Draw a rough sketch of the spectrum of 662 keV γ rays of ^{137}Cs source taken with a NaI(Tl) detector and explain the origin of each peak of the spectrum.

Exercise p-7.21: In what respect Ge(Li) detector is different from HPGe detector? Which is better and why?

Exercise p-7.22: Solid-state detectors are better than any other detectors for γ ray detection. Do you agree with the statement, if yes, give reasons of your answer?

Exercise p-7.23: What is meant by the resolution of a detector setup? On what factors the resolution of a system depends? Why the resolution of semi-conductor detectors is so good as compared to other detectors.

Exercise p-7.24: Write notes on: (i) Position sensitive counters, (ii) Pulse-shape discrimination and phoswich detectors, (iii) BGO as a detector for high energy γ rays.

Exercise p-7.25: How will you detect and determine the energy of a fast neutron?

Exercise p-7.26: Review the major international efforts on foot to detect neutrinos.

Exercise p-7.27: Describe in detail the working principle of SSNTD and some of its applications.

Exercise p-7.28: Derive an expression for the maximum energy that can be lost by a particle in an interaction with another particle.

Multiple choice questions

Note: It is possible that more than one alternative is correct in case of some questions. In such cases, all correct alternatives should be selected.

Exercise M-7.1: Which of the following radiations show exponential absorption?

(a) Mono-energetic α particles (b) mono-energetic β particles
(c) γ rays (d) mono-energetic neutrons

Exercise M-7.2: Cross-section for photoelectric effect in an atom is maximum for

(a) any bound electron (b) valance electron
(c) K-shell electron (d) M-shell electron

Exercise M-7.3: Auger electrons are emitted when

(a) a γ ray is converted into an electron
(b) excitation energy of an atom is given directly to the atomic electron
(c) α rays knockout an electron from the atom
(d) in γ decay internal conversion takes place.

Exercise M-7.4: The mass absorption coefficient of a certain γ ray in aluminium (density = 2700.0 kg/m^3) is 0.5 cm^2/g. The linear absorption coefficient of the γ ray is

(a) 0.5 m^{-1} (b) 1.35 m^{-1} (c) 0.185 cm^{-1} (d) 1.35 cm^{-1}

Exercise M-7.5: Maximum energy (in MeV) that a 0.511 MeV g ray can transfer to a free electron is

(a) 0.511 (b) 0.255 (c) 0.170 (d) 0.128

Exercise M-7.6: The Compton wavelength (in meter) of an electron is of the order of

(a) 10^{-15} (b) 10^{-12} (c) 10^{-9} (d) 10^{-6}

Exercise M-7.7: A heavy charged particle, such as proton, may also undergo collisions with the nuclei of the absorbing material just like the collisions it suffers with the atomic electrons of the absorbing medium, but we do not take energy loss by nuclear collisions in consideration while calculating the stopping power. Why?

(a) Nucleus is much smaller in size than the proton
(b) The charge on the nucleus is much larger than the electronic charge
(c) Compton wavelength is smaller than the nuclear dimensions
(d) Energy loss is inversely proportional to the mass of the target particle and nuclear mass is much larger than the mass of the electron; hence, negligible energy is lost in collisions with nuclei.

Exercise M-7.8: A proton and a deuteron of same velocity are moving in aluminium absorber, the ratio of the stopping power of proton to deuteron is

(a) 1 (b) 2 (c) 4 (d) 0.5

Exercise M-7.9: Approximate range (in meter) for a 5 MeV α particle in air is

(a) 0.05 (b) 1.0 (c) 0.5 (d) 0.005

Exercise M-7.10: The radiative energy loss is

(a) directly proportional to the kinetic energy of the particle
(b) inversely proportion to the kinetic energy of the particle
(c) directly proportional to the mass of the particle
(d) inversely proportional to the square of the mass of the particle.

Exercise M-7.11: Cerenkov radiations are emitted

(a) when a light charged particle moves in a medium with a velocity larger than the velocity of light in the medium
(b) when $\beta\mu$ for the particle < 1
(c) when $\beta\mu$ for the particle > 1
(d) when the Cerenkov radiations are emitted in a cone of semi-angle $\theta = \beta\mu$

Exercise M-7.12: A GM counter detects g rays primarily through

(a) Compton scattering with the counter gas
(b) photoelectric effect with counter gas
(c) photoelectric effect with the cathode of the counter
(d) pair production with the counter gas

Exercise M-7.13: In an experiment, the width at half maxima of a photo peak is 0.1 V and the peak occurs at 50 V. The percentage resolution of the detector is

(a) 20% (b) 2% (c) 0.2% (d) 0.02%

Exercise M-7.14: It is required to record the intensity of 10 MeV γ rays in cosmic ray background at some place. Which of the following scintillators will be best for such measurements?

(a) BGO (b) NaI(Tl) (c) plastic (d) Antharecene

Exercise M-7.15: Which of the following detectors is good for the permanent record of fission fragments?

(a) BGO (b) SSNTD (c) NaI(Tl) (d) MWPC

8 Nuclear Reactions

More than 90% of nuclear reactions at moderate energies are of binary type, that is, there are two particles in the initial system and two in the final system. A binary reaction may be represented by the following equation:

$$^{A_x}_{Z_x}X + ^{A_a}_{Z_a}a = ^{A_y}_{Z_y}Y + ^{A_b}_{Z_b}b \tag{8.1a}$$

Here, A_i and Z_i are, respectively, the atomic mass number and atomic number of the *i*-th particle, *i* being X, a, Y or b. The reaction expressed by Eq. (8.1) may be represented by X (a, b) Y in short. In a binary nuclear reaction, the two nuclei (or a nucleus and a nuclear particle) in the incident channel interact with each other through their nuclear fields and as a result rearrangement of nucleons and/or sharing of energy between them takes place in a very short time. The resulting final products of the reaction generally move out. The system, in general, is well defined before the reaction has taken place and after the reaction, but what happens in the very less time during which sharing of energy and redistribution of nucleons takes place is not exactly known. However, following quantities are found to be conserved in a nuclear reaction. We shall consider the reaction in Eq. (8.1b) as an example and will apply the conservation laws to see the restrictions put by these laws. At present, we are not concerned about the energies and whether the reaction is possible or not, these aspects will be discussed later.

$$^{14}_{7}N + ^{4}_{2}He \rightarrow \left[^{18}_{9}F^{*}\right] \rightarrow ^{17}_{8}O + ^{1}_{1}H \tag{8.1b}$$

As may be observed, Eq. (8.1b) represents the reaction in which the nucleus of the stable isotope of nitrogen $^{14}_{7}N$ interacts with an α particle ($^{4}_{2}He$) and finally the nucleus $^{17}_{8}O$ and a proton are formed. Exactly how this change occurred is not known; however, one way of understanding the reaction process is to assume that ^{14}N and ^{4}He first fuses into one composite nucleus $^{18}_{9}F^{*}$. The intermediate nucleus $^{18}_{9}F^{*}$, often called the compound nucleus, is in excited state and unstable, which quickly decays into ^{17}O and a proton. This is the approach of the so-called 'compound reaction mechanism'. Some properties of the nuclei involved in the reaction are detailed in table 8.1.

Table 8.1 *Some properties of the nuclei involved in the reaction*

Nucleus	Spin J (angular momentum)	Parity π	τ_3 (third component of isospin)
$^{14}_{7}N$	1	+	0
$^{4}_{2}He$	0	+	0
$^{18}_{9}F$	1	+	0
$^{17}_{8}O$	5/2	+	−1/2
$^{1}_{1}H$	1/2	+	1/2

8.1 QUANTITIES CONSERVED IN A NUCLEAR REACTION

The physical quantities conserved in nuclear reactions are discussed in the following sections:

8.1.1 Conservation of Total Charge (or of Proton Number *Z*)

$$(Z_X + Z_a)e = (Z_Y + Z_b)e \tag{8.1c}$$

Here, e is the charge of the proton.

In this case: $(7e + 2e = 9e = 8e + 1e)$; $\sum Ze$ is conserved.

8.1.2 Conservation of the Total Number of Nucleons

$$A_X + A_a = A_Y + A_b \tag{8.1d}$$

In this case: $(14 + 4 = 18 = 17 + 1)$; $\sum A$ is conserved.

As total charge (i.e. the number of protons) is conserved, and total number of nucleons is also conserved, the number of neutrons N is also conserved.

In this case: $(N = 7 + 2 = 9 = 9 + 0)$; $\sum N$ is conserved.

8.1.3 Conservation of Total Angular Momentum

Total angular momentum is the quantum mechanical sum of the angular momentums J_X and J_a of the individual particles in the incident channel and the orbital angular momentum ℓ^{Xa} of the relative motion of the two particles. Therefore, the total angular momentum in the incident channel J_i is $J_i = J_X + J_a + \ell^{Xa}$.

And the total angular momentum in the exit channel J_f is

$J_f = J_Y + J_b + \ell^{Yb}$, where J_Y, J_b are, respectively, the angular momenta of Y and b and ℓ^{Yb} is the orbital angular momentum of the relative motion of Y and b.

Hence, the conservation of angular momentum requires:

$$J_X + J_a + \ell^{Xa} = J_Y + J_b + \ell^{Yb} \tag{8.1e}$$

It may be pointed out that the sum of angular momentum referred in Eq. (8.1e) is the quantum mechanical sum and is quite different from the algebraic sum referred in Eqs. (8.1c) and (8.1d). The quantum mechanical sum $J_X + J_a$ may give many possible final values from $(J_X + J_a)$ to $(|J_X - J_a|)$ in steps of unity.

Let us calculate the total angular momentum in the incident channel J_i. It may be noticed that the values of J_X and J_a may be read from table 8.1, but what about ℓ^{Xa}? Two important points may be made in this regard. First, orbital angular momentum can have only integer values, 0, 1, 2 .... Second, at the lowest energy relative orbital angular moment between the two particles in incident channel may be zero (S wave interaction). As the energy in the incident channel increases larger values of orbital angular momentum may also become possible. Therefore, for simplicity we consider $\ell^{Xa} = 0$.

Total angular momentum in incident channel $J_i = 1 + 0 + 0$ (quantum mechanical sum) = 1.

It may be noted that the angular momentum of ^{18}F is also 1, which confirms to the conservation of angular momentum.

The angular momentum of final state $J_f = J_Y + J_b + \ell^{Yb} = 5/2 + 1/2 + \ell^{Yb} = (3, 2) + \ell^{Yb}$.

The quantum mechanical sum of 5/2 and 1/2 gives two possible values 3 and 2.

Let us consider the value of $J_Y + J_b = 3$.

Then, for the conservation of the angular momentum, we should have $3 + \ell^{Yb} = 1$ (the angular momentum in the incident channel).

This gives three possible values for ℓ^{Yb} = 4, 3, and 2.

If we consider the value of $J_Y + J_b = 2$; then we will have $2 + \ell^{Yb} = 1$.

The possible values of ℓ^{Yb} = 3, 2, and 1.

Thus, we see that in case the relative angular momentum for the incident channel is zero (S-state) then to conserve the total angular momentum the relative orbital angular momentum of ^{17}O and proton can have four possible values, 1, 2, 3, and 4.

It is left as an exercise to calculate the possible values of the relative orbital angular momentum if the relative orbital angular momentum in the incident channel is considered as 1.

8.1.4 Conservation of Parity

Parity of the initial system is conserved in nuclear reactions. It may be noted that parity that tells about the mirror symmetry of the system is not conserved in weak interactions.

$$\textit{Conservation of parity demands: } \pi^X \cdot \pi^a \cdot (-1)^{\ell^{Xa}} = \pi^Y \cdot \pi^b \cdot (-1)^{\ell^{Yb}}.$$

The parity of the initial system $\pi^i = \pi^X \cdot \pi^a \cdot (-1)^{\ell^{Xa}} = (+)\,(+)\,(-1)^0 = +\text{ve}$.

Here, π^X and π^a are, respectively, the parities of nucleus X (^{14}N) and particle a (^4He), and we have considered $\ell^{Xa} = 0$. Note that in case the relative orbital angular momentum ℓ^{Xa} is considered as 1, instead of zero than the parity of the initial system will become negative.

$$\text{The parity of the final system is} = \pi^f = \pi^Y \cdot \pi^b \cdot (-1)^{\ell^{Yb}} = (+)\,(+)\,(-1)^{\ell^{Yb}}.$$

Here, π^Y and π^b are, respectively, the parities of ^{17}O and ^1H both of which are positive from table 8.1.

We observe that the parity of the final system depends on the value of ℓ^{Yb}. If it is even, the parity of the final state is positive and if it is odd, the parity will be negative. While discussing the conservation of angular momentum we have observed that for the case when the orbital angular momentum ℓ^{Xa} in initial state was zero, the possible values of the orbital angular momentum ℓ^{Yb} were 1, 2, 3, and 4.Out of these four values obtained from the conservation of angular momentum, the odd values 1 and 3 are not allowed by the conservation of parity as these values will give the parity of the final state a negative value.

In the case when $\ell^{Xa} = 1$, the parity of the initial state will be negative and, hence only odd values of the orbital angular momentum ℓ^{Yb} will be allowed.

8.1.5 Conservation of Statistics

Nuclei are made of nucleons that individually follow Fermi–Dirac statistics, but if a nucleus has even number of nucleons it follows Bose–Einstein statistics, which is the statistics of a system that consists of two fermions. As the total number of nucleons is conserved in a nuclear reaction, the statistics is also conserved.

In our example, the X and a, both contain even number of nucleons they follow Bose–Einstein statistics. The statistics of the initial system is, therefore, Bose–Einstein. In the final system, there are two particles both of which have odd number of nucleons, hence follow Fermi–Dirac statistics, but the total statistics of the system is again Bose–Einstein.

8.1.6 Conservation of Isospin and the Third Component of Isospin

More and more evidence now suggest that isospin is conserved in nuclear reactions. The third component of isospin τ_3 is related to atomic number Z by the relation

$$Z = \tau_3 + A/2 \tag{8.1f}$$

As both Z and A are conserved in nuclear reactions, the third component of the isospin τ_3 is conserved. The values of τ_3 for the components of the initial and final systems are given in table 8.1. Conservation of τ_3 may be confirmed from these values.

8.1.7 Conservation of Linear Momentum

In nuclear reactions, both the nucleus X and the particle 'a' are generally charged. Therefore, to bring them close enough to be within the range of the nuclear force of each other it is essential that at least one of them is projected with a certain velocity on the other. Conservation of linear momentum demands that the total linear momentum of the initial system is equal to the total linear momentum of the final system. The principle of conservation of linear momentum will be used in deriving Q-equation.

8.1.8 Conservation of Total Energy

In nuclear reactions, the total energy of the initial system, which is the sum of the rest mass energies of the initial nucleus X and the particle 'a' and their kinetic energies E_X and E_a, is conserved. The energy conservation may be written as

$$E_i = M_X c^2 + M_a c^2 + E_X + E_a;\ E_f = M_y c^2 + M_b c^2 + E_Y + E_b \tag{8.1g}$$

And

$$E_i^t = E_f^t \tag{8.1h}$$

Here, M_X, M_a, M_Y, M_b, E_X, E_a, E_Y, and E_b are, respectively, the rest masses and kinetic energies of X, a, Y, and b. E_i^t and E_f^t represents the total energies of the initial and final systems. Equating total initial energy E_i^t to the final total energy E_f^t, and rearranging terms, we get

$$\left[(M_X + M_a) - (M_Y + M_b)\right]c^2 = \left[(E_Y + E_b) - (E_X + E_a)\right] \equiv Q \quad (8.2)$$

Or
$$(M_i - M_f)c^2 = (E_f - E_i) = Q$$

Equation (8.2) defines the Q-value of the reaction. It may be noted that the Q-value of a reaction is equal to the energy corresponding to the difference in the initial and final masses. Also, if the final mass (sum of the masses in the final state) is less than the initial mass then the Q-value will be positive and in that case the sum of the kinetic energies in the final state E_f $(= E_Y + E_b)$ will be more than the sum of the kinetic energies E_i $(= E_X + E_a)$ in the initial state. This means that in positive Q value reactions the energy corresponding to the difference in the masses of the initial and final states is converted into kinetic energy. As such, in a positive Q-value reaction, energy will be released in the reaction (in the form of the kinetic energies of Y and b) and the reaction will proceed of its own (provided X and a are within the range of the nuclear force of each other). Positive Q-value reactions are called *exoergic* reactions as energy is released in such reactions.

In case the total mass in the final state is more than the total mass in the initial state, that is $[(M_Y + M_b) > (M_X + M_a)]$, the Q-value of the reaction is negative. As is obvious, a negative Q-value reaction will not take place unless energy equivalent to the mass difference of the final and initial states is supplied to the system. The required energy may be supplied in the form of the kinetic energy $[E_X + E_a]$ in the initial state. In most of the real experiments, E_x is zero, that is, the heavier target X is stationary and the lighter particle 'a' is incident on X with some kinetic energy E_a. From the balance of energy, it appears that a negative Q-value reaction will take place if $E_a = \{E_f - (-Q)\}$ is supplied to the system. The minimum value of E_a will be when E_f $(=E_y + E_b)$ is minimum, that is zero and in that case Y and b will be formed in the final state with zero kinetic energies, that is at rest. However, this is not exactly true. It is because when 'a' is made to hit 'X' with kinetic energy E_a, it supplies some linear momentum p_i $[= \sqrt{2M_aE_a}\,]$ to the initial system. This linear momentum must be conserved. To conserve the linear momentum, Y and b must also move out with some velocities. Therefore, in actual experiments involving negative Q-value reaction, when particle 'a' is made to hit 'X' with some kinetic energy E_a, the kinetic energy E_a should be slightly more than $|-Q|$ so as to provide energy for the creation of additional mass and also to provide velocities to Y and b for momentum conservation. The minimum energy required to initiate a negative Q-value reaction is called the *threshold* energy and is given by

$$\text{Threshold energy } E_{\text{thre}} = |-Q|\left(\frac{M_Y + M_b}{M_Y + M_b - M_a}\right) \cong |-Q|\left(\frac{M_X + M_a}{M_X}\right) \quad (8.3)$$

Let us calculate the Q-value for the reaction represented by Eq. (8.1b) using the following data:

M_X = Mass of ${}^{14}_{7}N$ = 14.003074 u; M_a = Mass of ${}^{4}_{2}He$ = 4.002603 u; M_Y = Mass of ${}^{17}_{8}O$ = 16.999132 u; M_b = Mass of ${}^{1}_{1}H$ = 1.007825 u; considering 1 u = 931.47 MeV.

It is easy to see that the total mass in the final state is more than the total mass in the initial state. Hence, the reaction is a negative Q-value reaction.

$$\text{Also, } Q = (M_i - M_f) = -0.001273 \text{ u} = -1.186 \text{ MeV}$$

The threshold energy calculated using Eq. (8.3) comes out to be

$$E_{\text{thre}} = 1.525 \text{ MeV}$$

These calculations show that this reaction will take place if the α particle ($^4_2$He) is made to hit the stationary $^{14}_{7}$N nucleus at least with a kinetic energy of 1.525 MeV. If it is hit by a less energetic α particle reaction will not proceed.

8.2 QUANTITIES THAT ARE NOT CONSERVED IN NUCLEAR REACTIONS

Quantities such as the magnetic dipole moment and electric quadrupole moment are not conserved in a nuclear reaction. It is because the magnetic dipole moment depends on the distribution of current elements while the electric quadrupole moment on the distribution of charge, both of which undergo change in nuclear reaction.

8.3 THE *Q* EQUATION

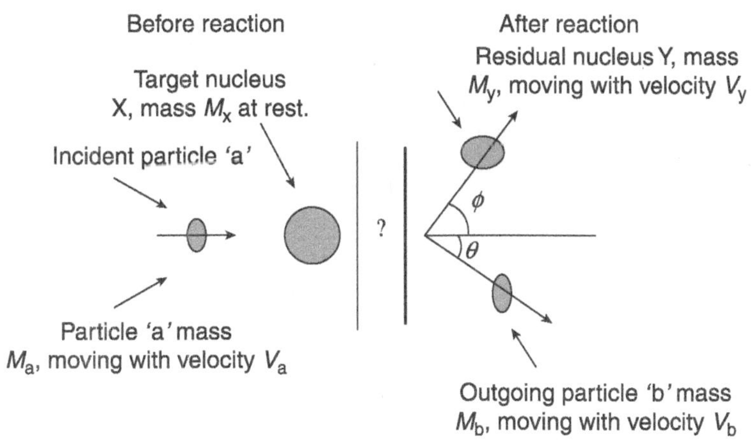

Figure 8.1 *Pictorial representation of a typical nuclear reaction*

The general layout of a typical nuclear reaction is shown in figure 8.1, where a nuclear particle or a light nucleus 'a' of mass M_a impinges with a kinetic E_a on a heavier nucleus X of mass M_X, which is at rest. If both 'a' and X are charged, the incident particle 'a' will face a Coulomb barrier as it approaches X. The maximum height of the barrier will be when 'a' touches X, that is when they are separated by the distance $(r_a + r_X)$ where r_a and r_X are, respectively, the radii of particle 'a' and the target nucleus X. Therefore, the Coulomb barrier height V_{Cou} will be given by

$$V_{\text{Cou}} = \frac{Z_a Z_X e^2}{4\pi\varepsilon_0 (r_a + r_x)} = 1.44 \frac{Z_a Z_X}{r_0 \left(A_a^{1/3} + A_X^{1/3}\right)} \text{ MeV} - \text{Fermi} \tag{8.4}$$

The magnitude of the Coulomb barrier may be calculated in MeV if the value of r_0 in Eq. (8.4) is substituted in Fermi. A reaction with positive Q-value will proceed if the energy E_a is larger than V_{Cou}. In the case where E_a is little less than V_{Cou}, the reaction may still proceed through quantum mechanical tunnelling. However, if E_a is much less than V_{Cou}, the reaction may not proceed because the incident particle 'a' will not be able to reach within the range of the nuclear field of X. It may be noted that a positive Q-value does not guarantee that the reaction will definitely take place it only states that if the interacting particles are within the range of their nuclear force then the reaction will occur. In the case of negative Q-value reaction, the reaction is energetically allowed if the incident kinetic energy E_a is equal or more than the threshold energy E_{thre}. However, again, the reaction will occur only if $E_a > E_{thre} > V_{Cou}$. However, presently we are interested only in the dynamics of the reaction and are not concerned with the Coulomb barrier.

Conservation of total energy (rest mass energy and kinetic energy) and the X and Y components of the linear momentum give:

$$E_Y + E_b - E_a = Q \tag{8.5}$$

$$\sqrt{2M_Y E_Y}\,\cos\varphi + \sqrt{2M_b E_b}\,\cos\theta = \sqrt{2M_a E_a} \tag{8.6}$$

$$\sqrt{2M_Y E_Y}\,\sin\varphi - \sqrt{2M_b M_b}\,\sin\theta = 0 \tag{8.7}$$

The excess energy available in the final state is shared between Y and b as their kinetic energies. Although the exact amounts of their energies depend on the angles θ and φ but Y being heavier moves with small velocity and often stops within the target foil that contains the target X. As such, in most of the experiments detectors are set at a given angle θ to detect and measure the energy of 'b'. If one eliminates φ and E_y from the Eqs. (8.5)–(8.7), we get

$$E_b\left(\frac{M_y + M_b}{M_y}\right) - E_a\left(\frac{M_y - M_a}{M_y}\right) - \frac{2\sqrt{M_a E_a M_b E_b}}{M_Y}\cos\theta = Q \tag{8.8}$$

Or $$\left(\frac{M_y + M_b}{M_y}\right)(\sqrt{E_b})^2 - \left\{\frac{2\sqrt{M_a E_a M_b}}{M_Y}\cos\theta\right\}\sqrt{E_b} - \left[E_a\left(\frac{M_y - M_a}{M_y}\right) + Q\right] = 0 \tag{8.9}$$

Equation (8.8) is called the Q equation. The following remarks may be made about this equation:

1. All quantities, that is E_b, E_a, and angle θ, are in laboratory frame and the treatment is non-relativistic.
2. The Q-equation has been derived without making any assumption about the reaction mechanism and, therefore, is valid for all those processes in which interaction of X and a leads to Y and b.
3. For $\theta = 90°$ the $\cos\theta$ term becomes zero.
4. It is true that for exact calculations the precise masses of X, a, Y, and b be used. However, in actual experiments where energies and other quantities are measured, there are always uncertainties of measurements. As such, often it is sufficient to replace exact masses by the corresponding mass numbers A_X, A_a, A_Y, and A_b, respectively.
5. Q-equation may be treated as a quadratic equation in $\sqrt{E_b}$ and may be written as Eq. (8.9) after rearranging terms.
6. For energetically possible reactions, $\sqrt{E_b}$ should be real and positive.

The roots of Eq. (8.9) may be written as

$$\sqrt{E_b} = B \mp \sqrt{B^2 + A} \tag{8.10}$$

where

$$B \equiv \frac{\sqrt{M_a M_b E_a}}{(M_b + M_y)} \cos\theta \tag{8.11}$$

$$A \equiv \frac{\left[M_y Q + E_a (M_y - M_a)\right]}{(M_b + M_y)} \tag{8.12}$$

8.3.1 Exoergic Reactions

We consider exoergic reactions for which Q-value is positive ($Q > 0$)

8.3.1.1 $E_a \cong 0$: Thermal neutron included reaction

If the incident particle 'a' and the target X are both charged particles, then even a positive Q-value reaction will not proceed unless the incident energy E_a is more or at least be of the order of the Coulomb barrier between them. However, a neutron does not face any Coulomb barrier and even thermal energy neutrons, energy $E_a = 0.025$ eV easily enter the nucleus and initiate nuclear reactions. A well-known example is the reaction ${}^{10}_{5}\text{B} + {}^{1}_{0}\text{n} \rightarrow {}^{7}_{3}\text{Li} + {}^{4}_{2}\text{He} + 2.8\,\text{MeV}$, which is used to detect thermal neutrons by BF_3 counters. For all practical purposes, the incident energy E_a may be taken as zero, that makes $B = 0$ (Eq. (8.11)) and, therefore,

$$E_b = \frac{M_Y Q}{(M_b + M_Y)} \tag{8.13}$$

It may be noted that the energy of the emitted particle 'b' is fixed and it does not depend on the direction of emission θ. Further, as the linear momentum of the initial system is zero (or nearly zero) the particles Y and b in the final system are emitted in opposite direction, that is $\theta + \varphi = 180°$, and $E_f = E_Y + E_b = Q = 2.8\,\text{MeV}$. Replacing atomic masses by their atomic numbers, we get $E_b = E_{He} = \frac{7 \times 2.8}{11} = 1.782\,\text{MeV}$; and $E_Y = E_{Li} = \frac{4 \times 2.8}{11} = 1.018$ MeV.

8.3.1.2 $E_a > 0$, Finite incident energy

In case of finite value of the incident energy, there are two possibilities:

1. $M_Y > M_a$

   In this case, A is always positive for exoergic reactions (see Eq. (8.11)). Therefore, the term $\sqrt{B^2 + A}$ is positive and larger than B; Hence, $\sqrt{E_b} = B + \sqrt{B^2 + A}$ and the other root $\sqrt{E_b} = B - \sqrt{B^2 + A}$, which gives a negative value is not possible. Thus, the energy of the outgoing particle 'b' is always single valued. The numerical value of E_b, however, depends on θ through Eq. (8.11). For $\theta = 90°$, $B = 0$ and therefore $E_b = A$.

2. $M_Y < M_a$

(a) If the mass of the particle 'Y' is less than the mass of incident particle 'a', A will be zero for $E_a = \frac{M_Y Q}{(M_a - M_Y)} = E_1$ (say) and will remain positive for $E_a < E_1$. Like the previous case, for positive A values the energy E_b will be single valued.

(b) For incident energy E_a larger than E_1, A will be negative. In that case, $\sqrt{B^2 + (-A)}$ will be less than B and hence both the roots of $\sqrt{E_b}$ will be positive and valid, $\sqrt{E_b} = B \mp \sqrt{B^2 + (-A)}$. This gives rise to the double values of E_b for a given angle. It means that at a given angle in lab frame there will be two groups of particle 'b' with energies $E_b = \left[B + \sqrt{B^2 - A}\right]^2$ and $\left[B - \sqrt{B^2 - A}\right]^2$

(c) Further, if $E_a > E_1$, for $\theta < 90°$, $\cos\theta$ is positive and, therefore, B is positive. At $\theta = 90°$, B becomes zero and if A is negative ($E_b > E_1$) then E_b becomes imaginary which is not possible. The same is true for $\theta > 90°$ when both B and A become negative. As a result, particle 'b' will be emitted only at angles below 90°, with two values of energies at each angle.

8.3.2 Endoergic Reactions

Reactions with negative Q-value are called endoergic reactions or threshold reactions. In endoergic reactions, the total mass in the final state is more than the total mass in the initial state. These reactions do not take place unless the kinetic energy in the input channel is more than a minimum value called the threshold energy E_{thre}. It is because at $E_a < E_{thre}$, the kinetic energy E_a is not enough for the creation of additional mass and for the conservation of linear momentum. E_{thre} corresponds to the minimum value of E_b when the particle 'b' is emitted in forward direction, that is $\theta = 0$. The minimum value for E_b or $\sqrt{E_b}$ may be obtained by setting $(B^2 + A) = 0$ and $\theta = 0$. These conditions when put in Eqs. (8.11) and (8.12) give

$$E_{thre} = E_{b_{min}}^{\theta=0} = -Q\left[\frac{M_Y + M_b}{M_Y + M_b - M_a}\right] \tag{8.14}$$

However,

$$M_a + M_X = M_Y + M_b + Q/c^2 \tag{8.15}$$

Therefore, Eq. (8.14) may be written as

$$E_{thre} = -Q\left[\frac{M_X + M_a - Q/c^2}{M_X - Q/c^2}\right] \tag{8.16}$$

In case $M_X >> Q/c^2$, Eq. (8.16) reduces to

$$E_{thre} = -Q\left[1 + \frac{M_a}{M_X}\right] \tag{8.17}$$

It may be noted that in centre of mass frame the final products Y and b are emitted with zero kinetic energy at reaction threshold. However, in lab Frame, the final products Y and b have non-zero kinetic energies due to the motion of the centre of mass. When the incident kinetic energy is sufficiently large the kinetic energy of the outgoing particle might have double values.

8.3.3 Relativistic *Q* Equation

So far only non-relativistic treatment has been worked out. It is because nucleons become relativistic at very high energies of the order of 1000 MeV. The relativistic Q equation may be obtained from the non-relativistic Eq. (8.8) by adding a relativistic correction term δr given by

$$\delta r \approx \frac{1}{2M_Y c^2}\left[E_a^2 + E_b^2 - E_Y^2 - \sqrt{M_a M_b E_a E_b}\left(\frac{E_a}{M_a} + \frac{E_b}{M_b}\right)\cos\theta\right] \quad (8.18)$$

In Eq. (8.18), the kinetic energy of the residual nucleus Y is generally very small and can be neglected.

Q equation may also be used to describe elastic scattering with $Q = 0$, and $M_b = M_a$, if the angular distribution of the incident particle is required to be calculated.

In the inelastic scattering, the residual nucleus is left in an excited state. If W is the excitation energy of the residual nucleus, then substitution of $Q = -W$, $M_Y = M_X$, $M_a = M_b$ may be used to describe the inelastic scattering using the Q equation.

Solved example S-8.1

The reaction ${}^3_1\mathrm{H}\left({}^2_1\mathrm{D}, {}^1_0\mathrm{n}\right){}^4_2\mathrm{He}$ with Q value of +17.6 MeV is often used to produce neutrons using accelerated beams of deuterons. What will be the maximum and minimum energies of the emitted neutrons if a beam of 0.3 MeV is used?

Solution. We use Eq. (8.8).

$$E_b\left(\frac{M_Y + M_b}{M_Y}\right) - E_a\left(\frac{M_Y - M_a}{M_Y}\right) - \frac{2\sqrt{M_a E_a M_b E_b}}{M_Y}\cos\theta = Q \quad (8.8)$$

And substitute $a \to {}^2_1\mathrm{D}; \mathrm{X} \to {}^3_1\mathrm{H}; \mathrm{Y} \to {}^4_2\mathrm{He}; \mathrm{b} \to {}^1_0\mathrm{n}; E_a \to 0.3\,\mathrm{MeV}, M_a \to 2; M_b \to 1, M_X \to 3, M_Y \to$ and get, $5E_n - 1.549\sqrt{E_n}\cos\theta - 71 = 0$. The maximum energy of the emitted neutron will be when θ is zero and minimum when θ is 180. Substituting these values, $(E_n)^{max} = 15.41\,\mathrm{MeV}$ and $(E_n)^{min} = 13.08\,\mathrm{MeV}$.

Solved example S-8.2

5 MeV protons hit a stationary target foil of ${}^{107}_{47}\mathrm{Ag}$ to produce the nucleus ${}^{108}_{48}\mathrm{Cd}^*$ in an excited state. From the following data, calculate (i) the excitation energy of the ${}^{108}\mathrm{Cd}$, (ii) the kinetic energy of ${}^{107}\mathrm{Ag}$ to produce the same excited state on hitting a stationary sample of hydrogen (protons), that is in the inverse reaction, (iii) the linear momentum of the proton and the ${}^{107}\mathrm{Ag}$, (iv) the maximum value of the orbital angular momentum ℓ for the grazing interaction of nuclei in the two situations. Given, $M_p = 1.007825$ u; $M^{107}{}_{Ag} = 107.009490$ u; $M^{108}{}_{Cd} = 108.009581$ u; 1 u = 1.66×10^{-27} kg; $r_0 = 1.2$ F, radius of proton 0.8 F.

Solution.

(i) The excitation energy U = kinetic energy of 5 MeV proton in CM + Q-value of (p + ^{107}Ag = ^{108}Cd) reaction.

$$Q = (M_p + M^{107}{}_{Ag}) - (M^{108}{}_{Cd}) = 0.007734 \text{ u} = 7.2040 \text{ MeV};$$

$$\text{CM-energy of proton} = 5\times\left(\frac{107}{107+1}\right) = 4.9537.$$

Hence, the excitation energy $U = 7.2040 + 4.9537 = 12.1577$ MeV.

(ii) The Q-value of the reaction will not change if the proton is at rest and ^{107}Ag is moving with some kinetic energy E_{Ag} to produce the same final state (i.e. Cd at same excitation). It means that for the same U, the CM energy of ^{107}Ag should be 4.9537MeV. Hence, $E_{Ag} \times (M_p/108) = 4.9537$ MeV or $E_{Ag} = 534.9996$ MeV.

(iii) $(MV)_{proton} = \sqrt{2M_p E_p^{kinetic}} = \sqrt{2\times5\times1.67\times10^{-27}\times1.6\times10^{-13}} = 5.17\times10^{-20}$ kg-ms^{-1};

$(MV)^{107}{}_{Ag} = \sqrt{2M_{107_{Ag}} E^{kinetic}_{107_{Ag}}} = 53.51 \times 10^{-20}$ kg-ms^{-1}

(iv) For the grazing case, the distance d between the two nuclei will be $(R_p + R^{107}{}_{Ag})$

$$d = 0.8\,\text{F} + r_0 \times 10^{-15}\,(107^{1/3})\,\text{m} = 6.68\,\text{F}.$$

Now, $\ell\hbar = \text{MV}\cdot d$ or $\ell = \frac{\text{MV}\cdot d}{\hbar}$. Using the values of MV and d calculated and considering $\hbar = 1.0544\times10^{-34}$, we get $\ell(\text{p}) = 3.37$ or simply 3 and $\ell(\text{Ag}) = 33.90$ or simply 34. We thus see that in the inverse reaction the kinetic energy of ^{107}Ag is much larger and the almost ten times angular momentum can be pumped in the system.

Exercise p-8.1: Calculate the Q value for $^2_1$H ($^2_1$H, $^1_0$n)$^3_2$He reaction from the following data and show that the energy of neutrons emitted at 90° with respect to the incident direction is given by $E_n = 2.45 + E/4$ MeV, where E is the energy of the incident deuteron.

$$M_n = 1.008665 \text{ u}; \; M^{21}{}_H = 2.014102 \text{ u}; \; M^3{}_{He} = 3.016030.$$

Exercise p-8.2: Show that the energy released in the following fission reaction is 200.6 MeV. $^{235}_{92}$U + $^1_0$n = $^{141}_{56}$Ba + $^{92}_{36}$Kr + 3 $^1_0$n. Given that the masses of nuclei are M($^1_0$n) = 1.008665 u; M($^{235}_{92}$U) = 235.043915 u; M($^{141}_{56}$Ba) = 140.913900 u; M($^{92}_{36}$Kr) = 91.897300 u.

Exercise p-8.3: Derive an expression for the minimum kinetic energy E_a of the incident particle so that the outgoing particle b just appears at angle θ.

Exercise p-8.4: Using Q equation show that in an inelastic collision the kinetic energy acquired by the struck particle when it is projected at an angle θ is $E_X = 4\,E_a\,\{(M_a M_X)/(M_a + M_X)^2\}\cos^2\theta$.

Exercise p-8.5: Calculate the Q value for the reaction: $^{30}_{14}\text{Si} + ^2_1\text{H} \rightarrow ^{31}_{14}\text{Si} + ^1_1\text{p}$ using the following data: $^1_0\text{n} \rightarrow ^1_1\text{H} + \beta^- + \bar{\nu} + 0.78\,\text{MeV}$; $^{31}_{14}\text{Si} \rightarrow ^{31}_{15}P + \beta^- + \bar{\nu} + 1.51\,\text{MeV}$, and $^{30}_{14}\text{Si} + ^2_1\text{H} \rightarrow ^{31}_{15}\text{P} + ^1_0\text{n} + 5.10\,\text{Mev}$.

8.4 THE PARTIAL WAVE ANALYSIS OF NUCLEAR REACTION CROSS-SECTION

We consider a beam of mono-energetic, charge and spinless particles (may be neutrons) of velocity v moving along the z-direction towards a target 'P'. We assume that the coordinate system is attached to the target at P. This beam moving along the z-direction may be represented as plane wave $e^{ikr\cos\theta}$. At large distance from P (asymptotically), the incident beam may be represented by partial waves

$$\psi_{\text{inc}} = e^{ikr\cos\theta} \to \frac{\pi^{1/2}}{kr}\sum_{\ell=0}^{\infty}\sqrt{2\ell+1}\; i^{\ell+1}\left\{\exp\left[-i\left(kr-\frac{\pi}{2}\ell\right)\right]-\exp\left[i\left(kr-\frac{\pi}{2}\ell\right)\right]\right\}Y_{\ell,0}(\theta) \quad (8.19)$$

Here, the wave number $k = \frac{1}{\hbar}\sqrt{2mE_{\text{kin}}}$ with E_{kin} the kinetic energy and m the mass of the particle. It is easy to show that $k = 0.2187\sqrt{mE_{\text{kin}}}$ F^{-1} if mass m is in unified mass unit u and the kinetic energy E_{kin} in MeV.

Expression (8.19) may be looked as the combination of partial waves of two kinds: one moving towards the origin, $\exp\left[-i\left(kr-\frac{\pi}{2}\ell\right)\right]$ and the other, $\exp\left[i\left(kr-\frac{\pi}{2}\ell\right)\right]$ moving out of the origin. Suppose now a nucleus is placed at the origin P. The nucleus located at P may interact with the partial waves falling within the range of its nuclear field and may change the amplitude and the phase of the outgoing partial waves. Note that the nuclear interaction will not disturb the waves approaching it, but only the partial waves falling within the range of its nuclear field and moving out after the interaction with the nucleus will be affected. It may be recalled that in the case of scattering we considered that the phase of the scattered wave is changed. Now, we are considering two things, the change of phase as well as the change in the amplitude of the outgoing partial waves. Reduction in the amplitude of the outgoing partial waves will mean that some incident particles have been removed from the beam giving rise to reaction. In the case of scattering, the particles were redistributed and are not removed; hence, the amplitudes of the partial waves remained unaltered. Let the amplitude of the ℓth outgoing partial wave be changed by the factor η_ℓ (figure 8.2). The factor η_ℓ may be complex and must at the most be equal to 1 or less than 1, that is

$$|\eta_\ell| \le 1 \quad (8.20)$$

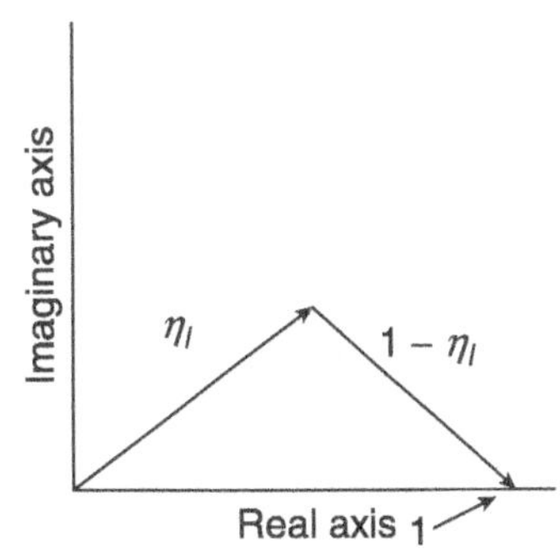

Figure 8.2 *The complex factor η_ℓ by which the amplitude of the outgoing wave has changed*

The modified wave ψ_{mod} after the interaction is

$$\psi_{\text{mod}} = \frac{\pi^{1/2}}{kr}\sum_{\ell=0}^{\infty}\sqrt{2\ell+1}\; i^{\ell+1}\left\{\exp\left[-i\left(kr-\frac{\pi}{2}\ell\right)\right]-\eta_\ell\exp\left[i\left(kr-\frac{\pi}{2}\ell\right)\right]\right\}Y_{\ell,0}(\theta) \quad (8.21)$$

The scattered wave ψ_{sc} is the difference of the modified wave and the incident wave, and may be written as

$$\psi_{sc} = \frac{\pi^{1/2}}{kr} \sum_{\ell=0}^{\infty} \sqrt{2\ell+1}\, i^{\ell+1} (1-\eta_\ell) \exp\left[i\left(kr - \frac{\pi}{2} \right)\ell \right] Y_{\ell,0}(\theta) \tag{8.22}$$

To calculate the number of particles scattered per second, we integrate Eq. (8.22) over the surface of sphere of radius r, and find out how many particles are crossing this surface per second.

Number of particles scattered per second is given by

$$N_{sc} = \left(\psi_{sc}\psi_{sc}^*\right) v \int |Y_{\ell,0}(\theta)|^2 r^2 \sin\theta \, d\theta \, d\phi \tag{8.23}$$

The first term in Eq. (8.23) gives the flux of the scattered wave, and as it is a mono energetic beam of chargeless particles each particle is moving with the velocity v and the flux is equal to the particle density $\psi\psi^* v$. The second integral term is unity as the spherical harmonic functions are normalized. Hence,

$$N_{sc} = \frac{\pi}{k^2}[2\ell+1]\, |1-\eta_\ell|^2 v \tag{8.24}$$

The ℓ-th partial wave scattering cross-section $\sigma_\ell^{sc} = \dfrac{N_{sc}}{\text{incident flux}} = \dfrac{\pi}{k^2}[2\ell+1]\, |1-\eta_\ell|^2$ (8.25)

It may be noted that the incident flux of each partial wave is v, as the particle density of each incident partial wave is 1.

The number of particles that have been absorbed and get lost from the incident ℓ-th partial wave is proportional to $(1 - |\eta_\ell|^2)$. The absorbed particles are lost because they initiate nuclear reactions. Following a procedure similar to the one carried out for scattering cross-section, the ℓ-th partial wave absorption cross-section, which is also the reaction cross-section may be given as

$$\sigma_\ell^{abs} = \frac{\pi}{k^2}[2\ell+1]\left(1-|\eta_\ell|^2\right) \tag{8.26}$$

At moderate and low energies, only the S-partial wave ($\ell = 0$) and a little bit of P-wave ($\ell = 1$) lie within the range of the nuclear field (≈2 F) of the target nucleus and, therefore, major contribution to the scattering and reaction cross-sections comes from the S-wave. Assuming that only the S-wave interacts with the target, the total modified wave function ψ_{mod} may be written as ψ^0_{mod} which is given by

$$\psi^0_{mod} = \frac{\sqrt{\pi}}{kr} i\left\{e^{-ikr} - \eta_0 e^{ikr}\right\} Y_{0,0} = \frac{u_0}{r} Y_{0,0} \tag{8.27}$$

Here, $$u_0 = \frac{\sqrt{\pi}}{k} i\left\{e^{-ikr} - \eta_0 e^{ikr}\right\} \tag{8.28}$$

It is now useful to define a quantity called logarithmic derivative of u_0 evaluated at $r = R$ (the radius of the target nucleus) as

$$f_0 \equiv \left\{ \frac{r}{u_0}\frac{du_0}{dr} \right\}_{r=R} = \frac{\left[(-ik\,R)e^{-ikR} - (ik\eta_0 R)e^{ikR}\right]}{(e^{-ikr} - \eta_0 e^{ikr})} \tag{8.29}$$

From Eq. (8.29), it is easy to get

$$\eta_0 = \left[\frac{f_0 + ikR}{f_0 - ikR}\right] e^{-2ikR} \tag{8.30}$$

Equation (8.30) gives the value of the factor η_0 by which the amplitude of the S-partial wave has changed as a result of nuclear interaction in terms of the logarithmic derivative f_0. The scattering and absorption (or reaction) cross-sections may now be found in terms of f_0, and k using Eqs. (8.25), (8.26), and (8.30).

8.5 REACTION MECHANISM

In a nuclear reaction, the initial state is defined by the experimental setup and the parameters of the final state may be determined by actual measurements. However, what happens during the small time interval Δt_{re} when the target nucleus and the incident particle are within the range of nuclear forces, exchanging energies and redistributing nucleons, is not known. Reaction time Δt_{re} may vary from $\approx 10^{-22}$ s to 10^{-16} s depending on the reaction mechanism. In absence of exact knowledge, models for the reaction mechanism have been developed. The first such attempt was made by Niels Bohr in 1932. It is called compound nuclear reaction mechanism or CN mechanism in short.

8.5.1 Compound Nuclear Reaction (or CN) Mechanism: Reactions Leading to Continuum

The basic assumption of this reaction mechanism is the formation of an excited composite system as a result of the absorption of the incident particle by the target nucleus. The composite system so formed lives long enough (10^{-16} s), during which large number of interactions between the nucleons of the target and the incident particle take place, giving rise to thorough mixing, sharing, and distribution of energy amongst many degrees of freedom, that ultimately leads to the establishment of a thermodynamic equilibrium. The equilibrated compound system forgets about its history of formation and is specified by its quantum parameters. Once thermal equilibrium is established, by chance excess energy is stored at a nucleon or a cluster of nucleons, which leaves the compound nucleus giving rise to its decay. Emission of particles from the excited compound nucleus may be compared to the evaporation of molecules from the surface of the boiling liquid. The formation and the decay of the compound nucleus are assumed to be two independent steps. A compound nucleus formed in a given quantum state, by whatever process, will decay in the same way irrespective of its way of formation. As such, the cross-section for a reaction (α, β), initiated by channel α, and leading to channel β, may be written as

$$\sigma(\alpha,\beta) = \sigma^{C}(E_C,\alpha)G_C(E_C,\beta) \tag{8.31}$$

Here, $\sigma^{C}(E_C,\alpha)$ is the cross-section for the formation of the compound nucleus at excitation energy E_C through channel α, and $G_C(E_C,\beta)$ the probability that the compound nucleus at excitation energy E_C decays through channel β. The symbol α contains all the quantum parameters, such as the spins J_X of the target nucleus X and J_a of the particle 'a', and the relative orbital angular momentum ℓ^{Xa} of X and a. Similarly, β, contains all parameters that completely define the final state of Y and b.

8.5.1.1 Excitation energy of the compound nucleus

Let as consider a real experiment situation. A nuclear particle ${}^{A_a}_{Z_a}a$ (mass M_a, atomic number Z_a, mass number A_a) moving with kinetic energy E_a^{lab} (in laboratory frame) hits a stationary target nucleus ${}^{A_x}_{Z_x}X$ (mass M_X, atomic number Z_X, mass number A_X) in ground state. The incident particle is absorbed by the target to form the compound nucleus ${}^{A_{cn}}_{Z_{cn}}CN^*$ of mass M_{cn}, atomic number Z_{cn} and mass number A_{cn}. The incident particle on absorption, will share its kinetic energy and nucleons with the target forming the compound nucleus. A part of the energy E_a^{lab} will be used to give recoil energy to the compound nucleus to conserve the linear momentum. The remaining part of the incident energy will be deposited in the compound nucleus as a part of its excitation energy. The compound nucleus (CN) is, therefore, formed in an excited state with excitation energy E_c. To indicate that a nucleus is in excited state a (*) mark is put as a superscript on the right hand side of the symbol of the nucleus. A simple way of calculating the excitation energy E_c of the compound nucleus is to treat the problem in centre of mass frame (denoted by CM-frame). Because in CM frame conservation of linear momentum need not to be considered explicitly and total kinetic energy E_a^{CM} of the incident particle in (CM) frame is available for the excitation of the compound nucleus. It is easy to show that

$$E_a^{CM} \text{ (kinetic energy of particle a in CM-frame)} = E_a^{lab}\left(\frac{M_x}{M_x+M_a}\right) \approx E_a^{lab}\left(\frac{A_x}{A_x+A_a}\right).$$

This energy E_a^{CM} is lost in giving excitation energy to the compound nucleus. However, there is one very important point that needs consideration. When incident particle was absorbed in the target nucleus and a bound compound nucleus was formed, some mass equivalent to the binding energy B_a^{cn} of the particle 'a' in the compound nucleus must have converted into energy. The energy released from the conversion of mass ($=|B_a^{cn}|$) is also available for the excitation of the compound nucleus. Therefore, the excitation energy E_c of the compound nucleus is the sum of E_a^{CM} and B_a^{cn},

$$E_c = E_a^{CM} + |B_a^{cn}| \approx E_a^{lab}\left(\frac{A_x}{A_x+A_a}\right) + B_a^{cn} \tag{8.31a}$$

The value of B_a^{cn} can be found in standard tables of binding energies. A little thinking will tell that $B_a{}^{cn}$ is also equal to the Q value of reaction ${}^{A_x}_{Z_x}X(a,\gamma)\,{}^{A_{cn}}_{Z_{cn}}CN$.

8.5.1.2 Spin distribution of compound nucleus

The spin J_c of the compound nucleus is the quantum mechanical sum of spins S_a, of incident particle, S_x of the target and the orbital angular momentumℓ.

$$J_c = S_a + S_x + \ell \tag{8.31b}$$

As Eq. (8.31b) represents a quantum mechanical sum, there may be several possible values of J_c even when each quantity on the RHS of the equation has a fixed value. In practice, however, S_a

and S_x have fixed value but the value of ℓ can have several possible integer values starting from 0, depending on the energy of the incident particle. Hence, the spin of the compound nucleus is not single valued instead there is a distribution of the values of J_c. In an actual experiment, a beam of incident particles hit a target foil that contains a large number of nuclides X. Large number of compound nuclei are simultaneously formed by the absorption of the incident particles by many nuclides X. These many compound nuclei so formed have a distribution of their spins J_c.

8.5.1.3 Maximum value of the orbital angular momentum ℓ_{max}

The maximum orbital angular momentum ℓ_{max} will be pumped by the incident particles in grazing incidence. The value of the impact parameter b for grazing incidence is equal to the sum of the radii R_a and R_x, respectively, of the incident particle 'a' and the target nucleus 'X'. If p is the linear momentum of the incident particle then

$$\ell_{max}\hbar = p \cdot b$$

$$\text{or} \quad \ell_{max} = \frac{p \cdot b}{\hbar} = \frac{b\sqrt{2M_a E_{kin}^{CM}}}{\hbar} = \frac{r_0\left(A_a^{1/3} + A_x^{1/3}\right)\left(\sqrt{2M_a E_a^{CM}}\right)}{\hbar} \tag{8.31c}$$

E_a^{CM} in Eq. (8.31c) is the centre of mass kinetic energy of the incident particle.

8.5.2 Cross-section for the Formation of the Compound Nucleus $\sigma^C(E_C, \alpha)$

We now attempt to derive an expression for $\sigma^C(E_C, \alpha)$. We are still considering the incident wave to be the S-wave of chargeless and spinless particles of energy E_{kin} incident on a nucleus X that has well-defined surface at $r = R$. We further make the following assumptions:

1. Once inside the nucleus, the kinetic energy of the incident particle increases as it now moves in the potential well (of depth V) of the target nucleus and, hence, the particle has a new wave number $K = \frac{1}{\hbar}\sqrt{2m\left(E_{kin} + V\right)}$.
2. The probability of re-emission of the particle from the compound nucleus with same energy E_{kin} is negligible. The event in which the incident particle is re-emitted from the compound nucleus with same energy is called *compound elastic scattering* to differentiate it from the normal elastic scattering event that takes place at the surface of the target nucleus and is called shape or potential elastic scattering. The validity of the assumption that compound elastic scattering is negligible depends on the excitation energy of the compound nucleus E_C. If the excitation energy is large, many closely spaced levels of the compound nucleus are simultaneously excited and overlap each other because of their larger relative widths. This results in two important consequences: (i) very large number of channels becomes open for the decay of the compound nucleus and, therefore, the probability of decay through compound elastic channel becomes very small and (ii) the incident particle of any incident energy E_{kin} is absorbed by the target nucleus. This is in contrast to the low-energy excitation of the compound nucleus where the excited levels are far apart and incident particle of only fixed values of E_{kin} may be absorbed to raise the compound

nucleus to an individual excited level. The low excitation region is called resonance region and will be discussed later.

With these assumptions the radial wave function u_0 may be written as

$$u_0 = e^{-iKr} \text{ for } r \leq R \tag{8.32}$$

However, now there is no corresponding outgoing term. Further, the logarithmic derivative f_0 becomes (see Eq. (8.29))

$$f_0 = -iKR \tag{8.33}$$

And
$$|\eta_0| = \frac{(K-k)}{(K+k)} \tag{8.34}$$

When the value of $|\eta_0|$ is substituted in Eq. (8.26), we get

$$\sigma^{\mathrm{C}}(E_{\mathrm{C}}, \alpha) = \sigma_0^{\mathrm{abs}} = \frac{\pi}{k^2}\left(1 - |\eta_0|^2\right) = \frac{4\pi K}{k(K+k)^2} \tag{8.35}$$

Equation (8.35) gives the desired cross-section for the formation of a compound nucleus by chargeless and spinless particles of orbital angular momentum $\ell = 0$.

At higher incident energies, partial waves with higher orbital angular momentum also contribute to the absorption cross section and in principle total absorption cross-section can be found by summing the contributions of each partial wave. If the incident particle is charged, the effect of the Coulomb barrier has to be taken. One way of doing this is by multiplying the absorption cross-section for each partial wave by a corresponding barrier transmission factor (Gamow factor). Similarly, if the outgoing particle 'b' is also charged a barrier transmission coefficient for the outgoing particle needs to be included. Another factor that has not been included in our derivation is the spin of the incident and of the outgoing particles. As will be shown later, the spin effects may be included through the statistical weights of the states.

8.5.3 Decay of the Compound Nucleus and the Branching Ratio $G_C(E_C, \beta)$

It has been assumed that the excitation energy E_{C} of compound nucleus is so high that a large number of levels of the compound nucleus are simultaneously excited and overlap. Also, at such high excitations, one does not talk of the individual level but of the level density, that is the number of levels per unit energy interval. This region of excitation is called continuum, in contrast to the excitations at lower energies where levels are well separated and far apart, called the resonance region. In continuum, a large number of exit channels, $\beta, \beta', \beta'', \ldots$ each with its own decay constant $\lambda_\beta, \lambda_{\beta'}, \lambda_{\beta''}, \ldots$ are open for the decay of the compound nucleus. The mean life τ_β of the decay channel β is related to the decay constant λ_β through the relation

$$\tau_\beta = \frac{1}{\lambda_\beta} \text{ and from the uncertainty principle, } \tau_\beta \Gamma_\beta = \hbar. \text{ Hence,}$$

$$\Gamma_\beta = \hbar \lambda_\beta \tag{8.36}$$

Here, Γ_β is the width of the level β.
Therefore,

$$\text{the branching ratio } G_C(E_C, \beta) = \frac{\Gamma_\beta}{\sum_{\beta'}^{\text{all open channels}} \Gamma_{\beta'}} = \frac{\Gamma_\beta}{\Gamma} \tag{8.37}$$

The reciprocity theorem co-relates the compound nucleus formation cross-section through different channels and their corresponding widths through the following relation:

$$k_\alpha^2 \frac{\sigma_C(E_C, \alpha)}{\Gamma_\alpha} = k_\beta^2 \frac{\sigma_C(E_C, \beta)}{\Gamma_\beta} = k_{\beta'}^2 \frac{\sigma_C(E_C, \beta')}{\Gamma_{\beta'}} = \ldots = F(E_C) \tag{8.38}$$

The proportionality constant $F(E_C)$ is a function of the excitation energy E_C only. It follows from Eq. (8.38) that

$$\Gamma_\gamma = \frac{k_\gamma^2 \sigma_C(E_C, \gamma)}{F(E_C)} \tag{8.39}$$

The expression for the branching ratio $G_C(E_C, \beta)$ may now be obtained by substituting the value of the width Γ_γ from 8.39 into Eq. (8.37)

$$G_C(E_C, \beta) = \frac{k_\beta^2 \sigma_C(E_C, \beta)}{\sum_\gamma k_\gamma^2 \sigma_C(E_C, \gamma)} \tag{8.40}$$

Equation (8.40) gives the desired expression for the decay probability of the compound nucleus through channel $\boldsymbol{\beta}$.

The cross-section $\sigma(\alpha, \beta)$ may now be given as:

$$\sigma(\alpha, \beta) = \sigma^C(E_C, \alpha) G_C(E_C, \beta) = \frac{4\pi K}{k(K+k)^2} \frac{k_\beta^2 \sigma_C(E_C, \beta)}{\sum_\gamma k_\gamma^2 \sigma_C(E_C, \gamma)}$$

As k and K in this expression depend on energy, the cross-section can be determined if the energy is known.

8.5.4 Special Feature of Continuum Reactions

Parity is conserved in nuclear reactions. However, in the region of continuum levels of odd and even parity may overlap to a large extent. It is because the incoming particle may have both even and odd orbital angular momentums and therefore, even and odd parities. Similarly, the outgoing particle may also carry both types of angular momentum and a definite state of the final nucleus may also be reached by both parity states in the compound nucleus. If a state of the residual nucleus is fed by both parity states of the compound nucleus, then the angular distribution of the reaction product is not symmetrically distributed around 90° in the centre of mass. This may be seen in many charged particle reactions.

In case the parity is strictly conserved, that is states of only one parity are involved, then the angular distributions of the reaction products in centre of mass are symmetrical about 90°. In the case of resonance region, a single level of the compound nucleus is populated, the parity is strictly conserved and the angular distribution of reaction products in the centre of mass is symmetrical about 90°, that is $\frac{d\sigma}{d\Omega}(\theta) = \frac{d\sigma}{d\Omega}(\pi - \theta)$.

8.5.5 Thermodynamic Approach to Continuum Reactions

The energy spectrum of particles emitted in reactions leading to the continuum has been found to have the same shape as that of the molecules evaporated from the surface of a boiling liquid. The energy distribution follow Maxwell–Boltzmann law and the intensity $I(E)$ of particles having energy E and $E + dE$ is given as

$$I(E) \propto E e^{-\frac{E}{\theta}} dE$$

Here, θ, which has the dimensions of energy, is called the 'nuclear temperature' and is related to the density of levels ω of the residual nucleus Y through the relation

$$\frac{1}{\theta} = \frac{dS}{dE}$$

With
$$S = \ln \omega(E)$$

Here, $\omega(E)$ is the density of levels of residual nucleus Y around the energy E. The equation involving S and $\omega(E)$ is just like the thermodynamical relation between the entropy and the density of final states, without the Boltzmann constant $\mathfrak{R}$ (In thermodynamics entropy $S = \mathfrak{R} \ln \rho$). The relation $\frac{1}{\theta} = \frac{dS}{dE}$ is the well-known equation of thermodynamics that defines the temperature in terms of entropy. Assuming equidistant single particle levels the level density $\omega(E)$ may be given as

$$\omega(E) \sim e^{2\sqrt{aE}}$$

The level density parameter 'a' plays an important role in compound nucleus reactions. Detailed analysis of particle spectra has indicated that for non-magic nuclei 'a' may be given as

$$a = \frac{A}{K}$$

where K is a constant and is often treated as a free parameter in computer codes written to carry out CN-calculations. It may have values from 6 to 10.

8.5.6 Reaction and Scattering in Resonance Region

The nuclear excited levels show a typical behaviour with the excitation energy. At low excitations, the levels are sharp and well separated from each other. As the excitation energy increases, the level separation decreases and the level width increases. Ultimately, at sufficiently high excitation, the levels are so closely spaced and widths are relatively large that there is considerable overlap in excited levels. This is called the continuum region and the statistical approach is appropriate to describe the system. For example, the cross-section σ_C is an average quantity over several levels of the compound nucleus.

In resonance region, one deals with a single isolated level of the compound nucleus, which may be reached by different incident particles of specific energies hitting the corresponding target nucleus. As shown in figure 8.3, the same levels in the compound nucleus ${}^{28}_{14}\text{Si}$ may be reached by ${}^{24}_{12}\text{Mg} + {}^{4}_{2}\text{He}$ reaction and also by ${}^{27}_{13}\text{Al} + {}^{1}_{1}\text{H}$ reaction. Many other reactions may be used but in each case the energy of the incident particle must vary within very narrow limits. The cross-section for the formation of the compound nucleus sharply increases attaining a maximum and again decreases sharply with in a very narrow interval of incident energy. The sharp peaks are called resonance peaks.

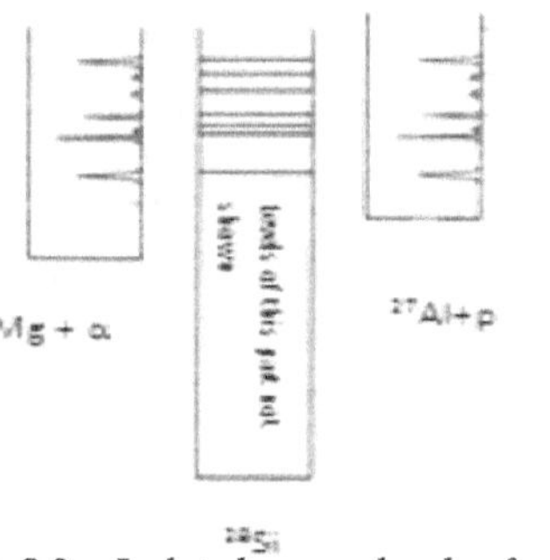

Figure 8.3 *Isolated energy levels of the compound nucleus ^{28}Si reached through ^{24}Mg + ^{4}He and ^{27}Al + ^{1}H reactions*

We shall now apply the concepts of the logarithmic derivative f_0 and of η_0 developed earlier to see how the resonance peaks can be reproduced.

Figure 8.4 shows the wave function u inside and outside the nucleus in three cases. In case (a), the tangent at the boundary $r = R$ is horizontal and $(du/dr)_{r=R}$ is zero. The amplitude of the wave function inside the nucleus is a maximum in this case. In case (b), the derivative $(du/dr)_{r=R}$ is positive and the amplitude of the wave function inside is small. In case (c), $(du/dr)_{r=R}$ is negative and the amplitude of the wave inside is again small. It is evident that the maximum amplitude of the wave penetrated inside the nucleus occurs when its derivative is zero at the boundary. Absorption of the incident wave and formation of a quasi-compound nucleus occurs only in situation (a) when $(du/dr)_{r=R} = 0$. $f_0 \equiv \left[\dfrac{r\,du_0}{u_0\,dr}\right]$ at $r = R$, therefore, passes through zero at some energy E_s and in a small energy interval around E_s one may write

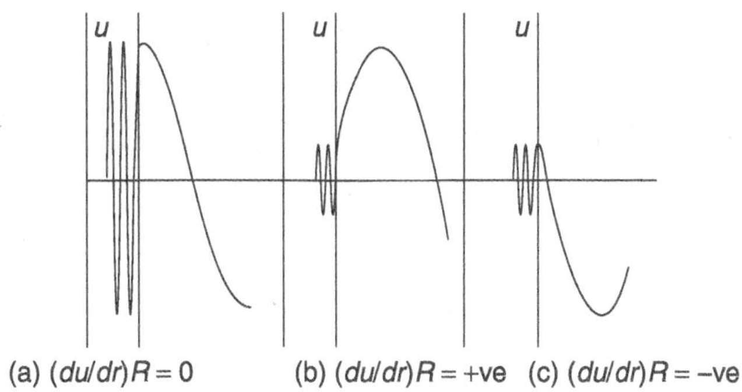

Figure 8.4 *Pictorial representation of the wavefunction u outside and inside the nucleus, and its derivative at the nuclear boundary r = R*

$$f_0 = -a(E - E_s) \tag{8.41}$$

Further, if f_0 is a purely a real quantity then from Eq. (8.30) $|\eta_0|^2$ becomes = 1 and from Eq. (8.26) the reaction or absorption cross-section $\sigma^{\text{abs}}_{\ell=0} = \dfrac{\pi}{k^2}\left[1-|\eta_0|^2\right]$ becomes zero. To have a non-zero value of the absorption cross-section, it is essential that f_0 must contain an imaginary term. As $|\eta_0| \leq 1$, the imaginary term must be of negative sign, hence

$$f_0 = -a(E - E_s) - ib \tag{8.42}$$

Here, both a and b are constants and it is assumed that Eq. (8.42) describes the behaviour of f_0 in a small energy interval around E_s.

Now, $$|1-\eta_0| = \left|1 - \frac{(f_0 + ikR)}{(f_0 - ikR)} e^{-2ikR}\right| = \left|\frac{e^{2ikR} - \dfrac{f_0 - ikR + 2ikR}{f_0 - ikR}}{e^{2ikR}}\right|$$

The term e^{2ikR} in the denominator will be multiplied by e^{-2ikR} on taking the modulus and will give 1. Hence, it may be dropped for simplicity, and therefore,

$$|1-\eta_0| = \left|e^{2ikR} - \frac{f_0 - ikR + 2ikR}{f_0 - ikR}\right| = \left|e^{2ikr} - 1 - \frac{2ikR}{f_0 - ikR}\right| \tag{8.43}$$

$$= |A_{\text{pot}} + A_{\text{res}}| \tag{8.44}$$

where $$A_{\text{pot}} = \left(e^{2ikR} - 1\right) = e^{ikr}\left[e^{ikR} - e^{-ikR}\right] = 2ie^{ikR} \sin kR \tag{8.45a}$$

and $$A_{\text{res}} = \frac{-2ikR}{f_0 - ikR} \tag{8.45b}$$

8.5.6.1 Significance of A_{pot} and A_{res}

It may be observed that A_{pot} does not contain f_0, and hence it does not depend on the properties of the interior of the nucleus. It represents the situation where the projectile does not penetrate the target nucleus as in the idealized situation where the nucleus is considered an impenetrable hard sphere. A_{pot} is responsible for the scattering from the surface of the nucleus, called potential scattering and hence the subscript 'pot'. On the other hand A_{res} depends on f_o and through it on the interior properties of the nucleus. A_{res} gives rise to resonance absorption.

We now substitute the value of f_0 from Eq. (8.42) in Eq. (8.45b) to get the value of A_{res} as

$$A_{\text{res}} = \frac{-2ikR}{-a(E - E_s) - ib - ikR} = \frac{2i\left(\dfrac{kR}{a}\right)}{(E - E_s) + i\left(\dfrac{b}{a} + kR/a\right)} \tag{8.46}$$

If A_{pot} is neglected in comparison of A_{res}, then from Eq. (8.25) it follows

$$\sigma_{\ell=0}^{\text{scat}} = \frac{\pi}{k^2}|1-\eta_0|^2 = \frac{\pi}{k^2} \frac{4(kR/a)^2}{(E - E_s)^2 + \left(\dfrac{b}{a} + kR/a\right)^2} \tag{8.47}$$

A plot of $\sigma_{\ell=0}^{\text{scat}}$ with E shows the shape of a resonance curve, increasing as E approaches E_s, attaining a maximum value at $E = E_s$ and again decreases as E increases beyond E_s. If for a moment b is set to zero, that is there is no absorption or reaction, then the full width at the half maxima $\Gamma_{\ell=0}^{\text{scat}}$ is given by

$$\Gamma_{\ell=0}^{\text{scat}} = \frac{2kR}{a} \tag{8.48}$$

The absorption cross-section for charge and spinless particles for S-wave interaction may be determined using Eq. (8.26) as

$$\sigma_{\ell=0}^{\text{abs}} = \frac{\pi}{k^2}\left(1-|\eta_0|^2\right) = \frac{\pi}{k^2}\left[1-\left|\frac{-a(E-E_s)-ib+ikR}{-a(E-E_s)-ib-ikR}e^{-2ikR}\right|^2\right]$$

Or

$$\sigma_{\ell=0}^{\text{abs}} = \frac{\pi}{k^2}\frac{4\left(\frac{b}{a}\right)\left(\frac{kR}{a}\right)}{(E-E_s)^2+\left(\frac{b}{a}+kR/a\right)^2} \tag{8.49}$$

The quantity $2b/a$ is called the reaction width and is represented by $\Gamma_{\ell=0}^{\text{rec}}$ and, therefore

$$\sigma_{\ell=0}^{\text{abs}} = \frac{\pi}{k^2}\frac{\Gamma_{\ell=0}^{\text{scat}}\,\Gamma_{\ell=0}^{\text{abs}}}{(E-E_s)^2+\frac{1}{4}\left(\Gamma_{\ell=0}^{\text{scat}}+\Gamma_{\ell=0}^{\text{abs}}\right)^2} \tag{8.50}$$

And

$$\sigma_{\ell=0}^{\text{scat}} = \frac{\pi}{k^2}\frac{\Gamma_{\ell=0}^{\text{scat}^2}}{(E-E_s)^2+\frac{1}{4}\left(\Gamma_{\ell=0}^{\text{scat}}+\Gamma_{\ell=0}^{\text{abs}}\right)^2} \tag{8.51}$$

Equation (8.50) shows that if $\Gamma_{\ell=0}^{\text{scat}}$ is zero (no scattering), the $\sigma_{\ell=0}^{\text{abs}}$ is also zero. It means that there is no absorption (or reaction) without scattering. However, Eq. (8.51) states that scattering takes place even if $\Gamma_{\ell=0}^{\text{abs}}$ is zero, which means that scattering without absorption is possible but not vice versa.

To determine the scattering cross-section far from the resonance energy E_s where absorption cross-section is zero, we use the expression

$$\begin{aligned}\sigma_{\ell=0}^{\text{scat}}\left(2|E-E_s|>>\frac{\Gamma}{kR}\right) &= \frac{\pi}{k^2}\left|\frac{\Gamma_{\ell=0}^{\text{scat}}}{(E-E_s)+1/2\left(\Gamma_{\ell=0}^{\text{scat}}+\Gamma_{\ell=0}^{\text{abs}}\right)}+2e^{ikR}\sin kR\right|^2 \\ &= \frac{4\pi}{k^2}\left|\frac{\Gamma}{2(E-E_s)+i\Gamma}+kR\right|^2\end{aligned} \tag{8.52}$$

At large distances from the resonance, the second term in expression (8.52) dominates and therefore,

$$\sigma_{\ell=0}^{\text{scat}}\left(2|E-E_s|>>\frac{\Gamma}{kR}\right) \approx 4\pi R^2 \tag{8.53}$$

$4\pi R^2$ is the scattering cross-section for the hard sphere. Figure 8.5 shows plot of $\sigma_{\ell=0}^{\text{scat}}$ as a function of energy in the resonance region.

It may again be pointed out that in the derivation so far chargeless and spinless incident particle and spinless target nucleus have been considered. To use the expressions for absorption and scattering cross-sections for neutron-induced reactions it is required to multiply the respective cross-sections by the statistical factor $F(J_C)$ given by

$F(J_C) = \dfrac{2J_C+1}{2(2J+1)}$, where the compound nucleus spin $J_C = J \pm 1/2$, J being the spin of the target nucleus.

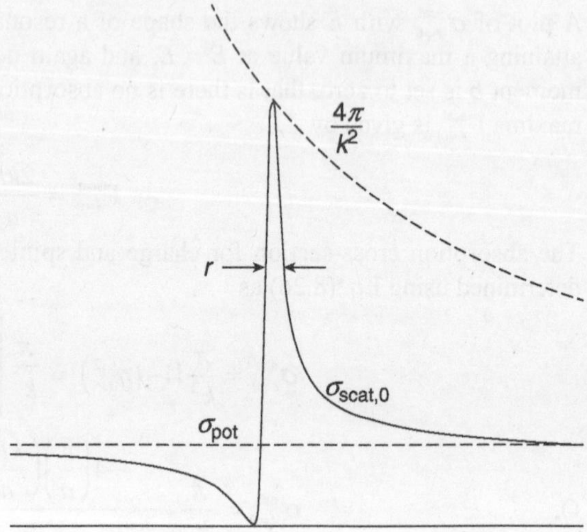

Figure 8.5 *Plot of the scattering cross-section for S-wave interaction in resonance region*

Equations (8.50) and (8.51) directly lead to the Breit–Wigner formula

$$\sigma^{\text{scat}} = \frac{g\left(\dfrac{\lambda}{2\pi}\right)^2 \pi \Gamma_{\text{scat}}^2}{(E-E_s)^2 + \left(\dfrac{\Gamma}{2}\right)^2}; \sigma^{\text{reac}} = \frac{g\left(\dfrac{\lambda}{2\pi}\right)^2 \pi \Gamma_{\text{scat}} \Gamma_{\text{reac}}}{(E-E_s)^2 + \left(\dfrac{\Gamma}{2}\right)^2} \tag{8.53a}$$

And
$$\sigma^{\text{total}} = \sigma^{\text{reac}} + \sigma^{\text{scat}} = \frac{g\left(\dfrac{\lambda}{2\pi}\right)^2 \pi \Gamma_{\text{reac}} \Gamma_{\text{scat}}}{(E-E_s)^2 + \left(\dfrac{\Gamma}{2}\right)^2} \tag{8.53b}$$

where Γ is the total width, $k^2 = \left(\dfrac{2\pi}{\lambda}\right)^2$ and the subscript $\ell = 0$ has been dropped. The factor g gives the statistical weight factor given by: $g = \dfrac{(2J_c+1)}{(2J_a)(2J_X+1)}$.

8.6 TEST OF THE COMPOUND REACTION MECHANISM: EXCITATION FUNCTIONS AND GHOSHAL'S EXPERIMENT

A graph showing the variation of cross-section with energy of the incident particle is called the excitation function. Excitation functions for many reactions induced by neutrons and charged particles in several targets spanning over the periodic table have been measured.

8.6.1 Stacked Foil Activation Technique

Stacked foil activation technique is often used to measure excitation functions for those reactions, which leave radioactive residual nuclei of measurable half-lives. In this technique, a stack consisting of several thin target foils and catcher energy degrader foils is irradiated with an accelerated beam of desired particles of desired energy. The energy of the incident beam decreases as it passes through the stack and successive foils of the stack are irradiated with incident particles of different energies. The energy of the incident ion falling on target foils can be adjusted to the desired value by taking catcher foils of suitable thicknesses. After irradiation for a time that depends on the half-life of interest, the stack is taken out and the activity induced in each foil as a result of irradiation is counted by a suitable and pre-calibrated detector. The intensity of the induced activity gives the cross-section for the reaction responsible for the production of the activity. Stacked foil technique is advantageous in many ways. First, it is non-destructive; second, cross-sections at different incident energies can be determined in a single irradiation and third, excitation functions for several reactions, each leaving a radioactive nucleus having different decay characteristics, can be determined in a single irradiation. Further, the technique requires very small amount of the material and is fairly accurate. The only limitation with the stacked-foil technique is that the residual nucleus must be radioactive of measurable half–life (figure 8.6). In those cases where the residual nucleus is stable or short-lived in-beam measurement of reaction products using particle identifying $\Delta E - E$ telescopes and so on have been done.

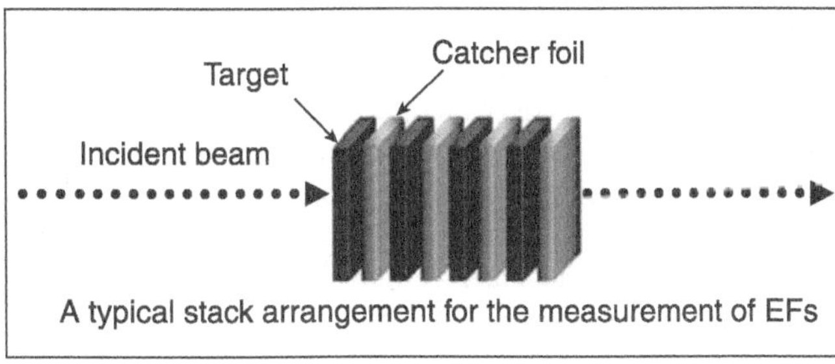

Figure 8.6 *Layout of stacked-foil technique*

The general characteristic of excitation functions for reactions leading to the continuum is a slow rise from the threshold energy reaching a maximum value and a decrease thereafter when threshold for other reactions is reached. Typical excitation functions for $(n, x\alpha)$, where $(x = 1, 2, 3)$ are shown in figure 8.7. As may be observed from the figure 8.7, excitation function for each statistical reaction starts from the threshold energy of the reaction, increases with the incident energy reaching a maximum value and then decreases with further increase in energy. The initial increase in the cross-section is due to the fact that more and more partial waves

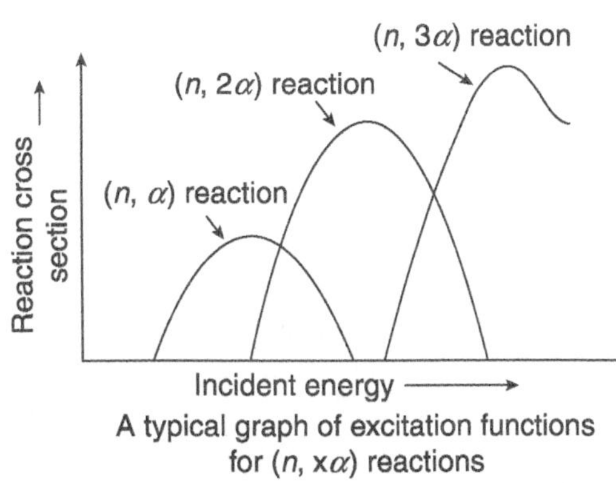

Figure 8.7 *Typical excitation functions for reactions leading to the continuum of levels in the compound nucleus*

of higher ℓ values contribute to the cross-section as the incident energy is increased. However, with the increase of the incident energy, reactions with higher threshold energy also become possible and, therefore, new exit channels open with energy. As a result, the excitation function for a given channel decreases with energy after attaining a maximum value due to the opening of new higher threshold reaction channels that take away a part of the total cross-section.

S.N. Ghoshal carried out some beautiful experiments to test the validity of compound reaction mechanism. In this set of experiments, same excited compound nucleus $^{64}_{30}Zn^*$ in the same range of excitation energy was produced through two different incident channels. Scheme of Ghoshal's experiment is shown in figure 8.8. The energies of the α and proton beams were adjusted such that the compound nucleus $^{64}Zn^*$ is formed in the same range of the excitation energy. Decay of the compound nucleus to the n, 2n, and np channels were recorded.

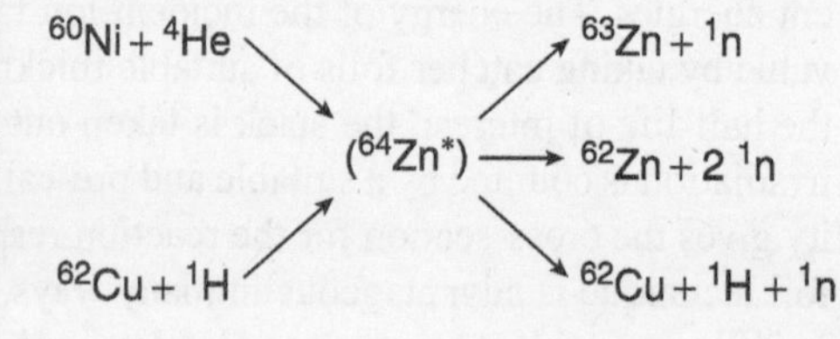

Figure 8.8 *Scheme of Ghoshal's experiment*

The pair of yield curves for the three exit channels are shown in figure 8.9, where the horizontal axis shows the energies of the α particle (bottom X-axis) and for the proton (X-axis at the top) to achieve the same excitation energy of the compound nucleus. It may be seen from these figures that the ratio σ(p, n) : σ(p, 2n) : σ(p, pn) = $\sigma(\alpha$, n) : $\sigma(\alpha$, 2n) : $\sigma(\alpha$, pn). This shows that the formation and decay of the compound nucleus is independent of each other and also that the same compound system formed by two different channel has the same branching ratio, within experimental errors, for the given decay channel.

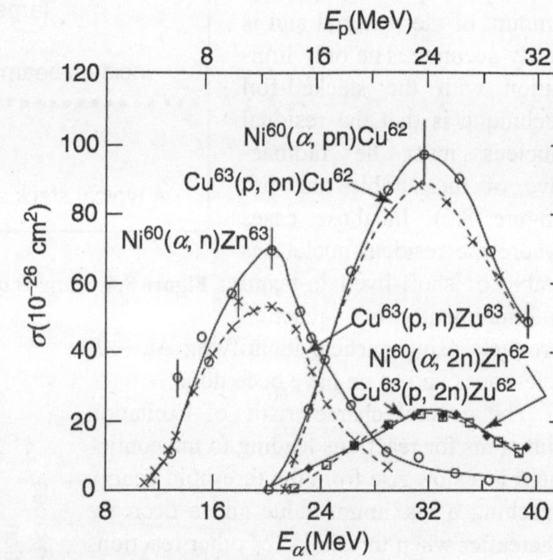

Figure 8.9 *Excitation functions of Ghoshal's experiment*

More detailed investigations of the energy spectrum of emitted charged particles in statistical nuclear reactions have further confirmed the compound reaction mechanism and have also indicated that the yields of reaction products is influenced by the angular momentum distribution of the compound nuclear states. It is possible to much the excitation energy of the compound nucleus by properly choosing the energies of the particles of the two incident channels; however, it is generally not possible to match the angular momentum distributions of compound nuclear states. For example, if J is the spin of the target nucleus then the compound nuclear states will have $J_C = J + 1/2 + \ell$ for the neutron- or proton-induced channels, while $J_C = J + 0 + \ell$ for αparticle-induced channels because of the spins of 1/2 for nucleons and 0 for α particle. It may, therefore, be observed that the spin distributions of compound nuclear states will be different for the nucleon and α particle induced channels. However, it is possible to apply corrections for the difference in J_C distributions using theoretical tools of angular momentum coupling. The compound reaction mechanism with experimental support stands on a firm footing. Further confirmation of the CN mechanism comes from the sum of the measured reaction cross-sections for all open channels, which compares well with the total absorption cross-section predicted by the CN mechanism.

In the resonance region, the measured widths of scattering and absorption resonance peaks have given information of considerable value about the compound nuclear levels and their properties. As such, the compound reaction mechanism has proved to be a tool of great value both in understanding the reaction mechanism and nuclear structure.

8.7 PRE-EQUILIBRIUM (OR PRE-COMPOUND) EMISSION IN STATISTICAL NUCLEAR REACTIONS

Figure 8.10 shows a typical energy spectrum of charged particles emitted in statistical nuclear reactions. The total spectrum may be divided into three parts. The broad peak at lower particle energies (region A) may be assigned to the compound nuclear process while the isolated peaks at higher energies corresponding to the low-lying discrete levels of the residual nucleus (part B) may be explained by the direct reaction mechanism. However, the part C of the spectrum showing particles of energy larger than what is predicted by the CN mechanism cannot be explained either by the CN or the direct reaction mechanisms. Both intuition and similar other experimental observations questioned the validity of the assumption of the CN mechanism that the compound nucleus decays only after the establishment of the thermodynamic equilibrium. It is quite likely that particles may be emitted from the intermediate states formed during the equilibration of the compound system.

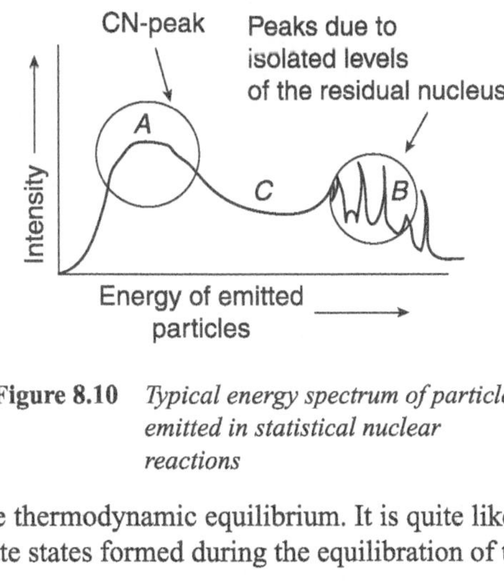

Figure 8.10 *Typical energy spectrum of particle emitted in statistical nuclear reactions*

This is termed as pre-equilibrium or pre-compound emission. Pre-equilibrium emission serves as a bridge between the compound reaction and direct reaction mechanisms. The important characteristic features of the pre-equilibrium emission are as follows:

1. Slowly descending tails of excitation functions
2. Larger number of high energy particles than what is predicted by CN mechanism
3. Forward peaked distributions of emitted particles
4. Skewed distribution of particles in angular momentum space

Several semi-classical models such as intra-nuclear cascade model, Harp-Millar and Bern model, exciton model, hybrid model, geometry-dependent hybrid model, and totally quantum mechanical models have been developed to explain the pre-equilibrium emission.

All semi-classical models assume that the initial state formed in the first interaction of the incident particle with the target nucleus is driven towards equilibrium through two body residual interactions. In exciton model (see figure 8.11), the intermediate states are defined in terms of the numbers of excited particles p and holes h. Each two-body residual interaction changes the exciton number $n\ (= p + h)$ by two. A state with given exciton number 'n' may move to a state of '$(n + 2)$' excitons or may revert back to a state of '$(n - 2)$' excitons, as a result of two-body residual interactions. However, the probability of transition to a higher exciton number state is always more than for the transition to a state of lower exciton number. It is because of the level density of states with higher exciton number is always more than that for the state with lower exciton number. For a given value of the exciton number, there are some particle–hole configurations that have particles in continuum that may be emitted. Probability of pre-equilibrium emission may be obtained by proper book-keeping of continuum to non-continuum configurations at each step of excitation. The probability of pre-equilibrium emission decreases with the complexity (i.e. the number of excitons) of the exciton state as in a state of higher complexity the energy is shared amongst larger number of excitons. Ultimately, the system attains a state of large exciton number, say N, such that the transition probability to $(N - 2)$ and $(N + 2)$ states become almost equal. This defines the complete equilibration of the compound nucleus.

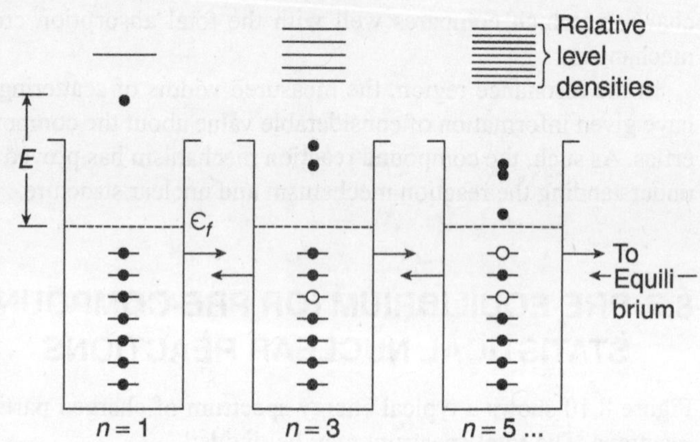

Figure 8.11 *Pictorial representation of Exciton model*

In nuclear reactions at moderate excitation energy, both the compound and the pre-compound emission occur simultaneously. The contribution of the pre-equilibrium component in the experimental data on excitation functions, energy, and angular distributions of emitted ejectiles

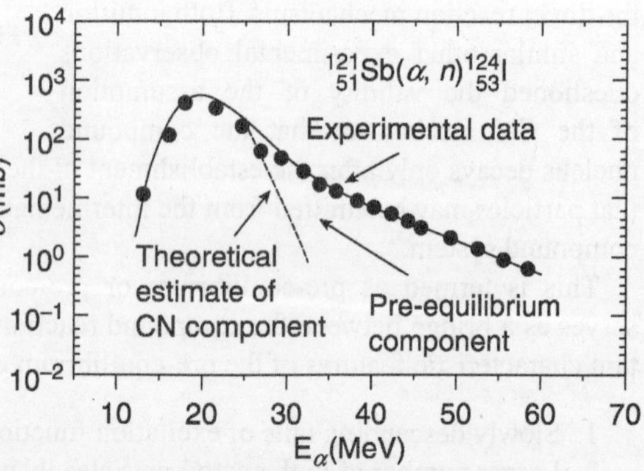

Figure 8.12 *Resolution of PE-contribution from measured excitation function*

may be obtained by subtracting the theoretically calculated contribution of the compound component and/or by finding out the enhancement in the number of higher energy ejectiles than what is predicted by the CN mechanism (figure 8.12). In angular distribution data, the presence of PE-emission is reflected in the larger number of emitted particle in forward direction than in the backward. Analysis of experimental data on PE-emission has indicated that the relative strength of PE-component increases with the increase in the energy of the incident particle. Further, at a given value of excitation energy per nucleon of the composite excited system (formed as a result of absorption of the incident ion by the target nucleus) the PE-emission is more in a bigger nucleus which has larger surface area and hence more nucleons at the surface. This finding has indicated that PE-emission, as expected, is a surface phenomenon.

Initial exciton number n_i $(=p_i + h_i)$, which is the number of excitons created in the first interaction between the incident particle and the target nucleus, is an important parameter that governs the relative strength of pre-equilibrium component over the compound component. In most of the computer codes based on exciton model of PE-emission, initial exciton number is treated as an adjustable free parameter. A small value of n_i means that the excited composite system will pass through a larger number of intermediate stages before attaining thermal equilibrium. As each intermediate stage contributes to PE-emission, a smaller value of initial exciton number results in larger PE strength. A larger value of n_i increases the relative strength of CN component over the PE-component. The pre-equilibrium fraction F_r, defined as the ratio of the PE-component to the total of (PE + CN) is found to increase with the excitation energy per nucleon as well as the number of nucleons at the surface of the composite system.

Feshback–Kerman–Kooni (FKK) quantum mechanical theory of pre-equilibrium emission assumes that a nuclear reaction takes place in series of stages corresponding to the interaction of the incident particle with the nucleons of the target nucleus. At each stage, nucleons are excited to higher states and may be emitted. This corresponds to pre-equilibrium emission. The theory distinguishes between two types of chains: the P-chains in which the particles after first interaction always remain in continuum and Q-chains in which the particles always remain in bound state after the first interaction. The P-chains dominate at higher energies and give rise to cross-sections peaked in the forward direction, while the Q-chains become important at lower energies giving rise to cross-sections symmetrical at 90° in centre of mass. The interactions in P-chain are called multistep direct (MSD) and those in Q-chains multistep compound (MSC) (figure 8.13).

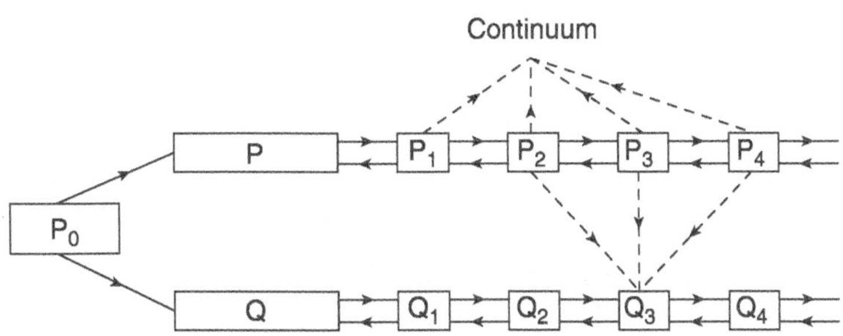

Pictorial representation of FKK model P-chain represents MSD process and Q-chain represents MSC process

Figure 8.13 *Pictorial representation of P and Q chains*

Both semi-classical and quantum mechanical theories of

pre-equilibrium emission have been successful to the same degree in describing the experimental data on pre-equilibrium emission. Recently, pre-equilibrium emission has been observed in heavy ion reactions at energies as low as 7 MeV per nucleon.

Solved example S-8.3

A plot of neutron cross-section versus energy for an even–even nucleus shows resonance at energy 200 eV in the total cross-section with a peak cross-section of 1400 b. The width of the peak is 20 eV. Determine the partial width for scattering at this resonance.

Solution. As the target nucleus is even–even, its spin is zero. Hence, the statistical weight factor $g = 1$. Now, from Breite–Wigner formula, the total cross-section at resonance $(E = E_s)$ is given by $\sigma_{total} = \dfrac{\pi\left(\dfrac{\lambda}{2\pi}\right)^2 \Gamma_{scat}\Gamma}{\dfrac{\Gamma^2}{4}} = \dfrac{\lambda^2\Gamma_{scat}}{\pi\Gamma}$ or $\Gamma_{scat} = \dfrac{\pi\Gamma\sigma_{total}}{\lambda^2} = \dfrac{3.14\times 20\ (\text{eV})\times 1400\times 10^{-24}\ (\text{cm}^2)}{\lambda^2} = \dfrac{8.792\times 10^{-20}\ (\text{eV}-\text{cm}^2)}{\lambda^2}$.

In the expression λ is the wavelength of a 200 eV neutron $= \dfrac{h}{\sqrt{2M_n E}}$.

$$\lambda = \frac{6.62\times 10^{-34}\ (J-s)}{\sqrt{2\times 1.008665\times 1.66\times 10^{-27}\times 200\times 1.6\times 10^{-19}}} = 2.86\times 10^{-10}\ \text{cm}$$

Substituting this value of λ, we get $\Gamma_{scat} = 1\text{eV}$.

Ans: The scattering width at the resonance is 1 eV.

Exercise p-8.6: Give a brief account of the compound reaction mechanism and derive expressions for the formation and the decay of the compound nucleus.

Exercise p-8.7: What is pre-equilibrium emission? What are its characteristic features?

Exercise p-8.8: Briefly discuss exciton model of pre-equilibrium emission.

Exercise p-8.9: Discuss Ghoshal's experiment as a test for CN-mechanism.

Exercise p-8.10: Drive expressions for the scattering and absorption cross-sections in resonance region for chargeless and spinless particles of relative angular momentum $\ell = 0$.

Exercise p-8.11: Show that scattering without reaction is possible whereas reaction without scattering is not.

8.8 HEAVY ION (HI) REACTIONS

Charged particles heavier than α particle are termed as heavy ions (HI). Initial interest in heavy ion (also called complex nuclei) reactions was due to the possibility of creating of super heavy nuclei (SHE) from their fusion. Another reason of interest in complex nuclei reactions is the population of high angular momentum states in such reactions. We know that the angular momentum carried by an incident particle depends on its linear momentum p (=mv) and the impact parameter. An accelerated heavy ion in peripheral interaction with another target heavy ion may impart large angular momentum to the fused system. States with angular momentum of 50 $\hbar$ or more, which cannot be reached in light ion reactions, can be easily produced in heavy ion reactions. With the view to study high spin states and many other interesting phenomena associated with HI reactions, large number of heavy ion accelerators has come up in the world. It is now possible to accelerate all ions starting from boron up to uranium. These HI-accelerators are capable of delivering sufficient HI-current and have very good energy resolution. In India, at present, there are three HI-machines, a cyclotron at Variable Energy Cyclotron Centre (VECC), Kolkata, a 15-MV Tandem Van-de-Graf accelerator, called 15-UD palletron at Inter University Accelerator Centre (IUAC), New Delhi and a 14-MV Tandem at TIFR-BARC, Bombay. To utilize these facilities to the fullest extent and for carrying out internationally competitive research work, many more auxiliary national facilities such as Heavy Ion Reaction Analyser (HIRA), Gas-filled Hybrid Recoil Analyser (HYRA), Indian National Gamma Analyser (INGA), and Charged Particle Detection Array (CPDA) have been developed. Information of considerable value regarding the HI reaction mechanism and nuclear structure has been obtained from experimental and theoretical work done in the country during the last two decades or so.

Quantum mechanical description of heavy ion interactions is quite involved as both the target and the incident ions are many-body quantum systems. However, on account of the large mass and momentum of the incident heavy ion, its de Broglie wavelength is very small. As such, it is possible to describe the trajectories of the incident ions classically and treat their interaction quantum mechanically through a potential V, which is the sum of Coulomb and nuclear potentials with centrifugal barrier (figure 8.14).

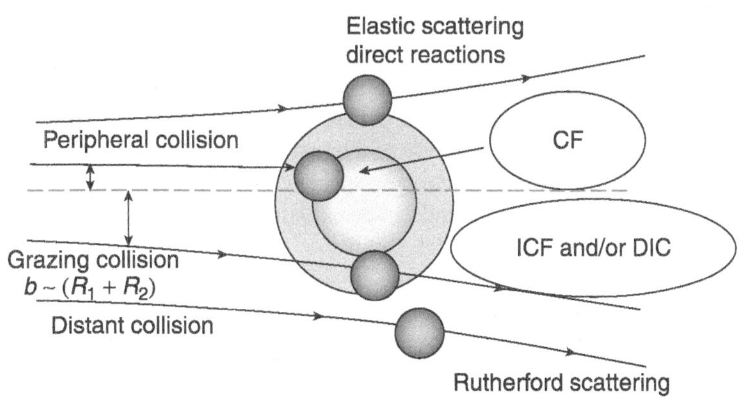

Figure 8.14 *Pictorial representation of heavy ion reactions*

The interactions between two heavy ions may be classified in terms of the distance of closest approach b between them (figure 8.14). If $b >> (R_1 + R_2)$, where R_1 and R_2 are the radii of the two ions, they may interact only through their Coulomb fields resulting in Rutherford scattering. For $b \cong (R_1 + R_2)$, the collision is called peripheral collision and elastic scattering is the dominant process. When the incident ion approaches the target ion more closely ($b < R_1 + R_2$) and both feel the presence of their nuclear fields, deep inelastic scattering and incomplete fusion of the incident ion with the target may take place. Still, smaller impact parameter b leads to central collisions and complete fusion of the incident and target ions.

As has already been emphasized, it is not possible to have a bear nucleus, particularly of a heavy atom. In heavy ion accelerators, a beam of the desired heavy atoms is first ionized producing a distribution of different charged states of these atoms. The atomic ions with different charged states are then passed through a suitable magnetic sector that separates out ions with different charges. Thus, a beam containing ions of a particular charge is further accelerated to the desired energy. In a tandem Van de Graff type accelerator, if V is the voltage at the terminal of the accelerator and q is the charge state of the accelerated ion then the ultimate energy of the ion coming out of the accelerator will be $(1 + q)$ eV.

8.8.1 Complete and Incomplete Fusion of Heavy Ions

Out of the many interesting phenomena associated with HI reactions the one which has got much attention during the last decade or so is the so-called incomplete fusion (ICF) at low energies. In heavy ion reactions at moderate and low energies, it is expected that the incident and the target ions will completely fuse (CF) in central and peripheral collisions making an excited compound nucleus. The compound nucleus so formed is likely to follow its normal course of decay as shown in figure 8.15.

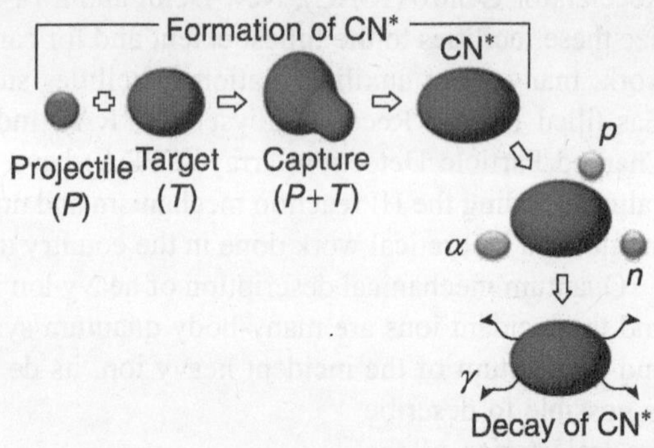

Figure 8.15 *Sequence of events in complete fusion of heavy ions*

However, what has been observed is the partial fusion, that is a part of the incident ion fuses with the target and the remaining part moves out with the velocity of the incident ion without any interaction as spectator. Incomplete fusion was found to occur in HI reactions at very high energies and can be explained in terms of the very small de Broglie wavelength of the incident ion and the small compressibility of nuclear matter at those energies. However, these arguments do not hold good at low energies. Further, it has been observed that incident ions such as $^{12}_{6}C$ and $^{16}_{8}O$ that have cluster structure exhibit large component of ICF even at low energies of the order of <10 MeV per nucleon.

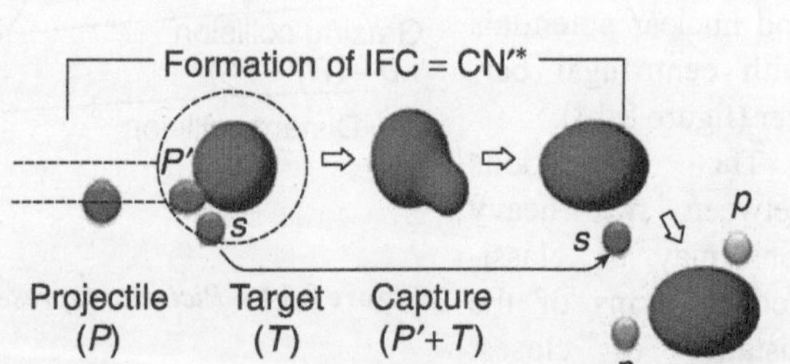

Projectile(P) = participant(P′) + spectator(S)
Incompletely fused composite = IFC
CN formed via ICF = CN′*

Figure 8.16 *Pictorial representation of in-complete fusion of heavy ions*

Figure 8.16 shows the expected sequence of events in ICF. It is assumed that the incident ion breaks into its component clusters when it enters the nuclear field of the target nucleus. One part of the incident ion fuses with the target nucleus forming the ICF-compound nucleus that first decay by particle emission and when that is not possible, the excitation energy is released in the form of γ rays. The un-fused part of the incident ion moves as spectator without any interaction.

As is evident from figures 8.15 and 8.16, the residues of both complete and incomplete fusion are present in the exit channel. It is also possible that both the CF and the ICF interactions produce the same residue. As an example, let us consider the reaction $^{16}O + {}^{169}Tm$. In the case of complete fusion, the compound nucleus $^{186}Ir^*$ will be formed that may decay by the emission of 2 αs, one proton and one neutron leaving the residual nucleus ^{175}Hf. The complete fusion reaction may be written as

$$^{16}O + {}^{169}Tm \rightarrow {}^{185}Ir^* \rightarrow {}^{175}Hf + 2\alpha np \tag{8.54a}$$

In addition, ^{16}O may break into (i) $^{12}C + {}^{4}He$ or into (ii) $^{8}Be + {}^{8}Be$. In case (i), if ^{12}C fuses with ^{169}Tm and ^{4}He goes out as spectator, then again ^{175}Hf may be produced by the following reaction:

$$^{12}C + {}^{169}Tm \rightarrow {}^{181}Re^* \rightarrow {}^{175}Hf + \alpha np \tag{8.54b}$$

In case (ii), when ^{16}O breaks into two ^{8}Be and one of them fuses with ^{169}Tm, the same residual nucleus ^{175}Hf may be produced by the following reaction:

$$^{8}Be + {}^{169}Tm \rightarrow {}^{177}Ta^* \rightarrow {}^{175}Hf + np \tag{8.54c}$$

We see that the same residual nucleus ^{175}Hf may be produced by complete fusion (Eq. (8.54a)) as well as through two in-complete fusion channels represented by Eqs. (8.54b) and (8.54c)). In an actual experiment, when a target of ^{169}Tm is irradiated by an accelerated beam of ^{16}O, residual nuclei ^{175}Hf formed by all the three processes will be present in the exit channel. It may, however, be noted that the ^{175}Hf nuclei that are product of CF (Eq. (8.54a)) will have largest linear momentum as in CF all the linear momentum of the incident ^{16}O ions is transferred to the compound nucleus $^{185}Ir^*$ and in turn to the CF residue. As such, the complete fusion residue ^{175}Hf will have largest kinetic energy. In case of the ICF of ^{12}C, only a part of the incident linear momentum will be transferred to the compound nucleus $^{181}Re^*$ and in turn to ^{175}Hf residues produced by ICF channel represented by Eq. (8.54b). The residue ^{175}Hf formed via ICF channel of Eq. (8.54b) will have smaller kinetic energy as compared to the same residues formed by CF channel (8.54a). In a similar way, the residue ^{175}Hf formed via the ICF channel Eq. (8.54c) will have least kinetic energy. If the residues formed in the irradiation of ^{169}Tm by ^{16}O beam are allowed to pass through some absorber material, the most energetic ^{175}Hf formed via CF reaction will travel the largest thickness, and the ICF products will travel less distances in the absorber. This essentially means that the recoil ranges of reaction products will be different in the absorber medium depending on the mode of their formation. As ^{175}Hf is radioactive, it may be identified through its characteristic decay properties such as the half-life and the γ radiations emitted by it. Figure 8.17 shows the recoil range distribution of ^{175}Hf residue in aluminium absorber. The observed distribution can be resolved into three distinct peaks corresponding to the CF and two ICF channels. As the area under the peaks is proportional to the probability of the process responsible for the peak, it is possible to determine the relative contributions of CF and ICF channels.

8.8.2 Fusion–Fission and Other Reactions

Fusion–fission reactions constitute another important component of HI interactions. Complete fusion of the two interacting heavy ions forms an excited composite system with high spin that may undergo fission. Investigation of such reactions provides an important tool to study the *spin dependence* of fission. Deep inelastic scattering and few nucleon transfer reactions have also been investigated using HI reactions.

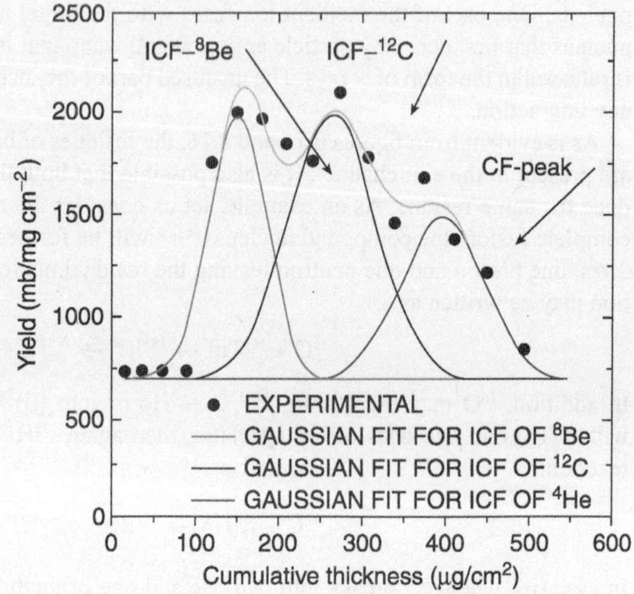

Figure 8.17 *Recoil range distribution of ^{175}Hf formed via CF and ICF channels*

Solved example S-8.4

A beam of $^{16}_{8}$O ions of charged state +6e is accelerated in a tandem Van de Graff accelerator with terminal voltage of 13 MV. In two separate experiments with this beam, thin foils of $^{27}_{13}$Al and $^{197}_{79}$Au were irradiated. For these two cases, calculate (i) kinetic energy of the beam, (ii) the Coulomb barriers, (iii) CM-energy of the incident ions, and (iv) the maximum possible value of the spin that may be imparted to the systems. Consider $r_0 = 1.2$ F.

Solution.

(i) Kinetic energy of the beam in lab.

$$E_{\text{lab}} = (6+1)\times 13\,\text{MeV} = 91\,\text{MeV}$$

(ii) Coulomb barrier for Al: $V_{\text{C}}^{\text{Al}} = 1.44\dfrac{8\times 13}{1.2\left(16^{1/3}+13^{1/3}\right)} = 22.32$ MeV

Coulomb barrier for Au: $V_{\text{C}}^{\text{Au}} = 1.44\dfrac{8\times 79}{1.2\left(16^{1/3}+197^{1/3}\right)} = 90.87$ MeV

(iii) Centre of mass energy for Al case: $E_{\text{Al}}^{\text{CM}} = 91\times\dfrac{13}{16+13} = 40.79\,\text{MeV}$

Centre of mass energy of Au case: $E_{\text{Au}}^{\text{CM}} = 91\times\dfrac{179}{16+179} = 83.53\,\text{MeV}$

(iv) We calculate $v/c = \beta$ for the two cases using $\beta = \sqrt{\dfrac{2E^{\text{CM}}\,(\text{MeV})}{m\,(\text{in MeV})}}$ and then calculate $J_{\max}$ for the two cases using the relation $J_{\max} = \dfrac{mc^2\,(\text{MeV})\times\beta\times b\,(\text{in Fermi})}{\hbar c\,(\text{in MeV}-\text{m})}$.

$\beta^{Au} = 10.59 \times 10^{-2}$ and $\beta^{Al} = 7.4 \times 10^{-2}$

$$J^{Al}_{max} = \frac{16 \times 931 \times 7.4 \times 10^{-2} \times 5.59 \times 10^{-15}}{197.3 \times 10^{-15}} = 31.23 \approx 31;$$

$$J^{Au}_{max} = \frac{16 \times 931 \times 10.59 \times 10^{-2} \times 8.35 \times 10^{-15}}{197.3 \times 10^{-15}} = 66.76 \approx 66$$

We thus see that up to $31\hbar$ and $66\hbar$ of angular momentum may be pumped in the two systems by the oxygen beam.

Exercise p-8.12: What additional information can be obtained from the study of heavy ion interactions that cannot be obtained from the light ion reactions?

Exercise p-8.13: Give a brief description of the classification of HI-reactions and discuss one important mode of reaction in details.

Exercise p-8.14: Write a note on stacked foil activation technique and discuss some of its advantages and limitations.

Exercise p-8.15: Describe in short the complete and incomplete fusion in heavy ion reactions.

8.9 OPTICAL MODEL APPROACH TO NUCLEAR REACTIONS

Quantum mechanically nuclear interactions are described in terms of a potential $V(r)$ that the projectile faces as it approaches the target nucleus. While describing elastic scattering it was assumed that the potential changes the phase of the outgoing partial waves without any loss of the incident flux, the incident flux was only redistributed. In nuclear reactions, a part of the incident flux is lost or absorbed. One way of including the reduction of incident flux is by multiplying the amplitude of each outgoing partial by a factor η_ℓ (<1). H. Feshback and co-workers in 1953 suggested that absorption of incident flux may also be achieved if the potential $V(r)$ has an imaginary part. Accordingly, the nuclear potential $V(r)$ may be written as

$$V_{total}(r) = V(r) + iW(r)$$

Here, $V_{total}(r)$ denotes the complex potential with real part $V(r)$ and the imaginary part $W(r)$. The effect of including the imaginary term is that the incident wave *penetrates into the nuclear region as an exponentially decaying wave*. As such, the nucleus behaves as a translucent body and the interaction of the incident wave with the target nucleus may be compared to the interaction of a light beam with a crystal ball, and hence, the name optical model.

A general optical model potential contains the following five terms:

$$V_{total}(r) = V_R(r) + V_I(r) + V_D(r) + V_S(r) + V_C(r) \tag{8.55a}$$

The first term $V_R(r)$ may be given by

$$V_R(r) = -Vf(r, R, a) = -\frac{V}{\left(1 + \exp\left(\frac{r-R}{a}\right)\right)} \tag{8.55b}$$

where V is the depth of the real part of the potential and $f(r, R, a)$ is the Woods–Saxon form factor given by

$$f(r,R,a) = \frac{1}{\left(1+\exp\left(\frac{r-R}{a}\right)\right)}$$

R being the radius of the real potential and 'a' the diffuseness parameter.

The second and third terms constitute the imaginary part of the potential and are given by

$$V_I(r) = -iWf(r,R_I,a_I) \text{ and } V_D(r) = 4i\,a_I W_D \frac{df(r,R_I,a_I)}{dr}$$

Here, W and W_D are the depths of the imaginary potentials with R_I and a_I their corresponding radius and diffuseness. The two imaginary terms are complementary of each other to the extent that $V_I(r)$ becomes important for high-energy incident particles, which may travel deep inside the nucleus and may give rise to absorption in the volume of the target nucleus. The $V_D(r)$ peaks at the nuclear surface and is important at lower energies giving rise to surface absorption.

The fourth term is the spin–orbit term that considers the polarization effects. A beam, which is initially un-polarized, that is contains particles with randomly oriented spins, gets partially polarized after the interaction. This effect is taken care by this term which is given as

$$V_S(r) = s\cdot\ell V_S \frac{df(r,R_S,a_S)}{dr}$$

The fifth and the last term is Coulomb term. It has different values inside and outside the Coulomb radius of the nucleus.

$$V_C(r) = \frac{Z_1Z_2e^2}{2R_C}\left(3-\frac{r^2}{R_C^2}\right) \quad \text{for} \quad r \le R_C$$

$$= \frac{Z_1Z_2e^2}{r} \quad \text{for} \quad r > R_C$$

Here, R_C is the Coulomb radius of the nucleus.

Figure 8.18 shows the experimental and theoretically calculated values of angular distribution of protons of 17 MeV centre of mass energy by various nuclei in the range of $Z = 26$–30. The five-term optical potential mentioned earlier has been used for these calculations.

The optical model has a limited set of adjustable parameters and is not capable to describe abrupt variations in the cross-sections, as it happens for isolated resonances. However, it can give a good description of the cross-sections in the presence of the oscillations of large width in the continuous region, as it treats these as wave phenomena.

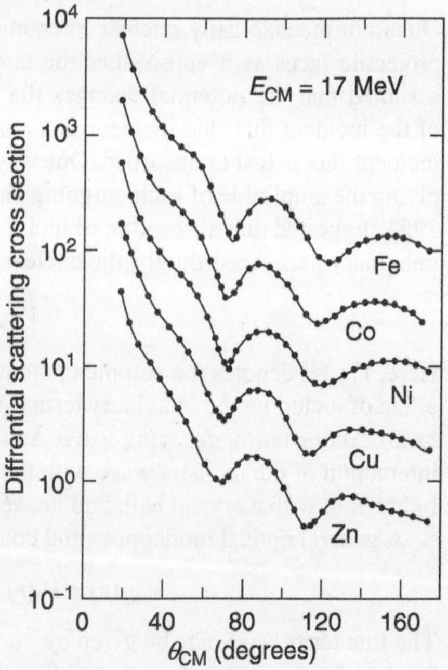

Figure 8.18 *Angular distribution of 17 MeV protons from nuclei with Z = 26–30*

8.10 DIRECT REACTION MECHANISM

Direct reactions are those in which the interacting nuclei make glancing contact and then separate out immediately. The incident projectile may exchange some energy and/or angular momentum or one or two nucleons are transferred from it or to it from the target. The reaction takes place at the nuclear surface or near to it with large impact parameter. Direct reaction products are strongly forward peaked. These reactions take place at short time scale that decreases with the increasing energy of the projectile. Time scale of direct reaction is of the order of 10^{-22} s, the time that a nucleon takes in crossing a typical nucleus. As only localized heating of the target nucleus takes place in direct reactions, only a few degrees of freedom are involved in these reactions in comparison to large degrees of freedom involved in compound nucleus reactions. Analysis of the experimental data has indicated that at moderate energies a given reaction proceeds both with CN and direct reaction mechanism. The relative strengths of the CN and direct processes depend strongly on the incident energy and also on other parameters.

There are two characteristic types of direct reactions. In the first type, the incident particle scatters in-elastically and the transferred energy is used to excite a collective mode of the nucleus. Rotational and vibrational bands can be studied in this way. The second type involves a modification in the nuclear composition. Examples are transfer of nucleons, as pick-up and stripping reactions. An important reaction of the latter kind is a knock-out reaction where the incident particle knocks out a particle of the target nucleus and continues in its path, resulting in three reaction products. Reactions with nucleon exchange can also be used to excite collective states. An example is a pick-up reaction where a projectile captures a neutron from a deformed target and the product nucleus is in an excited state belonging to a rotational band.

Direct reactions exhibit a peculiar form of angular distribution, which allows us to extract information on the reaction mechanism with the employment of simple models. Typical examples are the stripping reactions (d, n) and (d, p), where the angular distribution of the outgoing nucleon presents a forward prominent peak and smaller peaks at larger angles, with the characteristic aspect of a diffraction phenomena. As an example, let us consider the stripping reaction X(d, p)Y, where a deuteron of energy E_d hits a stationary and heavy target X that does not recoil, transfers a neutron to the target and is scattered in a direction θ with an energy equal to the incident energy plus the Q-value of the X(d, p)Y reaction. The initial and the final stages of the reaction are shown in figure 8.19a, b, while figure 8.19c shows the linear momentum conservation triangle, where K_d, K_p, and K_{trns} are, respectively, the linear momentums of the incident deuteron, the spectator proton, and the linear momentum carried by the neutron to the target nucleus.

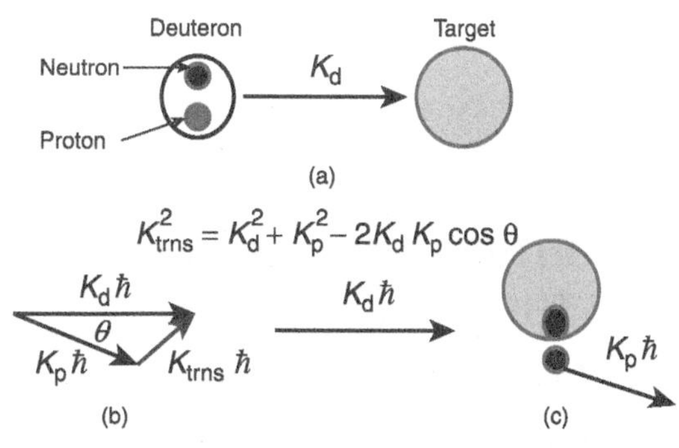

Figure 8.19 *(a) The initial state, (b) the final state, (c) linear momentum conservation triangle*

Suppose that the neutron is captured in a single particle state very near to the surface of the target nucleus at a distance R from the centre of the nucleus, where R is the radius of the nucleus, then the orbital angular momentum ℓ of the neutron in nucleus Y will be $K_{\text{trns}} \cdot R$. In semi-classical approach, the orbital angular momentum ℓ should be quantized, and therefore,

$$K_{\text{trns}} \cdot R = \sqrt{\ell(\ell+1)} \tag{8.56a}$$

Also from the linear momentum conservation triangle (figure 8.19c), we have

$$K_{\text{trns}}^2 = K_d^2 + K_p^2 - 2K_p K_d \cos\theta \tag{8.56b}$$

From Eq. (8.56b), it is clear that depending on the angle of emission θ of the proton, there may be different values of K_{trns}, and corresponding to each value of K_{trns} there will be a discrete value (Eq. (8.56a)) of orbital angular momentum ℓ. As such each angle θ for proton emission may be associated with the orbital angular momentum of the state in which the neutron has been captured by the residual nucleus Y. More specifically, for the case of ${}^{45}_{21}$Sc (d,p) ${}^{45}_{21}$Sc reaction, the Q-value is ≈ 6.5 MeV and for deuterons of about 7 MeV the value of K_d is ≈ 0.82 F^{-1}. The energy of outgoing proton is ≈ 7 + 6.5 = 13.5 MeV and the corresponding K_p ≈ 0.79 F^{-1}. Taking the radius of ^{46}Sc to be ≈ 6 F, one gets using Eqs. (8.56a) and (8.56b)

$$\theta \cong 0, \quad \ell = 0; \theta \cong 16°, \ell = 1; \; \theta \cong 30°, \; \ell = 2; \theta \cong 42°, \ell = 3$$

In actual experiment, there were peaks in the angular distributions of protons at 17° and at 42°, corresponding to orbital angular momentums of 1 $\hbar$ and 3 $\hbar$. Accordingly, the neutron has been captured in the p- or f-states. Also, according to the shell model, the 21st neutron of ^{45}Sc may go to the f-state. In this way, the simple semi-classical approach of stripping reactions may be used to probe the structure of residual nuclei.

Exercise p-8.16: Write a note on the optical model of nuclear reactions. In what respect is the optical model approach different from the usual way of phase shift analysis and of multiplying the amplitude of outgoing wave by a complex factor to take into account the absorption of incident wave.

Exercise p-8.17: Explain giving a suitable example how information about the energy levels of the residual nucleus can be obtained from the stripping reactions.

Exercise p-8.18: Bring out the differences between the compound, pre-equilibrium, and direct reaction mechanisms.

8.11 INTERACTION OF ELECTROMAGNETIC RADIATIONS WITH THE NUCLEUS

8.11.1 Coulomb Excitation

Coulomb excitation is an inelastic scattering in which the Coulomb field of a charged particle (a nucleus) excites the other nucleus. The Coulomb field of the projectile nucleus may be decomposed into its electric and magnetic multipoles such as E1, E2, M1, …, which carry well-defined angular momentum and parity. At low bombarding energies, E2 (quadrupole) excitations are more common, while at higher energies E1 excitations dominate. As an example of a low-energy reaction, let us consider the excitation of a quadrupole state in a head-on collision below the Coulomb barrier, that is for a situation in which the projectile decelerates as it approaches the target and stops before reaching in the range of the nuclear interaction, re-accelerating backwards after that. The differential cross-section is given by the product of the Rutherford differential cross-section at 180° and the probability of transition of the target from state i to state f, along the trajectory, measured by the square of A_{if}.

$$\left[\frac{d\sigma}{d\Omega}\right]_{(\theta=180)} = \left[\frac{d\sigma_{\text{Ruther}}}{d\Omega}\right]_{(\theta=180)} \times |A_{\text{if}}|^2 \tag{8.57a}$$

The square of A_{if} measures the transition probability from i to f and this probability should be integrated along the trajectory. A simple calculation for A_{if} can be carried out in the case of the excitation from the ground state $J = 0$ of a deformed nucleus to an excited state with $J = 2$. The perturbation P comes, in this case, from the interaction of the projectile charge $Z_a e$ with the quadrupole moment of the target nucleus. This quadrupole moment works as an operator that acts between the initial and final states, that is

$$P = \frac{1}{2}\frac{Z_a e^2 Q_{\text{if}}}{r^3} \tag{8.57b}$$

where r is the projectile–target separation distance and

$$Q_{\text{if}} = \sum_{\text{i}} \int \psi_{\text{f}}^* \left(3Z_{\text{i}}^2 - r_{\text{i}}^2\right)\psi_{\text{i}}\, d\tau \tag{8.57c}$$

Here, the sum extends to all protons at the positions $r_{\text{i}} = (x_{\text{i}}\ y_{\text{i}}\ z_{\text{i}})$ in the target. The amplitude is then given by

$$A_{\text{if}} = \frac{4Q_{\text{if}}E_{\text{a}}^2}{3Z_{\text{a}}e^2\hbar v_{\text{a}}Z_A^2} \tag{8.57d}$$

Finally, if the values of differential Rutherford scattering cross-section at $\theta = 180°$ and A_{if} are substituted in Eq. (8.57a), we may obtain

$$\frac{d\sigma}{d\Omega_{(\theta=180°)}} = \frac{M_{\text{a}}E_{\text{a}}\,|Q_{\text{if}}|^2}{18\hbar^2 Z_A^2} \tag{8.57e}$$

Equation (8.57e) indicates three important facts, which are as follows:

1. $\frac{d\sigma}{d\Omega_{(\theta=180°)}}$ is directly proportional to M_{a} the mass of the incident ion. It means that heavy ions are more suited for Coulomb excitation as compared to the light ions.
2. $\frac{d\sigma}{d\Omega_{(\theta=180°)}}$ is directly proportional to the kinetic energy of the projectile E_{a}.
3. $\frac{d\sigma}{d\Omega_{(\theta=180°)}}$ does not depend on the charge of the incident ion.

In case where the initial and final state wave functions are similar, as in case of the excitation to the first rotational level, the quadrupole operator Q_{if} becomes equal to the intrinsic value of the quadrupole moment. For such transitions, measured value of the differential Coulomb excitation cross-section at $\theta = 180°$ may be used to determine the quadrupole moment of the nucleus.

8.11.2 Photo-disintegration and Giant Resonance Excitations

When a nucleus is hit by a mono-energetic beam of γ rays (photons) the γ absorption cross-section depends strongly on the energy of the incident γ rays. If the energy of the γ ray exceeds the separation energy of nucleons sharp resonances corresponding to the nucleon emission appear in γ ray excitation function. However, on further increasing the energy of γ rays, a peculiar resonance is found to occur for every nucleus at around 25 MeV for light nuclei and around 15 MeV for heavier ones. The resonance energy E_{GDR} decreases as $80/A^{1/3}$. This resonance is rather broad with a width of about 3–5 MeV, and of about 7 MeV in some cases. These resonances that appear for all nuclei over the periodic table are called giant resonances and are due to collective motion of neutrons and protons in the nucleus. The time varying electric field of the photon is very effective in inducing a collective motion of protons against the neutrons in the nucleus.

Giant electric dipole resonance (GDR) is excited when angular momentum of 1 $\hbar$ is transferred to the nucleus, for example an even–even nucleus is excited to the 1^- state. The isospin state of the nucleus is also changed by $\Delta T = 1$, and due to that this is called the iso-vector resonance. The photons can also excite but less effectively the iso-vector resonance for which $\Delta T = 0$. However, electric monopole ($\Delta \ell = 0$) and electric quadrupole $(\Delta \ell = 2)$ resonances are produced in charged particle reactions. In electric giant quadrupole resonance, a nucleus, which in its ground state is spherical, oscillates between spherical and ellipsoidal shapes. If neutrons and

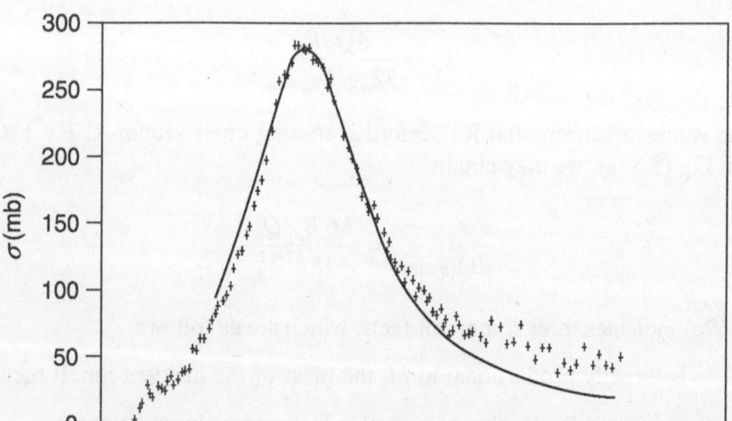

Figure 8.20 *Giant electric dipole resonance in the absorption of γ rays by ^{120}Sn*

protons oscillate in phase, the resonance is iso-scalar ($\Delta T = 0$) and if they oscillate out of phase, the resonance is iso-vector ($\Delta T = 1$). The case of electric monopole resonance is very typical; in this case, the shape of the nucleus does not change but its volume increases and decreases giving rise to what is called the 'breathing mode'.

Apart from the giant electrical resonances, there are also magnetic resonances. In electric resonance, the shape of the nucleus changes; however, in magnetic resonance, the nucleons with spin up vibrate with respect to the nucleons with spin down. The number of nucleons involved in such motion cannot be very large because of the Pauli's exclusion principle. Magnetic resonances can also be separated into iso-scalar when nucleons of spins in one direction oscillate with respect to the nucleons of spin directed oppositely. In magnetic iso-vector resonance, protons of spin up and neutrons of spin down vibrate against their counterparts of opposite spins. A third type of giant resonance called Gamow–Teller giant resonance appears in reactions of type (d, p) and (d, n) reactions. A typical electric giant dipole resonance observed in the absorption of γ rays by ^{120}Sn is shown in figure 8.20.

8.12 NUCLEAR REACTIONS USING RADIOACTIVE ION BEAMS

Research in nuclear physics initiated with the aim to understand the nature of the basic nuclear force that binds nucleons inside the nucleus has paid many beneficial returns to the society in the form of nuclear medicines, NMR imaging, accelerator-based radiation therapy, ion-implantation technique for modification of properties of solids, carbon dating technique to determine the age of archeologically important artefacts etc, to name a few. This is a beautiful example of how investments in basic science generate applications that are beneficial to the society.

Most of these applications have become possible by accelerating stable or long-lived isotopes numbering more than 250 in suitable accelerators. Further advancement in techniques has made it possible to accelerate radioactive unstable nuclides. The number of such radioactive candidates may be much larger ≈ 6000 or so. They are the nuclei that lie near the so-called drip lines in the chart of nuclides. It is expected that research with these new tools will also prove beneficial to the society.

The study of nuclear interactions initiated by unstable nuclei is important to understand the role played by such rare reactions in the early stages of the evolution of the universe and also in the birth and death of stars. These reactions are also expected to further enhance our knowledge of the true nature of nucleon–nucleon force.

As we know, the stability of a nucleus is determined by the N/Z ratio of the nucleus. For light stable nuclei, N/Z has a value around 1. In case of uranium, which is the heaviest nucleus of very large half-life, the ratio is ≈ 1.6. Light nuclei that have N/Z value much different than 1 and heavy nuclei with $N/Z > 1.6$ are unstable and are termed as exotic nuclei. These nuclei lie beyond the valley of stability near to the proton and or neutron drip lines. To create a radioactive ion beam (RIB) facility, it is necessary to produce desired radioactive (unstable) ions in sufficient number for a considerable period of time, separate them from other undesired nuclei, and accelerate them for experiments. All these steps need to be done quickly, before a large component of the radioactive beam decays out. Although there are several methods for producing exotic nuclei, the four important are as follows:

1. *Nuclear spallation reaction*: In this method, a very high energy proton or other stable light ion beam is made to hit a solid heavy target. As a result, many stable and exotic nuclei and nuclear particles are created. Often, the exotic nuclei of interest remain within the target,

particularly if the target is thick. Heating and other methods are used to extract the desired radioactive ions, they are then mass analysed and are further accelerated.

2. *Fusion of stable ions*: Exotic ions are also produced in the fusion–fission reactions of heavy stable ions. Again, separation of the desired unstable ion and further acceleration is required in this case as well.
3. *Self-fission and induced fission*: Exotic nuclei are continuously formed in the self-fission of ^{252}Cf and other self-fission nuclei. Strong ^{252}Cf self-fission source is often used to produce radioactive ion beams.
4. *Fragmentation of heavy ions*: When an high-energy uranium ion, called participant, undergo central collision with the target a part of it get scrapped off and forms a highly excited mixture of nucleons with some nucleons of the target nucleus. A part of the projectile called spectator flies away with the same velocity as the incident beam. Small spectator fragments that are formed in such collisions have N/Z values much different than 1, which is characteristic of light stable nuclei. Statistically, a large number of exotic fragments far away from the line of stability are produced in such collisions. These nuclei may be collected in secondary beams and can be further accelerated to induce reactions in targets. To induce fragmentation of the heavy ion it is essential that the projectile velocity is more than the velocity of nucleons inside the nucleus, much higher than the Fermi energy.

First, experiments using radioactive beams were carried out to study the matter distribution of nucleons. Usually, for stable nuclei, electron scattering provides information about charge distribution; however, such experiment in case of radioactive beams can only be done using collider beam accelerators. Information about matter density could also be obtained from the reaction cross-section obtained from the collision of radioactive beam with a fixed target nucleus. Experiments on matter distribution has shown that many neutron rich nuclei such as ^{6}He, ^{8}He, and ^{11}Li have unusually large radii, much larger than the value obtained by $R = r_0 A^{1/3}$. The large value of their radii is also reflected in the small binding energies of the order of 300 keV in the case of ^{11}Li and 900 keV in case of ^{6}He (instead of around 6–8 MeV) for the last neutron in such nuclei. These nuclei are called halo nuclei. Figure 8.21 shows the experimental values of radii for some halo nuclei.

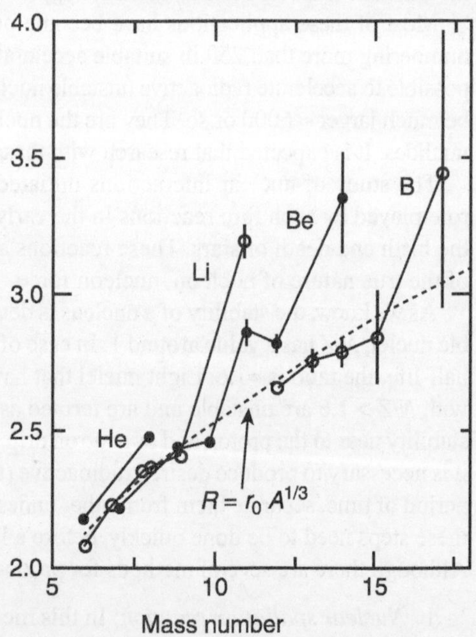

Figure 8.21 *Radii of some halo nuclei*

Exercise p-8.19: Write notes on: (i) radioactive ion beams and their applications, (ii) halo nuclei, (iii) giant resonances, and (iv) Coulomb excitation.

8.13 NUCLEAR REACTIONS RESPONSIBLE FOR NUCLEOSYNTHESIS

More than 500 different stable and unstable nuclides are found in the universe. It is interesting to know how these nuclides got synthesized in nature. According to the Big Bang theory of the creation of universe and the standard model of cosmology, light nuclides up to lithium were formed within the first 30 min of the Big Bang. The heavier nuclei got synthesized by nuclear reactions in stars and supernova. The synthesis of light elements after the Big Bang is called primordial or Big Bang synthesis and the other as nucleosynthesis in stars.

8.13.1 Primordial Nucleosynthesis

According to the Big Bang theory, the universe evolved out of the phase transitions accompanying expansion and cooling down of the infinite dense radiation and matter plasma created in the initial stage. The synthesis of nucleons/nuclei took place after about 1 μs of the Big Bang. After 1 μs, the temperature cooled down to about 10^{12} K corresponding to 100 MeV. Up to this time, gluons and quarks were free but now they got bound forming nucleons, the lightest members of baryon family and into mesons. Because of the identical densities of up and down quarks, nearly same number of neutrons and protons were produced at this stage. The temperature was still high for the formation of nuclei; however, the system rapidly underwent expansion and cooling till about a second of Big Bang. During the period 10^{-6} s to 1 s of Big Bang, neutrons and protons were in thermal equilibrium with each other through the following weak interactions:

$$\nu_e + \mathrm{n} \rightarrow \mathrm{p} + \beta^- \text{ and } \overline{\nu_e} + \mathrm{p} \rightarrow \mathrm{n} + \beta^+ \tag{8.58a}$$

After 1 s, the temperature has fallen to about 1 MeV (10^{10} K) and because of the larger mass of the neutron the balance of Boltzmann distribution shifted significantly in favour of the proton density $\rho(\mathrm{p})$ as compared to the neutron density $\rho(\mathrm{n})$, that is

$$\frac{\rho(\mathrm{p})}{\rho(\mathrm{n})} = e^{\left[\frac{(m_\mathrm{n} - m_\mathrm{p})c^2}{\Re T}\right]} \tag{8.58b}$$

where m_n and m_p are, respectively, the neutron and proton masses, and $\Re$ is Boltzmann constant.

The radiation density now is so small that conversion of proton into neutron via second equation of (8.58a) does not take place; the anti neutrino gets decoupled and moves out in the universe virtually without any further interaction.

During the period of 1 s to about 3 min after the Big Bang, the density of neutrons further decreased as the free neutron decayed with a half-life of about 12 min. However, during this time the synthesis of nuclei such as deuteron does not take place in spite of the positive Q-value of the reaction. It is because of the sufficiently high radiation density that induces photo-disintegration of deuteron immediately as they are formed.

$$\mathrm{n} + \mathrm{p} \rightarrow {}^2_1\mathrm{D} + \gamma + 2.22 \text{ MeV} \tag{8.58c}$$

$${}^2_1\mathrm{D} + \gamma \rightarrow \mathrm{n} + \mathrm{p} \tag{8.58d}$$

At the end of 3-min period, the temperature of the universe dropped to about 100 keV and the baryonic matter at this time contained about 88% protons and 12% neutrons. During the period from 3 min to 30 min after the Big Bang, the radiation density dropped sharply, making deuteron a stable nucleus with a very little probability of photodisintegration. Fusion of neutron and proton with deuteron created ${}^3_1\text{H}$ and ${}^3_2\text{He}$ nuclei through the endoergic reactions

$$ {}^1_0\text{n} + {}^2_1\text{D} \rightarrow {}^3_1\text{H} + \gamma + 6.26\text{ MeV} \tag{8.58e} $$

And
$$ {}^1_1\text{p} + {}^2_1\text{D} \rightarrow {}^3_2\text{He} + \gamma + 5.5\text{ MeV} \tag{8.58f} $$

The highly stable α particle ${}^4_2\text{He}$ got synthesized by the fusion reactions of two deuterons (d + d), proton with ${}^3_1\text{H}$ (p + ${}^3_1\text{H}$) and of neutron with ${}^3_2\text{He}$ (n + ${}^3_2\text{He}$). Fusion of ${}^3_1\text{H}$ and ${}^4_2\text{He}$ created ${}^7_3\text{Li}$ nuclei that got readily destroyed by the highly endoergic reaction ${}^7_3\text{Li}\,(\text{p},\alpha)\,{}^4_2\text{He}$.

$$ {}^4_2\text{He} + {}^3_1\text{H} \rightarrow {}^7_3\text{Li} + \gamma + 2.47\text{ MeV} \tag{8.58g} $$

$$ {}^7_3\text{Li} + \text{p} \rightarrow 2\,{}^4_2\text{He} + 17.35\text{ MeV} \tag{8.58h} $$

After about 30 min of Big Bang, the temperature and the kinetic energies of nucleons and the synthesized nuclei became so low that they could not overcome the Coulomb barrier and hence further fusion of protons and other nuclei came to a stop. As there is no stable nuclide with $A = 5$ and $A = 8$, further synthesis of heavier nuclides did not take place after about 30 min from the Big Bang when the abundance of the hadronic matter was about 76% hydrogen and 24% helium with a very small percentage of the admixture containing ^{3}He and ^{7}Li. Synthesis of heavier nuclides took place much later in the formation of stars and in their decay – supernova explosions.

8.13.2 Elemental Synthesis in Stars

Stars are formed as a result of the gravitational contraction of interstellar gas and dust which consists of hydrogen and helium. Gravitational contraction increases temperature at the centre of the star up to the level of fusion of hydrogen. In the steady-state, the energy produced by nuclear fusion is equal to the energy radiated by the star. In equilibrium, the gravitational pressure at the centre is compensated by the radiation from the fusion reactions. Fusion reaction changes the chemical composition of the star.

Sun is a star in its initial phase of evolution. Some of its important parameters are as follows: Mass of the sun $M_\odot = 2\times10^{30}$ kg; surface temperature = 6000 K; temperature at the centre = 1.5×10^7 K; pressure at the centre = 6×10^{11} atmosphere; density at the centre = 150×10^3 kg/m^3; composition = 70% hydrogen, 29% helium, and remaining 1% other heavier elements such as carbon and oxygen.

It is interesting to note that the source of energy of sun that is radiating huge amount of energy continuously for almost the last 4.9×10^9 years (assumed age of the sun) cannot be the fusion of hydrogen or of helium or of proton with helium as none of the end products of such fusion ${}^2_2\text{He}$, ${}^8_4\text{Be}$, or ${}^5_3\text{Li}$ are stable. The reaction responsible for the energy of the sun is of course the fusion of two protons but not through nuclear strong interaction that are fast but through the much slower nuclear weak interaction

$$ \text{p} + \text{p} \rightarrow {}^2_1\text{H} + \beta^+ + \nu_e + 0.4\text{ MeV} \tag{8.59a} $$

The cross-section for reaction (8.59) for protons of 1 MeV is very small about 10^{-23} barn. As such, the mean life of a 1 MeV proton due to the above reaction is about 10^{10} years in the sun. It is this slow source of energy production in the sun that decides its life. This also explains why the energy production in the sun is a slow process and not an explosion type like in a bomb.

The deuteron produced via reaction (8.59a) is almost immediately consumed by the reaction

$$^{2}_{1}D + p \rightarrow ^{3}_{2}He + \gamma + 5.49 \text{ MeV} \tag{8.59b}$$

$^{3}_{2}He$ further fuse to give $^{4}_{2}He$, which is one of the most stable structures

$$^{3}_{2}He + ^{3}_{2}He \rightarrow ^{4}_{2}He + 2^{1}_{1}H + 12.86 \tag{8.59c}$$

Or it may further fuse with $^{4}_{2}He$ to give $^{7}_{4}Be$

$$^{3}_{2}He + ^{4}_{2}He \rightarrow ^{7}_{4}Be + \gamma + 1.59 \text{ MeV} \tag{8.59d}$$

$^{7}_{4}Be$ may capture an electron to produce ^{7}Li

$$^{7}_{4}Be + e^{-} \rightarrow ^{7}_{3}Li + \nu_e + 2.47 \text{ MeV} \tag{8.59e}$$

In addition, ^{7}Be may also fuse with a proton to form the unstable ^{8}B, which immediately decays into two α particles and positron releasing large amount of energy

$$^{7}_{4}Be + p \rightarrow ^{8}_{5}B + \gamma + 0.135 \text{ MeV} \tag{8.59f}$$

And

$$^{8}_{5}B \rightarrow 2\,(^{4}_{2}He) + \beta^{+} + \nu_e + 18.09 \text{ MeV} \tag{8.59g}$$

Similarly, ^{7}Li may fuse with a proton to give two αs and 17.35 MeV of energy

$$^{7}_{3}Li + p \rightarrow 2\,(^{4}_{2}He) + 17.35 \text{ MeV} \tag{8.59h}$$

Reactions described by Eqs. from (8.59a) to (8.59h) constitute the p–p or hydrogen chain in which four protons essentially fuse together to form an α particle and release 26.7 MeV of energy. About 2% of this energy (≈0.5 MeV) is taken by neutrino. The p-p cycle may be represented by figure 8.22.

If the temperature in the interior of the star is higher and it contains heavier elements then another cycle called either CNO or Bethe–Weizsacker cycle may become important. The net result of this cycle is also the fusion of four protons and release of energy but this cycle is more efficient and faster than the p–p cycle. CNO cycle is shown in figure 8.23.

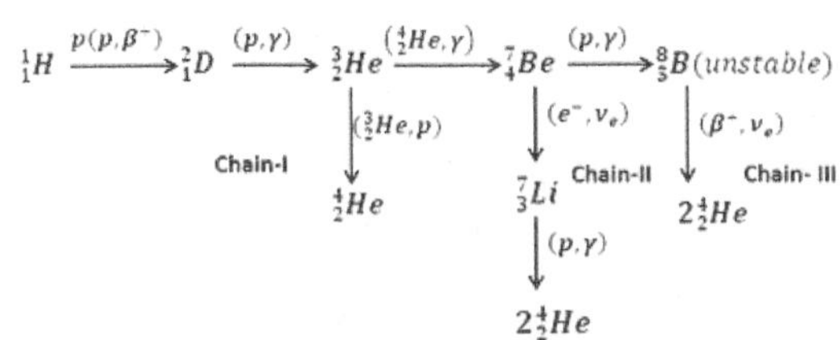

Figure 8.22 *p–p cycle chains*

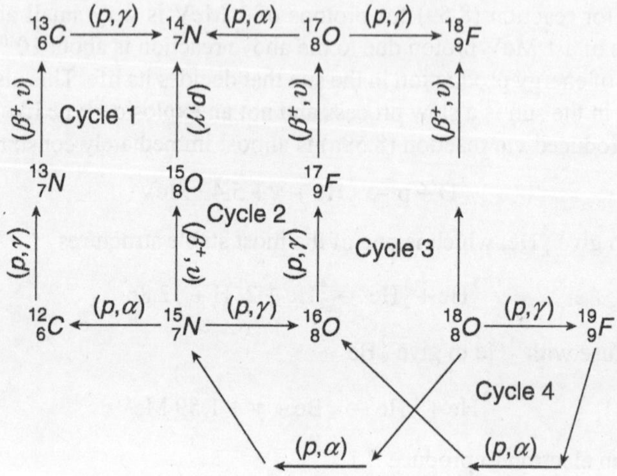

Figure 8.23 *CNO cycle chains*

After the consumption of all hydrogen at the core of the star the radiation pressure decreases and gravitational collapse of the star starts once again and the temperature in the interior increases. Further, fate of the star depends on its mass.

If the mass of the star M_{star} < 1.4 mass of the sun $M_{\odot}$, no further fusion in the star takes place. Its temperature decreases, and the gravitation collapse is balanced by the Fermi pressure of the degenerate electron gas. The star ends as a white dwarf. The mass value 1.4 $M_{\odot}$ is called *Chandrashekher limit* and is a very important astronomical constant. It decides the ultimate fate of the star.

Otherwise, if $M_{star} > 1.4\ M_{\odot}$, the gravitational collapse increases the temperature in the interior of the star simultaneously the hydrogen still present in the outer regions of the star starts burning due to fusion, the surface of the star expends and the colour of the star shifts to the red giving rise to the stage called red giant. In the interior of the star when temperature reaches beyond 10^8 K and density of the order of 10^8 kg m^{-3}, fusion of helium starts, which increases the temperature of the star and the radiation pressure in the interior. Burning of helium proceeds through

$$^{4}_{2}\mathrm{He} + {}^{4}_{2}\mathrm{He} \rightarrow {}^{8}_{4}\mathrm{Be} \qquad (8.60a)$$

Reaction (8.60a) is possible only at temperatures $>10^8$ K, because of the Coulomb barrier between two helium nuclei and the negative Q-value of about 90 keV. ^{8}Be fuses with another α particle forming ^{12}C in excited state.

$$^{8}_{2}\mathrm{Be} + {}^{4}_{2}\mathrm{He} \rightarrow {}^{12}_{6}\mathrm{C}^{*} \qquad (8.60b)$$

The net result of reactions shown by Eqs. (8.60a) and (8.60b) is

$$3\ ({}^{4}_{2}\mathrm{He}) \rightarrow {}^{12}_{6}\mathrm{C} + \gamma + 7.275\ \mathrm{MeV} \qquad (8.60c)$$

In a star such as the sun, helium burning may take 10^7 years or so. After the burning of the helium, star suffers further gravitational collapse and when the temperature in the interior of the star reaches a value $\geq 6 \times 10^8$ K, fusion of carbon takes off.

$$
\begin{aligned}
{}^{12}_{6}C + {}^{12}_{6}C &\rightarrow {}^{20}_{11}Ne + {}^{4}_{2}He + 4.62\ \text{MeV} \\
&\rightarrow {}^{23}_{11}Na + p + 2.24\ \text{MeV}
\end{aligned}
$$

$$
{}^{12}_{6}C + {}^{4}_{2}He \rightarrow {}^{16}_{8}O + \text{energy}
$$

$$
{}^{23}_{11}Na + {}^{4}_{2}He \rightarrow {}^{27}_{13}Al + \text{energy}
$$

As is indicated by these equations, fusion of carbon followed by the fusion of α particles results in the synthesis of heavier nuclei up to iron. No synthesis of heavier nuclei beyond iron takes place by the fusion of α particles as fusion beyond iron is endoergic. Burning of carbon and beyond up to the formation of iron takes only a few days. Once the fusion reactions in the star stops there is no source of thermonuclear energy in the star, the matter in the outer shell of the star collapses into the centre leading to a supernova explosion. If the mass of the star is up to two times the mass of sun, it ends up as a neutron star and if it is heavier, the supernova residue ends in a black hole. A black hole is a very dense and heavy body with extremely large gravitational field, so much that even the photons of light could not come out of the gravitational field of the black hole.

8.13.3 Synthesis of Heavier Elements

Heavier elements up to lead are synthesized essentially by the slow process (denoted by s-process) of neutron capture and β decay. Neutrons are generated in stars during the fusion by reactions of the type

$$
{}^{22}_{10}Ne + {}^{4}_{2}He \rightarrow {}^{25}_{12}Mg + n - 0.48\ \text{MeV}
$$

$$
{}^{13}_{6}C + {}^{4}_{2}He \rightarrow {}^{16}_{8}O + n - 0.91\ \text{MeV}
$$

Successive neutron capture and β negative decay results in the synthesis of nuclei along the valley of stability up to lead. Elements beyond lead and away from the valley of stability are formed in a process called the rapid or r-process where simultaneous capture of more than one neutron takes place. This leads to the production paths away from the valley of stability and beyond lead. The ultimate limit is put by the spontaneous self-fission.

The relative abundance of stable nuclides in our galaxy is shown in figure 8.24. The peaks from helium up to iron can be explained by the fusion processes in the star and almost constant distribution of elements heavier than iron by neutron capture and β^- decay. The peaks at magic numbers $N = 50$, 82, and 126 are due to the small cross-section for neutron capture.

Exercise p-8.20: Discuss the synthesis of light elements within 30 min of the Big Bang.

Exercise p-8.21: Draw a rough sketch of the relative abundance of elements in our galaxy versus their atomic mass number A and explain the peak in the distribution at $A \approx 56$. How can the smaller peaks at magic neutron number $N = 50$, 82, and 126 be explained?

Exercise p-8.22: Explain the synthesis of elements at the primordial stage.

Exercise p-8.23: How have heavy elements been synthesized in our galaxy? Give a brief outline of different reaction chains involved in this synthesis.

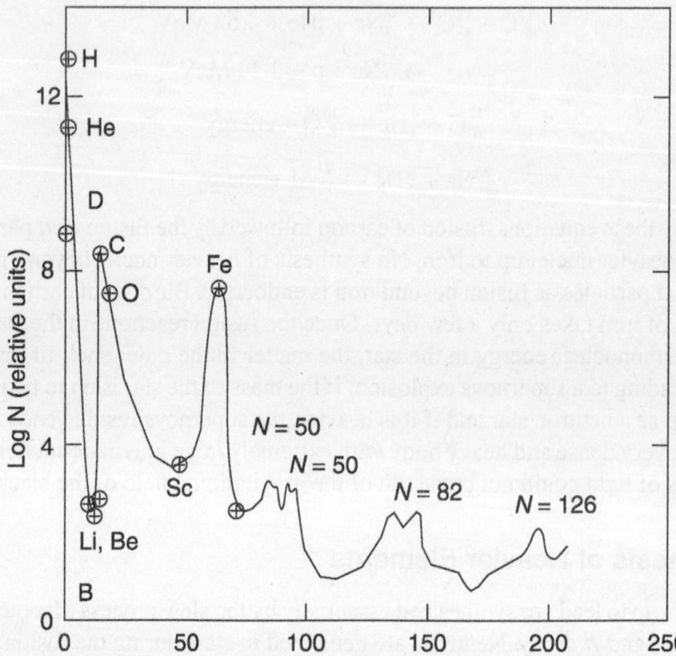

Figure 8.24 *Relative abundance of stable nuclides in our galaxy*

Multiple choice questions

Note: More than one alternative may be correct for some of the following questions. Tick (pickup) all the correct alternatives in such cases.

Exercise M-8.1: Which of the following quantity may not be conserved in strong nuclear reactions?

(a) Kinetic energy (b) Electric quadrupole moment
(c) Magnetic dipole moment (d) Parity

Exercise M-8.2: An initial excited state of spin $J_i^P = 2^+$ populates the final state of spin $J_f^P = 3^-$ on absorption of an α particle; the orbital angular momentum ℓ of α particle may have the value

(a) 1 only (b) 2 only (c) 5, 3, and 1 (d) 2 and 4

Exercise M-8.3: In the case of endoergic nuclear reactions, threshold energy of the reaction is approximately given by the relation:

(a) $|-Q|\left(1+\dfrac{\text{Mass of the projectile}}{\text{mass of the stationary target}}\right)$ (b) $Q\left(1-\dfrac{\text{Mass of the projectile}}{\text{mass of the stationary target}}\right)$

(c) $-Q\left(1-\dfrac{\text{Mass of the projectile}}{\text{mass of the stationary target}}\right)$ (d) $-Q\left(1+\dfrac{\text{Mass of the stationary target}}{\text{mass of the projectile}}\right)$

Exercise M-8.4: Largest number of degrees of freedom is excited in

(a) direct reactions (b) pre-equilibrium emission
(c) compound nuclear reactions (d) partial wave analysis

Exercise M-8.5: The amplitude of the nuclear wave function inside the nucleus is a maximum when the derivative of the wave function at the nuclear boundary has

(a) positive slope (b) zero slope (c) negative slope (d) infinite slope

Exercise M-8.6: The excitation functions for statistical nuclear reactions show a typical shape; the reaction cross-section first increases smoothly with the energy of the incident particle attains a maximum value and then decreases with further increase of the incident energy. The decrease of the cross-section with energy is due to

(a) opening of more and more reaction channels of larger negative Q-values
(b) participation of larger ℓ partial waves
(c) decrease of the de Broglie wavelength of the incident particle
(d) decrease of the reaction time

Exercise M-8.7: The radius R_H of a halo nucleus with A nucleons is

(a) $R_H >> r_0 A^{1/3}$ (b) $R_H = r_0 A^{1/3}$ (c) $R_H << r_0 A^{1/3}$ (d) $R_H^2 = r_0 A^{1/3}$

Exercise M-8.8: The giant resonance excited by the transfer of 1 $\hbar$ of angular momentum to a nucleus is called

(a) Giant Magnetic Dipole resonance (b) Giant Electric Dipole resonance
(c) Electric Monopole resonance (d) Electric Quadrupole resonance

Exercise M-8.9: The source of energy of stars is the fusion of protons by the nuclear

(a) strong force (b) electromagnetic force
(c) weak force (d) gravitational force

Exercise M-8.10: Synthesis of heavy stable nuclei in stars takes place by

(a) successive neutron capture and β decay
(b) successive proton capture and β decay
(c) successive α capture and neutron decay
(d) simultaneous capture of more than one neutron.

Exercise M-8.11: Elements up to what maximum mass number got synthesized at the primordial stage?

(a) 15 (b) 10 (c) 7 (d) 3

Exercise M-8.12: Approximate time after the Big Bang the primordial synthesis of element stopped is

(a) 1 h (b) 10 h (c) 1 day (d) 100 s

Exercise M-8.13: Stars lighter than *Chandrashekher limit* ends as a

(a) neutron star (b) red giant (c) black hole (d) white dwarf.

Exercise M-8.14: Fusion of α particle releases energy till the synthesis of

(a) ^{7}Li (b) ^{12}C (c) ^{40}Ca (d) ^{56}Fe

Exercise M-8.15: In a binary reaction it has been found that two groups of particles with different kinetic energies are emitted at angle θ. The maximum value of θ may be

(a) 0° (b) 90° (c) 180° (d) 360°

Exercise M-8.16: In the reaction a + X→ Y + b + Q, particle b will be emitted with two values of the kinetic energy at angles smaller than 90° if

(a) $M_a > M_Y$ and $E_a > (M_Y \cdot Q)/(M_a - M_Y)$
(b) $M_a < M_Y$ and $E_a > (M_Y \cdot Q)/(M_a - M_Y)$
(c) $M_a > M_Y$ and $E_a < (M_Y \cdot Q)/(M_a - M_Y)$
(d) $M_a < M_Y$ and $E_a < (M_Y \cdot Q)/(M_a - M_Y)$

Here, M_i stands for the mass of the i-th particle.

Exercise M-8.17: On absorption of a thermal neutron, a heavy nucleus divides into two parts of masses M and 1.5 M that travel with kinetic energies E_1 and E_2. The ratio of the velocities of the two fragments will be

(a) E_1/E_2 (b) E_1^2/E_2^2 (c) $E_1^{1/2}/E_2^{1/2}$ (d) None of the above

Exercise M-8.18 A 10 MeV α particle is made to hit a target $^{206}_{80}\text{Pb}$. Approximate percentage of the α energy consumed in giving motion to the centre of mass is

(a) 20 (b) 10 (c) 5 (d) 2

Exercise M-8.19: Peaks in the relative abundance of elements in our galaxy at higher magic numbers are due to

(a) low values of neutron capture cross-sections for these nuclides.
(b) high values of neutron capture cross-sections for these nuclei.
(c) low values of binding energies of magic nuclides.
(d) production of these nuclides in large amount at primordial stage.

Exercise M-8.20: 100 MeV neutrons are made to hit a nucleus for which the optical potential is $V = (-35 - 15i)$MeV. The kinetic energy (in MeV) of the neutron inside the target nucleus will be

(a) 65 (b) 85 (c) 115 (d) 135

9

Particle Accelerators

Particle accelerators are machines that impart large kinetic energy to charged particles. Charged particles of high energies are required for scattering experiments that provide information about the size and charge/matter distribution of subatomic systems such as nuclei and nucleons. Accelerated particles are also required for the generation of hitherto unobserved particles such as Higg's or the so-called 'God particles'. Energetic light and heavy ions are used to study the dynamics of nuclear reactions as well as the nuclear structure. Surface properties of solids can be manipulated by ion implantation using accelerated ions. Accelerators are finding applications in medicine as a tool of radiation therapy for the treatment of cancer and also in agriculture.

An accelerator setup can be divided into three broad parts: (1) ion source, (2) accelerating system, and (3) target and detectors. Each of these three components is more or less equally important and lot of research and efforts go into setting each. An ion source provides a nearly collimated initial ion beam of very low energy. The beam current essentially depends on the ion source. Development of suitable ion source for each special requirement is a full-fledged subject of research and is a highly technical job. However, at present it is possible to get ion sources for all ions starting from protons up to uranium. Even sources that may give negative ions up to uranium are now commercially available. The ion source is coupled to a system that accelerates the charged particles provided by the ion source. The other end of the accelerating system is coupled to a target and detector assembly. Target may be solid, liquid, or gas; it may be a biological sample and in some experiments target may itself be an accelerated beam. Detector set-up depends on the requirements of the experiment and varies from one experiment to the other. We shall, however, in this chapter study in some detail the accelerating system.

In all accelerating systems, charged particles from an ion source are accelerated due to the force of an electric field. A particle of charge Ze passing through an electric potential difference of V volt gets a push of kinetic energy $\Delta E = (ZeV)$ eV (electron volt). In electrostatic machines a high potential V is generated, which may be used once (such as in a Cockcroft–Walton and in a Van de Graaff generators) or two times as in Tandem Van de Graaff. Much higher energies may be obtained by using time-varying electric fields either in circular accelerators (cyclotron, synchrotron) or in linear accelerators. Circular accelerators employ magnetic fields to confine particles in orbits. Time-varying magnetic fields are also used to induce electric fields that accelerate charged particles (betatron).

9.1 ELECTROSTATIC ACCELERATORS

Electrostatic machines, also called direct current machines, may be classified into three types: (1) Cockcroft–Walton, (2) Van de Graaff, and (3) Tandem Van de Graaff.

9.1.1 Cockcroft–Walton Accelerator

John D. Cockcroft and Ernest Walton in 1932 built the first particle accelerator at Cavendish Laboratory, Cambridge, England (figure 9.1). They generated high voltage by the method of voltage multiplication using an AC voltage source. The principle of voltage multiplication is shown in figure 9.2. During the input half cycle, when the end B of the transformer secondary is positive (and A is negative) diode D_1 conducts and capacitor C_1 is charged via resistance R to the maximum value V_m of the input voltage. In the next half cycle, the positive input voltage at A and the potential at C_1 drives diode D_2, which charges capacitor C_2 to a potential $2V_m$. The ultimate output potential depends on the number of multiplication stages. The high voltage developed by employing multiple stages of voltage multiplication may be used to accelerate positively charged particles. Actual acceleration of charged particles is achieved in the accelerating tube. The accelerating tube is generally made of hollow cylindrical ceramic insulators. Three or four co-centric pieces of metallic tubes of decreasing diameter, called electrodes, are hung from the wall of the insulator ceramic tube as shown in figure 9.3. The first electrode is kept at the highest potential while the successive electrodes are provided decreasing potentials by a resistive potential divider chain. The last electrode is at ground potential. When positive ion beam passes through this accelerating column, it gains energy in steps at each gap between the electrodes. The electric lines of force in the gaps between the electrodes provide focusing of the beam along the axis of the accelerating column.

Figure 9.1 *The first accelerator*

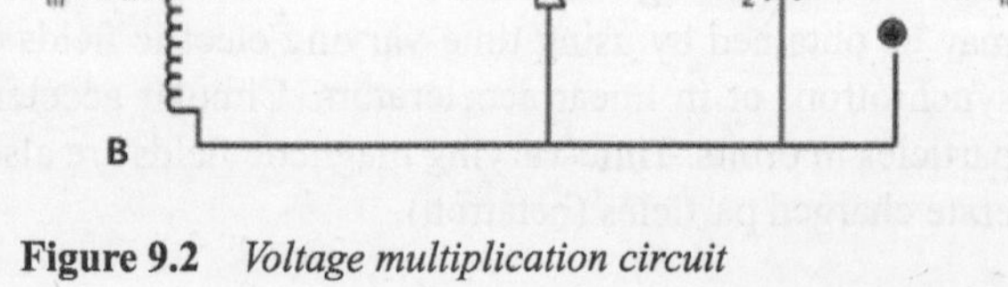

Figure 9.2 *Voltage multiplication circuit*

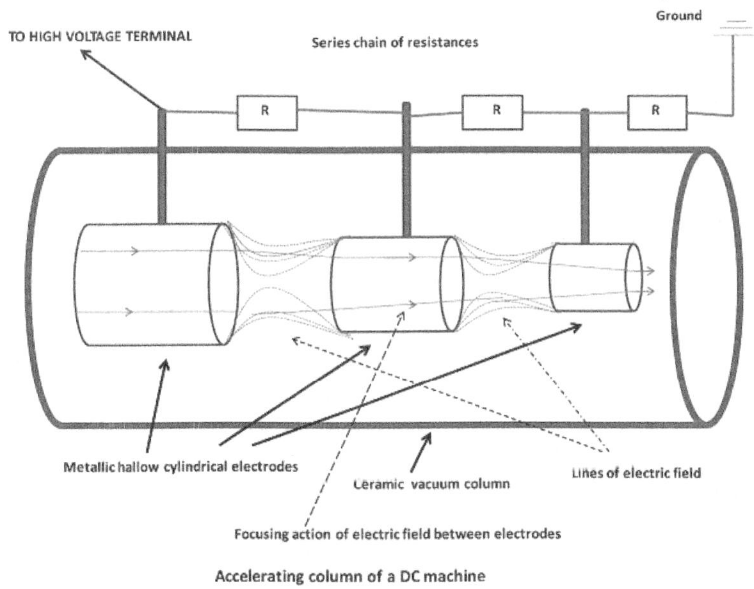

Figure 9.3 *Accelerating column*

Cockcroft and Walton carried out the first artificial transmutation by accelerated particles (protons).

$$ {}_{1}^{1}\mathrm{H} + {}_{3}^{7}\mathrm{Li} \rightarrow 2\left({}_{2}^{4}\mathrm{He}\right) $$

They were awarded Nobel Prize for their pioneering work in 1951. The voltage multiplication method could not go beyond about 1 MV because of the charge leaking through surfaces and in the atmosphere.

9.1.2 Van de Graaff Generator

Robert G. Van de Graaff, a post-doctoral fellow at Princeton, in 1929 built an 80-kV high-voltage generator based on the principle of developing high voltage through charging of a conductor of large capacitance. Later in 1933, a 7-MV accelerator was constructed at MIT. The problem of the leaking of charge, the main issue that puts an upper limit to high-voltage generation, was overcome by surrounding the high-voltage terminal with quenching gas at high pressure. Van de Graaff generators up to 20 MV have been built but terminal voltage up to 15 MV delivers stable current of accelerated particles. The principle of voltage generation in a Van de Graaff may be explained with reference to figure 9.4. As is

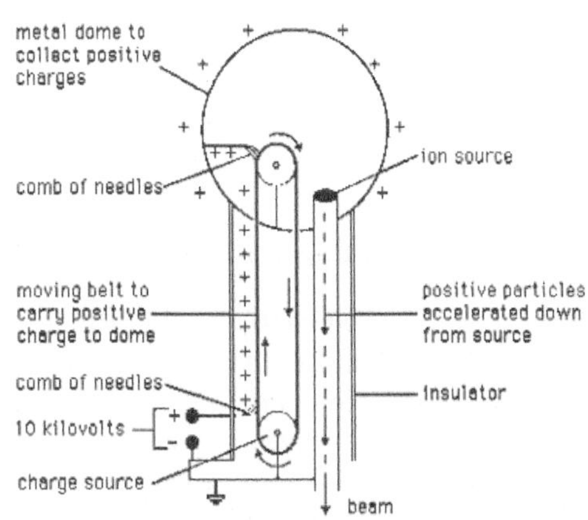

Figure 9.4 *Working principle of a Van de Graaff accelerator*

shown in the figure, a DC source (10 kV in the figure) of high voltage produces free positive and negative charges due to the sparking and ionization of the air between the two electrodes. The positive charges are picked up by the needle-comb arrangement and are sprayed on the non-conducting belt rotating between two pullies. The negative charges are grounded. The non-conducting belt carries positive charge away from the grounded negative charge and thus increases the electrostatic potential energy of the positive charge. At the other end, a second needle-comb arrangement picks up positive charges from the belt and transfers them to the big metallic spherical conductor, the metallic dome. It is interesting to know how a needle comb arrangement works. Needle-comb system is made of some metal such as copper. If some charge is picked up by the comb, it gets distributed all over the surface including needles. The tips of the needles are very pointed and have small surface area and capacitance, as a result the charge density at the tips of needles becomes large and tips become points of relatively high voltage. If there is a low potential surface near the needle tips, the charges from the tips got transferred to that surface. In this way, needle-comb arrangement picks up charge at the lower end and transfers it to the metallic dome at the other end. The dome is seamless, and has large electric capacitance C.

The potential of the dome-also called the high voltage terminal, increases as more and more charge is deposited on it ($V = Q/C$). In this way the mechanical energy spent in transporting positive charge is converted into electrostatic potential energy. Continuous running of the belt for some time generates sufficiently high voltage on the terminal. As the potential on the terminal increases, charges leak from the dome through corona discharge and also through surface leaking and ultimately a steady state is reached when there is no further increase in the terminal voltage. In machines where the high voltage terminal is exposed to the air, the maximum generated voltage is of the order of 10 MV. However, in new improved machines high voltage terminal is enclosed in a tank which contains some quenching gas like sulphur hexa fluoride (SF_6) at high pressure. The maximum terminal voltage of about 20 MV can be obtained from such machines. An ion source that provides desired positive ions is kept at the terminal voltage and the positive ions are accelerated through an accelerating column down to the ground terminal. The terminal voltage may be controlled to any desired level by a chain of variable resistances that provides a path for the leakage of charge from the terminal. The magnitude of the leakage current which can be controlled remotely, determines the value of the terminal potential. The potential gradient and the electric fields in the accelerating column are designed to provide beam focusing. Van de Graaff generators provide good quality beam of high intensity having good energy resolution and are extensively used in nuclear research.

9.1.3 Tandem Van de Graff

In a normal Van de Graaff, the high voltage is used only once to accelerate the positive ions. In the year 1951, Prof. Luis Alvarez suggested that the same terminal voltage may be used twice if initially a negatively charged ion is accelerated by attraction to the positive potential of the terminal and then the negative ion is converted to a positive ion by charge exchange within the high-voltage terminal. The positive ion may now be again accelerated from the terminal potential to the ground potential. In this way, the same terminal potential may be used twice: first to accelerate negative ion and then again to accelerate the charge exchanged positive ion. In the early stages, it was difficult to make negative ions of all desired elements; however, now it is possible to have negative ion sources that may deliver intense beams of negative ions of almost all elements from boron to uranium. A Van de Graaff generator that uses the terminal voltage twice for acceleration is called a Tandem Van de Graaff. Figure 9.5 shows the basic layout of a tandem Van de Graaff. Negative ions from a negative ion source, located outside the high-voltage terminal are steered to the accelerator terminal. Generally, a negative ion has one extra electron attached to the neutral atom and, therefore, gains energy of 1 eV when accelerated from ground potential to the terminal potential V of the tandem. The accelerated beam of negative ions is passed through a stripper material, which is generally carbon foil (or in same case may be a suitable gas) in the terminal. The stripper removes some electrons from each negative ion and makes them positively charged. Ions coming out after stripping have a distribution of positively charged states. The positive ion beam having distribution of charged states is passed through a magnetic sector. The parameters of the magnetic sectors may be set to select ions only of a particular charge state say +qe. These positive ions are now accelerated in the accelerating column, which in a Van de Graaff accelerator is housed in the quenching gas pressure tank along with the high voltage terminal. The total energy gained by a positive ion of charge state +qe is

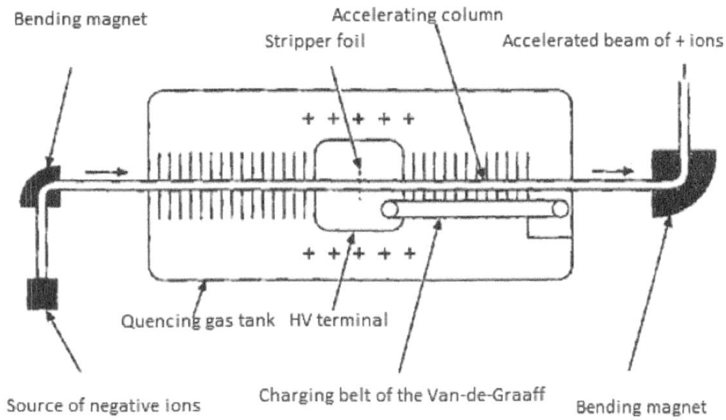

Figure 9.5 *Pictorial representation of a tandem Van de Graaff*

$$E = 1\text{ eV} + q\text{ eV} = (1 + q)\text{ eV}$$

where V is the potential of the high-voltage terminal. For example, let us suppose that the terminal voltage V is 12 MV and a ^{16}O beam of charge state 7^+ is accelerated, the total energy gained by the oxygen ion will be $(1 + 7) \times 12$ MeV = 96 MeV.

9.2 The Cyclotron

Another genius idea for the acceleration of charged particles was put forward by E. Lawrence in 1929. He suggested that the trajectory of a charged particle may be confined to a closed circular path by applying a uniform magnetic field normal to the plane of motion of the particle. As shown in figure 9.6, a particle of charge Q and mass m moving with velocity V_θ (in X–Y plane) under a uniform magnetic field B_Z along the Z-axis is constrained to trace a circular path of radius r. The centrifugal force F_{cen} in the circular motion is balanced by the magnetic force (Lorentz force) F_{M}.

However,

$$F_{\text{cen}} = \frac{mV_\theta^2}{r} \quad \text{and} \quad F_{\text{M}} = QV_\theta B_Z \tag{9.1a}$$

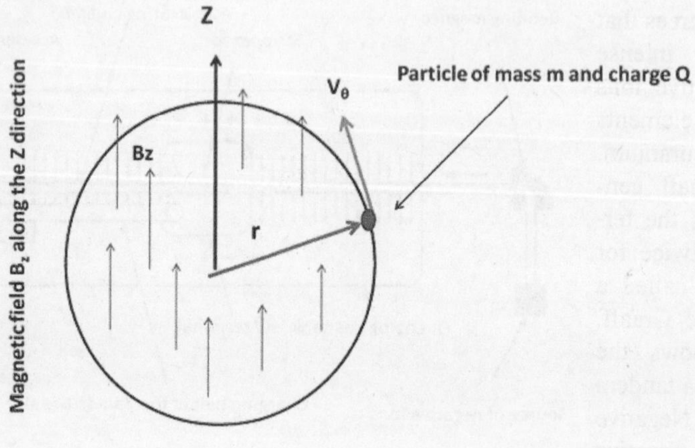

Figure 9.6 *Motion of a charged particle in a uniform magnetic field*

And, therefore,

$$\frac{mV_\theta^2}{r} = QV_\theta B_Z \tag{9.1b}$$

The angular frequency of revolution $\omega_{\text{rev}} = \dfrac{d\theta}{dt} = \dfrac{V_\theta}{r} = \dfrac{QB_Z}{m}$ (9.1c)

Equation (9.1c) states that the angular frequency of revolution ω_{rev} also called the cyclotron frequency depends only on the charge Q, mass m, and the strength of the magnetic field B_Z, and is independent of the energy of the particle. The kinetic energy of the charged particle is related to the radius r of the orbit by the relation

$$E = \frac{1}{2}mV_\theta^2 = \frac{1}{2}\frac{Q^2B_Z^2r^2}{m} \tag{9.1d}$$

It is evident from Eq. (9.1d) that the kinetic energy of a particle of given charge and mass in a fixed magnetic field B_Z is proportional to the square of the radius of its circular path, but from Eq. (9.1c) the frequency of revolution remains same for all particles irrespective of their energy. More energetic particle moves in a circle of larger radius such that it takes same time in completing one revolution as is taken by a similar particle of lesser energy which travels along a circle of smaller radius.

There are six important components of a cyclotron, which are as follows:

1. An electromagnet which can provide a uniform magnetic field of high intensity
2. There is enough gap between the pole pieces of the magnet where a cylindrical vacuum chamber may be fitted. The vacuum chamber is required to house the accelerating system of dees where beam of accelerated charged particles circle around. If the beam is allowed to move in air it will collide with gas molecules of the air and will lose intensity due to scattering.
3. A hollow metallic cylinder of small height (that may fit into the vacuum chamber) and of radius nearly equal to the radius of magnetic poles is cut into two halves making the two dees.

The dees are kept inside the vacuum chamber between the magnetic pole pieces with a little gap or separation between them.

4. An alternating (sinusoidal or some times square wave) electric potential of cyclic frequency ω_{ac} generated in a separate (RF) oscillator may be applied between the two dees through feeds provided in the vacuum chamber. When the frequency ω_{ac} is equal to the cyclotron frequency of the machine ω_{rev} the condition of resonance is met. The resonance frequency generally lies in the radio frequency (RF) range. It may be noted that for half cycle of the applied RF one of the dee is at a positive potential with respect to the other and during the next half cycle the polarities got reversed. Suppose a positively charged particle is introduced in the gap between the dees it will be repelled by the positive potential on the dee into the other dee, which is at a lower potential. As a result of this force of repulsion, the positively charged particle will gain energy and will move in the dee in a circle of larger radius. The charged particle does not experience any electrical force while moving inside the dees as electric fields do not exist inside metallic bodies. However, magnetic field that penetrates dees constrains the particle to move in a circular path. In the condition of resonance, when the particle comes out on the other end of the dee after half cycle the polarity of the RF field reverses and the particle finds that the dee through which it was travelling has acquired positive potential. The positively charged particle is then pushed again by the force of repulsion into the opposite dee. In this way, at each crossing of the gap between the dees the particle gets an injection of energy $\Delta E = QV$, where V is the instantaneous value of the RF potential at the instant of crossing the gap. With the gain in its energy at each gap crossing, particle moves in semicircles of successively increasing radius tracing a spiral path. As the particle crosses the gap two times in a cycle, the total number of gap-crossings is $2N$ if the particle has made N revolution and the energy of the particle has increased by $2NQV$. The kinetic energy of the particle is related to the radius of its motion through Eq. (9.1d).
5. A pair of deflecting plates is fixed near the periphery of the dees. A suitable DC potential may be applied to the pair of deflection plates, which deflects the beam of accelerated particles out of the cyclotron.
6. Ion source is an important part of any accelerator. Ions of desired species are produced by the ion source and are injected at the centre of the dees. Figure 9.7 shows the spiral path of accelerated charged particles as they gain more and more energy at each crossing of the gap between the dees. In some cyclotrons, there is only one dee, and AC potential is applied between the dee and a surface at ground potential.

9.2.1 Superconducting Cyclotron

The machine parameters that decides the highest energy that may be delivered by a given cyclotron are the strength of the magnetic field B_Z and the radius r (see Eq. (9.1d). It is difficult to produce uniform magnetic field of high strength over a large area (i.e. for large r) by conventional electromagnets. However, electromagnets made of wires of superconducting material through which high currents may be passed without much loss of power, may produce very high magnetic fields. Cyclotrons that employ superconducting magnets are compact and are capable of accelerating particles to very high energies.

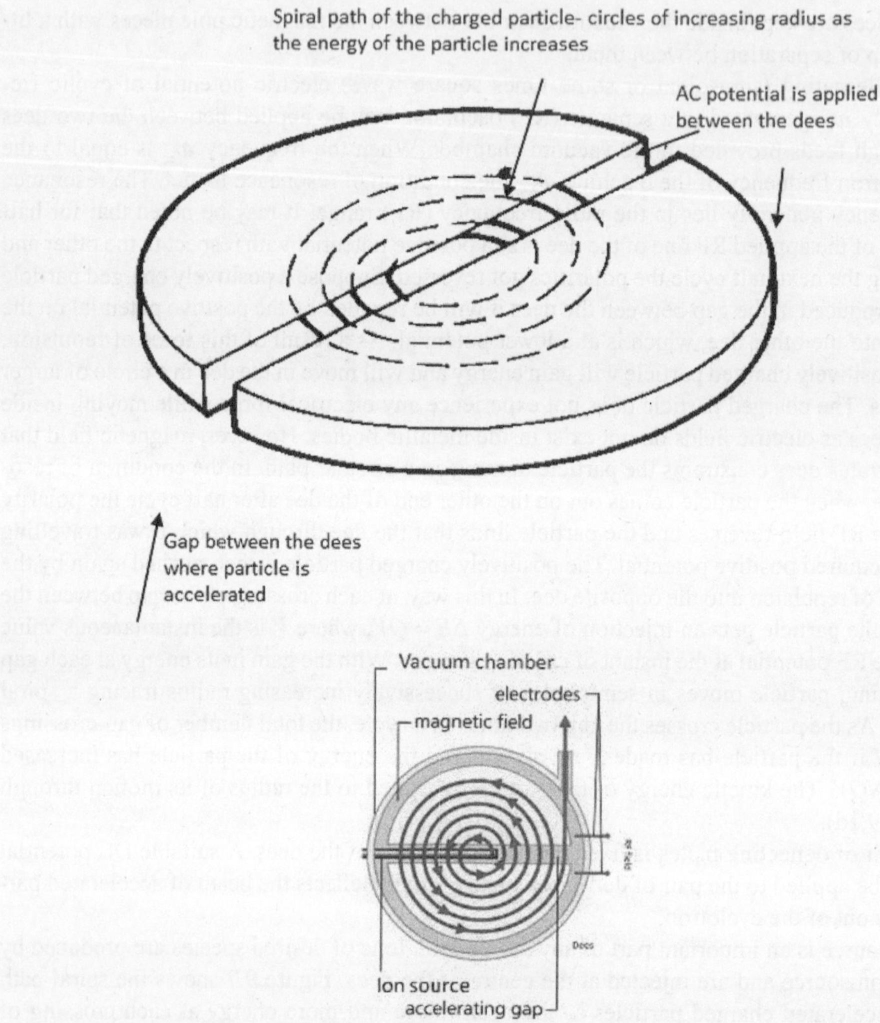

Figure 9.7 *Path of a charged particle through the dees of a cyclotron as it gains energy*

A simple cyclotron can accelerate particles only up to non-relativistic limits. As the energy of the charged particle increases, its velocity also increases and from the principle of relativity its mass changes. The particle mass m in motion with velocity V is related to its rest mass m_0 as follows:

$$m = \frac{m_0}{\sqrt{1-\frac{V^2}{c^2}}}, \text{ here } c \text{ is the velocity of light.}$$

$$= \gamma m_0, \text{ where } \gamma = \frac{1}{\sqrt{1-\beta^2}} \text{ and } \beta = \frac{V}{c} \tag{9.2a}$$

The change in the mass of the particle with energy is not appreciable till its kinetic energy is much less than its rest mass energy. A particle becomes relativistic when its energy approaches its rest mass energy. For example, the rest mass of an electron is 0.511 MeV and, therefore, a 0.3 MeV electron is quite relativistic and there is appreciable change in its mass. In addition, the rest mass of a proton (or neutron) is around 938 MeV and, therefore, relativistic mass variation for nucleons becomes important at 800 MeV or more.

In a cyclotron, the cyclotron frequency $\omega_{\text{rev}}\left(=\dfrac{QB_Z}{m}\right)$ depends on the mass of the particle (see Eq. (9.1c)). As the energy of the particle approaches relativistic limits, the mass increases with energy reducing the frequency of revolution. The resonance condition $\omega_{\text{rev}} = \omega_{\text{ac}}$ breaks down and further acceleration of the charged particle stops. As such, the relativistic variation of mass puts a limit on the energy of the particle to which it could be accelerated in a cyclotron.

9.2.2 Focusing of Particles in a Cyclotron: Weak Focusing

To get an intense beam of accelerated charged particles from a cyclotron, it is essential that particles remain confined to the median plane (a plane normal to the magnetic field and passing through the middle of the dees) and all particles of same energy move in an orbit of same radius. In cyclotrons with horizontal faces of magnetic poles, confining to the median plane means vertical focusing and the other requirement is radial focusing. Both vertical and radial focusing can be achieved if the magnetic field is such that it is large at the centre and decreases at periphery as shown in figure 9.8.

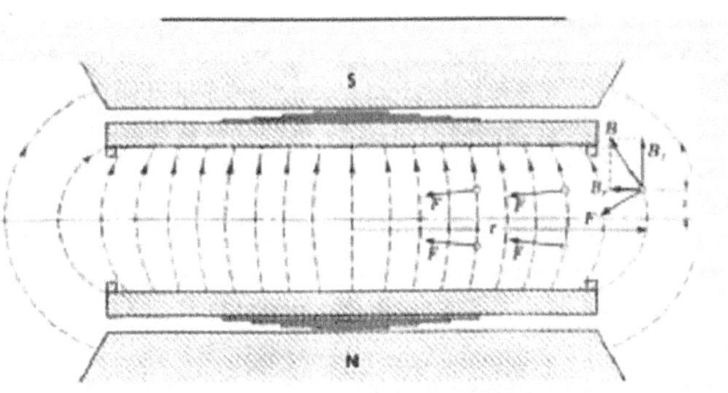

Figure 9.8 *A radially decreasing magnetic field provides both radial and vertical focusing*

As shown in figure 9.8, the magnetic force on the particle at any point away from the median plane may be resolved into two components: one pushing towards the median plane (i.e. in the Y direction) and the other towards the centre. The magnetic force towards the centre is counter-balanced by the centrifugal force, except for a small difference because of the non-uniform value of the magnetic field. This small unbalanced radial component of the force gives rise to small harmonic variations in the motion of the particle in the X-direction. The vertical component of the force is totally unbalanced; there is no force that counters it. As a result, it produces simple harmonic oscillation in vertical direction

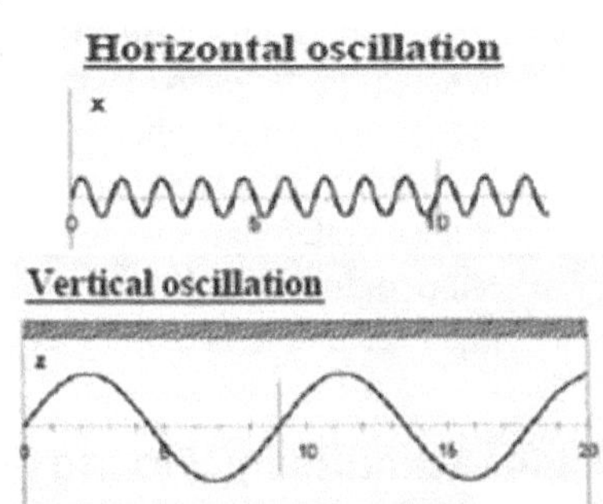

Figure 9.9 *Horizontal and vertical betatron oscillations*

(along Z-axis). The oscillations are of larger amplitude and low frequency in vertical direction and of smaller amplitude and high frequency in horizontal direction. These oscillations are called betatron oscillations and are shown in figure 9.9. This type of focusing is called weak focusing.

Considerable beam intensity is lost because of the betatron oscillations if the accelerated beam makes large number of rotations in the cyclotron. This coupled with the relativistic variation in the mass of the particle that slowly breaks the resonance condition: the first generation cyclotrons were designed such that the particles are taken out after about 100 revolutions. Stronger RF fields were used to get maximum ultimate energy within 100 revolutions. Thus, first generation cyclotrons employing weak focusing were able to accelerate particles in the range of 25–35 MeV.

9.2.3 Isochrone- and Synchrocyclotron

Particle acceleration in cyclotron is possible only if the resonance between the frequency of revolution and the frequency of the applied AC field is maintained. As particle gains more energy, the resonance breaks down because of the relativistic variation in the mass of the particle. The variation in the cyclotron frequency $\omega_{\text{res}}\left(=\dfrac{QB_Z}{\gamma m_0}\right)$, because of the relative variation in the mass, may be compensated in two ways to hold the condition of resonance.

1. The increase in the mass of the particle with energy is compensated by the increase of the magnetic field B_Z with r, such that $B_Z(r) = \gamma B_{Z0}$, where B_{Z0} is the value of the magnetic field at the centre. As $\gamma > 1$, this requires that the magnetic field should increase with r, which is against the requirements of median plane (weak) focusing. However, a different type of focusing called AVF or Thomas focusing can be employed. Thomas focusing will be discussed later separately. Isochronous cyclotrons employ magnetic field that increases with the radius to hold the resonance and provide high energy continuous beams of accelerated particles.
2. The other way to compensate for the relativistic increase in particle mass is to modulate (to decrease) the frequency ω_{ac} with time. Cyclotrons that employ RF frequency modulation are called synchrocyclotrons. The RF frequency at time 't' is given as $\omega_{\text{ac}}(t) = \dfrac{1}{\gamma}\omega_{\text{ac}}(0)$, where $\omega_{\text{ac}}(0)$ is the frequency of the applied AC field when a bunch of particles has just entered at the centre of the cyclotron. This particular bunch of particles will be accelerated till it reaches the periphery of the dees and is taken out. Suppose it takes time Δt for the bunch of particles to reach the periphery. After this time interval, the frequency ω_{ac} is again reset to the initial value $\omega_{\text{ac}}(0)$ and a new bunch of particles is introduced at the centre and a new cycle of frequency modulation starts. Thus, in a synchrocyclotron, pulses or bunches of accelerated particles are produced repeatedly after a certain interval of time. This mode of operation is called pulsed mode. In some experiments, such as those in which time of flight technique is employed, it is advantageous to have pulsed beam.

9.2.4 Azimuthally Varying Field (AVF) or Thomas Focusing or Sector Field Focusing

In cyclotrons employing weak focusing, a considerable intensity of the beam is lost by the large amplitude betatron oscillations in vertical direction. The intensity loss becomes serious if the particles are accelerated to high energies and make large number of revolutions. L.H. Thomas in 1938 suggested that strong vertical focusing can be achieved if the magnetic field varies with the azimuthal angle.

As shown in figure 9.10, in principle a magnetic field that varies with azimuthal angle may be produced by shaping the magnetic poles so that the gap between them alternately decreases and increases. The magnetic field lines are also shown in the figure. As shown in the figure, a charged particle moving above or below the meridian plane will experience strong vertical force directed towards the median plane both when it enters the high-intensity field region and also when it leaves it. Although there will be no focusing in the regions of low magnetic field (called valley), the focusing produced in the high field region (called the hill) is quite large. This type of focusing is also called sector focusing as the high-intensity magnetic fields may be produced by magnetic sectors. In this type of focusing, the trajectory of the particle is not circular but has smaller curvature in hill regions and larger curvatures in valley regions as shown in figure 9.11.

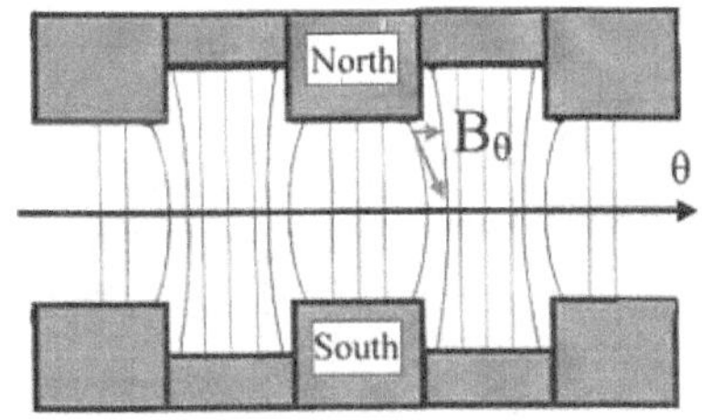

Figure 9.10 *Vertical focusing action of a magnetic field that varies with azimuthal angle*

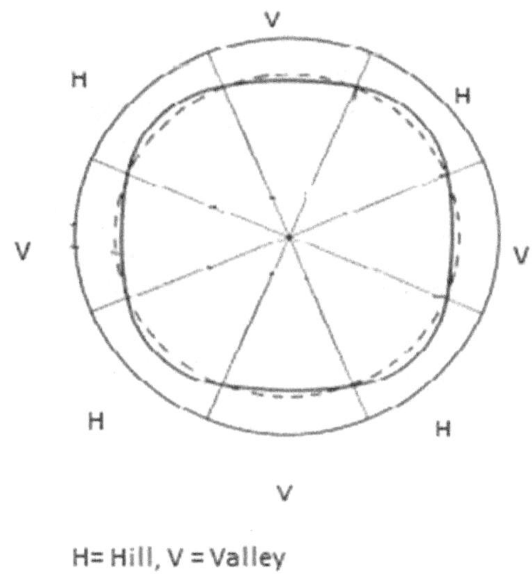

Figure 9.11 *Path of a particle in AVF focusing*

9.3 LINEAR ACCELERATOR (LINAC)

A linear accelerator is in a way a hybrid of an electrostatic accelerator and a cyclotron. Like cyclotron the accelerating potential in linear accelerator (called linac in short) is an RF field while the accelerating column is linear like that of an electrostatic accelerator. However, there is no magnetic field to confine the accelerated particles in circular orbits. Figure 9.12 shows the basic layout of a linear accelerator. A number of co-axial hallow metallic cylindrical pieces of increasing lengths, called drift tubes, are hung in an evacuated chamber. Alternate drift tubes are joined together and are coupled to a high-voltage RF oscillator. An ion source supplies a beam of charged particles that are already accelerated to a low energy in the ion source. Both positively charged particles such as protons and negatively charged particles such as electrons may be accelerated to high energies in a linac. Principal of operation is simple. It may be recalled that charged particles move with constant velocity when they pass through the drift tubes and get an injection of energy $\Delta E = QV_0$ (where Q is the charge of the particle and V_0 is the instantaneous potential of the RF wave at the moment the particle crosses the gap between the drift tubes) when they cross the gap. The phase of the RF signal and the lengths of the drift tubes are so adjusted that the particles find the polarity of the next drift tube appropriate for acceleration, whenever they reach a gap. The velocities of the particles increase as they cross gaps between drift tubes and gain energy.

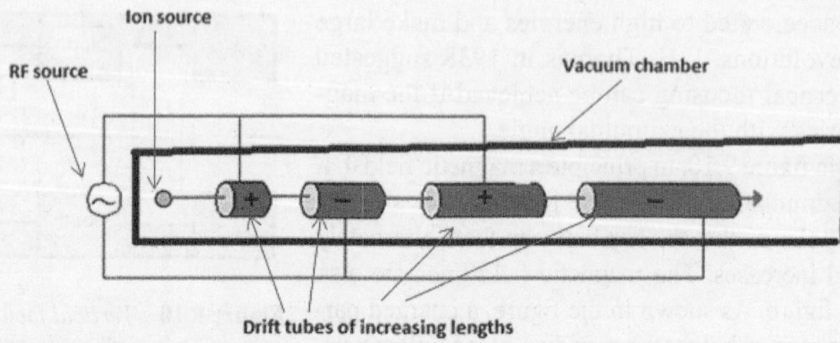

Figure 9.12 *Layout of a linear accelerator*

Let us assume that the frequency of the RF signal is f, and V_n and L_n are, respectively, the velocity of the charged particle in the n-th drift tube and the length of the n-th drift tube. To maintain resonance between the polarity of the RF signal required for the acceleration and the time particle takes in reaching the gap, it is required that the particle cover the length of each drift tube exactly in half time period (T) of the RF signal, that is

$$\frac{T}{2} = \frac{1}{2f} = \frac{L_n}{V_n} \tag{9.2b}$$

Or

$$L_n = \frac{V_n}{2f} \tag{9.2c}$$

However, the kinetic energy E_n of the particle when it is in n-th drift tube is

$$E_n = \frac{1}{2} m V_n^2 = nQV_0 + K \tag{9.2d}$$

Here, m is the mass of the particle. The constant K in Eq. (9.2d) considers the kinetic energy of the particle when it entered the first drift tube. From Eq. (9.2d), we obtain

$$V_n = \sqrt{\frac{2(nQV_0 + K)}{m}} \tag{9.2e}$$

Also, for RF $c = f\lambda$ and therefore,

$$f = \frac{c}{\lambda} \tag{9.2f}$$

Substituting the value of V_n from Eq. (9.2e) and of f from Eq. (9.2f) in Eq. (9.2c), we get

$$L_n = \frac{\lambda}{2c}\sqrt{\frac{2(nQV_0 + K)}{m}} \tag{9.2g}$$

Thus, the length of each drift tube and the total length of the accelerating column are proportional to the wavelength of the RF signal. Further, the length of each drift tube may be calculated using expression (9.2g). It may be shown that the total length of the accelerating column, that is the sum of the lengths of the drift tubes ignoring the lengths of the gaps and the initial energy, is given by

$$L = \left(\frac{1}{2f}\right)\left(\frac{2eV}{m}\right)^{1/2} \sum_{1}^{n} \sqrt{n} \tag{9.2h}$$

where f is the frequency of the RF potential of magnitude V applied at the gaps to accelerate particles of mass m. N is the number of drift tubes in the linac.

9.3.1 Acceleration of Electrons Using Linac

Acceleration of electrons to very high energy is quite simple in a linac. It is because the electron becomes relativistic at low energies (≈0.30 MeV) and moves almost with the velocity of light c which does not change further with the increase in the energy of the electron. As a result, in an electron linear accelerator the drift tubes are all of the same length.

Drift tubes of a linear accelerator may also be treated as wave guides filled with electromagnetic waves from the RF source. An electromagnetic wave in free space propagates such that both the electric field vector and the magnetic field vector are perpendicular to the direction of propagation. However, if an electromagnetic wave is made to travel in a region bounded by conducting walls (such as the drift tube or the waveguide/resonator), the boundary condition that the tangential component of the electric field and the normal component of the magnetic field at the conducting boundary must vanish, does not allow the electromagnetic wave to travel in the same way as it travels in free space. The boundary conditions can be fulfilled if either the electric field component or the magnetic field component travels in the longitudinal direction (say Z-direction) and the other component travels in the transverse direction (i.e. along the X-direction). The electromagnetic wave in which the electric field component travels in Z-direction and the magnetic component in X-direction is denoted by (TM) waves and may be used for the acceleration of charged particles along the Z-direction. Further, it may be shown that the phase velocity V_{ph} of the (TM) wave in the Z-direction is always more than the velocity c of light. For synchronization of the wave and the particle, it is essential that the V_{ph} and the particle velocity V_p are equal. It is, therefore, required to reduce the phase velocity in longitudinal direction. This is achieved by putting conducting disks at regular intervals in the wave guides. All modern electron accelerators are travelling wave devices. Most of these machines operate in pulse mode to reduce the power loss.

The size of the linac depends on the final energy and the particle type. The longest linac is at Stanford University, USA which is 2-mile long, while meter and a half long linacs are often used in hospitals for radiation therapy.

9.4 THE BETATRON

It is an accelerator that is used to accelerate electrons to high energies. It is based on the principle of induction. A time-varying magnetic field generates an electric field (by induction) that is used to accelerate the electrons. It consists of a hollow circular accelerating tube (also called doughnut shaped tube) with electron gun at one side and a target on the other, as shown in figure 9.13.

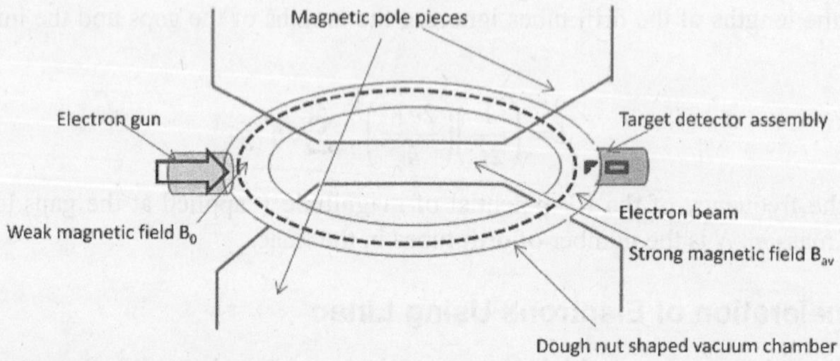

Figure 9.13 *Doughnut shaped betatron accelerating tube with electron gun and target*

The accelerating tube is placed between the pole pieces of an electromagnet. The strength of the magnetic field may be changed by changing the current through the magnet coils. However, the magnet is designed in such a way that the magnetic field intensity in the central region of the accelerating tube is twice as strong as at the location of the accelerating tube, at all instants of time when the field changes with time. The time-varying strong magnetic field in the central region induces strong electric field at the location of the accelerating tube, which accelerates the electrons. The magnetic field at the location of the electron orbit confines the electron beam in a circular path.

9.4.1 Betatron Condition for Acceleration

Let us assume that at the instant 't':

1. The average magnetic field in the central region = $B_{av}(t)$.
2. The magnetic field at the location of electron orbit = $B_0(t)$.
3. The radius of the electron orbit = r_0.

The magnetic flux ϕ_m linked with the circle of radius r_0, $\phi_m = \pi r_0^2 B_{av}(t)$.

If the magnetic field varies with time, it will induce an electric field $E(t)$ at the location of the electron orbit at $r = r_0$. Using Stokes theorem, we get

$$2\pi r_0 E(t) = \frac{d\phi_m}{dt} = \pi r_0^2 \frac{dB_{av}(t)}{dt}$$

Or

$$E(t) = \frac{r_0}{2}\frac{dB_{av}(t)}{dt} \tag{9.3a}$$

The induced electric field $E(t)$ will exert a force $F(t) = QE(t)$, on a particle of charge Q at $r = r_0$. Thus, the force experienced by the charged particle (electron)

$$F(t) = Q\frac{r_0}{2}\frac{dB_{av}(t)}{dt} \tag{9.3b}$$

However, force is equal to the rate of change of momentum; hence,

$$\frac{dp(t)}{dt} = F(t) = Q\,\frac{r_0}{2}\frac{dB_{av}(t)}{dt}$$

And

$$p(t) = Q\frac{r_0}{2}\,B_{av}(t) \qquad (9.3c)$$

The magnetic field $B_0(t)$ at the location of the orbit of the charged particle confines it in a circular path of radius r_0. This happens because the centrifugal force is balanced by Lorentz magnetic force, that is

$$\frac{mV^2(t)/r_0}{r_0} = QV(t)B_0(t)$$

Or

$$mV(t) = Qr_0B_0(t)$$

However, $mV(t)$ is the instant linear momentum $p(t)$ of the particle; hence,

$$p(t) = Qr_0B_0(t) \qquad (9.3d)$$

Equating (9.3c) and (9.3d), we get

$$2\,B_0(t) = B_{av}(t) \qquad (9.4)$$

The condition (9.4) says that the average central field should be twice as strong as the magnetic field at the location of the orbit at all instants. This condition must be satisfied at all instants as the fields change with time. This ensures the acceleration of the particle as well as their confinement in circular orbit. The strong central field and relatively weak magnetic field at the location of the accelerating tube are provided by designing the pole pieces and through the auxiliary correction coils of the electromagnet. The pole pieces have smaller gap in central region and larger gaps at the location of the doughnut tube. The electron gun provides a copious beam of electrons, which are accelerated by the induced electric field as long as the magnetic field is changing with time. The rate of acceleration is decided by the strength of the central magnetic field and the rate at which it changes with time. As is obvious, the acceleration takes place only during the variation of the magnetic field, which is increased up to a maximum, after that the accelerated beam is deflected by another pulsed magnetic field on to the target. Betatron provides a pulsed beam. Electrons up to a maximum 300 MeV energy have been accelerated by betatron. Betatron is better suited for the acceleration of electrons as they move almost with the velocity of light and there is no further change in their mass with energy once they are relativistic. A magnetic field that decreases with r provides both radial and vertical focusing.

9.5 THE SYNCHROTRON

9.5.1 Electron Synchrotron

One big drawback of betatron is the consumption of large power in producing time-varying strong central magnetic field that induces electric field for acceleration. This also makes betatron very bulky as it has a heavy magnetic core. An electron synchrotron has hollow circular doughnut

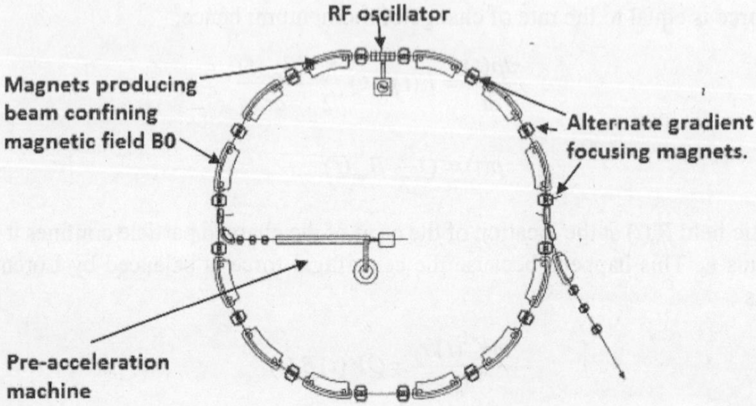

Figure 9.14 *Diagram showing the layout of a synchrotron*

shaped evacuated accelerating tube (such as betatron). Electrons from some other accelerator where they are already accelerated to relativistic energies (≈1 MeV) are injected in the machine. Electrons move in a circular orbit of radius r_0 and are confined to the circular path by the magnetic field B_0 at the location of the electron orbit. There is no strong time-varying magnetic field at the centre region of the accelerating tube. An RF field between a pair of electrodes (hollow metallic cylindrical pieces) placed around the electron orbit provides accelerating electric field. Electrons gain energy every time they pass through the gap between the electrodes. As electrons become relativistic at relatively low energy and move almost with the velocity of light, the radius of electron orbit does not change with the increase of energy. As such, the resonance or synchronization between the RF and the cyclotron frequency of motion is maintained. Electron synchrotrons may accelerate electrons up to several hundred BeV (=10^3 MeV) of energy. However, high-energy electrons moving in circular orbit have centripetal acceleration and lose energy by radiation. It may be shown that the energy lost by radiation ΔE_{rad} per revolution is given by

$$\Delta E_{rad} = \frac{8.85 \times (E)^4}{r_0} 10^{-14} \text{ MeV per revolution} \tag{9.5a}$$

where the energy E of the electron is in MeV and the radius of its motion r_0 is in metre. It is obvious that the energy gained per revolution must be larger than ΔE_{rad} if electrons have to be accelerated.

9.5.2 Proton Synchrotron

A proton synchrotron also works on the same principle as an electron synchrotron, except that a proton becomes relativistic at much higher energies and, therefore, the resonance condition does not hold good. The cyclotron frequency of a particle of m and charge q in magnetic field B_0 is given by

$$\omega_{cyc} = \frac{q B_0}{m} \tag{9.5b}$$

The resonance condition between the frequency of the applied RF field and the cyclotron frequency of the particle breaks down due to the increase of the mass of proton with energy. As such, in a proton synchrotron the applied RF frequency is modulated in accordance with the cyclotron frequency of the proton.

9.5.3 Alternate Gradient Focusing

A big problem with high-energy cyclic machines is the loss of beam intensity, which increases as the particle makes large number of revolutions to gain more energy. The intensity loss is due to the lack of vertical and radial focusing. Detailed analysis of the motion of charged particles in magnetic fields has indicated that a magnetic field that varies as $B(r) = B_0 r^n$, where n is a number can provide radial and vertical focusing but rather under opposite conditions. If n is large ($n > 1$), there is strong focusing in vertical direction (along Z-axis); however, there is de-focusing in radial direction. In addition, a negative large value of n gives radial focusing but no vertical focusing. It can, however, be shown that strong focusing both in vertical and in radial planes may be obtained by using a large number of magnets around the circumference of the particle orbit each having the same magnitude of the field B_0 at the central orbit, but alternately positive and negative n values. This is called the principle of alternate gradient focusing. The focusing action may be understood in terms of the following optical analogy. A combination of two lenses, a convergent lens, and another divergent lens of same focal length separated by a distance always produce a net focusing effect as shown in figure 9.15.

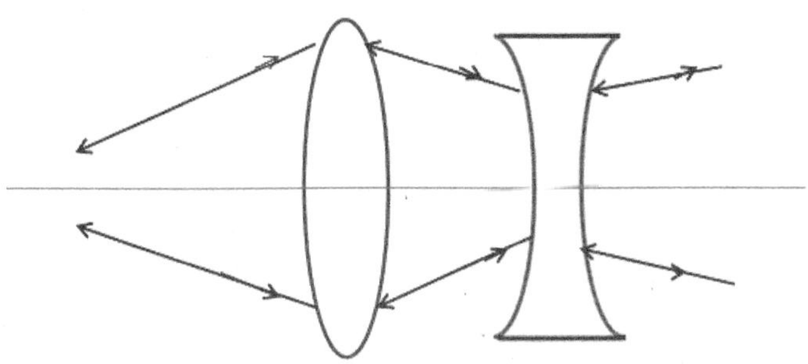

A combination of a divergent and a convergent lens of equal focal lengths always produces a convergent combination

Figure 9.15 *A combination of convergent and divergent lenses of equal focal lengths always produces net focusing*

9.6 PARTICLE COLLIDER MACHINES

Accelerators are used to study the structure of subatomic particles as well as to create and study the properties of hitherto unknown elementary particles. The energy available for actual interaction in collisions of accelerated particles is of paramount importance for the production of elementary particles. More is the available energy; elementary particles of heaver mass and of larger binding energies may be produced. We know that when a 10 MeV proton hits another stationary proton, a maximum of only 5 MeV of energy is available for the production of new particles or for the excitation of the system. The rest of the energy is used up as the kinetic energy of the products to conserve the incident linear momentum. At higher energies when relativistic effects become important, the available energy becomes still less; the available energy in relativistic collision with a stationary target is only $(2Mc^2E)^{1/2}$. In addition, if two particles of same mass moving with the same energy E in opposite directions are made to collide, $2E$ energy becomes available for the production of elementary particles or for the excitation of the composite system. Further, larger energies correspond to higher temperatures. Study of matter at very high temperatures is important from the perspective of the Big Bang theory of the creation of universe. Collider accelerators have been developed with these objectives in mind. Colliders are basically synchrotrons in which two beams of identical particles or particles and antiparticles are accelerated simultaneously but moving in opposite directions. The beams are then made to collide with each other creating matter at very high temperature. Colliders using P–P, e^-e^+, and P–P′ beams are presently in use.

9.7 SOME INTERNATIONALLY IMPORTANT ACCELERATORS

9.7.1 Large Hadron Collider (LHC)

Hadron refers to the composite elementary particles made up of quarks and held together by strong nuclear force. Neutron and proton are examples of hadron. The large hadron collider (LHC) is presently the world's highest energy accelerator of hadrons. It has been commissioned by the European Organisation for Nuclear Research (CERN). It is a synchrotron that is constructed in a circular tunnel of circumference 27 km long, 50 m to 175 m below the surface of mountainous terrain at the boarder of France and Switzerland near the city of Geneva in Switzerland. The synchrotron is designed initially to accelerate up to 7 TeV two beams of protons in opposite directions and to make them collide. Later, beams of lead nuclei will be accelerated up to 574 TeV. The collider tunnel contains two adjacent parallel beam lines that intersect at four points, each having accelerated proton beams moving in opposite directions. Six special detectors have been setup near the four points of intersection to study different processes.

9.7.1.1 Experimental setups of LHC

1. **ATLAS:** It is a toroidal LHC apparatus that is designed to investigate the origin of mass and to look for extra dimensions.
2. **CMS:** It is a Compact Muon Solenoid that is a part of the general detector setup and is dedicated to hunt for Higg's boson and dark matter.
3. **ALICE:** A large ion collider experiment is the detector setup meant for the detailed study of quark gluon plasma state of matter. This state of matter is likely to be produced in the collision of very high energy lead nuclei.
4. **LHCb:** Large Hadron Collider Beauty is the detector setup that will investigate as to what happened to the antimatter that was created along with matter in equal amount at the time of Big Bang. At present, there is only matter and almost no antimatter in the world.
5. **TOTEM:** A setup designed to study total cross-section for elastic scattering, diffraction, and dissociation processes.
6. **LHCf and MoEDAL:** Detectors will look for events in the forward direction, exotic events and monopoles.

9.7.2 Fermi Lab Tavatron

Tavatron was world's second highest energy synchrotron accelerator which was commissioned in 1983 at Fermi National laboratory just east of Batavia, Illinois, USA. It accelerated protons and antiprotons up to energies of 1 TeV and, hence, the name Tavatron. The main achievement of the accelerator was the discovery in 1995 of top quark that was predicted by the standard model of particle physics. The accelerator stopped operation on 30 September 2011 by the time more powerful LHC become operational.

9.7.3 CERN Super Proton Synchrotron (SPS)

This is the second largest machine in CERN's accelerator complex measuring nearly 7 km in circumference. It is a synchrotron that may accelerate proton, antiproton, electron, and positron and also heavy ions up to 400 GeV energy. Experiments code name UA1 and UA2 carried out during 1981–1984 using proton–antiproton beams led to the discovery of W and Z bosons. Carlo Rubbia and Simon van der Meer were awarded Nobel Prize for these discoveries in 1984. SPS is now used to inject high-energy beams to LHC for further acceleration.

9.7.4 Relativistic Heavy Ion Collider at BNL

It is one of the only two operating relativistic heavy ion colliders, at Brookhaven National Laboratory, in Uptown, New York, USA. RHI is a storage intersecting ring collider, which has six intersection points. Unique feature of the machine is that it may produce collisions between polarized protons. Heavy ion collisions at high energies are likely to produce state of matter that existed a few moments after the Big Bang. Study of primordial matter is of great importance from the perspective of creation of universe.

9.7.5 RIKEN-RIBF

Riken is Japan's multidisciplinary laboratory complex where a high-current radioactive ion beam facility is being developed to study exotic nuclei.

9.7.6 SLAC – Stanford University Linear Accelerator

Two mile (3.2 km) long radio frequency linear accelerator at United States Department of Energy National Laboratory run by the Stanford University is the world's straightest and largest linear accelerator that may accelerate electrons and positrons up to 50 GeV. Research work at these laboratories has won three Nobel Prizes in physics for discoveries of Charm quark, quark structure of nucleons, and tau-lepton.

Solved example S-9.1

In a synchrocyclotron, the frequency of the dee voltage at the beginning of an acceleration sequence is 20 Mc/s. What must be the final frequency if the deuterons in the pulse have energy of 469 MeV. The rest mass energy of deuteron is 1876 MeV.

Solution. As the mass of the particle increases with the increase of its energy, resonance frequency must be decreased to keep the resonance. If the initial frequency was f_0 and the final frequency is f, then $f = \frac{f_0}{\gamma}$. Therefore, we have to find out the value of γ.

$$\gamma = \frac{1}{\left(1-\frac{v^2}{c^2}\right)^{1/2}} = \left(1-\frac{v^2}{c^2}\right)^{-1/2} = \left(1+\frac{1}{2}\frac{v^2}{c^2}-\ldots\right) \approx \left(1+\frac{1}{2}\frac{m_0 v^2}{m_0 c^2}\right) = \left(1+\frac{\text{Kinetic energy}}{\text{rest mass energy}}\right)$$

$$\gamma = \left(1+\frac{469}{1876}\right) = 1.25. \text{ Hence, } f = f_0/1.25 = 20/1.25 = 16 \text{ Mc/s.}$$

Solved example S-9.2

A linac has five drift tubes and is driven by a 100 Mc oscillator. If protons are injected in the first drift tube with energy 50 kV and gain 100 kV in each gap calculate (a) the output energy after the fifth drift tube (b) total length of the accelerating section, ignoring the gap lengths.

Solution. (a) As there are five drift tubes, there are only four gaps. Hence, the total energy at the end of the fifth drift tube = 50 keV + 4 × 100 keV = 450 keV.

(b) The length of the accelerating column in a Linac of n tubes is given by $L = \left(\frac{1}{2f}\right)\left(\frac{2eV}{m}\right)^{1/2} \sum_{1}^{n} \sqrt{n}$ where f is the frequency of the applied RF field, V the voltage of the RF, m the mass of the particle to be accelerated and n the number of drift tubes. It is to be kept in mind that all these quantities are in MKS units.

$$\text{Therefore, } L = \left(\frac{1}{2\times 100\times 10^6}\right)\left(\frac{2\times 1.6\times 10^{-19}\times 100\times 10^3}{1.67\times 10^{-27}}\right)^{1/2} \sum_{1}^{4} \sqrt{n} = 13.43\,\text{cm}$$

Exercise p-9.1: Protons are accelerated in a cyclotron to 31 MeV energy. If the radius of the cyclotron dee is 50 cm, calculate the strength of the uniform magnetic field and the cyclotron frequency.

Exercise p-9.2: The charging belt of a tandem Van de Graaff is 25 cm wide moving with a speed of 1 m/s charges the terminal with a 250 mA current. Calculate the electric field strength near the belt.

Exercise p-9.3: Discuss the working of an electron linear accelerator. Why conducting ring are inserted in the path of the electromagnetic waves?

A linear accelerator produces 100 pulses per second of 5×10^{10} electrons with a final energy of 3 GeV. Calculate the average beam current and output power of the accelerator.

[**Hint:** Average current = charge carried per second = $100 \times 5 \times 10^{10} \times 1.6 \times 10^{-19} = 0.8\ \mu$A. The power = voltage × current = 3 GV × 0.8 × 10^{-6} W = 1.8 KW.]

Exercise p-9.4: Explain how electrons get accelerated in a betatron. Derive the necessary conditions that the central magnetic field and the magnetic field at the orbit must satisfy.

Exercise p-9.5: What were the limitations of the first generation cyclotrons? How were these overcome in later cyclic machines?

Exercise p-9.6: Discuss the role of the stripper in the working of a Tandem Van de Graaff.

Exercise p-9.7: What are collider accelerators? What is their special property? Give some details of the LHC.

Head-on collisions are observed between protons each moving relative to a fixed observer with the velocity corresponding to 10^{10} eV. If one of the proton is to be at rest with respect to the observer, what need to be the energy of the other proton to produce the same collision energy as before. The rest energy of the proton is 10^9 eV.

[**Hint:** As it is a head-on collision, the two protons are moving in opposite direction. Therefore, the observer is in the CM frame and the total collision energy is $E = (2 \times$ rest energy of protons $+ 2 \times$ kinetic energy of motion$) = (2 \times 10^9 \text{ eV} + 2 \times 10^{10} \text{ eV}) = 22 \times 10^9 \text{ eV} = 22$ GeV.

If one of the proton is stationary and the other is moving with kinetic energy E_1 then the collision energy $= (2 \times$ rest energy of proton $+ 2 \times$ rest energy of stationary proton $\times E_1)^{1/2} = E = 22$ GeV or $(2 \times 10^9 \text{ eV} + 2 \times 10^9 \times E_1 \text{ eV}) = 484$ GeV or $E_1 = 242$ GeV.

However, the rest energy of the proton is 10^9 eV $= 1$ GeV; therefore, the required energy is $242 - 1 = 241$ GeV.]

Multiple choice questions

Note: Some questions may have more than one correct alternative. Tick all correct alternatives in such cases.

Exercise M-9.1: The terminal voltage of the Van de Graaff accelerator is controlled by
(a) speed of the belt
(b) power of the DC charging source
(c) series chain resistances
(d) size of the high-voltage terminal

Exercise M-9.2: Strong focusing in the *Z*-direction may be achieved if the magnetic field in a synchrotron varies as
(a) $B(r) = B(0)r^{-5}$ (b) $B(r) = B(0)r^{-1}$ (c) $B(r) = B(0)r^{0}$ (d) $B(r) = B(0)r^{2}$

Exercise M-9.3: In a circular ring type, accelerator machine radial focusing will occur for
(a) $B(r) = B(0)r^{-5}$ (b) $B(r) = B(0)r^{-1}$ (c) $B(r) = B(0)r^{0}$ (d) $B(r) = B(0)r^{2}$

Exercise M-9.4: Which of the following accelerator(s)/(inherently) give pulsed beam?
(a) Betatron
(b) proton synchrocyclotron
(c) Van de Graaff
(d) Cockcroft–Walton accelerator

Exercise M-9.5: Pressurized quenching gas in van de Graaff machines is used to
(a) increase maximum terminal potential
(b) increase the ion current
(c) increase radiation shielding
(d) decrease power consumption

Exercise M-9.6: Which detector system associated with LHC is designed to study matter properties few moments after Big Bang?
(a) ATLAS (b) ALICE (c) CMS (d) TOTEM

Exercise M-9.7: The frequency of the RF field of a cyclotron is N times the cyclotron frequency. The machine will accelerate charged particles when

(a) N is even
(b) N is odd
(c) only when $N = 1$
(d) for all integer values of N

Exercise M-9.8: The time variation of the magnetic field in a betatron suddenly stops. The bunch of electrons under acceleration will

(a) stop moving
(b) keep moving and accelerating
(c) keep moving with constant energy
(d) starts deceleration

Exercise M-9.9: Particles with energy E_1 are injected in the first drift tube of 11 tube system. If the energy gain per gap is E_2, then the final energy at the exit of the 11th tube will be

(a) $E_1 + 10E_2$
(b) $10E_2 - E_1$
(c) $E_1 + 11E_2$
(d) $11E_2 - E_1$

Exercise M-9.10: In a colliding beam experiment, identical particles of rest energy 1 GeV moving with energy corresponding to 5 GeV collide head-on. The collision energy (GeV) in centre of mass is

(a) 12
(b) 10
(c) 8
(d) 6

10

Nuclear Energy

Energy that is defined as the capacity to perform work may be broadly classified into two types: the kinetic energy and the potential energy. The energy possessed by a moving body is kinetic energy, a very useful form of energy, as it can be made to do work. From kinetic theory, we know that temperature of a body is a measure of the kinetic energy of its constituents. It is the kinetic energy of the H_2O molecules in high temperature steam that gives it the capability to run turbines and generate electricity. Electricity is another very useful form of energy primarily because of the ease of its transmission from one place to another. Energy is also locked in all fields in the form of potential energy of the field. A body at a higher elevation possesses larger gravitational potential energy than a body at a lower height. In a similar way, energy is also stored in the nuclear field of the nucleus in the form of nuclear potential energy. It is possible to unlock the nuclear potential energy through nuclear reactions. A significant part of the energy stored as the potential energy of nucleons in the nuclear field of some nuclei may be converted into the kinetic energy by two nuclear reactions, the fission and the fusion. In this chapter, we discuss both these processes in more detail.

10.1 ENERGY FROM NUCLEAR FISSION

The process of fission was discovered in 1939 by Otto. Hahn and F. Strassemann when they observed that the heavy uranium nucleus breaks into two nearly equal halves if hit by neutrons. The possibility of fission and the energy released in the process may be easily understood from the variation in the average binding energy of nucleons with the size of the nucleus. Figure 10.1 shows the variation of the binding energy per nucleon (B/A) as a function of mass number A. The overall average value of the binding energy per nucleon is ≈8.0 MeV. Deviations from this overall average value give rise to the release of energy in fission and fusion of nuclei. For example, if a heavy nucleus with A around 200 breaks into two smaller parts each part with $A \approx 100$, the difference in initial and final binding energies, which is ≈200 MeV, is released. Fission of a heavy nucleus results into two fission fragments (FF), few (two to three) fission neutrons, and prompt γ rays. The fission fragments are highly excited neutron rich nuclei and they decay by β emission accompanied with neutrino and γ rays. The energy (≈200 MeV) released in fission is distributed amongst the fission products approximately in the following order:

Kinetic energy of fission fragments (FF)	≈ 82.5%
Prompt γ rays	≈ 3.5%
Fission neutrons	≈ 2.5%
β particles from FF	≈ 3.5%

γ rays from FF	$\approx 3.0\%$
Neutrinos	$\approx 5.0\%$

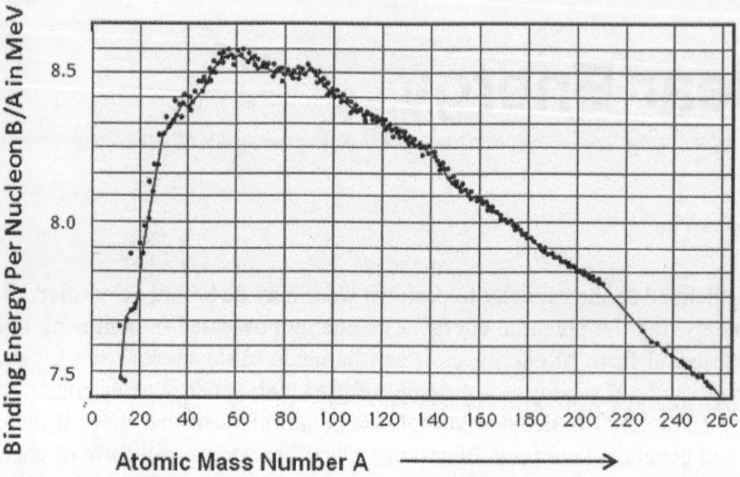

Figure 10.1 *Binding energy per nucleon as a function of atomic mass number A*

As may be seen, a very large part of the energy is released in the form of the kinetic energy of fission fragments. Fission fragments and other fission products lose their kinetic energy in collisions with the walls of container vessel, which in turn gets heated.

Bohr and Wheeler, on the basis of liquid drop model, showed that fission is the outcome of the interplay between the Coulomb energy and the nuclear surface energy. When a big nucleus is deformed its Coulomb energy decreases and the surface energy increases. As long as the increase in the surface energy is more than the decrease in the Coulomb energy, the deformed nucleus returns back to the equilibrium shape. However, if the deformation in the shape of the nucleus is so large that the change in Coulomb energy over rides the change in surface energy, the nucleus breaks into two smaller fragments ready to separate under the Coulomb repulsive force. However, the Coulomb barrier, also called fission barrier, restricts their separation. Tunnelling through the barrier becomes significant if the fission barrier is not too high above the kinetic energy of the FF inside the nucleus. Fission may be induced by supplying energy, called activation energy, to an otherwise stable nucleus. Further discussion on the topic is included in chapter 5 under the liquid drop model.

Nuclides that undergo fission by thermal neutrons are called fissile nuclides and they are all odd-A nuclides (^{233}U, ^{235}U, and ^{239}Pu). These nuclei are either stable or have very large mean life for self-fission. It is because they have a fission barrier of about 6–8 MeV height. Let us consider the thermal fission of ^{235}U as a typical example. When ^{235}U absorbs a thermal neutron, a compound nucleus ^{236}U* is formed, which is excited. Although the kinetic energy of a thermal neutron is negligible ($\approx$0.025 eV), its separation energy from the nucleus ^{236}U is about 7 MeV and, therefore, the excitation energy it supplies to the compound nucleus is around 7 MeV. As such, the compound nucleus ^{236}U* formed by the absorption of a thermal neutron has an excitation energy, which is larger than the height of the fission barrier. As a result, ^{236}U* undergoes fission and divides into nearly two equal parts (called fission fragments), a few (two or three) fast neutrons and releasing around 200 MeV of energy. Fission fragments are neutron-rich unstable nuclei and within a very short time (fraction of a second) they decay into more stable residual nuclei via β negative decay. One typical property of fission fragments is that, generally, they are not the same nuclei in all fission events. For example, three different fission modes of 236U* are as follows:

$$^{235}U + n\ (\text{thermal}) \longrightarrow 236U^* \longrightarrow \begin{cases} ^{144}Ba + {}^{90}Kr + 2\ n\ (\text{fast}) + \text{about } 200\ \text{MeV} \\ ^{141}Ba + {}^{92}Kr + 3\ n\ (\text{fast}) + \text{about } 170\ \text{MeV} \\ ^{139}Te + {}^{94}Zr + 3\ n\ (\text{fast}) + \text{about } 197\ \text{MeV} \end{cases}$$

It may thus be observed that each fission of a parent atom produces a different set of fission product nuclides. However, while an individual fission is not predictable, the fission products are statistically predictable. The amount of any particular isotope produced per fission is called its yield, typically expressed as percent per parent fission; therefore, yields total to 200% not 100%.

While fission products include every element from zinc through the lanthanides, the majority of the fission products occur in two peaks. One peak occurs at about (expressed by atomic number) strontium to ruthenium while the other peak is at about tellurium to neodymium. A typical mass distribution of product nuclei is shown in figure 10.2.

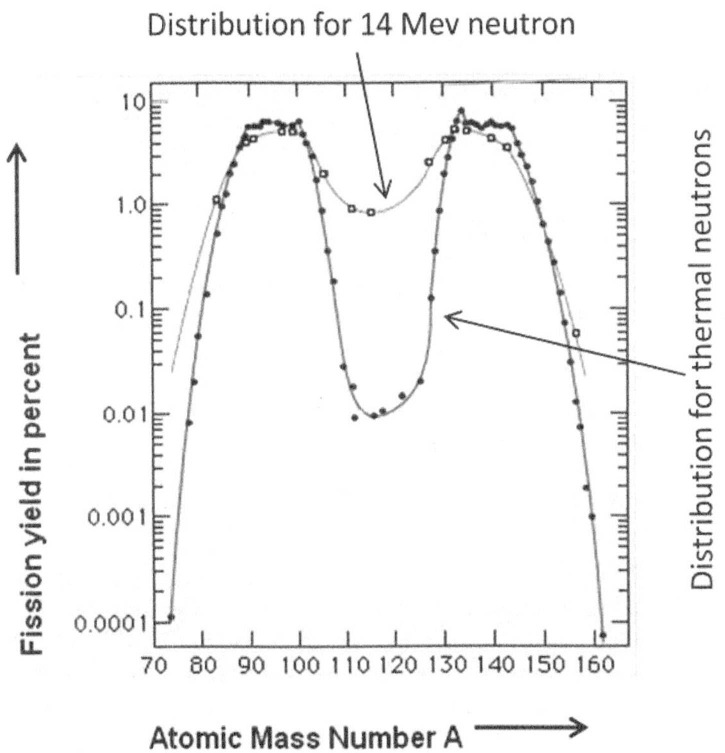

Figure 10.2 *Fission yield distribution of fission products for thermal and 14 MeV neutron fission of* ^{235}U

As may be observed in figure 10.2, symmetrical fission in which two identical fission products are formed is very much unlikely for thermal neutron induced fission. However, symmetrical fission becomes more probable as the energy of the incident neutron increases and for 14 MeV neutrons the fission of ^{235}U is almost symmetrical. From the theoretical perspective, the main difference between thermal neutron induced fission and the fission induced by high-energy neutrons is the participation of only $\ell = 0$ partial waves in the former, and of higher orbital angular momentum partial waves in the later. As such, the asymmetric mass distribution in fission appears to depend on the total angular momentum involved in the fission process.

10.1.1 Materials and Reactions Important from Fission Point of View

Heavy nuclides and nuclear reactions important from the fission viewpoint are classified as follows:

1. Fissile material: ^{233}U, ^{235}U, ^{239}Pu, and ^{241}Pu that may undergo fission by thermal neutrons.
2. Fertile material: ^{232}Th and ^{238}U that may be converted into fissile material by inducing nuclear reactions.
3. Self fission (SF) nuclides: Nuclides ^{241}Pu and ^{252}Cf that undergo spontaneous fission are called self-fission nuclides.
4. Nuclear reaction that converts a fertile material into a fissile material is called the breeding reaction. Two important breeding reactions are as follows:

$$^{232}_{90}\text{Th} + {}^{1}_{0}\text{n (Thermal)} \xrightarrow{(\text{n},\gamma)} {}^{233}_{90}\text{Th} \xrightarrow{\beta^-,\, t_{1/2}=22.4\text{ m}} {}^{233}_{91}\text{Pa} \xrightarrow{\beta^-,\, t_{1/2}=27\text{ d}} {}^{233}_{92}\text{U} \tag{10.1a}$$

$$^{238}_{92}\text{U} + {}^{1}_{0}\text{n (Fast)} \xrightarrow{(\text{n},\gamma)} {}^{239}_{92}\text{U} \xrightarrow{\beta^-,\, t_{1/2}=23.5\text{ m}} {}^{239}_{93}\text{Np} \xrightarrow{\beta^-,\, t_{1/2}=2.3\text{ m}} {}^{239}_{94}\text{Pu} \tag{10.1b}$$

10.2 FISSION CHAIN REACTION

A good source of energy must supply energy at a constant rate for a long time. To use fission as a source of energy, it is essential that a large number of nuclei undergo fission successively one after the other for a long time. This requires fission chain reaction. The possibility of self-sustained chain reaction in fission is due to the emission of fast neutrons in fission called fission neutrons. The fission of a fissile nucleus $^{A}_{Z}X$ by absorption of a thermal neutron may be represented by the following general equation:

$$^{A}_{Z}\text{X} + {}^{1}_{0}\text{n(thermal)} \rightarrow ({}^{A+1}_{Z}\text{X})^{*} \rightarrow \text{FF}_1 + \text{FF}_2 + \nu\, {}^{1}_{0}\text{n(fast)} \tag{10.2a}$$

In Eq. (10.2a), $[^{A+1}_{Z}X]^*$ is the intermediate excited compound nucleus (CN) formed as a result of the absorption of a thermal neutron by the fissile nucleus $^{A}_{Z}X$. In general, the compound nucleus (CN) may undergo fission or it may revert to the ground state by emitting γ rays. The fission probability of the (CN) is proportional to its fission cross-section σ_f^{CN}, which in turn depends on the energy of the incident neutron. Similarly, the probability of decay of the CN by γ emission is proportional to the cross-section σ_γ^{CN}. FF_1 and FF_2 are the two fission fragments and ν is the average number of fast neutrons emitted at the moment of fission. These fast neutrons emitted in fission play the key role in fission chain reaction. It is because these neutrons may be used to produce fission in other fissile nuclei, and neutrons emitted in the second generation fission may initiate third generation of fission and in this way the chain reaction may build up. It may, however, be realized that all fast neutrons ν generated in fission cannot initiate fission in other fissile nuclei. It is because some of the fission neutrons absorbed by the fissile nuclei may lead to γ emission from the CN instead of fission. This is called radiative capture of the fission neutron. An important parameter, η, defined as the number of fast neutrons generated per thermal neutron absorbed, is given by

$$\eta = \frac{\nu \sigma_f^{CN}}{\sigma_f^{CN} + \sigma_\gamma^{CN}} \tag{10.2b}$$

The factor η defined by Eq. (10.2b) considers the loss of neutrons by radiative capture. A larger value of η means higher chances of sustained chain reaction. As may be observed from Eq. (10.2b), apart from ν the value of η also depends on the CN-fission and γ emission cross-sections σ_f^{CN} and σ_γ^{CN}. Both these cross-sections are sensitive to neutron energy. Typical values of these parameters for the three fissile nuclei and for ^{238}U and are given in table 10.1.

^{238}U, which is about 99.3% in natural uranium, has negligible fission cross-section for low-energy neutrons but has very large cross-section (resonances) for the radiative capture (^{238}U (n, γ) reaction) of 5–100 eV neutrons. It essentially means that 5–100 eV energy neutrons if come in contact with ^{238}U nuclei are readily absorbed and are lost from the system. Let us consider the possibility of self-sustained fission chain reaction in a lump of natural uranium. Suppose that a stray thermal neutron present in the environment fission out a ^{235}U nucleus in the lump. To establish a self-sustained fission chain reaction in a sample of natural uranium (99.3% ^{238}U and 0.7% ^{235}U)it is, therefore, essential that: (1) the fast fission neutrons produced in fission events may not be allowed to be lost to the surrounding and (2) the fast fission neutrons be thermalized rapidly before being captured by ^{238}U and lost. Loss of neutrons from the lump may take place in two ways: (1) they leak out from the surface of the lump into the surrounding without producing fission in other ^{235}U nuclei. (2) Fast fission neutrons undergo scattering with ^{238}U and ^{235}U nuclei present in the lump, lose their energy and when their energies are of the order of 5–100 eV they are absorbed by ^{238}U nuclei before acquiring thermal energies (0.025 eV). The loss of neutrons through surface may be minimized by reducing the surface area of the lump. For a given mass, the spherical shape has minimum surface area and, therefore, if the lump is of spherical shape the neutron loss due to leakage through its surface may be minimized. However, to reduce the

Table 10.1 *Typical fission parameters of some nuclei*

Fissile nucleus	Neutron energy	σ_f^{CN}	σ_γ^{CN}	ν	η
^{233}U	0.025 eV (thermal)	525 b	70 b	2.51	2.21
^{235}U	0.025 eV (thermal)	582 b	108 b	2.47	2.08
^{239}Pu	0.025 eV (thermal)	729 b	300 b	2.91	2.06
^{238}U	Below 1 MeV Between 5 eV and 100 eV	0 0	Small Large resonances		

possibility of radiative capture by ^{238}U, the energy reduction (also called thermalization or energy moderation) of fast fission neutrons may be done rapidly and outside the assembly (where there are no ^{238}U nuclei). As a matter of fact, self-sustained fission chain reaction in a lump of natural uranium is not possible. However, if about 5 kg of pure ^{239}Pu or 15 kg of highly enriched ^{235}U are taken in a spherical shape and the spherical lump is further surrounded by neutron reflectors, fission chain reaction may start and the fissile nuclei fission out rapidly generating large amount of energy in a very short time. This uncontrolled fission chain reaction is used in making fission bombs. The minimum mass of the enriched fissile material required for sustained chain reaction is called the critical mass. In fission bombs, an amount of enriched fissile material more than the critical mass is taken into two parts that are kept separated from each other. The mass of each individual part is less than the critical mass and, therefore, is safe. A remote controlled system may bring the two parts of the enriched material in contact making the joint mass more than the critical mass and thus initiating the uncontrolled fission chain reaction. Once fission chain reaction starts, it cannot be stopped or controlled till whole of the enriched fissile material is consumed.

10.3 NUCLEAR FISSION POWER REACTOR

In a nuclear reactor, fission chain reaction is used in a controlled way. This is achieved by the use of moderator and control rods (also called arrestor rods). Moderator is a material that rapidly reduces the energy of fast fission neutrons. Low-Z materials, such as water (H_2O), heavy water (D_2O), and graphite (carbon), are commonly used as moderators.

10.3.1 Properties of a Good Moderator

A good moderator should rapidly reduce the energy of fast fission neutrons without absorbing them. The energy of (two to three) neutrons, emitted in the fission of a ^{235}U nucleus by thermal neutron, may vary between 8 MeV and 0.1 MeV. However, the average value of the energy of fast fission neutrons may be taken to be 2 MeV. A moderator is then required to reduce this 2 MeV of average neutron energy to the average thermal energy of 0.025 eV in the shortest possible time so that the neutron is not lost. Elastic scattering with the nuclei of the moderator is the dominant process by which a neutron loses energy. The energy loss by elastic scattering in each individual collision depends on the initial energy E_i of the neutron, the mass of the moderator nucleus and the angle of scattering. Therefore, the energy loss in each collision is different and, therefore, the final energy E_f of the neutron after each collision is different. If the average values of E_i and E_f are denoted by $\overline{E}_i$ and $\overline{E}_f$, respectively, then it has been found that the quantity ($\ln \overline{E}_i - \ln \overline{E}_f$), called the *average logarithmic energy decrement*, denoted by Greek letter ξ, remains nearly constant for all collisions with the nuclei of a given moderator. The value of ξ is given by the relation

$$\xi = 1 + \frac{(A-1)^2}{2A} \ln \frac{(A-1)}{(A+1)}$$

Here, A is the mass number of the moderator nucleus.

For $A > 10$, the value of ξ reduces approximately to $\xi_{A>10} = \dfrac{2}{(A+2/3)}$.

The average number of collisions N^C in which the initial energy E_i of the neutron falls to a value E_n may be given by

$$N^C = \frac{1}{\xi} \ln \frac{E_i}{E_n}$$

If σ_s denotes the microscopic scattering cross-section per nucleus (dimensions of area) and N the number of scattering nuclei per unit volume of the moderator (number per unit volume), then the macroscopic scattering cross-section, denoted by Σ_s, may be written as

$$\sum_s = N\sigma_s$$

Σ_s has the dimension of inverse length (cm^{-1} or m^{-1}).

The slowing down time, t, in which the initial energy E_i of the neutron becomes E_f as a result of elastic scattering with the nuclei of the moderator is given by

$$t = \frac{\sqrt{2m}}{\xi \sum_s} \left[E_i^{-1/2} - E_f^{-1/2} \right]$$

The *slowing down power* of a moderator is defined as ($\xi\Sigma_s$) and the *moderating ratio* as $\left[\frac{\xi\Sigma_s}{\Sigma_{abs}}\right]$. Here Σ_{abs} is the macroscopic absorption cross section.

The approximate number of collisions for various light elements required to thermalize fast neutrons ($\approx$2MeV) are given in table 10.2.

Table 10.2 *Average number of collisions required for thermalization of fast fission neutrons in different moderators*

Material	H_2O Light water	D_2O Heavy water	He Helium	Li Lithium	C Graphite
No. of collisions	18	25	43	67	114

Although light water as a moderator is better than all others, it also absorbs neutrons. Heavy water is costly. Graphite being economical and easily available is often used as a moderator.

10.3.2 Properties of the Control Material

Control rods are made of a material that absorbs thermal neutrons without producing fission or creating another neutron. Commonly used materials for control rods are hafnium, cadmium, and boron all of which have large cross-section for the absorption of thermal neutrons. Figure 10.3 shows the basic structure of a controlled fission chain reaction. If fission is induced in a ^{235}U nucleus by stray neutron it will build through the fast fission neutrons that are thermalized by the moderator. First generation neutrons lead to a higher number of second generation neutrons, which in turn will lead to still larger number of third generation neutrons and so on. More and more nuclides fission out as the fission chain builds up. As a result, energy is released at a faster rate. If it is required to stop further build up of fission chain, control roads are inserted in the system, which absorb thermal neutrons and reduce the rate of fission of fissile nuclei, thereby reducing the rate of energy production. When energy production rate starts dropping down below the desired level, the control rods are withdrawn. In this way, by inserting and withdrawing control roads in succession, energy production rate of the reactor may be maintained at the desired level.

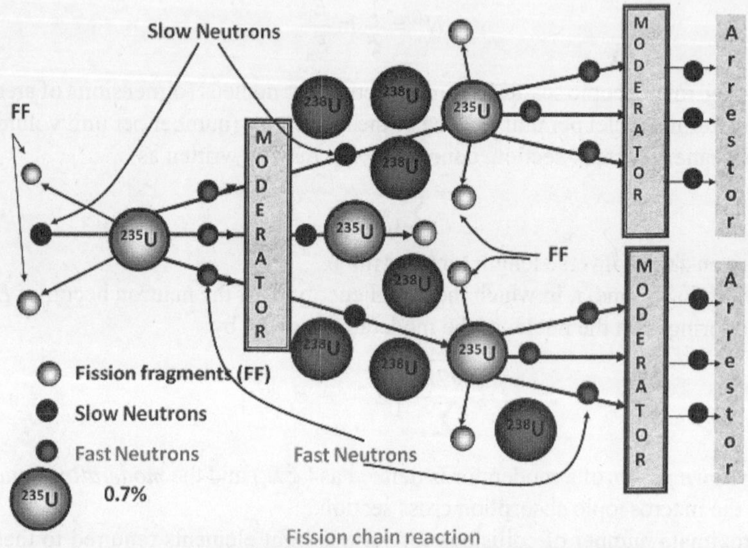

Figure 10.3 *Pictorial representation of fission chain reaction*

The simplest way to make a fission reactor or a chain-reacting pile, as it was called, is to make lattice of alternate blocks of uranium and graphite (see figure 10.4). The fast fission neutrons are produced in the uranium block by the fission of ^{235}U and they move out to the graphite block where they got thermalized. It may be noted that in alternate graphite and uranium lattice arrangement, thermalization of fast fission neutrons takes place in graphite lattice and not in uranium lattice. This reduces the probability of radiative capture of fast neutrons by ^{238}U nuclei. Thermal neutrons initiate fission in other ^{235}U nuclei in the next uranium block. In this way, a self-sustained fission chain reaction builds up in the pile.

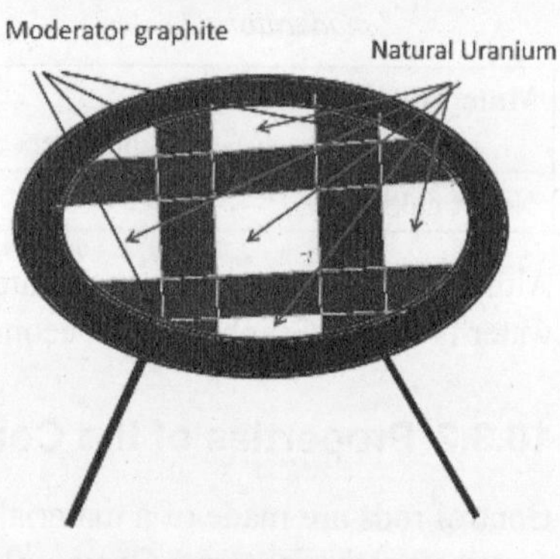

Figure 10.4 *Basic layout of a pile reactor with alternate lattices of natural uranium and moderator graphite*

10.3.3 The Neutron Multiplication Factor *k*

The fission chain reaction is characterized by the neutron multiplication factor k, which is defined as the ratio of the number of fission neutrons in n-th generation to $(n-1)$th generation. If k is less than one, it will mean that the reaction chain will not grow but stop. In reactor terminology, $k < 1$ means a sub-critical system, $k = 1$ a just critical system, and $k > 1$ a super critical system.

If it is desired that a chain reacting pile or any nuclear reactor generates energy at a constant rate, the neutron multiplication factor k should be 1. Assuming that the reactor is infinitely large so that no neutrons leak through the reactor walls, we may calculate the value of k_∞, the neutron multiplicity factor for an infinite reactor, by following the life cycle of the neutron, that is, what happens to the neutrons as they go from one generation to the next.

- We begin with N thermal neutrons.
- Each of this will produce ν fast neutrons per fission. However, all of the ν neutrons will not produce fission, in particular some are captured. If capture of thermal neutrons is considered, we are left of η the mean number of fission neutrons produced per thermal neutron (Eq. 10.2b) instead of ν. The value of η for some nuclei is given table 10.1. Let us calculate the value of η for natural uranium. The value of ν for natural uranium is the same as for the isotope ^{235}U, as isotope ^{238}U has negligible fission cross-section for thermal neutrons. The radiative capture cross-section for ^{238}U is 2.75 b (barn). Keeping in mind that the relative abundance of ^{235}U and ^{238}U in natural uranium is 0.7% and 99.3%, respectively, and using the expression for η as given by Eq. (10.2b), we have

$$\eta_{\text{nat U}} = 2.47\frac{\frac{0.7}{100}\times 582+\frac{99.3}{100}\times 0}{\frac{0.7}{100}\times 108+\frac{99.3}{100}\times 2.7} = 1.33$$

- *Using enriched uranium*: We see that the value of η for natural uranium is very nearly 1 if it is assumed that no loss of thermal neutrons from leakage occurs. However, to ensure that the system remains critical, instead of natural uranium, enriched uranium with about 3% of ^{235}U is used. The value of η for 3% enriched uranium is $\approx$1.84.
  Moderation: Out of our original N neutrons some have been absorbed and some cause fission so we now have ηN fast neutrons, which we must arrange to lose energy and become thermal neutrons – to take advantage of the high fission cross-section for thermal neutrons.
- This is achieved in the collisions with nuclei in the moderator. However, in the course of their random walk through the pile they may encounter ^{238}U nuclei, which have a small cross-section (~1 b) for fission by fast neutrons. We have to introduce another factor ε – the fast fission factor, which considers this effect.
- Typical value of ε for natural uranium is $\approx$1.03. The number of fast neutrons is now $N\eta\varepsilon$.
- While losing energy from the range of MeV to 0.025 MeV (thermal energy) the neutron will pass through 2–100 eV energy range in which ^{238}U has large number of resonances with very large cross-sections ($\approx$1000 b)for radiative capture. Unless the neutrons are shielded from these resonances we will lose all our potential thermal neutrons. How to avoid these resonances? The answer lies in making the moderator thickness sufficiently large so that neutrons are thermalized in the moderator and are of energy less than 5 eV when they come out of the moderator. In case the moderator is graphite, 19-cm thick graphite layer is sufficient to thermalize neutrons of average energy of 2 MeV. This does not totally eliminate the problem. Some neutrons are still lost by absorption in resonances. This is accounted by introducing the resonance escape probability p.

- A typical value of p is 0.9.
- Now, the number of neutrons left after thermalization is $N\eta\varepsilon p$.
- While the thickness of the moderator must be sufficient to thermalize the fast neutrons it must not be too large. It is because some neutrons are also lost by absorption in moderator medium as well as in other reactor components. To consider these neutron losses, a factor f-called thermal utilization factor is introduced, which gives the fraction of the thermal neutron actually available for fission.
- Typically f is about 0.9.
- Overall number of neutrons that survive thermalization is $N\eta\varepsilon pf$.
- The neutron multiplicity factor $k_\infty = \frac{N\eta\varepsilon pf}{N} = \eta\varepsilon pf$ (10.3a)
- This is called the *four factor formula*.

The value of the neutron multiplicity k_∞ has been derived assuming that the reactor is of infinite dimensions and, therefore, no leakage of thermal or fast neutrons from the surface of the reactor takes place. However, the real reactors are of finite dimensions. Hence, corrections for neutron loss from the finite dimensions of the reactor both for thermal (or slow) neutrons and for fast neutrons need to be applied.

If L_f and L_s are, respectively, the fractions of the fast and slow neutrons lost then for a real reactor

$$k_{\text{real}} = (1-L_f)(1-L_s)\eta\varepsilon pf \tag{10.3b}$$

Eq. (10.3b) is called the *six factor formula*.

For an infinite reactor that uses natural uranium as fuel, the approximate values of various factors are

$$\eta_{\text{nat U}} = 1.33;\ \varepsilon = 1.03;\ p = 0.9;\ f = 0.9 \text{ and, therefore, } k_\infty = 1.33 \times .1.03 \times 0.9 \times 0.9 = 1.11$$

The factors $(1 - L_f)$ $(1 - L_s)$, however, reduce k_{real} to less than 1. It is, therefore, required to use fuel that is slightly (about 3%) enriched in ^{235}U.

10.4 THE DELAYED NEUTRONS

So far we have been discussing about fast fission neutrons that are emitted at the moment of fission. However, there is another type of neutrons, called delayed neutrons, which are emitted sometime after the fission from the excited fission fragments as they acquire stability. The fission fragments are generally neutron-rich nuclei that decay by β negative emission with a certain half-life. A nucleus ^A_ZX on β negative decay leaves the residual nucleus $_{Z+1}^{A}Y$ in states of different excitations, depending on the original excitation energy and the energy carried away by the β^- particle and the anti-neutrino. The excitation energy of some of the residual nuclei is more than the separation energy of the last neutron. Neutrons are emitted from such excited β^- residues. The emission of these neutrons is delayed by the mean life time of the β decay. The number of delayed neutrons is about 0.6% of the fission neutrons and they are emitted with time delays of 0.2–50 s.

10.5 CONTROLLING REACTOR POWER OR REACTIVITY: IMPORTANT ROLE OF DELAYED NEUTRONS

The instantaneous power of a reactor depends on the number of fissions taking place at the instant. There is a delicate balance between the neutron production and absorption in a reactor. The power level of the reactor depends on the number of fissile nuclei undergoing fission at a given instant, which in turn is proportional to the number of neutrons available at that instant. It is not possible to design a reactor in which the number of neutrons remains constant for successive generations. Instead, extra fissile fuel is loaded and the additional neutrons, whenever they increase beyond the required power level, are removed from the system by control rods. Control rods are made up of neutron absorbing materials such as cadmium with silver and indium, boron, and hafnium. Control rods can be inserted or withdrawn from the reactor core (active volume where fission takes place) by remote control.

A slight variation of neutron multiplicity factor k results in the rapid, almost exponential growth or decay of neutrons. For example, it can be shown that if k changes from 1 to 1.01 the number of neutrons will grow by approximately 22,000 in one second. This is a very rapid growth and it may not be possible to control the reactor power level. Fortunately, delayed neutrons come for rescue. Delayed neutrons lengthen the time scale of neutron growth and make it possible to insert or withdraw control rods to regulate the power level. Reactors are designed in such a way that the system remains super critical if both the fast and the delayed neutrons are present and becomes subcritical if delayed neutrons are absorbed by the control roads. Without delayed neutrons reactivity control in reactors would not have been possible.

10.6 ESSENTIAL ELEMENTS OF A REACTOR

1. **Fuel** – pellets of UO_2 (generally, 1 cm diameter by 1.5 cm long) are arranged in tubes to form fuel rods, which are usually formed into fuel assemblies in the core.
2. **Moderator** – usually water but may be graphite or heavy water.
3. **Control rods** – made with neutron-absorbing material so that inserting or withdrawing the rods controls or halts the growth of the reaction rate.
   Note: Secondary shutdown systems involve adding other absorbers of neutrons, usually in the primary cooling system.
4. **Coolant** – liquid or gas circulating in the core to carry away heat. In light water reactors, the coolant acts both as moderator and coolant.
5. **Pressure vessel** – usually a robust steel vessel of tubes containing the core and moderator/coolant but it may be a series of tubes holding the fuel and conveying the coolant through the moderator.
6. **Steam generator** – part of the cooling system where the reactor heat is used to make steam to drive the turbines.
7. **Containment** – structure round core to protect it from intrusion and to protect the outside from radiation in case of a major malfunction.

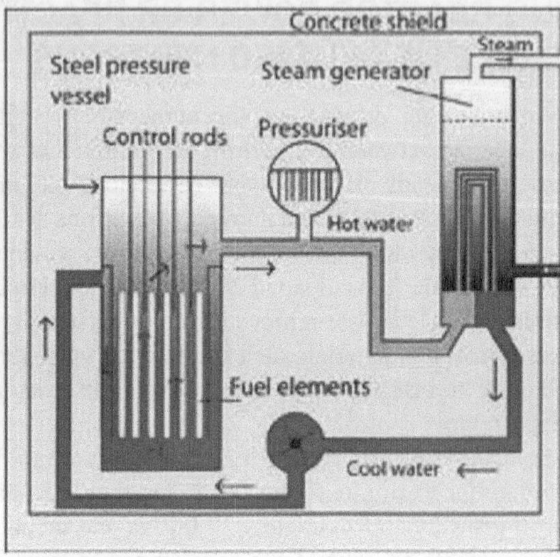

Figure 10.5 *Working diagram of a nuclear fission reactor*

Figure 10.5 shows the working diagram of a nuclear fission reactor.

10.7 CLASSIFICATION OF FISSION REACTORS

Fission reactors may be classified in several ways:

1. *On the basis of the energy of neutrons that drive the fission chain*: (i) thermal (neutron) reactors and (ii) fast (neutron) reactors. As thermal reactors are economically more viable, most of the reactors in operation are of this type. Fast reactors are costly and require enriched fuel. However, the radioactive waste of fast reactors has relatively less component of long-lived radioactive nuclei.
2. *Classification by moderator material used in thermal reactors*: Following materials may be used as moderators: deionized light water, heavy water, and graphide. Light-water reactors use ordinary water to moderate and cool the reactors. If the temperature of the water increases, its density drops, and fewer neutrons passing through it are slowed enough to trigger further reactions. This negative feedback stabilizes the reaction rate. Graphite and heavy water reactors tend to be more thoroughly thermalized than light-water reactors. Due to the extra thermalization, these types can use natural uranium/un-enriched fuel.
3. *Classification by coolant*:

   (a) Water-cooled reactors:

   i. Pressurized water-cooled reactors: Most of the commercial water moderated and water-cooled reactors use pressurized water. As a matter of fact, the reactor vessel is a pressure tank.

   ii. Boiling water reactors: In these reactors, water at high pressure is circulated through the fuel. Heat produced as a result of fission is taken away by water, which starts boiling. The boiling water then transfers heat to another cold water loop where steam is generated to run turbines. As such, there are two different cooling loops and, therefore, the chances of the leakage of radioactive water coming directly from the fission zone in loop one, is minimized.

iii. Liquid metal-cooled reactors: Water, because of its moderating properties, cannot be used as a coolant in a fast neutron breeder reactor. Liquid sodium is often used in such reactors.

iv. Gas-cooled reactors: Inert gases such as helium, carbon dioxide, nitrogen have been used for cooling of reactor.

4. *Classification by use*: Reactors may be classified according to their use as:
   (a) Power plants
   (b) Propulsion, foremost nuclear marine propulsion
   (c) Various proposed forms of rocket propulsion
   (d) Other uses of heat, for example H_2 production for use in a hydrogen economy
   (e) Desalination
   (f) Heat for domestic and industrial heating
   (g) Production reactors for transmutation of elements
   (h) Breeder reactors. Fast breeder reactors are capable of enriching uranium during the fission chain reaction (by converting fertile U-238 to Pu-239), which allows an operational fast reactor to generate more fissile material than it consumes. Thus, a breeder reactor, once running, can be re-fuelled with natural or even depleted uranium.
   (i) Creating various radioactive isotopes, such as americium for use in smoke detectors, and cobalt-60, molybdenum-99, and others, used for imaging and medical treatment.
   (j) Production of materials for nuclear weapons such as weapons-grade plutonium
   (k) Providing a source of neutron radiation and positron radiation (e.g. neutron activation analysis and potassium–argon dating)
   (l) Research reactors: Typically, reactors used for research and training, materials testing, or the production of radioisotopes for medicine and industry. These are much smaller than power reactors or those propelling ships, and many are on university campuses.

10.7.1 Modern Technologies

1. *Pressurized Water Reactors (PWR)*: These are reactors cooled and moderated by high pressure liquid (even at extreme temperatures) water. They are the majority of current reactors, and are generally considered the safest and most reliable technology currently in large-scale deployment.
2. *Boiling water reactors (BWR)*: These reactors use water both as coolant as well as moderator. The pressure on water is slightly less so that it boils in the reactor vessel. Generally, there is a single loop and the boiling water directly generates steam for turbines. They are less safe as the chances of radioactive water leaking are high.
3. *Pressurized heavy water reactors (PHWR)*: This is a Canadian design of reactor called CANDU. Pressurized heavy water is used both as coolant and moderator. The reactor vessel is not a single big size vessel as in PWR, but consists of a large number of small-sized pressure tubes that are used to keep the fuel. This is one of the most efficient types of reactors.

10.8 PROBLEMS ASSOCIATED WITH FISSION REACTORS

10.8.1 Core Meltdown

Fission reactors use self-sustained fission chain reaction that builds up with the number of neutrons in the system. The neutron multiplication factor k is >1 and the system is supercritical. Control rods are used to maintain the reactivity and the power level of the reactor. Any malfunctioning of control rod and associated mechanical systems that may result in non-removal of extra neutrons (not absorbed by the control roads) will lead to very rapid growth of neutrons and the chain reaction. The reactor will thus behave like un-controlled bomb and very high temperatures will be generated inside the reactor vessel. The metallic claddings of fuel rods may get melted and the melted lump of fuel may settle at the bottom of the reactor vessel. In the case of enriched fuel, it is also possible that the total mass of the molten fuel at the bottom may increase beyond the critical mass that may further accelerate the fission process. In extreme case, the reactor vessel may itself melt. Very high pressures may get generated within the containment walls and they may collapse realizing highly radioactive gases and matter in the atmosphere. This is referred as core meltdown. A similar scenario may develop if the heat extraction system that removes heat from the reactor core and uses the extracted heat to produce steam fails.

Two major and one minor accident involving complete and partial meltdowns have happened so far. In 1986, the core of a reactor at Chernobyl, Ukraine (in Russia at that time) suffered complete meltdown. It was an old reactor and had no containment walls. Large amount of radioactive material leaked into the atmosphere. Partial core meltdown of reactors at Fukushima Daiichi reactor complex in Japan occurred in 2011 just after the complex was hit by large Tsunami waves. A relatively smaller meltdown accident occurred at the Three Mile Island, USA in 1979.

10.8.2 Production of Weapon-Grade Fissile Material

Most of the present day reactors use natural or enriched uranium as reactor fuel that has large percentage of ^{238}U isotope. Substantial amount of $^{239}_{94}$Pu, a weapon-grade fissile material, is produced in such reactors by the following reaction:

$$^{238}_{92}\text{U} + ^{1}_{0}\text{n} \rightarrow ^{239}_{92}\text{U} + \gamma$$

$$^{239}_{92}\text{U} \rightarrow ^{239}_{93}\text{Np} + \beta^- + \nu^-;$$

$$^{239}_{93}\text{Np} \rightarrow ^{239}_{94}\text{Pu} + \beta^- + \nu^- \quad (10.4)$$

The three-step reaction shown in Eqs. (10.4) in which fertile material (^{238}U) is converted into fissile material is called breeding reaction. As the fission cross-section for ^{239}Pu is large (729 b) it is often used to make fission bombs and is termed as weapon-grade fissile material. As more than 400 fission reactors are in operation in the world, large amount of weapon grade fissile material has already accumulated in the world.

10.8.3 Radioactive Nuclear Waste

The spent fuel from reactors also called reactor waste contains highly radioactive materials. Some of these radioactive materials have very long lives. Nuclear waste is hazardous because of the emission of radiations and also because it is toxic. No fool-proof method of its safe disposal has been found as yet. In some cases, the waste after extracting unspent fuel is first kept in large water tanks for cooling and for the decay of short-lived radioactive products. Later, the waste containing long-lived radio isotopes is imbedded in glass matrix, hermitically sealed in stainless steel containers and buried deep in arid areas where there is no underground water. Accumulation of radioactive waste is a big problem.

10.8.4 Limited Supply of Uranium Fuel

Supply of natural uranium, the main source of reactor fuel, is limited. Like coal or oil the presently available uranium ore is likely to deplete in the next 50 years or so.

10.9 FAST AND THERMAL BREEDER REACTORS

10.9.1 What is a Breeder Reactor?

In a normal fission reactor, some fissile fuel burns, energy is generated and some fissile material is also produced as a by product. For example, in a reactor that uses natural uranium {a mixture of ^{238}U [(99.3%), fertile material] and ^{235}U [(0.7%), fissile material]}, some nuclides of ^{239}Pu are also synthesized via the breeder reaction shown by Eq. (10.1b). $^{239}_{94}$Pu is a fissile material. However, the conversion factor, which is the ratio of the number of fertile nuclides ($^{239}_{94}$Pu) produced to the number of fissile nuclides ($^{235}_{92}$U) burnt is less than 1 in normal fission reactors. The conversion factor for uranium fuel reactors is around 0.6. A reactor that is specifically designed such that the conversion factor is at least 1 or greater than 1 is called a *breeder* or *breeder reactor*. A breeder reactor produces more fissile material than what it consumes. After running for some duration, say T_D sufficient fissile material is accumulated in a breeder that may be enough to run a new reactor. The time T_D is called the doubling time and depends on the overall efficiency of the conversion process. The breeder reactor that employs reaction (10.1a), using thorium as fertile material, is called a *slow (or thermal) breeder*, as it utilizes slow or thermal neutrons for breeding, while the reactor that employs breeding reaction (10.1b) based on the conversion of fertile ^{238}U is called a *fast breeder*, because fast neutrons initiate breeding reaction.

10.10 ADVANTAGES OF FISSION ENERGY

10.10.1 Non-polluting Source of Energy

Nuclear energy is the most environment-friendly source of energy. Emission of obnoxious gases such as NO_2, NO, and greenhouse gas such as CO_2 are almost totally avoided. Because of the confinement of radioactive waste and relatively small amount of fuel required to run a reactor, both the ground and atmospheric pollution is negligible. Production and use of nuclear energy does not add to global warming.

10.10.2 Fuel Economy

A very small quantity of nuclear fuel is required to produce large amount of energy. Table 10.3 gives the approximate amount of electrical energy (in kilo watt hour – kWh) that may be produced

Table 10.3 *Energy density of different materials*

1 kg of material	Approximate amount of electrical energy in kWh
Fire wood	1
Fossil fuels	3–5
Natural Uranium	50×10^3
Plutonium	60×10^5

from 1 kg of different fuels. As may be seen from the table, the energy density of nuclear fuel is the highest.

10.10.3 Land Requirement, Installation Time and Cost

Nuclear fission reactor technology is time tested and well developed. The land area requirements are comparable to the conventional fossil fuel power plants of comparable size. Although the initial coast of a nuclear plant is more, in long run it is either comparable or economical compared to other sources of energy.

10.11 ACCELERATOR-DRIVEN ENERGY AMPLIFIER

The two main drawbacks of a fission reactor are (1) the production of high activity and long life radioactive waste and (2) the possibility of radioactive fallout if the control and/or the cooling mechanisms fail. It is, however, worth noting that different long-lived radioactive nuclides formed during the fission of ^{235}U may also undergo fission and thus burn out producing more energy and short-lived radioactive products if there are sufficient neutrons of different energies (other than thermal energies) that are suitable for the fission of these long-lived radio isotopes. Even ^{239}Pu, a by-product of present day fission reactors, may be made to burn producing more energy if the number of thermal neutrons in a reactor is more than what is required for sustaining the fission chain reaction of ^{235}U. The basic problem with a conventional nuclear power reactor is that the total neutron economy of the system is based on prompt and delayed fission neutrons. It means that if it is required to increase or decrease the number of neutron in the reactor, it can only be achieved by changing the fission rate. No other external factor can change the neutron flux. Further, the system is kept slightly supercritical (k is only slightly greater than 1) and when the number of neutrons increases it is brought back to the original level by inserting control rods. As such, there are never surplus thermal and other energy neutrons that may produce fission of long-lived radio nuclides and of weapon grade materials produced in the system. If k is kept much larger than 1, the control mechanism fails. This happens because the control mechanism is based on delayed neutrons. If k is much larger than 1, the number of fast fission neutrons becomes so large in comparison to the delayed neutrons that the fission chain will grow even if delayed neutrons are absorbed by control rods.

To overcome these problems, a new concept is taking shape in terms of the *accelerator based energy amplifiers*. A high-energy proton beam of about 800 MeV from an accelerator is made to hit a heavy material such as a lead target. A large number of neutrons, called spallation neutrons, are produced in the interaction of proton beam with the target. Spallation neutrons have energy distribution that is further modified by their scattering from the lead block. As such, there are zones around the point of interaction, which have neutrons of different energies. If heavy nuclei are introduced in zones with neutrons of appropriate energy they may undergo fission producing energy. Not only the nuclear waste produced in such a system will automatically burn out, nuclear waste accumulated at various fission reactor centres may be charged in the system to burn and produce energy. All weapon-grade material accumulated the world around may also be used to produce energy by putting the material in the system. The main advantage of the system is that the neutron flux in the system is governed by the strength and energy of the proton beam and not by the fission rate. Further, neutrons of all energies are available for the fission of different isotopes. The power level of such a system will not be governed by the fission rate instead it will be controlled by parameters of the proton beam. And the fission process may be immediately stopped by switching off the proton beam, just like switching off a light bulb. It is obvious that initially considerable energy will be required to accelerate the proton beam but in principle much higher amount of energy may be generated and subsequently a part of the generated energy may be fed back to run the system. Figure 10.6 shows the layout of the accelerator-based energy amplifier.

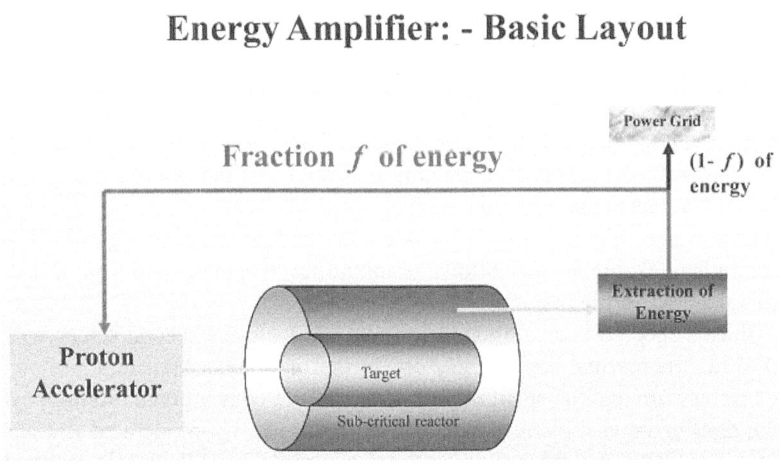

Figure 10.6 *Layout of an accelerator-based energy amplifier system*

An artist's view of an accelerator-based energy amplifier is shown in figure 10.7. Prototypes of these systems are being constructed and tested at many international laboratories.

10.12 INDIA'S THREE-STAGE PROGRAM FOR HARNESSING NUCLEAR ENERGY

Homi J. Bhabha initiated the nuclear energy program of India in 1950. Keeping in mind the fact that the known reserves of uranium in India (less than 2% of global uranium reserves) are much less than the reserves of thorium (about 30% of the global thorium reserves), India's nuclear energy program is a three-stage program.

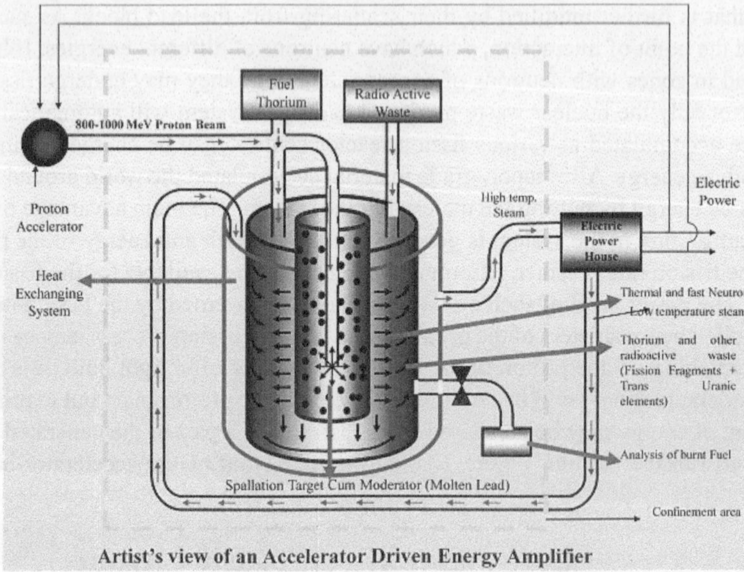

Figure 10.7 *Artist's view of an accelerator-driven energy amplifier*

1. In the first stage, energy will be produced by conventional fission reactors (a combination of pressurized heavy water reactors (PHWR) and the boiling heavy water reactors (BHWR)) that use natural uranium as fuel.
2. The second stage, which goes side by side with the first stage, enough fissile material will be accumulated from the conventional uranium-based reactors and from fast breeder reactors that use ^{238}U fertile material.
3. In the third stage, ^{239}Pu accumulated from the first and the second stages will be used as a fuel to run the thermal breeder reactors. In thermal breeder reactors, fission fuel will be ^{239}Pu, but thorium and the spent fuel (rich in ^{238}U) of conventional uranium cycle reactors may be distributed as a blanket around ^{239}Pu fuel rods. Un-moderated fast neutrons from the fission in ^{239}Pu may further breed the fissile material ^{239}Pu, while moderated thermal neutrons will initiate fission in ^{239}Pu fuel and will also breed ^{233}U from ^{232}Th.

So far around 4% of total electric power in India is produced by conventional reactors. Prototype fast breeder reactor has been developed and it is expected that in near future the second stage of the nuclear energy program will be completed. The third stage options are still open. It is possible that advanced nuclear technologies based on accelerator-driven energy amplifiers may supplement the thermal breeder reactors.

10.13 POSSIBILITIES OF CONTROLLED FUSION

Figure 10.1 showing the variation of B/A with A depicts that energy is liberated when two light nuclei fuse together. It is well known that thermonuclear fusion is the source of energy of stars such as sun. Uncontrolled fusion bomb called hydrogen bomb was first made in 1951.

Huge amount of money and efforts have since gone in developing a controlled fusion reactor for power generation. The interest in harnessing nuclear fusion energy is due to the fact that no radioactive waste is generated in it. Nuclear fusion source is, therefore, a clean source having no environmental and radiation concerns.

10.13.1 Nuclear Reactions Important for Fusion

Several endoergic nuclear fusion reactions in principle may be used for generating power. However, the three important ones are as follows with the average kinetic energy of the reaction products given in bracket:

$$\mathrm{D + D \rightarrow T\ (1.01\ MeV) + p\ (3.03\ MeV)} \tag{10.5a}$$

$$\mathrm{D + D \rightarrow {}^{4}_{2}He\ (0.82\ MeV) + {}^{1}_{0}n\ (2.45\ MeV)} \tag{10.5b}$$

$$\mathrm{T + D \rightarrow {}^{4}_{2}He\ (3.52\ MeV) + {}^{1}_{0}n(14.06\ MeV)} \tag{10.5c}$$

Largest amount of energy of about 17 MeV per fusion is, however, released in the third reaction. There is another advantage in using T + D reaction. Fresh supply of tritium ($\mathrm{T = {}^{3}_{1}H}$) may be obtained if neutrons produced via Eq. (10.5c) are made to hit ^{6}Li producing tritium and helium along with the release of additional 4.8 MeV of energy as given by the following equation:

$$\mathrm{{}^{6}_{3}Li + {}^{1}_{0}n \rightarrow {}^{3}_{1}T + {}^{4}_{2}He + 4.8\ MeV} \tag{10.5d}$$

In view of this, T + D fusion reaction is considered to be particularly suitable for fusion energy generation machines.

10.13.2 Requirements for Sustained Fusion

The fusion of tritium and deuterium nuclei will take place only if they come close enough to be within the range of their nuclear fields. Both nuclei being positively charged repel each other and have to overcome the Coulomb barrier before undergoing fusion. One way to impart large kinetic energies to the nuclei to overcome Coulomb barrier is to heat a mixture of the two gases. It is estimated that a temperature of about 40 million kelvin is required for the nuclei to come within the range of nuclear field and initiate fusion. However, at such high temperatures the gases become plasma. Plasma is the fourth state of matter. Plasma contains positive ions (nuclei of deuterium and tritium) and electrons moving at high speed. The intensity of plasma is measured in terms of its electron density: number of electrons per unit volume, generally denoted by n. Plasma contains only charged particles and, therefore, if left alone, it diffuses out with time. The fusion reaction will become a viable source of energy only if the rate of energy production by fusion is either more or at least equal to the rate at which energy is supplied to the system to maintain the plasma. Plasma loses energy primarily through two processes: Bremsstrahlung (radiation loss of accelerated charged particles) and conduction of heat through walls of the conductor. The rate of energy generation, H, by fusion in case of ^{3}T(D, n)^{4}He reaction may be given by

$$H = n^2 t \times (\sigma_{\mathrm{td}}\ v)_{\mathrm{average}} \times Q_{\mathrm{td}} \times \tau$$

Here, n is the number density of tritium (or of deuteron) ions in the plasma, σ_{td} the cross-section for the $^3T(D, n)^4He$ reaction, v the velocity of the ions, Q_{td} the Q-value of the reaction and τ the confinement time. The rate of energy loss from the plasma H', due to Bremsstrahlung and conduction may be estimate as,

$$H' = 6n\mathfrak{R}T$$

where n, $\mathfrak{R}$, and T are, respectively, the number density of deuteron, Boltzmann constant, and the temperature of the plasma. For sustained production of fusion energy, H must at least be equal to or more than H', Hence,

$$\frac{n \times (\sigma_{td} v)_{average} \times Q_{td} \times \tau}{6\,kT} \geq 1$$

This puts some restrictions, called Lawson criterion, on the plasma density n and the time of confinement τ. Using the realistic values of the parameters in the expression for T + D reaction, the Lawson criteria reduces to

$$n\tau > 1.7 \times 10^{20}\ \text{m}^{-3}\ \text{s}.$$

Unless the plasma density and the plasma confinement time fulfil this condition, fusion will not produce more energy than it consumes. The main problem with controlled fusion is that so far it has not been possible to achieve Lawson criteria in experimental setups.

An artist's view of a T + D fusion reactor is shown in figure 10.8. A mixture of deuterium and tritium in gaseous state is kept in a plasma cell which is heated to produce plasma at high temperature sufficient for the fusion to take place. Neutrons of ≈ 14 MeV created in the fusion cell are slowed down as they pass through the plasma cell wall. Slowed down neutrons hit the lithium blanket to produce tritium which replenishes the loss of it in fusion. Energy produced in fusion and tritium breeding is taken out and a part of it is utilized for power generation.

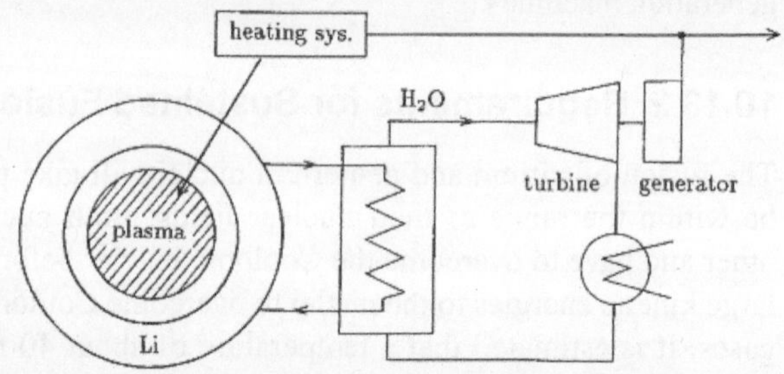

Figure 10.8 *An artist's concept of a fusion reactor*

The main problem faced in fusion power generation is the confinement and insulation of plasma. As the temperature of plasma is very high it will melt the walls of the containers in which it is kept if it is not kept away and insulated from walls. Further, being conducting energy will be lost if plasma is not kept insulated.

10.13.3 Magnetic Confinement

For the plasma fuel to produce energy it must be confined in a small volume away from any structural material so that it does not lose energy. In free plasma, the trajectories of plasma particles are random and the particles run away. However, if a current is passed axially through the plasma it gets enclosed by a magnetic field which pushes the charged plasma particles inwards. This is called *pinch effect.* In this type of plasma confinement, called open-ended confinement, plasma leaks through the ends. To avoid end-leaks, toroidal magnetic field is produced using torus windings all along the donut-shaped plasma tube (figure 10.9a). *Tokamak,*

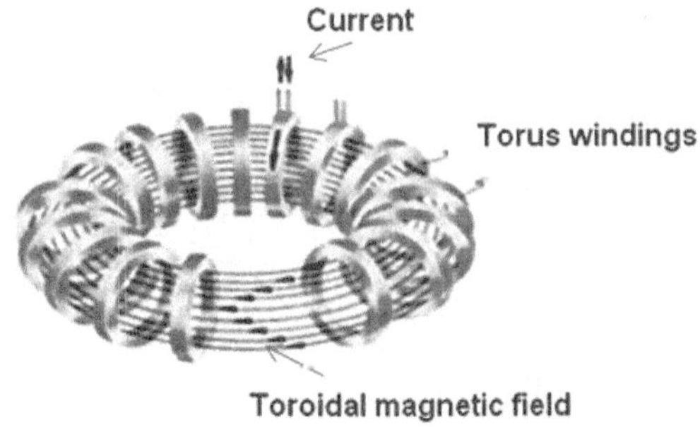

Production of Toroidal magnetic field by Torus windings

(a)

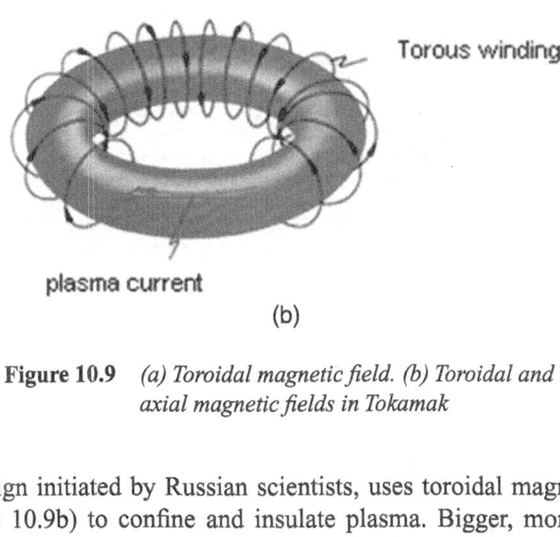

(b)

Figure 10.9 *(a) Toroidal magnetic field. (b) Toroidal and axial magnetic fields in Tokamak*

a fusion reactor design initiated by Russian scientists, uses toroidal magnetic field along with axial current (figure 10.9b) to confine and insulate plasma. Bigger, more powerful and with improved design tokamaks are being built at various places in the world. The design of a big tokamak at the Princeton University, USA is shown in figure 10.10. With a view to further increase the confinement time a combination of helical and toroidal coils is used in systems called *stellarator*. Combination of toroidal and helical windings generates magnetic field that produces three-dimensional confinement of the plasma, increasing the confinement time (figure 10.11).

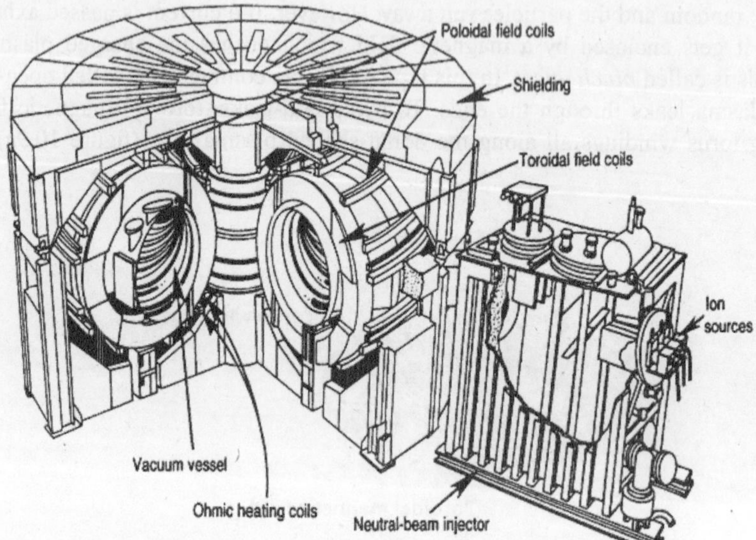

Figure 10.10 *A view of tokamak fusion test reactor (TFTR) at Princeton University, USA*

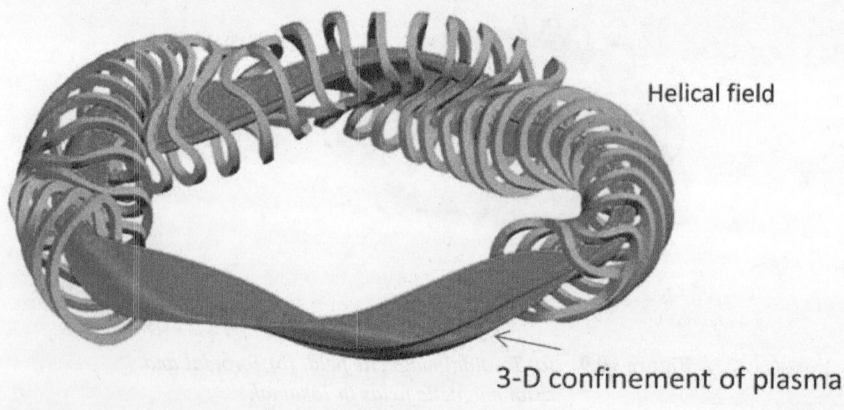

Figure 10.11 *Three-dimensional confinement of plasma by the combination of toroidal and helical windings in stellarator*

10.13.4 Inertial Confinement

The characteristic of inertial confinement is that the extremely high-density plasma is produced by means of an intense energy driver, such as a laser or particle beam, within a short period so that the fusion reactions can occur before the plasma starts to expand. Magnetic confinement plays no part in this process, which has come to be called inertial confinement. For fusion conditions to be reached by inertial confinement, a small solid deuterium–tritium pellet must be compressed to a particle density 10^3 to 10^4 times that of the normal solid pellet particle density $ns = 4.5 \times 10^{22}$ cm^{-3}. One cannot expect the laser light pressure or the momentum carried by the particle beam to compress the solid pellet: they are too small. A more feasible method of compression involves irradiating the pellet from all side, as shown in figure 10.12a. The plasma is produced on the surface of the pellet and is heated instantaneously. The plasma expands immediately. The reaction force of the plasma jet accelerating outward compresses the inner pellet inward like a spherical rocket. This process is called *implosion*. The study of implosion processes is one of the most important current issues, and theoretical and experimental researches are being carried out intensively.

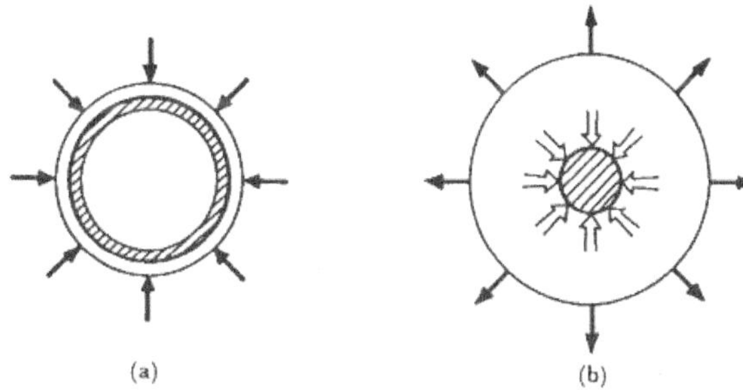

Figure 10.12 *(a) Implosion of T + D pallet by irradiation of laser or particle beams from all sides. (b) Further implosion by outgoing plasma jets*

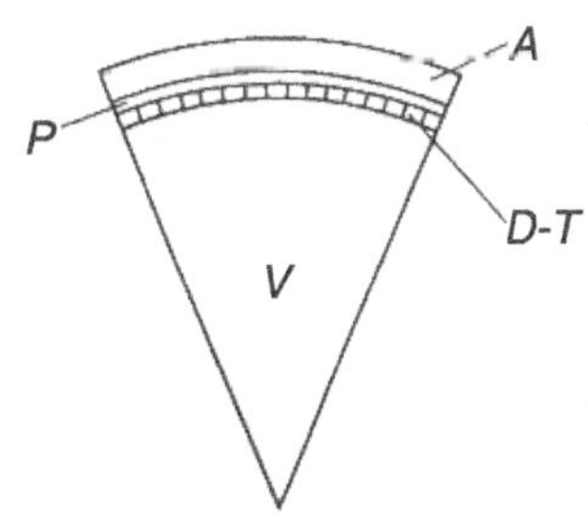

Figure 10.13 *Proposed structure of T + D pellet: A = Ablator, P = Pusher, D – T: Solid fuel, V = Vacuum*

A typical structure of a pellet is shown in figure 10.13. Outside the spherical shell of deuterium–tritium fuel, there is a pusher cell, which plays the role of a piston during the compression; an ablator cell with low-Z material surrounds the pusher cell and the fuel. The heating efficiency is equal to the conversion ratio of the driver energy to the thermal energy of the compressed core fuel. The heating efficiency depends on the interaction of the driver energy with the ablator, the transport process of the particles and the energy, and motion of the plasma fluid. The driver energy is absorbed on the surface of the ablator, and the plasma is produced and heated. Then, the plasma expands and the inner deuterium–tritium fuel shell is accelerated inward by the reaction of the outward plasma jet. The implosion takes place at the centre.

In spite of huge global efforts, it has not been possible to produce energy from fusion on a commercial scale in a controlled way. It is estimated that in the next 10 years or so some viable solution for high-density plasma confinement may emerge paving way for harnessing fusion energy.

10.14 GLOBAL STATUS OF NUCLEAR ENERGY

About 17% of world's energy is presently produced by about 439 nuclear reactors located in some 30 countries. Another 34 reactors are under construction, 74 planned and 162 proposed. Almost 20% energy needs of USA, 34% of Europe, and 4% of India are fulfilled by nuclear means. China, which presently uses only 1.9% nuclear energy, is planning to commission another 28 reactors and has proposal for 54 more. Countries in the middle-east and Africa that were till now totally dependent on oil are now planning for nuclear energy. However, at present all nuclear energy is derived from nuclear fission. With the rise of world population particularly in developing countries, energy needs in developing countries is likely to multiply many folds. Considering the land requirements, environmental pollution, global warming, commissioning time, cost and so on nuclear option appears to be the most favoured one. However, accidents in nuclear reactors at Chernobyl (1986) and after huge Tsunami at Fukushima, Daiichi, Japan (1911) has somewhat retarded the growth of nuclear energy.

Conventional energy sources apart from being highly polluting, produce green house gases adding to global warming. Generation of hydroelectric power involves large scale shifting of population for constructing water reservoir, which also adversely affect the flora and fauna of the place. Also the known reserves of conventional energy sources are depleting fast. The so-called renewable sources such as solar, wind, sea waves, and geothermal energy may contribute significantly up to about 25% of total global requirements. In spite of that nuclear option is expected to remain the leading option for energy. Development of safer fission reactors such as accelerator-driven energy amplifier and some likely breakthrough in fusion technology may bring back nuclear energy in forefront.

10.15 HEAVY AND SUPER HEAVY (SHE) ELEMENTS: TRANSURANIC/ACTINIDE AND TRANS-ACTINIDE NUCLIDES

Actinides form a family of elements with similar properties. Within actinides there are two groups: (a) transuranium (or transuranic) elements that follow uranium in periodic table and (b) transplutonium that follow plutonium in the periodic table. All actinides are generally rare and are not found in nature in much abundance. They are generally produced by neutron bombardment of heavy nuclei in reactors or by the bombardment of accelerated charged particles from accelerators. The actinides are generally short lived with their half lives decreasing with the atomic number. Actinides have 89–103 protons and 117–159 neutrons. A list of transuranic elements is given in table 10.4.

Many but not all of these elements have been synthesized in laboratories.

Table 10.4 *Atomic number and name of transuranic elements*

List of the transuranic elements	
Actinides	**Transactinide elements**
• 93 neptunium Np	• 104 rutherfordium Rf
• 94 plutonium Pu	• 105 dubnium Db
• 95 americium Am	• 106 seaborgium Sg
• 96 curium Cm	• 107 bohrium Bh
• 97 berkelium Bk	• 108 hassium Hs
• 98 californium Cf	• 109 meitnerium Mt
• 99 einsteinium Es	• 110 darmstadtium Ds
• 100 fermium Fm	• 111 roentgenium Rg
• 101 mendelevium Md	• 112 copernicium Cn
• 102 nobelium No	• 113 ununtrium Uut*
• 103 lawrencium Lr	• 114 flerovium Fl
	• 115 ununpentium Uup*
	• 116 livermorium Lv
	• 117 ununseptium Uus*
	• 118 ununoctium Uuo*

10.15.1 Super Heavy Elements

Nuclides having magic number of protons and or neutrons are found to be exceptionally stable in comparison to the other nearby nuclei. This is explained by the shell model of the nucleus. Generally speaking the shell structure is the outcome of confinement of a number of identical particles in a limited space by a central potential. Strong spin–orbit coupling provides additional binding to nuclei at shell colours. Doubly magic nuclei such as ^{40}Ca and ^{208}Pb show remarkable stability. Theoretical calculations made using the microscopic–macroscopic theory have indicated that there might be another island of stability near the second doubly magic region around $Z = 112$–124 and $N = 184$. The nuclei lying in the second island of stability are called *super heavy nuclei* or *super heavy elements* (SHE). Traditionally, nuclei lying above the last actinide ($Z = 103$, lawrencium), starting with rutherfordium are termed as SHE. These nuclei represent the very top end of the periodic table and are all expected to be alpha emitters and or self-fissioning nuclei with short half-lives. However, the mean lives of some of them are expected to be of the mean age of the earth, and therefore, some of them may be present in the environment in traces. Production and study of physical and chemical properties of SHE is of great interest from the point of view of understanding the limits of stability in mass and charge. Attempts are being made to produce SHE by the fusion of accelerated heavy ions with actinide nuclei obtained from reactors. Several laboratories in the world are trying to produce and study the properties of super heavy nuclei.

10.16 NUCLEAR WEAPONS

Armaments that use nuclear fission or fusion energy for destructive explosion are called nuclear weapons. These devices may be atom bombs, hydrogen bombs, or their smaller versions, which may be carried by continental or inter-continental missiles. The destruction power of a nuclear device is several orders of magnitude larger than the conventional non-nuclear explosives. In an explosive device, energy is produced at a high rate and once triggered the energy production cannot be stopped. Nuclear explosive devices are, therefore, uncontrolled fission or fusion reactors.

An atom bomb is like a fission reactor without control elements. Although ^{235}U is a fissile material, its fission cross-section for thermal neutrons is small as compared to the fission cross-section for ^{239}Pu. As in an explosive device energy should be generated at a fast rate, ^{239}Pu is better suited as compared to ^{235}U, for making nuclear explosives. It is for this reason that ^{239}Pu is called the weapon grade material. Further, the critical mass, the minimum mass of the fissile material required to have a self-sustained chain reaction, for ^{239}Pu is not very large and, therefore, it is frequently used for making atom bombs. Although the exact construction details of the atom bomb are not known, it is speculated that in an atom bomb the fissile material, slightly more than the critical mass is kept in two pieces initially separated from each other. The assembly is covered from all sides by neutron reflectors to further increase the fission rate. The whole system is then kept in a thick and hard casing. A remote control device brings the two pieces of the fissile material in contact, increasing the total mass beyond the critical value. Due to the fission chain reaction, which is further accelerated by neutrons reflected from the casing, large amount of heat energy is produced in a very short time. As the system is enclosed in a hard cover, everything inside is converted into gas at high temperature and the pressure becomes several times the atmospheric pressure. The outer cover suddenly breaks down under the high temperature and pressure, releasing the high pressure gas. The gas now suddenly expands generating a shock wave. Everything within the initial burst of high temperature-high pressure gas melts down; the shock wave (the wave containing a front of large pressure difference) expands and destroys everything in its path.

A hydrogen bomb is a nuclear explosive device where most of the energy is obtained from the fusion of deuterons and of tritons. Exact constructional details of a hydrogen bomb are also not known but it is speculated that it contains a fission device at the centre, which is covered by a blanket of lithium deuteride. The assembly is enclosed in a tamper, a thick and hard outer layer of fissionable material that holds the contents together in order to obtain a large explosion (figure 10.14).

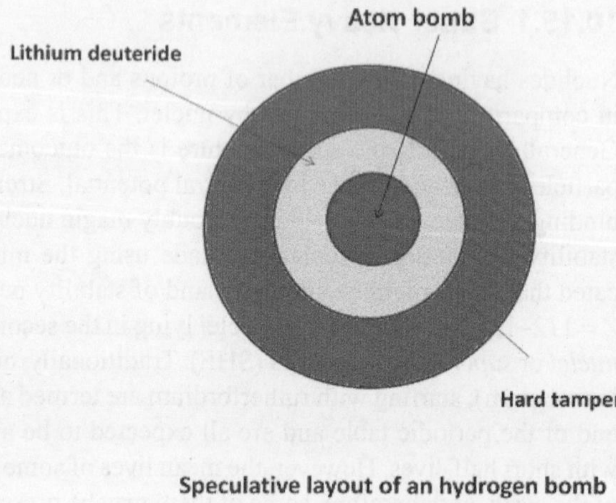

Figure 10.14 *Speculative layout of an hydrogen bomb*

To explode the device, first the atom bomb is fired. Neutrons emitted from the central fission device causes lithium to turn into tritium and helium releasing energy according to the following reaction:

$$^{6}_{3}\text{Li} + ^{1}_{0}\text{n} \rightarrow ^{3}_{1}\text{H} + ^{4}_{2}\text{He} + 4.8\,\text{MeV}$$

Atomic explosion at the centre also generates high temperatures of the order of 50×10^6 K that are sufficient to initiate the following fusion reactions:

$$^{3}_{1}\text{H} + ^{2}_{1}\text{H} \rightarrow ^{4}_{2}\text{He} + ^{1}_{0}\text{n} + 14.5\,\text{MeV}$$

And $$^{2}_{1}\text{H} + ^{2}_{1}\text{H} \rightarrow ^{3}_{2}\text{He} + ^{1}_{0}\text{n} + 4.5\,\text{MeV}$$

Additional neutrons produced from the fusion reactions further accelerate the fission at core and initiate fission in the fissile material of tamper. Very large quantity of heat is thus produced in a very short time within the tamper before it brakes resulting in the rapidly expanding shock and heat wave.

A hydrogen bomb is said to be a clean bomb as very little radioactive dust (produced from the central and tamper atomic devices) is spread with the shock wave.

10.16.1 Neutron Bomb

If the tamper of the hydrogen bomb does not contain any fissionable material, the amount of radioactive dust is further reduced. However, large amount of neutrons are produced, which may destroy every form of life without really destroying structures. This type of bomb is called a neutron bomb.

10.16.2 Cobalt Bomb

When the tamper material of the hydrogen bomb is loaded with cobalt metal, some of the stable cobalt nuclei are converted into radioactive ^{60}Co, which emit high energy γ rays (of about

1 MeV) with half-life of about 5.26 years. The dust containing radioactive cobalt spreads with winds over a large area, which ultimately settles down on the ground. Thus, a very large area is badly contaminated with long-lived ^{60}Co. Living organism/human are continuously irradiated by high energy γ rays, which may produce cancer and other deceases. Thus, cobalt bomb is the most lethal weapon of long term damage and mass destruction.

Solved example S-10.1

Assuming that in each fission of ^{235}U, 200 MeV of energy is released and 5% is lost in neutrinos, estimate the amount of ^{235}U burnt out to produce 1×10^{11} kWH of energy in one year. The efficiency of the reactor is 50%.

Solution. Let n nuclei of ^{235}U fission out per second. The energy released per second H

$H = n \times 200 \times 10^6 \times 1.6 \times 10^{-19} \times 0.95$ (5% loss by neutrinos) $\times 0.5$ (efficiency) $= 1.52 \times 10^{-11} \times n$ Joule per second

Now, to generate 100×10^{11} kWH of energy per year the energy H' per second will be $H' =$

$$\frac{100 \times 10^{11} \times 10^3 \times 3600 \text{ (seconds in one hour)}}{3.15 \times 10^7 \text{ (number of seconds in one year)}} = 1.143 \times 10^{11} \text{ Joule per second.}$$

Equating H' to H, we get $n = 1.143 \times 10^{11}/1.52 \times 10^{-11} = 0.713 \times 10^{22}$.

Thus, per second 0.713×10^{22} nuclei need to be burnt. Therefore, in 1 year $0.713 \times 10^{22} \times 3.15 \times 10^7 = 2.246 \times 10^{29}$ nuclei have to burnout. Now, in 1 g of natural uranium there are

$$\frac{\left(\frac{0.7}{100}\right) \times 6.02 \times 10^{23}}{238} = 1.77 \times 10^{19} \text{ nuclei of } ^{235}U$$

The amount of natural uranium that will contain 2.246×10^{29} ^{235}U nuclei is

$$= \frac{2.246 \times 10^{29}}{1.77 \times 10^{19}} = 1.269 \times 10^{10} \text{ g} = 1.269 \times 10^4 \text{ ton of natural uranium.}$$

Ans: 1.269×10^4 ton of natural uranium per year will be required.

Solved example S-10.2

Calculate the plasma confinement time for T + D reaction from the following data, assuming that the Lawson criteria is just fulfilled.

Q-value = 17.62 MeV; $(\sigma_{td}.v)_{average} = 10^{-22}$ m^3 s^{-1}; number density of triton = number density of deuteron = 7×10^{18} m^{-3}; kT = 10^5 eV.

Solution. For Lawson criteria to be just fulfilled

$$\frac{n_t^2 \times (\sigma_{td} v)_{average} \times Q_{td} \times \tau}{6 n_d kT} = 1$$

Therefore,

$$\tau = \frac{6 n_d kT}{n_t^2 \times (\sigma_{td} v)_{average} \times Q_{td}} = \frac{6 \times 10^5}{7 \times 10^{18} \times 10^{-22} \times 17.62 \times 10^6}$$

$$= 48 \text{ second}$$

Ans: Plasma confinement time = 48 s

Exercise p-10.1: Discuss the possibility of self sustained fission chain reaction in a lump of natural uranium. Estimate the average number of collisions and the corresponding slow down time required to reduce the neutrons of average energy 2 MeV to average thermal energy(= 0.025 eV) in graphite moderator. The macroscopic scattering cross-section for graphite is 0.385 cm^{-1}.

Exercise p-10.2: What is η (eta) and what is its significance? Determine the value of η for natural uranium.

Exercise p-10.3: Give the need, properties and actual materials that may be used as moderators and arrestors. Assuming that the average energy released per fission of ^{235}U is 200 MeV, calculate the amount of ^{235}U required per day for a 100 MW reactor.

Exercise p-10.4: Define the neutron multiplication factor k and derive four-factor and six factor formulae for it.

Exercise p-10.5: Clearly differentiate between prompt and delayed neutrons involved in nuclear fission. Why delayed neutrons are important in reactor technology?

Exercise p-10.6: What is the difference between a reactor and a bomb? Write a short note on the classification of fission reactors.

Exercise p-10.7: Compare the advantages and disadvantages of fission energy.

Exercise p-10.8: What is an accelerator driven energy amplifier? In what respect is it better than a conventional reactor for energy production?

Exercise p-10.9: Discuss India's strategy of trapping nuclear energy.

Exercise p-10.10: Write a detailed note on the possibilities of controlled fusion and the efforts being done in this direction.

Suppose the fusion reaction ${}_1^2D + {}_1^2D \rightarrow {}_2^3He + {}_0^1n + 3.18\,MeV$ is initiated by deuterons of negligible kinetic energy, calculate the kinetic energy of the emitted neutron.

Exercise p-10.11: What is meant by inertial confinement? Give an account of the possible method of achieving sustained fusion using inertial confinement.

Exercise p-10.12: What are the main differences between an atom bomb and a hydrogen bomb? What are neutron and cobalt bombs and what are their special characters?

The fusion of deuterons takes place when they are 50 F apart. Calculate the minimum kinetic energy that deuteron must have to initiate fusion and if this energy is supplied by the thermal energy then the temperature of the deuteron [$k = 1.38 \times 10^{-23}$ J/K].

Multiple choice questions

Note: Some questions may have more than one correct alternative. Select all correct alternatives, if they are more than one for the answer.

Exercise M-10.1: The energy released (in Joule) in the complete fission of ^{235}U nuclei in 1 gram of natural uranium is approximately

(a) 2.55×10^{-8} (b) 2.55×10^{8} (c) 55.2×10^{7} (d) 5.52×10^{-7}

Exercise M-10.2: A minimum of 5% of the total energy released in the fission of ^{235}U nuclei cannot be used, because it is the energy associated with

(a) β negative particles (b) β positive particles
(c) fission neutrons (d) neutrinos

Exercise M-10.3: Fast fission neutrons take on an average 120 collisions in getting thermalized in a moderator medium. The most likely moderator is

(a) graphite (b) lead (c) light water (d) heavy water

Exercise M-10.4: In a conventional fission reactor, it is necessary to keep the thickness of moderator material more than a minimum value. This ensures that neutrons after thermalization have energy

(a) >5 eV (b) <5 eV (c) <0.025 eV (d) >0.025 eV

Exercise M-10.5: The ratio of neutron multiplication factor k of a reactor at the instants when control rods are in and when they are out is

(a) 1 (b) >1 (c) <1 (d) may have any value

Exercise M-10.6: The percentage yield of fission products depends on

(a) excitation energy of the compound nucleus
(b) kinetic energy of incident neutrons
(c) amount of fissile material
(d) total angular momentum involved

Exercise M-10.7: Accelerator driven energy amplifier are better than conventional fission reactors because

(a) fission fragments also undergo fission and release energy
(b) no weapon-grade fissile material is accumulated
(c) fission does not take place in them
(d) neutrons of only very high energy are produced

Exercise M-10.8: In inertial confinement of plasma, important role is played by

(a) magnetic field (b) jet action (c) pellet density (d) laser irradiation

Exercise M-10.9: A hydrogen bomb is based on

(a) only fusion (b) only fission
(c) both on fission and fusion (d) none of these

Exercise M-10.10: Which, in the long term, is the most devastating among the

(a) atom bomb (b) neutron bomb
(c) cobalt bomb and (d) (trinitrotoluene) TNT bomb?

Exercise M-10.11: Assuming Q-value for D + D fusion reaction to be 4.0 MeV and that the fission is initiated by deuterons of negligible energy, the kinetic energy (in MeV) of the emitted neutron will be

(a) 1 (b) 2 (c) 3 (d) 4

Exercise M-10.12: A deuteron of 1 MeV has a temperature (in Kelvin)

(a) 7.7×10^{11} (b) 7.7×10^{9} (c) 7.7×10^{7} (d) 7.7×10^{5}

Exercise M-10.13: The ratio of the temperature of a proton and a deuteron of 1 MeV energy is
(a) 0.5 (b) 0.75 (c) 1.0 (d) 2.0

Exercise M-10.14: The two prominent modes of energy loss by plasma are
(a) Compton scattering and pair production (b) Compton scattering and conduction
(c) Bremsstrahlung and photo electric effect (d) Bremsstrahlung and conduction

Exercise M-10.15: The three-dimensional confinement of plasma may be achieved by
(a) axial magnetic field
(b) solenoidal magnetic field
(c) helical magnetic field
(d) combination of toroidal and helical magnetic fields

11

Fundamentals of Elementary Particles

11.1 ELEMENTARY PARTICLES

Elementary particles are the building blocks of matter at the most fundamental level, at the smallest scale of size. It is a remarkable fact that matter at subatomic level is made up of tiny bits with relatively huge empty space between them. It is still more remarkable that these tiny bits are few in numbers, which are replicated again and again in extremely large numbers to make all the matter around us.

11.1.1 Leptons, Mesons, Baryons, and hadrons

The story of elementary particles in a way begun with the discovery of electron by J.J. Thomson in 1897, naming the nucleus of hydrogen as proton by Rutherford around 1910, and establishing the presence of neutron by Chadwick in 1932. It is not known who really discovered photon but it was Planck who in 1900 made the first description of the quanta of energy while trying to explain the energy distribution of black body radiations. It was Gilbert Lewis in 1926 who suggested the name photon for the quanta of radiations. In the initial stages, photon forced itself on the unreceptive scientific community but soon it found its natural place in the quantum field theory as the mediator of electromagnetic field. Extending the analogy of electromagnetic field to the strong nuclear field, Yukawa in 1934 predicted a mediator for the short-range strong nuclear field of mass around 300 times the mass of electron. As the expected mass of this quanta of strong field was intermediate between the mass of the electron and the mass of nucleon, the quanta was called meson. With the same logic, light-weight particles such as electrons were called leptons and nucleons baryons, meaning heavy weights. Although initially mu-lepton discovered in cosmic rays was considered to be the Yukawa particle, it was soon realized that it cannot be a quanta of strong nuclear field as it has very little interaction with strong field. Ultimately, in 1947 Powell discovered π-meson the true Yukawa particle. Elementary particles such as baryon and mesons that participate in strong interactions are called hadrons.

11.1.2 Particle and Antiparticle

A major development of great scientific significance happened in 1927 when Dirac developed the following expression for the energy E of a relativistic electron of mass m_e and momentum p.

$$E^2 = p^2c^2 + m_e^2c^4 \tag{11.1}$$

The solution of Eq. (11.1) has a negative energy solution associated with each positive energy solution. The solution with negative energy was interpreted as the description of an antiparticle. Discovery of positron in 1931 by Anderson firmly established that the duality of Dirac's equation is a manifestation of a very profound principle of quantum field theory that says '*for every particle there is an antiparticle*'. Discovery of antiproton and antineutrons, respectively, in 1955 and 1956 at Berkeley Bevatron put the particle–antiparticle principle on firm footing. It may not be out of place to mention that some charge-less particles such as photon and π-zero meson are their own antiparticles.

11.1.3 Principle of Cross-symmetry

With the firm establishment of the particle–antiparticle duality, an important principle of particle physics called the '*Principle of crossing symmetry*' emerged, according to which *if the reaction $X + Y \rightarrow A + B$ is possible than all other reactions obtained by taking any of the particles to the other side and replacing it by its antiparticle will also be a possible reaction, provided it is allowed by the mass–energy consideration.* As such, $X \rightarrow A + B + \bar{Y}$; $X + Y + \bar{A} \rightarrow B$; $X + Y + \bar{B} \rightarrow A$, reactions are all possible if the total initial energy (including mass energy and kinetic energy) is more than the total energy in the final state. As an example, let us write the equation of Compton scattering: scattering of a γ ray by a stationary free electron

$$\gamma + e^- \rightarrow \gamma + e^-$$

If electron on the right hand side is brought to the left hand side (so that it becomes e^+) and the γ on the left hand side is taken to the right hand side, we get

$$e^- + e^+ \rightarrow \gamma + \gamma$$

This equation is the well-known equation representing electron–positron annihilation.

11.1.4 Conservation of Baryon Number

In mid-fifties, Cowan and Reins conclusively proved the existence of the elusive neutrino, which was earlier postulated by Pauli to explain the continuous nature of the energy spectrum of β particles emitted in natural radioactive decay.

A parallel development was the discovery of strange particles. In 1947, Rochester and Butler discovered K^0 that decays into π^+ and π^-. However, in 1949 Powell found a charged version of K particle K^+ that decayed into three pions. $K^+ = \pi^+ + \pi^+ + \pi^-$. As some properties of K-particles were similar to mesons, they were included in the meson family. In due course of time, many more mesons such as η, ϕ, ω, and ρ were discovered. In 1950, Anderson and his group found another particle named lambda Λ that decays to proton and π-meson. Lambda belonged to the baryon family. To explain the stability of proton, Stuckelberg proposed the conservation of baryon number. Baryon number denoted by A is assigned a value +1 for baryons and −1 to antibaryons. The law of baryon number conservation states that '*In any physical process baryon number is conserved*'.

11.1.5 The Strange Particles and the Law of Associate Production

In following years, many more heavy baryons such as Σ, Ξ, and Ω were discovered. The newly discovered particles had a strange property: they are produced by strong interactions in a short time of the order of 10^{-23} s but decay slowly (10^{-10} s) via weak interactions. Gell-Mann and Nishijima assigned these particles a strangeness quantum number S that is conserved in strong interactions but not in weak interactions. A. Pais studied strange particles in detail and propounded the law of associate production of strange particles. The law of associate production of strange particle states '*Strange particles are produced in pair*'. For example,

$$\pi^- + p^+ \rightarrow K^+ + \Sigma^-;\ \pi^- + p^+ \rightarrow K^0 + \Sigma^0;\ \pi^- + p^+ \rightarrow \Lambda + \Sigma^0$$

In these reactions $S = 1$ for K, $S = -1$ for Σ and Λ; $S = 0$ for π and p+.

However, there is no such restriction on the decay of strange particles. The following decays in which strangeness quantum number S is not conserved are allowed:

$$\Lambda \rightarrow p^+ + \pi^-;\ \Sigma^+ \rightarrow p^+ + \pi^0;\ \Sigma^+ \rightarrow n^0 + \pi^+$$

It may not be out of place here to mention that there is no law that conserves the number of mesons.

11.2 THE EIGHT-FOLD WAY

A large number of elementary particles were discovered and classified as leptons, mesons, and baryons by 1960. It may be recalled that Russian chemist Mendeleev arranged elements in a periodic table on the basis of their chemical properties. A similar attempt was made by Murray Gell-Mann in 1961 to arrange baryons and mesons in geometrical patterns, called the eight-fold way. Y. Neeman has also proposed a similar scheme independently. The lightest eight baryons may be arranged in an octet as shown in figure 11.1.

It may be seen that particles of same charge fall on diagonal lines. The strangeness S is same for particles along horizontal lines.

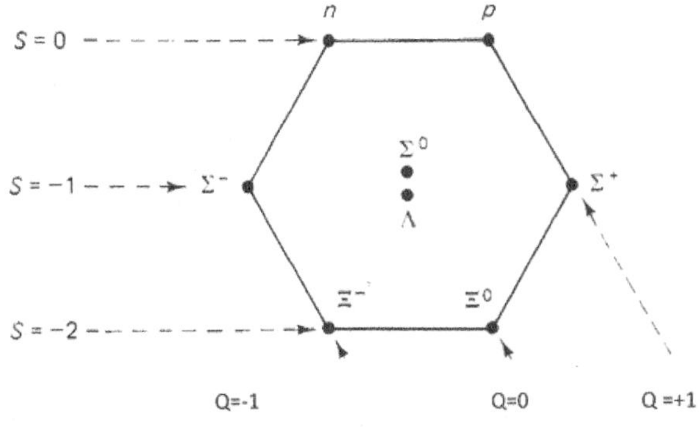

Figure 11.1 *The baryon octet*

Eight lightest mesons also formed an octet as shown in figure 11.2.

Ten heavier baryons in the eight-fold way were arranged in a triangular array called the baryon decuplet shown in figure 11.3.

An interesting fact about the heavy baryon decuplet is that omega minus baryon at the bottom was not discovered by the time Gell-Mann arranged baryons in eight-fold way. Gell-Mann not only predicted a meson with strangeness −3, charge −1 but also suggested how to detect it. Later, Ω^- was discovered in 1964 exactly the way as suggested by Gell-Mann.

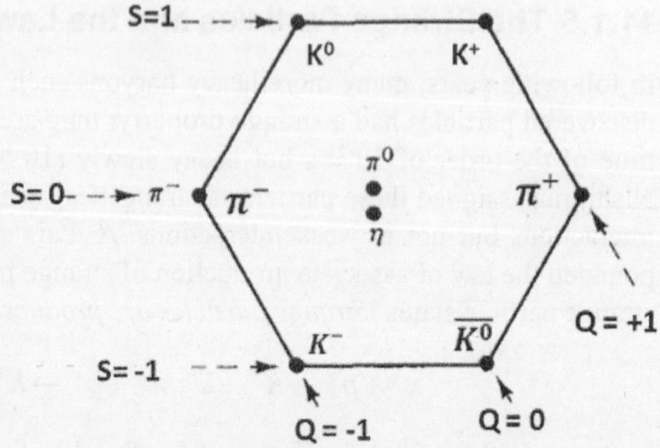

Figure 11.2 *Lightest meson octet*

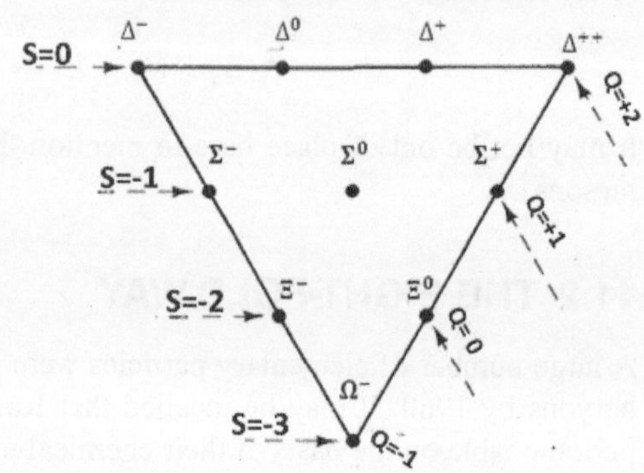

Figure 11.3 *Heavy baryon decuplet*

The mathematical basis of the eight-fold way is a kind of field theory called the 'special unitary group of arrays of size 3 × 3', denoted by SU(3). Some special conditions reduce the number of the components from 9 to 8. In principle, SU(3) is a generalization of the theory of isospin. SU(2) deals with the three-dimensional isospin while SU(3) considers eight-dimensional space.

11.3 QUARK MODEL OF HADRONS

To provide a sound basis to the eight-fold way, Gell-Mann in 1964 proposed that all hadrons are themselves made of still smaller entities, which he named as quarks. He proposed three types of quarks: up-quark denoted by u, down-quark denoted by d, and strange-quark denoted by s. Each of them has an anti-quark: $\overline{u}, \overline{d}$, and $\overline{s}$. In a way, the type or kind of a quark, also called the quark flavour is a quantum parameter such as spin and isospin that may be associated with different quarks. Quarks are particles with fractional charge and 1/2 $\hbar$ spin. Further, properties of these quarks and antiquarks are given in table 11.1.

According to the quark theory:

1. Baryons are the bound states of three quarks qqq.
   The quark structure of proton is uud while that of neutron is udd.
2. Mesons are the bound states of quark antiquark $q\overline{q}$.

Tables 11.2 and 11.3, respectively, list the properties and quark structures of some baryons and mesons.

Table 11.1 *Properties of quarks and antiquarks*

Property → Quark ↓	Charge in unit of e	Strangeness S
Up u	2/3	0
Antiup $\bar{u}$	−2/3	0
Down d	−1/3	0
Antidown $\bar{d}$	1/3	0
Strange s	−1/3	−1
Antistrange $\bar{s}$	1/3	1

Table 11.2 *Quark structure and properties of some baryons*

Baryon	Charge Q in units of e	Strangeness S	Quark structure qqq
Δ^{++}	2	0	uuu
Δ^{+}	1	0	uud
Δ^{0}	0	0	udd
Δ^{-}	−1	0	ddd
Z^{*+}	1	−1	uus
Σ^{*0}	0	−1	uds
Σ^{*-}	−1	−1	dds
Ξ^{*0}	0	−2	uss
Ξ^{*-}	−1	−2	dss
Ω^{*-}	−1	−3	sss

Table 11.3 *Quark structure and properties of mesons*

Meson symbol	Charge Q in unit of e	Strangeness S	Quark structure $q\bar{q}$
π^0	0	0	$u\bar{u}$
π^+	1	0	$u\bar{d}$
π^-	−1	0	$d\bar{u}$
η	0	0	$d\bar{d}$
K^+	1	1	$s\bar{u}$
K^0	0	1	$d\bar{s}$
K^-	−1	−1	$s\bar{u}$
$\bar{K}^0$	0	−1	$s\bar{d}$
η'	0	0	$s\bar{s}$

The quark structure of some of the above-mentioned multiplets may be identical, for example the quark structure of the proton and Δ^+ is the same uud. However, they are different because of the way the quarks are bound in them. In a similar way, π^+ meson and ρ^+ (rho plus) meson both have the structure $u\bar{d}$. It follows from these examples that the same quarks may generate different particles depending on their way of coupling. This is just like in the excited states of atoms or nuclei, which have the same basic structure as their ground states.

11.3.1 Quark Model and Pauli's Exclusion Principle

The quark structures of baryons, such as uuu, uud, ddd, and sss, appear to violate Pauli's exclusion principle as quarks are half-spin particles and not more than one particle can be accommodated in the same state. To overcome this difficulty, O.W. Greenberg in 1964 suggested another quantum number called *colour* to be assigned to each of the quark. The colour quantum number may have three different values or states: the *red*, the *green*, and the *blue*. In uuu, if there are three quarks each of different colour then there will be no problem of the violation of Pauli's exclusion principle. It may be emphasized that like strangeness, charge, spin, and others colour is also a quantum number; a red quark has zero blue colour and zero green colour. An anti-red quark has anti-red colour. When a red and an anti-red quark join together, the total structure becomes white or colourless. An important law of the colour scheme is:

All naturally occurring particles are colourless.

This law states why mesons have a quark and an antiquark and why baryons are made of three quarks or antiquarks. The colour property of a particle may be compared to the electrical charge, which can have two types: positive and negative. The colour charge may have three types: red, green, and blue.

11.4 NEW QUARKS

Since 1967 two new quarks have been identified: Charm quark, denoted by c and the Beauty quark denoted by b. These quarks were required to explain the existence of the very heavy meson J or ψ discovered in 1974 independently by C.C. Ting and Burton Richter. There are indications that another flavour of quark named Truth, denoted by T may also exist.

Today, we have six leptons: electron (e^-), electron neutrino (ν_e); μ lepton (μ), μ-leptonic neutrino (ν_μ); τ lepton (τ), and τ-leptonic neutrino (ν_τ). There are anti particles for each of the above leptons.

There are six flavours of quarks: up-quark (u), down-quark (d), strange quark (s), Charm quark (c), Beauty (or bottom) quark (b), and Truth-quark (or top quark) (t). Each flavour of quark has its antiparticle. A part from the usual quantum numbers such as strangeness, spin, and charge. Quarks and anti-quarks have another quantum number called colour. There can be three values of colour: red, green, and blue. All naturally occurring particles are colourless.

11.5 INTERMEDIATE VECTOR BOSONS

Around seventies, scientist Glashow, Weinberg, and Salam proposed the theory of electroweak interactions, unifying the electromagnetic and weak fields. The theory described the electroweak interaction through the exchange of three types of mediating bosons jointly called intermediate vector bosons: W^+, W^-, and Z^0. The two charged and one uncharged bosons were subsequently discovered in 1982–1983 at CERN by the research group led by C. Rubbia.

11.6 GLUONS

According to the present understanding, the strong force at the quark level is communicated by the exchange of particles called gluons. There are eight types of gluons. As gluon carry colour charge like quark, it cannot exist as a free particle. However, gluon balls, which contain few gluons such that the gluon ball becomes colourless, may be detected in experiments.

11.7 STANDARD MODEL

The standard model describes matter and the interactions due to electroweak and strong forces in a unified structure. Gravity is not included in standard model. According to the standard model, the matter and interactions may be described in terms of 61 elementary particles: 12 leptons + 36 quarks + 12 mediating quanta (8 gluon + 3 intermediate vector bosons + photon) + at least one Higg's particle. The quarks and leptons are sub-grouped into three generations of increasing mass. Particles in a given generation usually differ in their colour charge, while particles in the second and higher generations mimic them. The three generations and properties of leptons and quarks are listed in table 11.4.

The symbol Q stands for the electric charge in units of the electron charge. Lepton numbers corresponding to electron, μ and τ are represented by the corresponding subscript to L. Capital letters U, D, S, … represent the up-quark number, the down quark number, the strange quark number, and so on. Masses of leptons or quarks increase with its generation.

Table 11.4 *Three generations of Leptons and Quarks*

Leptons						Quarks							
Generation	**Symbol**	Q	L_e	L_μ	L_τ	**Symbol**	**Q**	**D**	**U**	**S**	**B**	**C**	**T**
Generation I	E	−1	1	0	0	D	−1/3	−1	0	0	0	0	0
	ν_e	0	1	0	0	U	2/3	0	1	0	0	0	0
Generation II	μ	−1	0	1	0	S	1/3	0	0	1	0	0	0
	ν_μ	0	0	1	0	C	2/3	0	0	0	0	1	0
Generation III	τ	−1	0	0	1	B	−1/3	0	0	0	−1	0	0
	τ_τ	0	0	0	1	T	2/3	0	0	0	0	0	1

11.8 HIGG'S FIELD AND HIGG'S PARTICLE

According to the quantum field theory, the vacuum of space–time continuum is filled with relativistic fields of different kinds such as field of strong force, field of electroweak force, and field of gravitational force. Disturbances in these fields consisting of energy packets behave like particles and mediate the field. Different fields generate particles of different types. In 1961, Sheldon Glashow proposed force carriers for weak interaction. They were W and Z bosons. These bosons required masses to reproduce the theoretical results obtained earlier by Fermi and were successful in explaining β decay. Massive carriers for weak interaction were quite different from the massless photon, the carrier particle of electromagnetic field and gluon, the mediator of strong field. The complete theory of electroweak field includes four force carriers: two charged carriers (W^+, W^-) and two uncharged particles (Z and γ photons) that mix at low energies – that is they evolve into each other as they travel. The electroweak theory successfully explained the experiments at low energy. However, at higher energies it could not hold. To fix the problem at high energy, Peter Higg's proposed theories in which a force carrier could acquire mass due to a new type of field. In 1967, Steven Weinberg and independently Abdus Salam incorporated this effect into Glashow's electroweak theory producing a consistent, unified electroweak theory. It included a new particle, dubbed the Higg's boson, which, when included in the scattering calculations, completed a new theory – the Standard Model – which made correct predictions even for very high-energy interactions. The way the Higg's field gives masses to the W and Z particles, and all other fundamental particles of the Standard Model (the Higg's mechanism), is subtle. The Higg's field, which like all fields lives everywhere in space, is in a different phase than other fields in the Standard Model. Because the Higg's field interacts with nearly all other particles, and the Higg's field affects the vacuum, the space (vacuum) particles travel through affects them in a dramatic way: It gives them mass. The bigger the coupling between a particle and the Higg's, the bigger the effect, and thus the bigger the particle's mass. This situation with the Higg's has a direct analogy with the freezing lake. At high enough temperatures, the Higg's field does not condense, which means that it takes on a constant value everywhere, and the W and Z are effectively massless. Lower temperatures can cause a transition in which the Higg's doublet condenses, the W and Z gain mass, and it becomes more difficult for them to move through the vacuum, as it is for fish in the lake, or boats on the surface when the lake freezes. In becoming massive, the W and Z absorb parts of the Higg's field. The remaining Higg's field has quantized vibrations that we call the Higg's boson that are analogous to vibrations on the lake itself. Recent experiments at the LHC (Large Hadron Collider) have confirmed the existence of Higg's bosons. Standard model shows that at very high energies $\approx 10^{15}$ TeV the strong, electromagnetic, and weak forces collapse into a single force. Attempts are now being made to develop a theoretical framework, which may also include gravitational force leading to the unification of all known forces and to TOE – theory of everything.

11.9 CONSERVATION LAWS

Conservation laws are reflections of symmetries in the nature. Conservation of linear momentum (and hence of energy) and of angular momentum are the outcome of translational and rotational symmetries of the universe. Some conservation laws, such as the conservation of energy, the conservation of linear and angular momentum, and the conservation of total electrical charge, are universal. These conservation laws also hold good at the level of interactions between elementary particles. It is, however, observed that some reactions seemingly allowed by the universal conservation laws do not occur. For example, consider the following reaction in which a proton decays into a positive and a chargeless pion.

$$P \rightarrow \pi^{+} + \pi^{0} \tag{11.2}$$

All quantities such as the total energy, linear momentum, and angular momentum could be conserved in reaction (11.2) if the two pions move out in opposite directions with appropriate energies. Further, the reaction should go fast with the speed characteristic of strong interactions as only hadrons are taking part in the reaction. However, the decay of proton has not been observed ever. This indicates that there is(are) some other quantity(ies) the non-conservation of which prohibits the process to occur. This and many other similar reactions, which are allowed by the universal conservation laws but do not actually occur, have given rise to what are called the '*family particle number conservation*' laws.

11.9.1 Conservation of Baryon Number *B*

The law that prohibits the decay of proton is the conservation of baryon number law. Just like the electric charge Q that may be assigned the value 0, ±1, ±2,... (in unit of the electronic charge e) to all elementary particles, a baryon number denoted by B, with possible values 0, 1, and −1, may be assigned to all elementary particles. Baryons have $B = 1$, anti-baryons $B = -1$, and for all other particles $B = 0$. The baryon conservation requires that the sum of the baryon number on the right hand side of a reaction must be equal to the sum of the baryon numbers on the left hand side. Proton is the lightest baryon and, therefore, all heavier baryons end up into proton. Proton is, therefore, stable. Reaction shown by Eq. (11.2) is forbidden by the conservation of baryon number law. The following reactions are allowed as baryon number is conserved in these reactions.

$$n \rightarrow p + e^- + \bar{\nu}_e \,;\; \Delta^+ \rightarrow p + \pi^0;\; \Delta^{++} \rightarrow p + \pi^+$$

11.9.2 Conservation of Lepton Numbers: Neutrino Oscillations

The lepton number conservation rule may be divided into two parts:

1. *Conservation of total Lepton number:* All leptons are assigned a lepton quantum number $L = 1$ and the anti-leptons the lepton number $L = -1$. The lepton number of all other particles is zero. The conservation of lepton number law demands that the total lepton quantum number in an allowed transition is conserved. The law holds for weak interactions that involve leptons.
2. In this scheme, three different leptons: electron (e^-), muon (μ), and tau-lepton (τ) along with their corresponding neutrinos, respectively, ν_e, ν_μ, and ν_τ are assigned a lepton number each as detailed in table 11.5. The lepton number of the antiparticle is same as that of the particle but with opposite sign. For example, the electron lepton number L_e for both $\bar{\nu}_e$ and e^+ is −1. The lepton number of all other particles is zero. The conservation law states that the total lepton number of each kind L_e, L_μ, and L_τ is separately conserved in a reaction.

Table 11.5 *Lepton numbers assigned to different leptons*

Lepton	Electron lepton number L_e	Mu lepton number L_μ	Tau lepton number L_τ
Electron e^-	1	0	0
ν_e	1	0	0
Muon μ	0	1	0
ν_μ	0	1	0
Tau lepton τ	0	0	1
ν_τ	0	0	1

The statement (1) of the law always holds, however, with the discovery of neutrino oscillations, statement (2) of the law is in doubt. Recent experiments have indicated that *neutrinos of one kind change into the other kind while in flight*. This is termed as *neutrino oscillations*. If neutrino oscillation is a real phenomenon then conservation of each kind of neutrino may not hold.

The decay of a μ-negative lepton into an electron may be given by Eq. (11.3a) if only statement (1) holds and by Eq. (11.2b) if statement (2) also holds.

$$\mu^- \rightarrow e^- + \bar{\nu} \tag{11.3a}$$

$$\mu^- \rightarrow e^- + \bar{\nu}_e \tag{11.3b}$$

11.9.3 Conservation of Strangeness Quantum Number

Some hadrons are created by strong interactions and decay via weak interactions. They are called strange particles. Further, it has been observed that strange particles are created in pairs. All hadrons are assigned a strangeness quantum number S that may have values either −1, zero or +1. *In strong interactions, the strangeness quantum number S is conserved but S is not conserved in weak or electromagnetic interactions.* The strangeness quantum number S for some hadrons is given in table 11.6.

Table 11.6 *Strangeness quantum number S, baryon number B, isospin I and the third component of isospin I_3 for some hadrons*

Hadrons	Strangeness S	Baryon no. B	Isospin I	Third component I_3
N P	0	1	1/2	−1/2 +1/2
π^0 π^+ π^-	0	0	1	0 +1 −1
Σ^0 Σ^- Σ^+	−1	1	1	0 −1 1
K^0 K^+	+1	0	1/2	−1/2 +1/2
K^- $\bar{K}^0$	−1	0	1/2	−1/2 +1/2
Ξ^0 Ξ^-	−2	1	1/2	+1/2 −1/2
Λ^0	−1	1	0	0
Ω^-	−3	1	0	0
η	0	0	0	0

11.9.4 Conservation of Isospin and the Third Component of Isospin

Hadrons Σ^0, Σ^-, and Σ^+ have nearly same mass but different charge states. It appears as if they are three different states of the same entity that has split due to electromagnetic interaction. There are many more examples of such splitting into doublets and triplets. A pair of quantum numbers I and I_3, respectively, called the isospin quantum number and the third component of the isospin, are assigned to each member of such families of hadrons. These quantum numbers for some hadrons are listed in table 11.6. The third component I_3 is related to the charge of the hadron by the following relation:

$$Q \text{ (in unit of e)} = I_3 + \left(\frac{S+B}{2}\right) \tag{11.4}$$

Here, $(S+B)$ is called the hypercharge and is denoted by Y.

The isospin I is conserved in strong interactions while I_3 is conserved both in strong and electromagnetic interactions.

It may be pointed out that I_3 behaves like a scalar quantity and follows algebraic summation, whereas isospin I behaves as an angular momentum and follow laws of addition of angular momentum.

The above-mentioned conservation laws along with the principle of crossing symmetries, discussed earlier, are tools to determine whether a given reaction is allowed or not.

Exercise p-11.1: Consider thc following reactions and determine whether each of the reaction is allowed; if yes, through what type of interaction, and if not then why?

(a) $\Xi^- \rightarrow \Sigma^- + \pi^0$
(b) $n + p \rightarrow \bar{n} + \bar{p}$
(c) $\Lambda^0 \rightarrow n + \pi^0$
(d) $K^- + p \rightarrow \Lambda^0 + \pi^- + \pi^+$
(e) $K^- + \Xi^0 \rightarrow \Omega^- + \pi^0$

Exercise p-11.2: What are the values of I, I_3, S, B, Y, and Q for particle Δ^0 which is a member of the isospin multiplet $\Delta^{++}, \Delta^+, \Delta^0$, and Δ^-

Exercise p-11.3: What are the possible values for the isospin of deuteron?

Exercise p-11.4: Assuming that a particle Z is created by the following strong reaction, assign the values of Q, S, B, and I_3 for particle Z.

$$K^+ + p \rightarrow Z + \Lambda^0$$

Exercise p-11.5: Discuss which of the following decays for the particle Z produced in p-11.4 are allowed and through what type of interaction. Also assign charges to the particles.

(i) $Z \rightarrow p + \pi$; (ii) $Z \rightarrow K + K$; (iii) $Z \rightarrow p + \bar{p} + \pi + \pi$

Exercise p-11.6: In the following reaction, particle Y could be a

$$Y \rightarrow \pi^0 + \mu^+ + \nu_\mu$$

(a) p (b) ρ^+ (c) K^+ (d) π^+

Exercise p-11.7: The reaction products of the collision of a π^+ and a (proton) p^+ may be

(a) $p^+ + p^+ + \bar{n}$ (b) $p^+ + p^+ + n$ (c) $p^+ + \bar{p}^+ + n$ (d) $\bar{p}^+ + \bar{p}^+ + \bar{n}$

Exercise p-11.8: Which of the following equations represents the decay of μ^- lepton?

(a) $e^- + \nu_e + \bar{\nu}_\mu$ (b) $e^- + \bar{\nu}_e + \nu_\mu$ (c) $e^- + \bar{\nu}_e + \bar{\nu}_\mu$ (d) $\bar{e}^- + \bar{\nu}_e + \bar{\nu}_\mu$

11.10 THE FUNDAMENTAL FORCES OF NATURE

Having discussed the elementary particles, it is now time to discuss the fundamental forces of nature that keep elementary particles together to make matter that we see around us. The fundamental forces may be categorized into four types:

1. Gravitational
2. Weak
3. Electromagnetic
4. Strong

Although it is difficult to define the relative strengths of different forces because they act on different types of particles and the strength of the force depends strongly on the distance between the particles, yet roughly speaking if the strength of the strong force is taken as 1, the strength of the electromagnetic force will be 10^{-2}, that of weak force 10^{-8}, and that of the gravitational force only around 10^{-45}. Once again it may be emphasized that these numbers comparing the strengths of different forces may not be taken seriously, and slightly different numbers may be found in some other literature for the reasons already mentioned.

In the quantum field theory, force is derived from the field and is mediated by field particle. The theory assumes that the space time continuum is filled with four types of fields, the strong, the electromagnetic, the weak, and the gravitational. Fields in a way may be compared to a pound. If a pebble is thrown in the pound a disturbance is created at the location where the pebble has fallen and it moves out in the form of a wave that carries energy. The quantum field theory describes the disturbance produced in a field as a field particle or field quantum that carries energy and brings about the interaction. For example, when electrically charged particles are placed in an electromagnetic field, the field at the location of the charged particles get disturbed creating the electromagnetic field quanta photon. The electromagnetic field quantum photon is then exchanged between the charged particles and in this way the charged particles interact with each other. The creation of the field quantum does not break the universal law of energy conservation. It is because of the Heisenberg's uncertainty principle, which states that the uncertainty ΔE in energy and the uncertainty in time Δt are related by $\Delta E \cdot \Delta t = \hbar$. It means that if the a small time interval Δt during which the field quantum is created and exchanged and the energy of the particle satisfy the uncertainty principle, there is no breakdown of the conservation of energy law. The field quanta so produced are called virtual particles and are only the messengers of the field. As such, according to the quantum field theory each field has its own field quanta. The mediating field quantum of gravitational field is called *graviton*. The graviton has not yet been discovered. Important characteristics of the four interactions are summarized in table 11.7.

As may be seen from table 11.7, weak decays have long life times $>10^{-13}$ s and small cross sections, typically $\approx 10^{-39}$ cm^2. Charged leptons experience both weak and electromagnetic interactions, while neutrino only weak interactions. Strangeness and parity are not conserved in weak decays. Weak decays are divided into three classes, depending on the extent to which leptons are involved in the process.

1. *Leptonic decays:* The decay products are only leptons: $\mu^- \rightarrow e^- + \bar{\nu}_e + \nu_\mu$
2. *Semi-leptonic decay:* Both hadron and leptons are involved in the decay. The strangeness may change by zero or integer value.
   (a) $n \rightarrow p + e^- + \bar{\nu}_e$ with $|\Delta S| = 0; K^+ \rightarrow \pi^0 + e^+ + \nu_e$ with $|\Delta S| = 1$
3. *Non-leptonic decays:* No leptons are involved in these decays and parity and strangeness are not conserved. Decay follows the selection rules: $\Delta S = \pm 1$, and $\Delta I = \pm 1/2$, for example $\Lambda \rightarrow p + \pi^-$.

Another important point is that *in weak decays the flavour of quark changes in contrast to the strong and electromagnetic interactions in which flavour is conserved.* This is indicated in the following reaction for the neutron decay in which a d-quark is converted into u-quark.

$$n \rightarrow p + e^- + \bar{\nu}_e \text{ or udd} \rightarrow \text{udu} + e^- + \bar{\nu}_e$$

Electromagnetic and weak interactions are two aspects of a single interaction called electroweak interaction. The corresponding charges are related by θ_w, the Weinberger angle; $\sin^2\theta_w = 0.2319$ and $\frac{M_W}{M_Z} = 0.88$.

Table 11.7 *Characteristics of fundamental interactions*

Interaction→	**Gravitational**	**Weak**	**Electromagnetic**	**Strong**
Quanta of the field with spin and parity $J^{\mp}$	Graviton?, 2+	$W^{\mp}, 1^-; Z^0, 1^+$	Photon, 1^-	Gluon, 1^-
Mass	0	80, 90 GeV	0	0
Coupling constant	$\frac{G_N M^2}{4\pi\hbar c} \approx 5\times10^{-40}$	$\frac{G(Mc^2)^2}{(\hbar c)^3}$	$\alpha = \frac{e^2}{4\pi\hbar c} = \frac{1}{137}$	≤ 1
Relative strength	10^{-45}	10^{-8}	10^{-2}	1
Time scale	–	$>10^{-13}$ s	10^{-18} to 10^{-20} s	10^{-23} s
Range	∞	10^{-18} m	∞	$\leq 10^{-15}$ m
Source	Mass	Weak charge	Electric charge	Colour charge

11.10.1 Cabibbo–Kobayashi–Maskowa (CKM) Matrix

The probability of transition from quark q to quark q′ is proportional to the square of the magnitude of the transition matrix element $V_{qq'}$. Cabibbo, Kobayashi, and Maskowa proposed the following transition matrix for quark transitions:

$$V_{ij} = \begin{pmatrix} V_{ud} & V_{us} & V_{ub} \\ V_{cd} & V_{cs} & V_{cb} \\ V_{td} & V_{ts} & V_{tb} \end{pmatrix}$$

The diagonal elements V_{ud}, V_{cs}, and V_{tb} of the transition matrix that corresponds to transitions within a family are only a few percent short of unity. Hence,

The transitions u → d, c → s, and t → b are Cabibbo favoured.

Elements V_{us}, V_{cd}, and V_{ts} are small but not zero.

The transitions u → s, c → d, and t → s are Cabibbo suppressed transitions.

Matrix elements V_{ub} and V_{td} are nearly zero;

Transition u → b and t → d are Cabbibo forbidden.

11.11 NUCLEON RESONANCES AND HYPERONS

Earlier than 1960 nucleons (neutron and protons) were considered to be elementary particles, having no further structure. However, after 1960 the quark model of hadrons indicated that nucleons are made up of three quarks and, therefore, have structures. With the availability of accelerated pion beams, it became possible to study nucleon structure through pion–nucleon scattering experiments. π – N Scattering data indicated peaks or bumps at fixed excitation energies. These peaks in π–N scattering cross-sections are called *nucleon resonances* and corresponds to the excited states of nucleons. Nucleon (or baryon) excited states or resonances are generated because of the flip of the spin of a quark or due to a different value of the relative orbital angular momentum of the quark when the resonance state decays. The resonances behave as short-lived particles and decay into a pair of nucleon and meson. The decay properties of the resonance states are determined from the widths and other characteristics of the peaks. The research group called particle data group (PDG) has made a compilation of the observed nucleon resonances. The resonance states are designated according to the symbol format N(M)L_{2I2J}. Here, N indicates that the state is an excited nucleon state, M is particle's (the excited resonance state that behaves as a particle) approximate mass in unified mass unit, L the relative orbital angular momentum of the nucleon–meson pair when the resonance decays, and I and J are, respectively, the states isospin and total angular momentum. As the isospin of nucleons is defined as 1/2, the factor $2I$ is always 1. The term $2J$ is always odd. L, the orbital angular momentum, can have integer values 0, 1, 2, 3 … that are also designated, respectively, as S, P, D, F… In case of nucleon resonances, it is quite common to drop N and reverse the order, that is $L_{2I2J}(M)$. For example, a proton may be represented as N(939)S_{11} or as S_{11}(939). A few important nucleon resonances are listed in table 11.8.

Table 11.8 *Some nucleon resonances*

Symbol resonance	J^P	Mass average (MeV/c²)	Full width	Common mode(s) of decay
$N(1520)D_{13}$	$3/2^-$	1520	115	$N + \pi$
$N(1535)S_{11}$	$1/2^-$	1535	150	$N + \pi$; or $N + \eta$
$N(1650)S_{11}$	$1/2^+$	1650	165	$N + \pi$
$N(1675)D_{15}$	$5/2^-$	1675	135	$N + \pi + \pi$; or $\Delta + \pi$
$N(1680)F_{15}$	$5/2^+$	1685	130	$N + \pi$
$N(1700)D_{13}$	$3/2^-$	1700	100	$N + \pi + \pi$
$N(1710)P_{11}$	$1/2^+$	1710	100	$N + \pi + \pi$

11.11.1 Hyperons

Hyperons are nuclear particles like baryons. Like all baryons, they are composed of three quarks. The term hyperon is generally used for a baryon containing one or more strange (s) quarks, as opposed (for example) to the proton and neutron, which contain only up (u) and down (d) quarks. The strange quark being unstable, hyperons decay into lighter baryons (such as protons or neutrons) plus mesons, with typical lifetimes of approximately 1/10 of a nanosecond. At high energies, these lifetimes are sufficient for a hyperon to travel several meters before decaying, as the hyperon can be moving at very nearly the speed of light they experience the time-dilation effect of special relativity. This long decay distance makes it possible to carryout experiments on the decay of hyperons.

11.11.2 Anti-matter Puzzle and CP Violation

One problem with the current Big Bang theory of the creation of universe is the almost complete absence of antimatter from the present world. According to the Big Bang theory equal amounts of matter and antimatter must have been produced at the time of Big Bang. There was no satisfactory answer to the question as to what happened to the antimatter? The inequality in the amounts of the matter and the antimatter may, however, be explained if charge conjugate and parity (CP) invariance is broken. Standard model allows for CP-violation in weak nuclear interaction. The concept of CP-violation was put forward by the Russian scientist Andrei Sakharov. Hyperon beams are very suitable to test the CP-non-conservation. Detailed experiments done at FERMI LAB, using hyperon/anti-hyperon beams have conclusively proved CP-violation in weak interactions.

11.12 DISCOVERY OF MUON AND π-MESON

As has already been discussed, the detection of charged particles depends on the ionization losses suffered by the charged particle while passing through the material of the detector. During 1930, charged particles present in cosmic rays were studied by cloud chambers. Charged particles passing through the cloud chamber produced tracks made up of droplets of super saturated water vapours present in the chamber. If a magnetic field is also present across the cloud chamber the tracks of charged particles got curved, curvature r depending on the strength B of the magnetic field and the momentum of the charged particle. According to Lorentz force law, the component of the momentum of the charged particle in the plane perpendicular to the magnetic field p (MeV/c) = 0.300×10^{-3} B (gauss) $\times$ r (cm), where r is the curvature of the track in cm. Thus, by measuring the track curvature it is possible to deduce the momentum of the particle. In addition, the energy of the particle can be measured by measuring the range of the particle (the total length of the track) in cloud chamber. As the energy loss per unit path length, also called stopping power, depends on the nature of the particle, it is possible to get valuable information about the characteristics of the particle by measuring range and momentum of the particle from cloud chamber data.

In 1937, Anderson and S.H. Neddermeyer made energy loss measurements by placing a 1-cm thick platinum plate in the cloud chamber. By measuring the curvature of the track on both sides of the platinum plate, they were able to determine the momentum loss suffered by the particle while passing through the platinum plate. As they observed particles in the 100–500 MeV/c momentum range, if the particles were electrons or positrons, they were highly relativistic and their energy was given simply by $E = pc$. According to the Bethe–Heitler theory, the particles should have lost in the plate an amount of energy proportional to their incident energy. Moreover, the particles with this energy should have produced electromagnetic shower. What Neddermeyer and Anderson observed was quite different. They found that the particles could be separated into two classes. The particles of the first class behaved just as the Bethe–Heitler theory predicted. The particles of the second class, however, lost nearly no energy in the platinum plate: They were highly penetrating. Moreover, they were not associated with electromagnetic showers. As the Bethe–Heitler theory predicted large energy losses for electrons because they were light and could easily emit radiation, Neddermeyer and Anderson were led to consider the possibility that the component of cosmic rays that did not lose much energy consisted of particles heavier than the electron. In addition, the particles in question could not be protons because protons of the momentum observed would be rather slow and would ionize much more heavily in the cloud chamber than the observed particles, whose ionization was essentially the same as that of the electrons. Neddermeyer and Anderson gave as their explanation

> *There exist particles of unit charge with a mass larger than that of a normal free electron and much smaller than that of a proton ... [That they] occur with both positive and negative charges suggests that they might be created in pairs by photons.*

While the penetrating component of cosmic rays had been observed by others before Neddermeyer and Anderson, the latter were able to exclude the possibility that this component was due to protons. Moreover, Neddermeyer and Anderson observed particles of energy low enough to make the application of the Bethe–Heitler theory convincing. At the time, many doubted that the infant theory of quantum electrodynamics could be trusted at very high energies. The penetrating component of cosmic rays could be ascribed to a failure of the Bethe–Heitler theory when the penetrating particles were extremely energetic. Neddermeyer and Anderson provided evidence for penetrating particles at energies for which the theory was believed to hold.

At nearly the same time, Street and Stevenson reported similar results and soon improved upon them. To determine the mass of the newly discovered particle, they sought to measure its momentum and ionization at the same time. As the ionization is a function of the velocity, the two measurements would in principle be able to determine the mass. However, the ionization is weakly dependent on the velocity except when the velocity is relatively low, that is, when the particle is near the end of its path and the ionization increases dramatically.

To obtain a sample of interesting events, Street and Stevenson used counters in both coincidence and anticoincidence: the counters record only if a charged particle passed through them and the apparatus was arranged so that the chamber was expanded to create super saturation and a picture taken only if a particle entered the chamber (coincidence) and was not detected if the particle exits (anticoincidence). As such, only those events in which the particle entered the cloud chamber and got stopped in the chamber were recorded. This method of triggering the chamber was invented by Blackett and Occhialini. In addition, a block of lead was placed in front of the apparatus to screen out the showering particles. In late 1937, Street and Stevenson reported a track that ionized too much to be an electron with the measured momentum, but travelled too far to be a proton. They measured the mass crudely as 130 times the rest mass of the electron, an answer smaller by a factor 1.6 than later, improved results, but good enough to place it clearly between the electron and the proton.

11.12.1 The Yukawa Particle?

In 1935, before the discovery of the penetrating particles, Hideki Yukawa predicted the existence of a particle of mass intermediate between the electron and the proton. This particle was to carry the nuclear force in the same way as the photon carries the electromagnetic force. In addition, it was to be responsible for β decay. As the range of nuclear forces is about 1 F, the mass of the particle predicted by Yukawa was about 200 MeV/c^2. When improved measurements were made, the mass of the new particle was determined to be about 100 MeV/c^2, close enough to the theoretical estimate to make natural the identification of the penetrating particle with the Yukawa particle. How could this identification be confirmed? In 1940, Tomonaga and Araki showed that positive and negative Yukawa particles should produce very different effects when they came to rest in matter. The negative particles would be captured into atomic-like orbits, but with very small radii. As a result, they would overlap the nucleus substantially. Considering that the Yukawa particle was designed to explain nuclear forces, it would certainly interact extremely rapidly with the nucleus, being absorbed long before it could decay directly. In addition, the positive Yukawa particles would come to rest between the atoms and would decay. The lifetime of the penetrating particle was first measured by Franco Rassetti who found a value of about 1.5×10^{-6} s. Improved results, near 2.2×10^{-6} s, were obtained by Rossi and Nereson, and by Chaminade, Freon, and Maze. Working in Italy during World War II, Conversi, Pancini, and Piccioni investigated further the decays of positive and negative penetrating particles that came to rest in various materials. Using a magnetic focusing arrangement that Rossi had developed, Conversi, Pancini, and Piccioni were able to select either positive or negative penetrating particles from the cosmic rays and then determine whether they decayed or not when stopped in matter. The positive particles did indeed decay, as predicted by Tomomaga and Araki. When the absorber was iron, the negative particles did not decay, but were absorbed by the nucleus, again in accordance with the theoretical prediction. However, when the absorber was carbon, the negative particles decayed.

This meant that the Tomonaga–Araki prediction as applied to the penetrating particles was wrong by many orders of magnitude: these could not be the Yukawa articles.

11.12.2 The Final Confirmation

Shortly thereafter, D.H. Perkins used photographic emulsions to record an event of precisely the type forecast by Tomonaga and Araki. Photographic emulsions provide a direct record of cosmic ray events with extremely fine resolution. The event in question had a slow negative particle that came to rest in an atom, most likely a light atom like carbon, nitrogen, or oxygen. After the particle was absorbed by the nucleus, the nucleus was blasted apart and three fragments were observed in the emulsion. This single event apparently showed the behavior predicted by Tomonaga and Araki, contrary to the results of the Italian group. The connection between the results of Conversi, Pancini, and Piccioni and the observation of Perkins was made by the Bristol group of Lattes, Occhialini, and Powell in one of several papers by the group, again using emulsions. Their work established that there were indeed two different particles, one of which decayed into the other. The observed decay product appeared to have fixed range in the emulsion, that is, it appeared always to be produced with the same energy. This indicated that the decay was into two bodies and not more. Because of inaccurate mass determinations, at first it was believed that the unseen particle in the decay could not be massless. Quickly, the picture was corrected and completed: the pion, decayed into a muon (the names given by Lattes et al.), and a very light particle, presumably Pauli's neutrino. The particle (which Perkins had likely seen) was much like Yukawa's particle except that it was not the origin of β decay, since β decays produce electrons rather than muons. The particle (which Anderson and Neddermeyer had found) was just like an electron, only heavier. The pion has two charge states, + and – that are charge conjugates of each other and which yield + and – muons, respectively, in their decays. In modern parlance, bosons (particles with integral spin) like the pion that feel nuclear forces are called mesons. More generally, all particles that feel nuclear forces, including fermions like the proton and neutron are called hadrons. Fermions (particles with half-integral spin) like the muon and electron that are not acted by these strong forces are called leptons. While a negative pion would always be absorbed by a nucleus upon coming to rest, the absorption of the negative muon was much like the well-known radioactive phenomenon of K-capture in which an inner electron is captured by a nucleus while a proton is transformed into a neutron and a neutrino is emitted. In heavy atoms, the negative muon could be absorbed (because it largely overlapped with the nucleus) with small nuclear excitation and the emission of a neutrino, while in the light atoms it would usually decay, because there was insufficient overlap between the muon and the nucleus.

11.13 MEASUREMENT OF THE SPIN AND PARITY OF PIONS

11.13.1 Measurement of the Spin of π-meson

11.13.1.1 Spin of the charged pion

The spin of the charged pion was determined from the following reversible reaction:

$$p + p \leftrightarrow \pi^{+} + d \qquad (11.5)$$

The theory of detailed balance requires that the transition probabilities for the forward and the backward reactions in case of reaction represented by Eq. (11.5) should be same. Now, the transition probabilities are proportional to the number of possible final states times the square of the linear momentum in centre of mass. Therefore, for the forward reaction (p + p → π^+ + d), the transition probability P_{for} may be written as

$$P_{\text{for}} \propto (2s_\pi + 1)(2s_\text{d} + 1)p_\text{d}^2 \tag{11.6}$$

Here, s_π, s_d, and p_d are, respectively, the spins of π-meson, spin of the deuteron, and the linear momentum of deuteron (or of π-meson) in the centre of mass frame.

Similarly, for the backward reaction $\pi^+ + \text{d} \rightarrow \text{p} + P$, the transition probability P_{back} may be written as

$$P_{\text{back}} \propto \frac{1}{2}\,(2s_\text{p} + 1)^2\, p_\text{p}^2 \tag{11.7}$$

Here, s_p and p_p are, respectively, the spin of the proton and the linear momentum of the proton in the centre of mass frame. The factor of 1/2 appearing in Eq. (11.7) is due to the Pauli exclusion principle that restricts states with identical quantum numbers for the two protons. Further, in CM frame the total linear momentum is zero, hence $|p_\text{d}| = |p_\text{p}|$.

The ratio $\frac{P_{\text{for}}}{P_{\text{back}}} = \frac{2\,(2s_\pi + 1)(2s_\text{d} + 1)}{(2s_\text{p} + 1)^2}$ is a function of the spin of the π meson s_π, the spins of deuteron s_d and of proton s_p. The spins of deuteron and of protons are known, and, therefore, the ratio of the forward to backward reaction rates depends on the spin of the π meson.

The transition rates for the forward and backward reactions were actually measured and gave the value of the spin of charged pion as zero.

11.13.1.2 Spin of the neutral pion

The neutral pion decays into two γs,

$$\pi^0 \rightarrow \gamma + \gamma \tag{11.8}$$

Now, the minimum spin of a γ is 1, and hence, the spin s_{π^0} may either be zero or it should be equal/larger than 2. As the spin of charged pions is zero, the spin of neutral pion must also be zero.

11.13.2 Parity of Pions

11.13.2.1 Parity of charged pions

To determine the parity of a system of particles, both the intrinsic parity of the constituents and the spin and angular moment of the system have to be considered. Pion parity can be deduced from the reaction

$$\pi^- + \text{d} \rightarrow n + n \tag{11.9}$$

in which a slow negative pion is captured by the deuteron to form an exotic atom. Capture occurs at overlap of the pion wave function with the deuteron, which is only large for the s state. Therefore, we know that the initial state has $\ell = 0$. As $s_{\pi^-} = 0$ and $s_d = 1$ the initial state has total angular momentum $J = \ell + s = 1$.

$$\text{The parity of initial system } (\pi^- + \text{d}) = P_{\text{ini}} = \left(P_{\pi^-}\right)\left(P_\text{d}\right)\left(-1^\ell\right).$$

Here, $\left(P_{\pi^-}\right)$ is the parity of negative pion and (P_d) the parity of deuteron and $(-1)^l$ is the parity of the relative motion with angular momentum ℓ. As has already been mentioned, the orbital angular momentum ℓ of the state in which deuteron captures a π^- meson is zero or S, $\ell = 0$ state, hence (-1) ℓ term is simple 1. Also, the parity of deuteron is even, that is $P_d = +1$ (or even) and therefore $P_{\text{ini}} = \left(P_{\pi^-}\right)$, the parity of π^- meson.

As parity is conserved in strong interactions, the parity of the initial system must be equal to the parity P_{final} of the final system. The final system consists of two neutrons. P_{final} is equal to the parity of the wave function ψ_n, of the two neutrons.

The wave function of the two neutrons may be written as the multiplication of two terms

$$\psi_\text{n} = \phi(\text{space})\alpha(\text{spin}) \tag{11.10}$$

In Eq. (11.10), ϕ and α are, respectively, the space and spin parts of the two neutron wave function.

The spin part of the wave function has the form $\alpha(S, S_z)$, where S and S_z are the total spin and its z-component. The following four combinations of spins are possible:

$$\alpha(1,1) = \uparrow\uparrow \tag{11.11a}$$

$$\alpha(1,0) = \frac{1}{\sqrt{2}}(\uparrow\downarrow + \downarrow\uparrow) \tag{11.11b}$$

$$\alpha(1,-1) = \downarrow\downarrow \tag{11.11c}$$

$$\alpha(0,0) = \frac{1}{\sqrt{2}}(\uparrow\downarrow - \downarrow\uparrow) \tag{11.11d}$$

The first three states are symmetric under spin-label interchange but the last is anti-symmetric. Therefore, the symmetry of the spin function is $(-1)^{S+1}$.

The symmetry of the spatial function ϕ is $(-1)^L$. Therefore, the total symmetry is $(-1)^{L+S+1}$. Identical fermions must have anti-symmetric wave function ψ_n and, therefore, the $L + S + 1$ must be odd. It means that the parity of the final state P_{final} must be odd, which means that $P_{\text{final}} = (-1)^1$. However, $P_{\text{final}} = P_{\text{ini}} = (P_{\pi^-})$, the pion must have the parity $= (-1)^1$.

11.13.2.2 Parity of neutral pion

The parity of the neutral pion was determined from the analysis of the rear decay.

$$\pi^0 \rightarrow \left(\text{e}^- + \text{e}^-\right)\left(\text{e}^+ + \text{e}^-\right) \tag{11.12}$$

From the detailed analysis of reaction (11.12), the parity of neutral π meson was also found to be $(-1)^1$.

11.14 THE CPT OR LÜDERS–PAULI THEOREM

The quantum field theories have been quite successful in describing the strong, electromagnetic and weak interactions. A fundamental ingredient in quantum field theories is the CPT invariance. CPT invariance means that all interactions and all laws of nature remain unchanged (invariant) when subjected to the combined operations of (1) particle–antiparticle interchange (the so-called charge conjugation, represented by C), (2) reflection of the coordinate system through the origin (replacing the space vector $\left(\vec{r}\right)$ by $\left(\vec{-r}\right)$, which is called the parity operation and is denoted by P), and (3) reversal of time (which means that the system goes back in time from the final state to the initial state, time reversal is denoted by T). In the language of operators, it may be said that the CPT operator commutes with the Hamiltonian of the system. It may further be realized that the order or the sequence of C, P, and T operations is not important. The operations may be performed in any order, PTC, TCP, CTP, and so on, and they are all entirely equivalent. Although, for the first time the CPT theorem evolved implicitly in the work of J. Schwinger around 1951–1952, a more explicit proof of the theorem was given by G. Lüders and W. Pauli in 1954, and, therefore, the theorem is also called the *Lüders–Pauli theorem*. At about the same time, and independently, the theorem was also proved by John Stewart Bell. These proofs are based on the validity of Lorentz invariance and the principle of locality in the interaction of quantum fields. The CPT theorem states that any Lorentz invariant in local quantum field theory with Hermitian Hamiltonian must have CPT symmetry.

Operator C, representing charge conjugation, is a mathematical operation that transforms a particle into an antiparticle, for example, changing the sign of the charge. Charge conjugation implies that every charged particle has an oppositely charged antimatter counterpart, called the antiparticle. The antiparticle of an electrically neutral particle may be identical to the particle, as in case of neutral π meson and the photon, or it may be distinct, as with the antineutron. Parity operator P implies space inversion, that is, the three space dimensions x, y, and z become, $-x$, $-y$, and $-z$. The parity conservation essentially means that the left and right and up and down are indistinguishable in the sense that a nucleus emits off decay products up as often as down and to left as often as right.

Invariance under time implies that whenever a motion is allowed by the laws of physics, the reversed motion is also an allowed one.

11.14.1 The Implication

The implication of CPT invariance or symmetry is that a mirror image of our universe – with all objects having their positions reflected by an imaginary plane (Parity operation), all momenta reversed (time inversion), and all matter replaced by antimatter (charge inversion) – would evolve under exactly the same physical laws as in our universe. The CTP operation turns our universe to its mirror image. CPT symmetry is regarded as a fundamental property of all physical laws. To preserve CPT symmetry, every violation of the combined symmetry of two of its components such as CP, must have a corresponding violation in the third component, such as T. Thus, violations in T symmetry are often referred to as CP violations.

It follows, in particular, from the CPT theorem that

1. Particles and the corresponding antiparticles have equal masses and equal lifetimes.
2. The charge of the particle and the corresponding antiparticle differ only in sign as do the magnetic moments.
3. The interaction of a particle and the corresponding antiparticle with the gravitational field is identical.
4. In those cases where interaction of particles in the final state is negligible, the energy spectra and angular distributions of the decay products are the same for particle and the corresponding antiparticle, but the projections of the spins are of opposite sign.

11.14.2 Non Conservation of Parity (P-violation)

An apparent lack of the conservation of parity in the decay of the charged K mesons into two and three π mesons prompted Chen Nig Yang and Tsung-Dao Lee to examine the experimental foundation of parity. In 1956, they showed that there was no evidence supporting parity invariance in weak interactions. Experiments conducted by C.S Wu in 1957 verified decisively that parity is not conserved in nuclear weak interactions like the β decay.

11.14.3 CP Invariance in Nuclear Weak Interactions

Further studies, however, showed that the charge conjugation symmetry is also broken in weak interactions, perhaps, to compensate for the non-conservation of the parity. The breaking of charge conjugation symmetry was demonstrated by examining the spins of the electrons and positrons emitted in the decays of positively charged and negatively charged muons, respectively. As such, it was believed, in general, that simultaneous breaking of the charge conjugate symmetry and of the parity symmetry (cancelled each other out exactly) leaves the CP symmetry invariant. This in turn will keep T symmetry invariant and the CPT-theorem intact.

11.14.4 Breaking of CP Symmetry

Experiments carried out in 1964 demonstrated that the electrically neutral K^0-meson that was initially thought to decay into three π mesons, decayed a fraction of time into only two π-mesons, thus violating CP symmetry. Assuming that the CTP invariance is universal and always holds for quantum fields, break down of CP symmetry implies break down of T symmetry as well.

To test the invariance of the combined CP symmetry, it is necessary to define an elementary particle state, which is either even or odd under CP-symmetry operation. Unfortunately, neutral kaon cannot be assigned a well-defined CP symmetry. It is because the CP operation transforms it into its antiparticle and, therefore, changes the identity of the wave function. To see the effect of the CP operation, the identity of the particle should not change, which in turn means that the identity of the wave function should not change. In the language of quantum mechanics, one may state that K^0 and $\bar{K}^0$ are not eigenstates of CP operation. The simplest eigenstates of CP operation are the states produced by the mixture of K^0 and $\bar{K}^0$

$$K_1^0 = \frac{1}{\sqrt{2}}\left(K^0 + \bar{K}^0\right) \text{ and } K_2^0 = \frac{1}{\sqrt{2}}\left(K^0 - \bar{K}^0\right) \tag{11.13a}$$

The wave functions of these states keep their identity under the CP operation. It may be shown that

$$\text{CP}\left(K_1^0\right) = +K_1^0 \text{ and CP}\left(K_2^0\right) = -K_2^0$$

As such, K_1^0 has even (+1) CP symmetry and K_2^0 the odd (−1) CP symmetry.

11.14.5 The Status of $K^0, \bar{K}^0$, K_1^0, and K_2^0

Neutral kaons ($K^0, \bar{K}^0$) are produced via strong interactions and are, therefore, eigenstates of parity operation with odd intrinsic parity along with a definite assignment of strangeness, whereas the particles that decay via weak interactions are K_1^0 and K_2^0 that are the eigenstates of the CP operation with +1 (even) and −1 (odd) CP symmetry, respectively. As in combined CP operation, the CP symmetry is conserved, K_1^0 decays into two pions (the final system has even symmetry under CP operation) and K_2^0 into three pions (the three-pion final state has odd symmetry under CP operation). In a simple language, it may be stated that K^0 and $\bar{K}^0$ interact only with nuclear strong field, through which they are produced. Neutral kaons do not sense the nuclear weak field, through which they decay. It appears a little contradictory, because if neutral kaons decay via nuclear weak field then they must sense the nuclear weak field also. However, what really happens may be better understood if we look to the mean life times of K_1^0 and K_2^0, which are as follows:

$$\text{Mean life time of } K_1^0 = 0.9 \times 10^{-10}\text{ s; mean life time of } K_2^0 = 520.0 \times 10^{-10}\text{ s.}$$

It may be observed that the mean life of K_2^0 is almost 520 times the mean life of K_1^0. Now, let us consider the following reaction that produces K^0 (and not $\bar{K}^0$) due to the conservation of strangeness:

$$\pi^- + p \rightarrow \Lambda^0 + K^0 \tag{11.13b}$$

At the instant of its creation by strong force (that drives reaction (11.13b)) K^0 is a mixture of K_1^0 and K_2^0 in equal amounts. However, because of the short mean life of K_1^0, it decays out (into two pions) and K_2^0 of longer mean life stays on. As such, the relative component of K_1^0 in a beam of K^0 decreases with time. Thus, after a sufficiently long time ($\approx 10 \times 10^{-10}$ s), the beam will have mostly K_2^0, which will decay into three pions.

An interesting aspect of reaction (11.13b) is that initially at the instant when K^0 was created, it was purely the particle K^0 but as time passes K^0 turns into K_2^0 $\left(= \frac{1}{\sqrt{2}}\left(K^0 - \bar{K}^0\right)\right)$ which has almost 50% $\bar{K}^0$, the antiparticle. To test whether this is true, one may look to a reaction that is initiated only by $\bar{K}^0$. One such reaction is

$$\bar{K}^0 + \text{N} \rightarrow \pi + \Lambda \tag{11.13c}$$

Reaction (11.13c) may be initiated only by $\bar{K}^0$ and not by K^0 because of the conservation of strangeness quantum numbers on the two sides of the equation. Now, if a beam which initially had only K^0 particles is allowed to decay the K_1^0 component and then made to hit a target containing neutrons, reaction (11.13c) will start to show up. The rate of reaction (11.13c) will increase with time as the K_1^0 component drops out. This has been experimentally verified. From the frequency or the rate of reaction (11.13c), the variation in the relative intensity of $\bar{K}^0$ component with time can be determined. This variation in the intensity is dependent on the mass difference between K_1^0 and K_2^0. Experiments have indicated a mass difference of 3.5×10^{-6} eV/c. As the mass of K^0 is 498 MeV/c, the mass difference between K_1^0 and K_2^0 is as small as of 1 part in 10^{14}.

11.14.6 Breaking of CP Symmetry in Nuclear Weak Interactions

To check whether CP symmetry is strictly and exactly conserved in weak interactions, Christenson, Cronin, Fitch, and Turlay took a beam of $K_2{}^0$ (the K^0 beam after travelling some distance in vacuum contains mostly $K_2{}^0$ as the $K_1{}^0$ component of shorter mean life decays out) and looked for the two-pion decay. If CP is exactly conserved in weak interactions then $K_2{}^0$ with odd CP symmetry should decay only into three pions and no event of two-pion decay should be observed. However, they recorded about 50 events (out of a total 23,000 events) in which two-pion decay of $K_2{}^0$ was observed. This number 50 was much more than the number expected from the background of the decay of residual $K_1{}^0$ left in the beam. This conclusively proved that a state ($K_2{}^0$) of odd CP symmetry may sometime go into a state of even-CP symmetry (a two-pion final state is even symmetry state) and the CP symmetry is not exactly preserved in weak interactions.

As CP symmetry is not exactly conserved in weak interactions, it is obvious that particles K_1^0 and K_2^0 do not represent the true eigenstates of CP symmetry, though initially we assumed them to be the eigenstates of weak interaction. The true eigenstates of CP symmetry are the admixtures of K_1^0 and K_2^0, denoted by K_L, the long-lived and K_S, the short-lived kaons, respectively, and defined in the following way:

$$K_L = K_2^0 + \varepsilon K_1^0; \text{ and } K_S = K_1^0 - \varepsilon K_2^0 \text{ with } \varepsilon \approx 0.002 \tag{11.14}$$

Although the non-conservation of CP symmetry in weak interaction has many important theoretical consequences, one important outcome is that it is possible to make a distinction between the right and left because of the breaking of CP symmetry in weak interactions.

Exercise p-11.9: What are leptons and their families? Discuss the law of lepton conservation in nuclear weak interactions. Is the decay $\tau^- \to \mu^- + \nu_\tau$ possible, if yes through what interaction, and if not then why?

Exercise p-11.10: Write a detailed note on mesons and their quark structure.

Exercise p-11.11: Give an account of the discovery of π-meson.

Exercise p-11.12: Discuss with necessary details a method for measuring the spin of π-meson.

Exercise p-11.13: What are intermediate vector mesons? Discuss their significance in electroweak interactions

Exercise p-11.14: Describe the eight-fold way of classifying elementary particles.

Exercise p-11.15: Write notes on:
(a) Particles and antiparticles
(b) Principal of cross symmetry
(c) Higg's field and elementary particle mass
(d) Nuclear resonances and hyperons.

Exercise p-11.16: Give a brief description of the fundamental forces of nature.

Exercise p-11.17: What are hadrons? What are their important properties? Discuss the quark structure of hadrons.

Exercise p-11.18: State CPT theorem and discuss the breakdown of CP symmetry in weak interactions.

Indicate how the following quantities will transform under P and T operations,
(i) r (ii) momentum p (iii) $\sigma = r X p$

Exercise p-11.19: The spin of ρ^0 is 1 and that of π-mesons is zero. On the basis of the symmetry of the final state total wave function decide which of the following decay is allowed: (a) $\rho^0 \rightarrow \pi^+ + \pi^-$ (b) $\rho^0 \rightarrow \pi^0 + \pi^0$

Exercise p-11.20: A K^0 meson of momentum p decays into π^+ with momentum p_1 and π^- with momentum p_2. The π mesons are emitted in opposite directions. Find the magnitude of p in terms of the masses of the mesons, if $p_1 = 2p_2$.

11.15 UNITS IN PARTICLE PHYSICS: THE NATURAL UNITS

Generally, in physics, we use SI system of units where the three quantities: mass, length, and time are fundamental quantities having kilogram, metre, and second as their basic units. All other quantities have derived units. The fundamental units, 1 kg and 1 m of the MKS system are defined in terms of the mass and length of prototypes kept at the bureau of international standards in Paris, France. The unit of time is defined, again with reference to the caesium clock, a prototype of which is kept in standards laboratories. The fundamental quantities of the SI system are defined on the so-called 'human' scale. However, the 'human scale' is not very convenient for particle physics. For example, the mass of proton is 1.67×10^{-27} kg, the radius of proton is 0.8×10^{-15} m and the spin of proton is 1/2 $\hbar$, which cannot be expressed in a convenient way in SI units. Particle physics uses the theory of relativity and quantum mechanics. In relativity, the velocity of light c enters in all expressions. Similarly, in quantum mechanism the Plank's constant h and rationalized Planck's constant $\hbar$ appear in most of the relations. People working in the field of particle physics have, therefore, devised a system of units that is based on the naturally occurring quantities in the theory of relativity and the quantum mechanics. The system of units used by particle physics workers is called the 'natural unit system' or simply natural units. In this system, the velocity of light in vacuum c and the rationalized Planck's constant is taken equal to unity

$$c \equiv \hbar \equiv 1.$$

With this substitution, the relativistic expression for energy, $E^2 = p^2c^2 + m^2c^4$, looks more simple as

$$E^2 = p^2 + m^2$$

The natural system of units may be considered as a system where the fundamental quantities are energy, velocity, and angular momentum instead of mass, length, and time. The dimensions and units for energy, velocity, and angular momentum in natural units are tabulated in table 11.9.

In natural units, mass, length, time, and so on are derived quantities having the dimensions and units shown in table 11.10.

In practice, one does not write unit of c or unit of angular momentum $\hbar$ and simply put 1. Thus, in natural units the unit of mass is GeV, of length GeV^{-1}, and of time also GeV^{-1}.

Now, 1 kg mass $= 1 \times c^2 = 1\text{ kg} \times (3 \times 10^8\text{ m})^2 = 9 \times 10^{16}$ Joule $= \dfrac{9\times10^{16}\text{ J}}{1.6\times10^{-19}}$ eV $= 5.625 \times 10^{35}$ eV

Or 1 kg mass $= 5.625 \times 10^{26}$ GeV/c^2.

Similarly, $\hbar = 1.055\times10^{-34}$ J-s, but in natural units $\hbar = 1$,

Therefore, 1.055×10^{-34} J-s $= 1$, or $\dfrac{1.055\times10^{-34}\text{ J}}{1.6\times10^{-19}\text{ J/eV}}$ s $= 1$, or 1 s $= \dfrac{1.6\times10^{-19}\text{ J/eV}}{1.055\times10^{-34}\text{ J}}$

Or 1 s $= 1.5 \times 10^{24}$ GeV^{-1}

Let us calculate the value of 1 m in natural units.

We know that $\hbar c = 197.3$ MeV-F $= 0.1973$ GeV^{-1} 10^{-15} m.

However, in natural units, $\hbar c = 1 = 0.1973$ GeV^{-1} 10^{-15} m.

Hence, 10^{-15} m $= \dfrac{1}{0.1973}$ GeV^{-1} or 1 m $= \dfrac{10^{15}}{0.1973}$ GeV^{-1} $= 5.068 \times 10^{15}$ MeV^{-1} or 1 m $= 5.068 \times 10^{15}$ MeV^{-1}.

Table 11.9 *Units and dimensions of fundamental quantities of natural units*

Quantity	Dimensions	Units
Energy	[E]	1 GeV $=1.602 \times 10^{-10}$ J
Velocity	[V]	1′ natural velocity unit c′
Angular momentum	[W]	1′ natural angular momentum unit $\hbar'$

Table 11.10 *Units and dimensions of mass, length and time in natural units*

Quantity	Dimensions	Units
Mass	[EV^{-2}]	1 GeV (unit of c)$^{-2}$
Length	[VWE^{-1}]	1 (unit of velocity c) 1 (unit of angular momentum $\hbar$) GeV^{-1}
Time	[WE^{-1}]	1 (unit of angular momentum $\hbar$) GeV^{-1}

Exercise p-11.21: Show that 1 mb is equal to 2.57 GeV^{-2}.

Exercise P-11.22: Write expressions for (i) Compton wavelength of an electron (b) Bohr radius a_0 for hydrogen atom in ground state.

11.16 INTRODUCTION TO FEYNMAN'S DIAGRAMS

In the year 1948, American physicist Richard Feynman invented a method of describing an interaction by two-dimensional diagrams and to calculate the quantum mechanical probability of the transition using these diagrams. Originally, Feynman devised this method for application in the field of quantum electrodynamics (QED); however, the method of Feynman's diagrams was soon extended to almost all branches of nuclear and particle physics, including the QCD (quantum chromo dynamics). To a very large extent, the credit of laying down the rules for drawing Feynman's diagram goes to Feynman's younger associate Freeman Dyson. Feynman and Dyson showed that in the case of electromagnetic interactions, calculations done using the technique of Feynman's diagrams give results correct up to fifth place of decimal. It may, however, be mentioned that the complete description of a transition requires not only one but a large number of Feynman's diagrams of increasing complexity, but in case of the electromagnetic interactions the relative contribution of complex diagrams decreases rapidly with the complexity of the diagram on account of the very small value of the coupling constant α (=1/137). Feynman has himself warned that the use of this method for calculating transition probabilities in case of strong interactions may lead to totally wrong results because of the large value of the coupling constant for strong interactions and significant contributions from complex diagrams. As an introduction to the method of Feynman's diagrams, in this section we shall just draw these diagrams for some interactions and will not try to calculate the transition probabilities.

Elements of Feynman's diagram: Since Feynman's diagrams are two-dimensional plots of the progress (and not trajectories) of interacting particles and the mediating quanta, it is required to show the different particles and quanta with different symbols. Convention generally followed in this regard is as follows:

1. The *X*-direction in Feynman diagrams, generally, represents space and the *Y*-direction time. However, this is not a hard and fast rule; one may represent space along the *Y*-direction and the time along the *X*-direction as well.
2. According to the quantum field theory, interactions occur because of the exchange of field quanta. The field quanta of electromagnetic interactions are photons. In electromagnetic interaction, virtual photons are exchanged between two charged particles when they interact. In Feynman's diagram, photons are represented by a wiggly line like a sinusoidal curve.

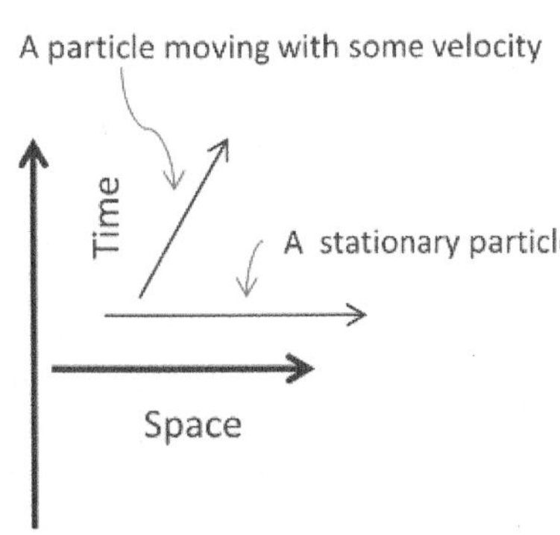

Figure 11.4 *Representation of Space and time in Feynman diagrams*

3. Fermions in Feynman's diagram are represented by a solid straight line with an arrow head indicating the direction of motion with time. Antiparticles of fermions are also indicated by solid straight lines but with the direction of arrow head reversed.
4. Electroweak interactions are communicated by three field quanta: Z^0 and $W^{\pm}$ bosons. These quantas of weak field are indicated with broken/dotted lines in Feynman diagrams.
5. The field quanta of strong field are quarks that are represented by springs.
6. The points in Feynman's diagram where field quanta are emitted and absorbed or two elements join together are called vertices. A Feynman's diagram that has more vertices is more complex. The order of the diagram represents it complexity and the simplest diagram is called the first-order diagram.

These elements are shown in figure 11.5.

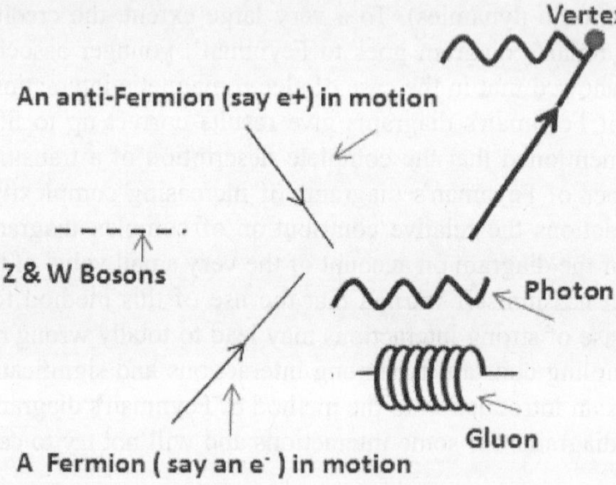

Figure 11.5 *Elements of Feynman's diagrams*

Drawing of Feynman's diagrams for some interactions is explained in the following solved examples.

Solved example S-11.1

Draw Feynman's diagram showing Coulomb repulsion between two electrons or draw Feynman's diagram for the electromagnetic interaction: $e^- + e^- \rightarrow e^- + e^-$.

Solution. As shown in figure S-11.1, an electron moving towards point A emits a virtual photon at vertex A and as a result of emitting virtual photon recoils back as scattered electron along AC. Another electron moving along B′B, absorbs the virtual photon at vertex B and got scattered in direction BD. Thus, the mutual repulsion between two electrons can be explained by this Feynman's diagram.

Feynman's diagram shown in figure S-11.1 is least complex diagram in which there are only two vertices A and B and only one virtual photon has been exchanged. More complicated diagrams may have four vertices if two virtual photons are exchanged and six if three and so on.

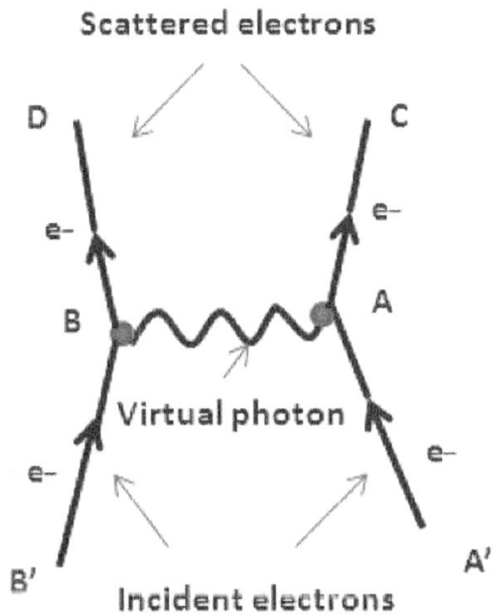

Figure S-11.1 *Feynman's diagram for repulsion between two electrons*

However, in electromagnetic interactions maximum contribution to the transition probability comes from the least complex diagram with minimum vertices. Least complex diagram is also called the first-order diagram.

Solved example S-11.2

Draw the first-order Feynman diagram for the interaction: $e^- + e^+ \rightarrow 3\gamma$

Solution.

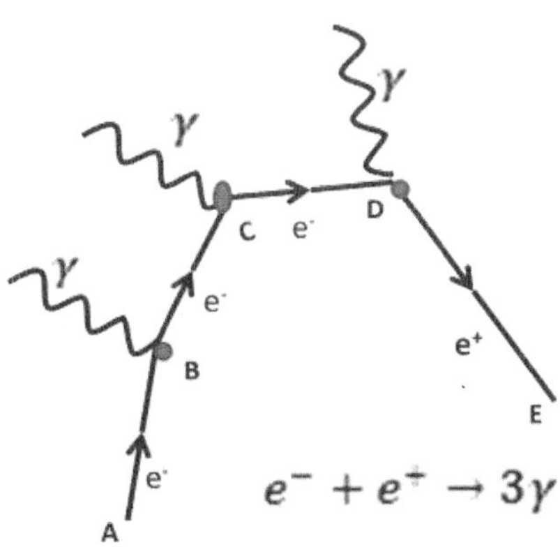

Figure S-11.2

As shown in figure S-11.2, an electron moving along AB emits a real photon, recoils in the direction BC, emits another real photon, and recoils in the direction CD. At vertex D, electron combines with a positron (antiparticle) and annihilates into the third γ.

Solved example S-11.3

Draw first order Feynman diagrams for the following:

(a) Decay of μ^- lepton ; (b) Pair production by γray (c) Compton scattering (d) Quark level decay $\Lambda \rightarrow p + e^- + \bar{\nu}_e$ and (e) Quark level interaction $p + \pi^- \rightarrow K^0 + \Lambda$.

Solution. The Feynman diagrams are as follows:

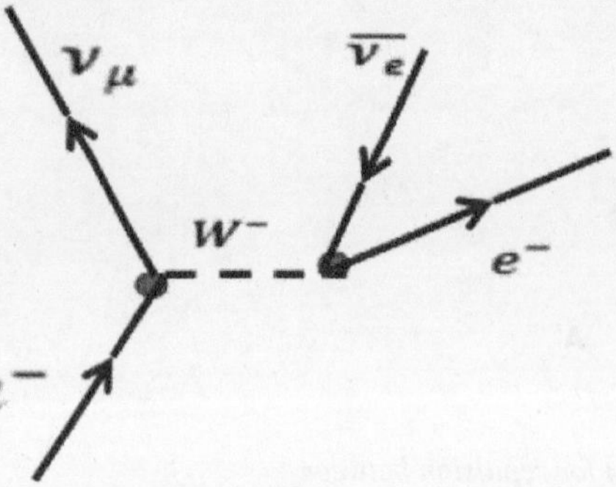

Figure S-11.3(a) *μ^- decay*

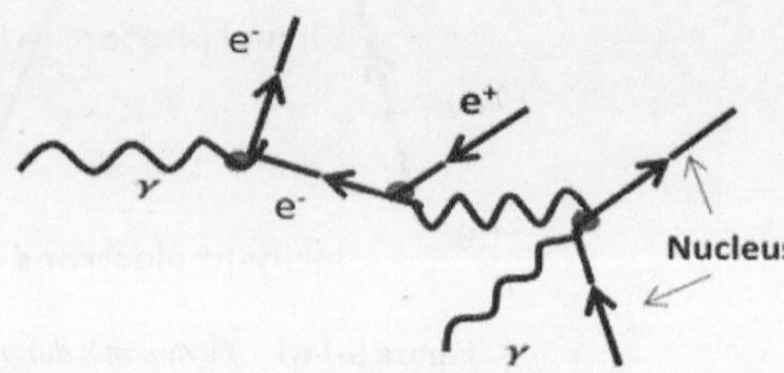

Figure S-11.3(b) *Pair production*

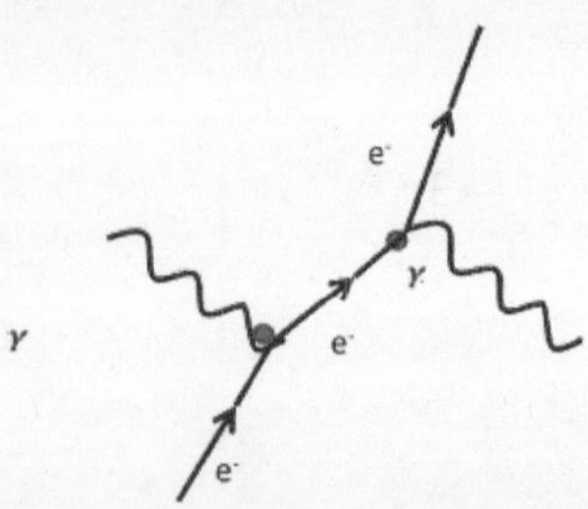

Figure S-11.3(c) *Compton scattering*

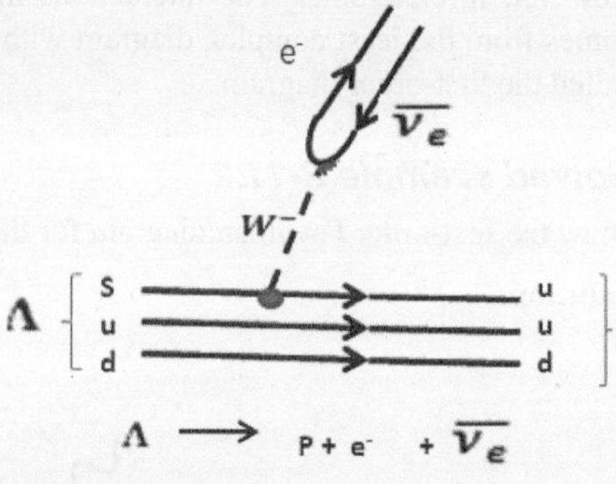

Figure S-11.3(d)

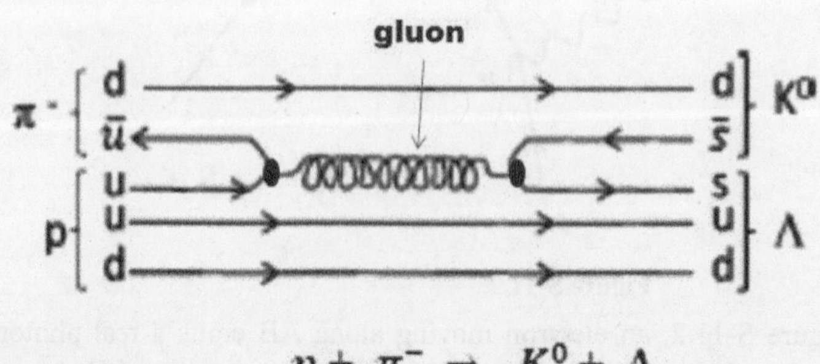

$$p + \pi^- \rightarrow K^0 + \Lambda$$

Figure S-11.3(e)

Exercise p-11.23: Explain the interactions shown by the two Feynman diagrams.

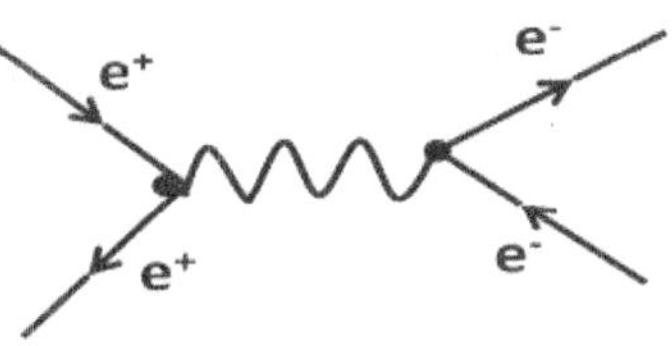

Figure p.11.23(a)

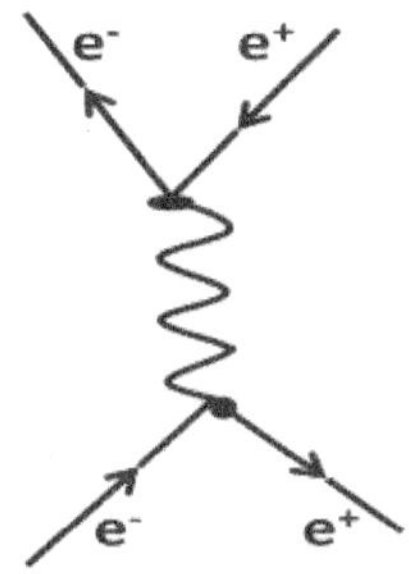

Figure p.11.23(b)

Multiple choice questions

Note: In some of the following questions more than one alternative may be correct. Select all the correct alternatives in such cases.

Exercise M-11.1: Missing particle in the reaction $K^- + p \rightarrow K^+ + ?$ is

(a) n (b) π^0 (c) Σ^- (d) Δ^-

Exercise M-11.2: In the decay $\mu^+ \rightarrow e^+ + X + Y$, particles X and Y are, respectively,

(a) $\nu_e, \bar{\nu}_\mu$ (b) $\bar{\nu}_e, \bar{\nu}_\mu$ (c) $\bar{\nu}_e, \nu_\mu$ (d) ν_e, ν_μ

Exercise M-11.3: The eigenstates of the weak field are

(a) K_S and K_L (b) K_1^0 and K_2^0 (c) K^0 and $\bar{K}^0$ (d) None of these

Exercise M-11.4: $(K_1^0 - \varepsilon K_2^0)$ defines

(a) K_2^0 (b) $\bar{K}_2^0$ (c) K_S (d) K_L

Exercise M-11.5: The resonance N(1675)D_{15} decays via

(a) $N + \eta$ (b) $\Delta + \Delta + \pi$ (c) $\Delta + \pi$ (d) $N + \pi + \pi$

Exercise M-11.6: The particle with quark structure 'ddd' is

(a) Δ^0 (b) Δ^- (c) Δ^+ (d) Δ^{++}

Exercise M-11.7: Which of the following quark structure(s) represent(s) a hyperon?

(a) uuu (b) uud (c) uds (d) sss

Exercise M-11.8: The particle that has strangeness number, baryon number, and isospin all zero is

(a) Λ^0 (b) π^+ (c) Ξ^- (d) η

Exercise M-11.9: Select the Cabibbo favoured quark transitions from the following:

(a) $u \rightarrow b$; (b) $c \rightarrow d$ (c) $u \rightarrow d$ (d) $t \rightarrow d$

Exercise M-11.10: Cabibbo forbidden transitions are

(a) $u \rightarrow b$ (b) $t \rightarrow d$ (c) $c \rightarrow d$ (d) $c \rightarrow s$

Exercise M-11.11: In natural unit system, the units of time are
(a) MeV^{-2} (b) GeV^{-2} (c) GeV^{2} (d) GeV^{-1}

Exercise M-11.12: The value of 1 second in natural system of units is
(a) 1.5×10^{24} GeV^{-1} (b) 1.5×10^{-24} GeV^{-1} (c) 1.5×10^{24} GeV^{-2} (d) 1.5×10^{24} GeV^{2}

Exercise M-11.13: Which elements of CKM matrix are negligible?
(a) V_{ub} (b) V_{ut} (c) V_{cs} (d) V_{td}

Exercise M-11.14: In natural unit system, the units of area are
(a) GeV^{2} (b) GeV (c) GeV^{-1} (d) GeV^{-2}

Exercise M-11.15: If σ and Σ, respectively, denote the microscopic and the macroscopic absorption cross-sections for a given radiation in a material, their units in natural unit system will respectively be
(a) GeV^{-2}, GeV (b) GeV^{-1}, GeV^{-2} (c) GeV^{-2}, GeV^{-1} (d) GeV^{-2}, GeV^{2}

12

Cosmic Rays

The subject of cosmic rays is unique in modern physics for the minuteness of the phenomena, the delicacy of observations, the adventurous excursions of the observers, the subtlety of the analysis, the grandeur of inferences.

Bruno Rossi

British physicist Abraham Bennet in the year 1787 invented an instrument called Gold-leaf electroscope (figure 12.1). The electroscope was used to test if a body is charged or not. When the metallic disc at the top of the electroscope was brought in contact with some charged body, the two gold leaves of the instrument go wide apart because of mutual repulsion, whereas when the electroscope is discharged the two leaves come close together. A charged electroscope may be discharged in two ways: (1) by connecting the top disc of the electroscope to a ground terminal or (2) by bringing some source of ionizing radiations, for example a radioactive source, near the electroscope. The radiations from the radioactive source ionize the gas around the gold leaves, making the gas conducting and thus allowing the charges on the gold leaves to flow out. It is interesting to see how a simple instrument devised to record the state of charge of a body ultimately became the tool for a big scientific discovery of cosmic rays, leading to the award of a Nobel Prize.

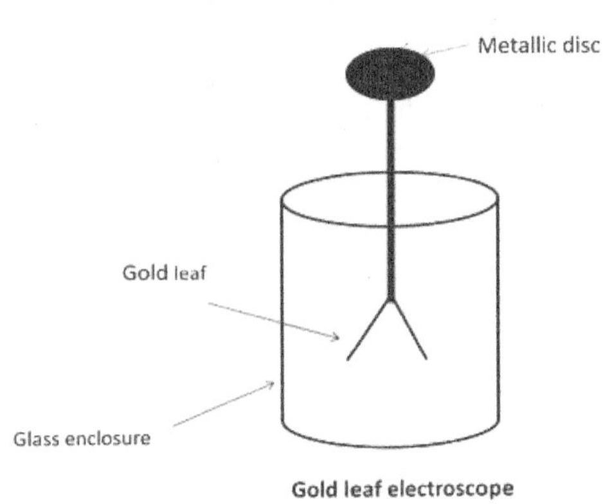

Figure 12.1 *Gold-leaf electroscope*

It was a common observation that a charged electroscope slowly discharges, of its own, even when no source of ionizing radiations is brought near it. The observation of spontaneous slow discharge of a charged electroscope, no matter where it is placed and in absence of any source of ionizing radiations, was puzzling scientist for quite some time. Initially, it was thought that the spontaneous discharge of the electroscope is due to the background radioactivity that is present in the ground, the walls, and other structures around the electroscope.

12.1 DISCOVERY OF COSMIC RAYS

Although the spontaneous discharge of a charged electroscope, even when apparently no source of ionizing radiations was in the vicinity, was hinting towards the presence of some mysterious invisible, ionizing radiations everywhere on the earth, the first direct evidence of such radiations (later called cosmic rays) was presented in 1912 when V.F. Hess after his balloon flight from Aussig in Austria to Pieskow in Germany covering altitudes up to 16,000 ft, published in a German research journal 'Phisikalische Zeitschrift' the results of his investigations on the discharge of electroscopes in flight, as follows:

The results of my observations are best explained by the assumption that a radiation of very great penetrating power enters our atmosphere from above.

R.A. Millikan and his collaborators at the California Institute of Technology carried out further experiments, during the period 1923–1926, both under water and at high altitudes and supported the findings of Hess. It was Millikan who named these radiations *cosmic rays*. Hess received Nobel Prize for his pioneering work on the discovery of cosmic rays in 1936, almost 25 years of his balloon flight experiments. Pfotzer, in 1936, made a detailed study of the variation in cosmic ray flux with the altitude on earth. He observed that the cosmic ray flux peaks at the altitude of about 15 km (that corresponds to the atmospheric pressure of about 170 mm Hg) above sea level and then decreases sharply.

12.1.1 Cosmic Rays as High Energy γ Rays

At the time when Hess announced his discovery of cosmic rays, the only known radiations were α particles, β particles, and γ rays (or photons), that are emitted in radioactive decay. Out of these three radiations, the γ rays had highest penetration power. Initially, it was assumed that cosmic rays consist of high-energy γ rays as γ rays have highest penetration power and they were also detected in experiments carried out at the sea level. It was argued that high-energy γ rays (of original cosmic rays) undergo Compton scattering with the atoms present in the atmosphere, producing lower energy scattered γ rays and electrons that were detected at the sea level. From the measured energy distribution of Compton scattered γs at sea level, Millikan theoretically calculated back the energy of initial cosmic ray γs. Millikan showed that the observed γ-ray distribution at ground level may be explained if it is assumed that the cosmic ray γs entering at the top of the atmosphere have three distinct energy groups of 26, 110, and 220 MeV. Millikan, while searching for the origin of these three energy groups of cosmic ray γs, speculated that interstellar space is filled with very dilute hydrogen gas and atoms of heavier elements might continuously evolve by the process of fusion of hydrogen atoms. When four hydrogen atoms fuse to make a helium atom, energy of 28 MeV is released. Similarly, synthesis of nitrogen and oxygen atoms from the fusion of 14 and 16 hydrogen atoms, respectively, may release about 100 MeV and 120 MeV of energy. Extending his argument further, Millikan showed that about 220 MeV of energy will be released if 28 atoms of hydrogen fuse together to synthesis silicon atom. As nitrogen, oxygen, and silicon are some of the most abundant elements in the solar system, their synthesis from hydrogen in intergalactic space appeared to be a fascinating idea and conversion of the energy released in their synthesis into γ rays appeared to be a valid hypothesis to explain the three groups of γs in cosmic rays. Thus, for a short duration of time, scientists believed that original cosmic rays are high-energy γ rays, as suggested by Millikan and even called these cosmic ray γs as the '*birth cry*' of elemental synthesis in intergalactic space.

12.1.2 Cosmic Rays as Charged Particles

However, around 1929, a very sensitive and efficient device called Geiger–Muller counter (in short called GM counter) was developed for the detection of radiations. Coupled with the development of vacuum-tube technology of coincidence counting, GM counter almost totally replaced the older electroscopes, which were earlier used for the detection of cosmic ray radiations. Bothe and Kohlhorster, two German scientists, carried out some coincidence experiments on the absorption of cosmic rays reaching the ground. The experimental setup of Bothe and Kohlhorster is out lined in figure 12.2. The coincidence unit gives an output pulse if the two input pulses from the two GM counters occur within a time interval of Δt of each other, which is called the resolution time of the system. If the resolution or resolving time is large, there are more chances that two different radiations may simultaneously pass through the two counters, by chance, and give a false coincidence pulse. This is called chance coincidence to distinguish it from the true coincidencc when a single radiation passing through the two counters, one after the other, gives a coincidence output pulse. Bothe and Kohlhorster used different thicknesses of gold sheets between the two GM counters and found that almost 76% of the cosmic ray radiations at ground level can penetrate up a thickness of 4.1 cm of gold. This was not possible if the cosmic ray radiations falling at the top of the atmosphere were high energy γ rays, as postulated by Millikan. Later, more detailed and multiple coincidence counting experiments in which the possibility of chance

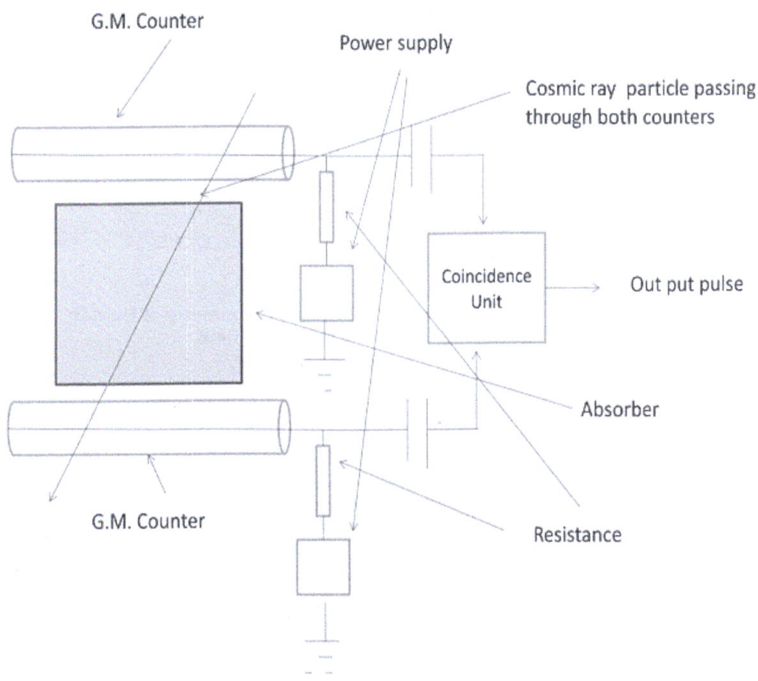

Figure 12.2 *Bothe and Kohlhorster's coincidence setup*

coincidences was almost negligible were carried out by Bruno Rossi. Rossi's experiments further confirmed that some of the cosmic radiations at ground level may penetrate almost 100 cm of lead. These experiments ruled out Millikan's hypothesis that cosmic rays are γ ray photons of high energy, because the electrons produced from the Compton scattering of initial cosmic ray γs with atoms of the molecules in the atmosphere could not penetrate such huge thicknesses of matter. These experiments lead to the hypothesis that cosmic ray radiations (falling on the top of the atmosphere of the earth from all the sides) are high-energy charged particles instead of γ rays. Confirmations to the charged particle hypothesis have been later obtained through rocket and satellite flights, which went much above the top of earth's atmosphere.

Information of considerable value regarding the nature of the cosmic rays reaching the top of earth's atmosphere was obtained even before the invention of men made satellites and rockets. A hint that charged particles entering the atmosphere of earth will suffer deflection due to the magnetic field of earth was given as far back as 1903 by C. Störmer, a Norwegian scientist who was studying the occurrence of auroral activity in the northern sky of Norway. Störmer made a detailed study of the motion of charged particles in the magnetic field of earth. The magnetic field of earth resembles the magnetic field of a dipole and extends well above the earth's atmosphere. However, the strength of earth's magnetic field is small (about 0.32 gauss at magnetic equator) and the geomagnetic poles do not coincide with geographical poles. The strength of the magnetic field is greatest at poles and weakest at equator. The magnetic axis of earth lies at an angle of 11° with the axis of the earth's rotation and the centre of earth's magnetic dipole is almost 320 km from the centre of the earth. The intensity of the magnetic field of earth decreases with the increase of altitude and at an altitude of 1600 km it is almost half of its value at the earth's surface. Earth's atmosphere extends to almost 350 km but the density of the atmosphere decreases rapidly with altitude. Under these conditions, it is very much likely that the high energy cosmic ray radiations are affected by the magnetic field of the earth before they undergo interaction with the molecules, atoms, or nuclei present in the atmosphere (figure 12.3).

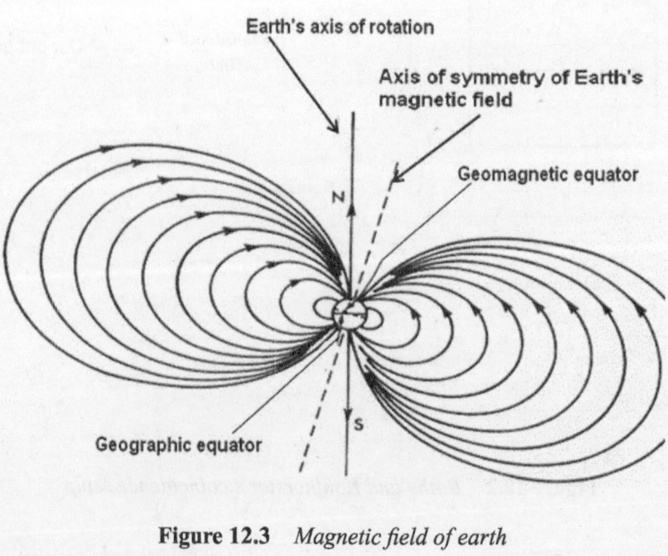

Figure 12.3 *Magnetic field of earth*

In case the cosmic ray radiations falling at the top of the atmosphere are photons or high energy γ rays, they will not be affected by Earth's magnetic field. However, if they are charged particles, then they will be subjected to a force F, called Lorentz force, given by the expression

$$\vec{F} = q(\vec{V} \times \vec{B})$$

Here, q is the charge of the particle, V its velocity, and B the strength of the magnetic field. The magnetic force acts normal to both the velocity of the particle and the direction of the magnetic field. As a result of the magnetic force, charged particles are constrained to move in circular paths. A positively charged particle circling the earth under the influence of earth's magnetic field moves from east to west as shown in figure 12.4. On the other hand, if the particle is negatively charged it will move the other way round, from west to east.

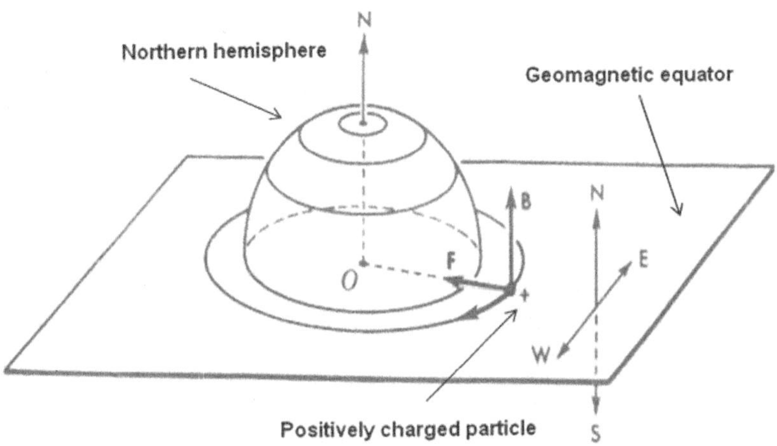

Figure 12.4 *Motion of a positively charged particle under the influence of earth's magnetic field*

Bruno Rossi, taking ideas from the work of Störmer, made detailed calculations for the trajectories of charged particles of different energies being projected on the magnetic field of the earth from all directions. As a matter of fact, Rossi looked at the problem of charged particle trajectories in the reverse way. He actually selected a point on the surface of the earth, selected the energy of the particle and its direction of approach, and then calculated the particle path in reverse direction. In this way, he was able to trace back two distinct types of trajectories:, one that were approaching a particular location on the earth from outer space and the other that were approaching that location from some other point on the earth (see figure 12.5). Obviously, the later type of trajectories, called

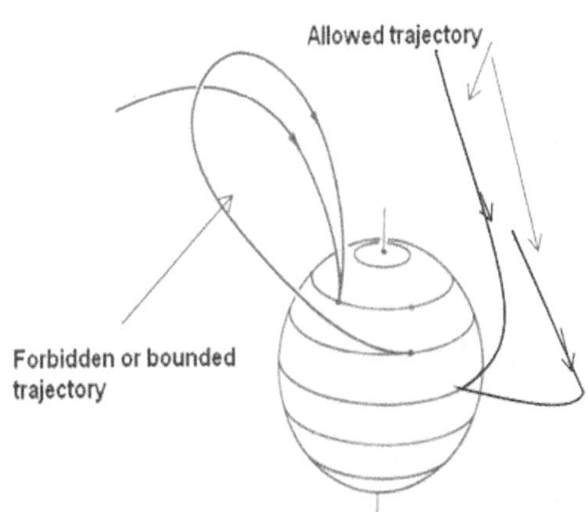

Figure 12.5 *Allowed and forbidden (bounded) trajectories at a point on the surface of the earth*

bounded trajectories, were not allowed for cosmic ray radiations as these trajectories do not originate from outer space. From the analysis of charged particle trajectories, Rossi arrived at the following conclusion:

For each point on the earth and for positively charged particles of any energy, there exists a cone (called Störmer cone), with the axis pointing towards east such that no cosmic ray particle can enter this cone from the outer space. The half angle of the forbidden Störmer cone depends on the geomagnetic latitude of the place and the energy of the incident particle.

12.2 THE EAST–WEST ASYMMETRY

As the forbidden Störmer cone opens towards east at each point on the surface of the earth (if cosmic ray particles are positively charged), and there is no forbidden direction towards the west; this is bound to lead to the asymmetry in the cosmic ray intensity recorded at a given point from the west and the east directions. This is called the east–west effect. It may, however, be noted that no such asymmetry in the number of cosmic ray particles from east and west will be observed if there are equal number of particles with both negative and positive charges in cosmic rays hitting the top of the atmosphere, whereas if cosmic ray particles are only negatively charged, Stormer cone will open towards west. Experiments carried out by Thomas H. Johnson (Mexico City, 29° geomagnetic latitude, 2250 m above sea level); L.W. Alvarez and Compton also working at the same place and Sergio De Benedetti working with B. Rossi (in Asmara, Eritrea 11° N geomagnetic latitude, 2370 m above sea level) independently observed east–west asymmetry using GM counter cosmic ray telescopes directed 45° to the vertical. The cosmic ray intensity was found to be 10–26% more from the west than from the east. This conclusively proved that almost all or a substantial number of cosmic rays hitting at the top of the earth's atmosphere are positively charged particles.

The half-angle of the Stormer cone as a function of the energy of the charged particle and the geomagnetic latitude λ of the place is shown in figure 12.6. The following observations may be made from the inspection of curves in figure 12.6.

1. At a fixed value of λ, the half angle of Stormer cone increases with the decrease of the energy of the particle. It means that at a given place on the surface of earth, cosmic ray particles below a certain cutoff energy E_{cut} do not reach; they are cutoff (or deflected away from the earth) by the earth's magnetic field.

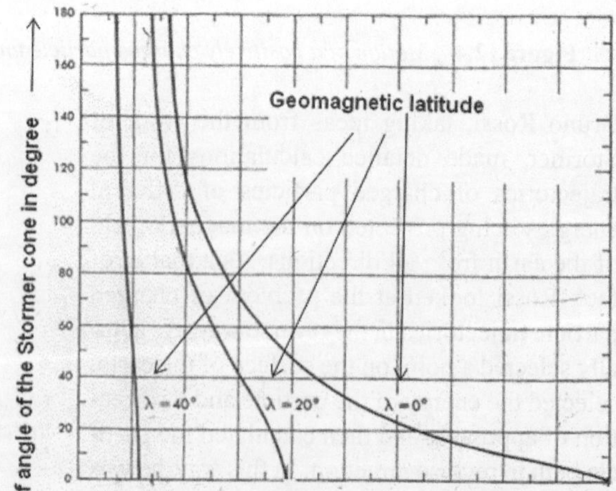

Figure 12.6 *Semi angle of Stormer cone as a function of particle energy and geomagnetic latitude*

2. The cutoff energy E_{cut} decreases with the increase of λ.
3. The half angle of Stormer cone becomes zero for all particles above the energy E_{cut}.

As such, only cosmic ray particles of energies greater than 60 BeV can reach at the magnetic equator of the earth ($\lambda = 0$), while at geomagnetic latitude 20°, all cosmic ray particles with energies greater than about 24 BeV may reach. It is obvious that at magnetic poles maximum number of cosmic ray particles, including particles of lowest energies may reach. It may be noted that in this discussion interaction of cosmic ray particles with atmospheric molecules, atoms, and nuclei have not been considered.

12.3 THE LATITUDE AND LONGITUDE EFFECTS

As the half angle of Störmer cone depends both on the geomagnetic latitude of the place of observation and the energy of the cosmic rays, it is obvious that the cosmic ray flux reaching at different latitudes (from the east side) will be different in magnitude, further the energy distribution of cosmic ray particles at different latitudes will also be different. This is called the *latitude effect*. Compton and his collaborators, including P.S. Gill from India, made a very detailed study of the variation in the intensity of cosmic rays on the ground and the sea. His group carried out 17 sea voyages, carrying GM counter-based cosmic ray telescopes, to study the variation in cosmic ray intensity at different geomagnetic latitudes. On the basis of these measurements, Compton developed maps showing cosmic ray intensity at different places on the surface of earth. He also drew curves connecting points of equal cosmic ray intensity and called these curves '*isocosms*'. From cosmic ray intensity maps developed by Compton, following observations about the cosmic ray intensity may be made:

1. The graphs of equal cosmic ray intensity, called isocosms, follow geomagnetic parallels more closely than they do geographical parallels. This observation clearly showed that the variation in cosmic ray intensity at the sea level on the earth is primarily due to the geomagnetic effect.
2. In 1937, Compton observed that the cosmic ray intensity at a given place varies slightly with the variation of the mean atmospheric temperature of the place. The intensity was found to decrease with the increase of the average atmospheric temperature.
3. As average atmospheric temperature at equator is larger than at the pole, the variation of the average atmospheric temperature tends to produce an *apparent latitude effect*, which is about one-third of the total latitude effect. The two-third of the latitude effect was truly due to the geomagnetic effect or due to the magnetic field of the earth.
4. At a given latitude, the latitude effect increases with the altitude. For example, at geomagnetic latitude of 80° and at the sea level, the cosmic ray intensity is about 15% more as compared to the value at geomagnetic equator ($\lambda = 0$), while at the same latitude of 80° but at altitudes of 2000 m and 4360 m, the cosmic ray intensities are, respectively, 22% and 33% of their values at geomagnetic equator.
5. Cosmic ray intensity was also found to vary with the longitude of the place. This variation in the intensity, called *longitude effect*, was attributed to the fact that the geomagnetic poles and geographic poles are not coincident.

Geomagnetic field essentially cuts down the number of particles with lower energies to reach the particular geomagnetic latitude. This is shown in figure 12.7, where the percentage flux of

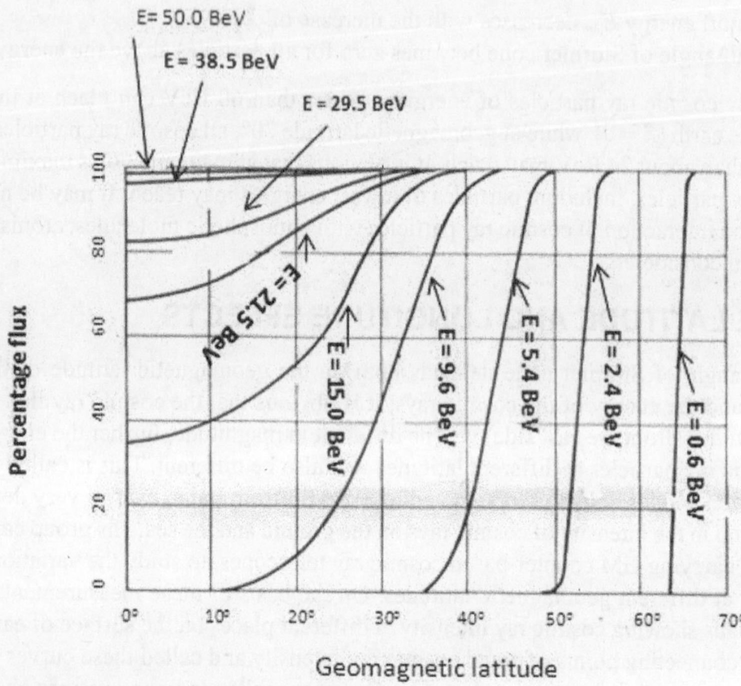

Figure 12.7 *Variation in the flux percentage of cosmic ray particles of different energies as a function of geomagnetic latitude*

cosmic ray particles of different energies reaching different geomagnetic latitudes is shown. As expected, the percentage cosmic ray flux at given latitude increases with the energy of the cosmic radiations. For example, in the geomagnetic belt of about 10° north to 10° south, cosmic ray particles of energies 9.6 BeV and less are totally cutoff and less than 40% flux of cosmic ray particles with energy 15 BeV reaches this region of the belt. Flux percentage increases for higher energy particles and particles with energies about 50 BeV or more reach this belt with 100% flux.

It is obvious that for given geomagnetic latitude, the latitude effect will be more pronounced at higher altitudes where the atmospheric thickness that cosmic ray particles travel is less and the density of atmospheric matter is also small. Therefore, at higher altitudes, probability of interaction of cosmic ray particles with atmospheric matter is small.

All these conclusions about the variations in cosmic ray intensity were based on the assumption that the atmosphere changes the nature of the initial cosmic radiations falling on the top of the atmosphere, but the directionality of the initial cosmic rays is maintained, to a large extent, during their interactions with atmospheric atoms and molecules.

Summing up, it may be stated that the geomagnetism produces two distinct effects: the east–west asymmetry and the increase of cosmic ray flux in going from geomagnetic equator to pole, called the latitude effect.

Experimental verification of east–west asymmetry and the latitude effect tells that cosmic rays hitting the top of the atmosphere are not γ rays and that they are predominantly positively charged particles. These particles may be protons or positron or even heavier nuclei as the present analysis does not make a distinction between a proton and a positron. It is because at such relativistic energies (of the order of BeV) the rest mass of the particle becomes negligible in comparison to their relativistic mass.

12.4 SOME IMPORTANT DETECTOR SETUPS USED IN THE STUDY OF COSMIC RAYS

As has been mentioned, gold leaf electroscope was the main detector that was used in the early part of research on cosmic rays. The electroscope in fact detected the average flux of ionizing radiations at a place. As the gas-filled counters were developed, ionization chambers were also used to detect and measure the average flux of cosmic radiations. However, GM counter soon replaced electroscopes and ionization chamber in view of its capability of detecting individual radiation. GM counters in coincidence (figure 12.8) makes cosmic ray telescope. In a telescope, two or more GM counters are arranged one above the other and events in which pulses from both the counters are detected within a small time interval Δt are recorded. These events are called coincidence events and occur when the same particle passes through the two counters. However, sometime it is possible that two different particles may pass one through counter-1 and the other through counter-2 within time Δt. An event of this type is called a chance coincidence and is not the genuine coincidence that happens when a single particle passes through both the counters. It is obvious that the probability of chance coincidence will be small if the time interval Δt, called the resolving or resolution time, is small. Present day coincidence circuits have resolution times of the order of micro or even milli-micro seconds. As such, the probability of chance coincidence in such setups is very small. As shown in figure 12.8, the field of view of the telescope depends on the relative separation of the counters. The coincidence counter telescopes may be used to study the directional variation of cosmic rays. A coincidence counter setup shown in figure 12.2 may be used to study the absorption of cosmic rays in a given absorber, such as lead, gold, or aluminium. In the same setup, if the two counters are in anti-coincidence,

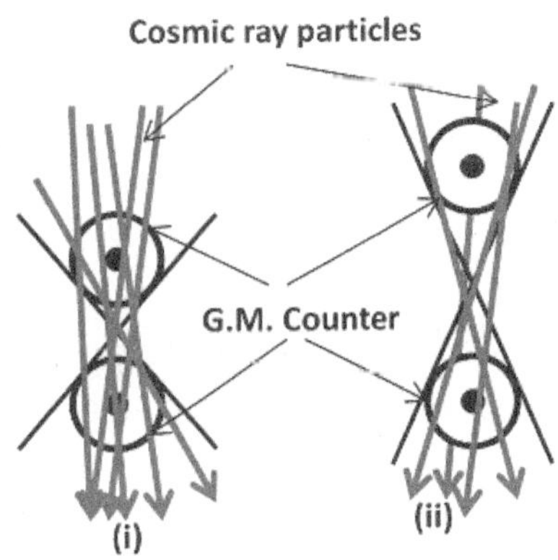

Figure 12.8 *Coincidence setup with wide and narrow field of views*

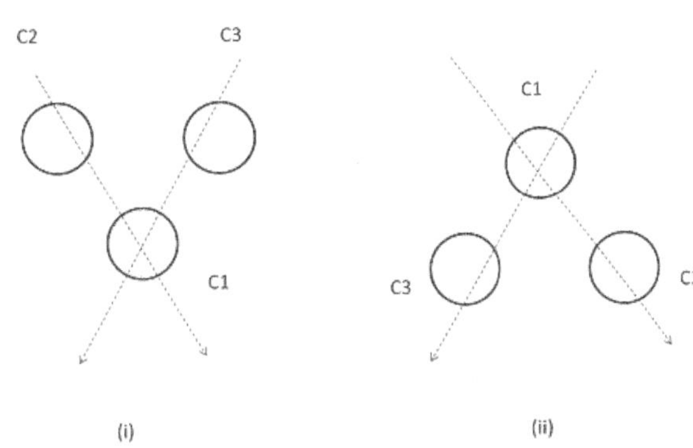

Figure 12.9 *Coincidence arrangement for detecting shower particles*

which means that the cosmic ray particle passes through the upper counter, enters the absorber and does not come out to reach the second counter, one may determine the number of events in which the particle is completely absorbed in the absorbing material.

Cosmic rays interact with the atoms and nuclei present in the atmosphere and produce cosmic ray showers in which large numbers of secondary particles are produced simultaneously. Coincidence counter-arrangement shown in figure 12.9 may be used to detect showers. The coincidence pulse will be generated only when at least two different particles pass through the triangular arrangement of counters. Several other arrangements of counters in coincidence and anti-coincidence with each other may be used to identify different types of events.

Another very versatile detector is the cloud chamber. Cloud chamber works on the principle of condensation of super-saturated vapours around the charged particles and ions. A cloud chamber consists of a cylindrical container of glass that is filled with saturated vapours. There is an arrangement for the sudden expansion of the cloud chamber volume and of taking a photograph of the chamber at the moment of its expansion. This is called the triggering of the chamber. As a result of the expansion of the chamber, the saturated vapours inside become super-saturated and condense as droplets around the charges present in the volume of the chamber. Thus, tracks of charged particles passing through the cloud chamber at the moment of triggering become visible in photographs as a sequence of droplets. The cloud chamber may be placed in an external magnetic field so that the trajectories of charged particles got bent in a certain direction because of the force exerted by the magnetic field on the particle. From the direction of the bending and the radius of curvature of the track, it is possible to determine the charge and the energy of the particle. Cloud chambers that are triggered by the pulse from a GM counter placed above the chamber may be used to take the photograph of a specific particle that has generated the pulse in the counter (figure 12.10a). In a similar way, arrangement for recording the event in which the cosmic ray particle has passed through counter-1, the cloud chamber, and counter-2 is shown in figure 12.10b. Figure 12.11 shows the setup for recording shower particles produced in the absorber. If the cloud chamber is placed in an external magnetic field the tracks recorded in the cloud chamber will show curvature.

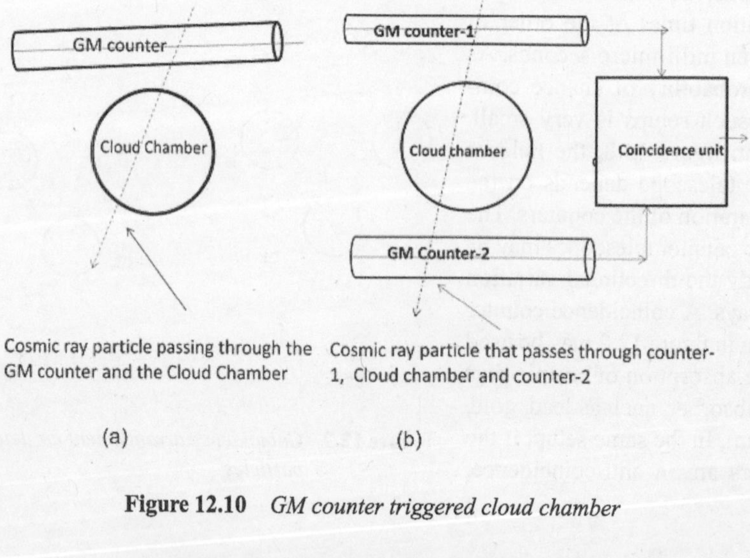

(a) (b)

Figure 12.10 *GM counter triggered cloud chamber*

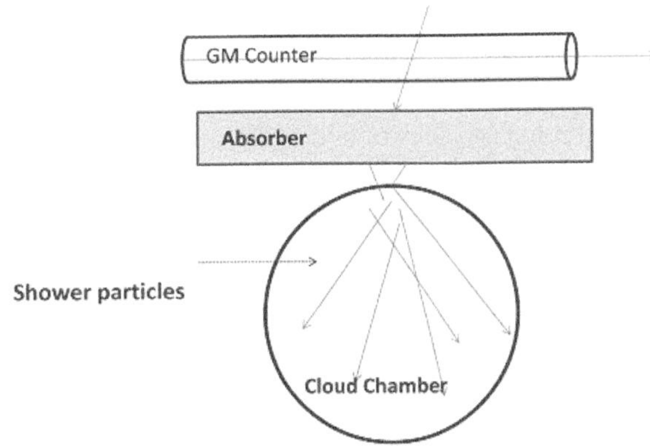

Figure 12.11 *Arrangement to detect shower particles by cloud chamber*

Nuclear emulsion plates developed by Kodak and other photographic film manufacturers have also been extensively used in detection and recording interaction events initiated by cosmic rays. Nuclear emulsion basically contains AgBr grains, which are affected by the passage of the charged particles. On chemical treatment by the suitable developer and fixer, silver grains are formed along the path of the charged particles. The particle trajectory appears as a black line when viewed under a high-resolution microscope. The energy loss by the ionizing charged particle may be estimated from the density of grains in the trajectory of the particle. As energy loss per unit path length depends both on the velocity and the charge of the particle, it is possible to determine the characteristics of the particle from emulsion records. The biggest advantage of nuclear emulsion detector is the high density of the detector material, as a result of which large number of cosmic ray interactions may occur in detector itself. This is a big advantage over GM counters and cloud chambers where the density of gas inside the detector is relatively small. Further, nuclear emulsions do not need any electronics and are stable and not very bulky. It is, therefore, easy to carry them in flights. It is also possible to load or introduce small amount of some desired element in nuclear emulsion so that the interaction of cosmic ray particle with the nuclei of the element may be studied.

In modern experimental setup, scintillation and solid-state detectors are frequently used. Charged particle and/or γ ray passing through scintillation detector generates an electronic pulse which may be analysed to determine the nature and the energy of the radiation. These detectors have high detection efficiency and better resolution for differentiating radiations of different energies and charges. It is also possible to record the energy spectra of incident radiations using these detectors. Scintillation and solid-state detectors are fast can handle large counting rates and can be used in coincidence and anticoincidence modes. However, generally these detectors require low temperature (liquid nitrogen temperatures) and detailed electronics for the recording and analysis of the data.

Cerenkov radiation detectors have also been used to detect the light charged particles of relativistic energies, such as electrons and positrons, which are produced as a result of the interaction of primary cosmic rays with the nuclei present in the atmosphere.

12.5 PRIMARY COSMIC RAYS

The primary cosmic rays are those entering the upper atmosphere, the cosmic rays of the interstellar medium. Secondary cosmic rays are those produced by the interactions of the primary rays in the atmosphere or in the earth. Also, products of cosmic ray interactions in the interstellar medium (e.g., spallation products from cosmic ray – cosmic ray collisions) are labelled as secondary cosmic rays. Cosmic rays can be of either galactic (including solar) or extragalactic origin.

12.5.1 Composition of Primary Cosmic Rays

The primary cosmic rays consist of the following charged particles:

1. Proton (nucleus of hydrogen)	about 90%
2. Alpha particles (nucleus of helium)	about 9%
3. Light nuclei (nuclei of lithium, beryllium, and boron)	about 0.25%
4. Heavier nuclei	≤1.0%
5. Electrons	≤1.0%
6. High energy photons	≤0.1%

The percentage of light nuclei in cosmic rays, though very small, is much larger than the natural abundance of light elements in universe. The percentage of light elements in universe is almost one-billionth of their percentage in cosmic rays. It appears that these light nuclei in cosmic rays are produced by the fragmentation of carbon and oxygen nuclei present in the galactic space when they are hit by high-energy cosmic ray protons.

12.5.2 Energy Distribution of Particles in Primary Cosmic Rays

The energies of primary cosmic rays may vary from a lower value of about 1 MeV to some 10^{15} MeV. However, the relative number of higher energy particle decreases sharply with the energy of the particle. According to a rough estimate, the number of cosmic ray particles above energy E falls off as $E^{-1.5}$. The peak of the distribution in energy is in the range of 100 MeV to 1 GeV. The intensity of cosmic rays of energy 1 GeV/nucleon or greater is about 1/cm^2 s sr. An approximate formula is

$$I_N(E) \sim 1.8\left(\frac{E}{\text{GeV}}\right)^{\alpha} \text{ nucleon per square cm. per second per steradian (sr).}$$

where E is the total energy per nucleon and $\alpha = -2.7$. The energy density corresponding to this is about 1 eV/cm^3. This can be compared to the energy density of stellar light that has a value 0.3 eV/cm^3.

12.6 PASSAGE OF PRIMARY COSMIC RAYS THROUGH THE ATMOSPHERE

The composition of the cosmic rays at the sea level on the earth, also called *local cosmic rays*, is quite different from the composition of the primary cosmic rays falling at the top of the atmosphere. It happens because a large number of positively charged primary cosmic ray particles interact with the molecules, atoms, and nuclei present in the atmosphere producing secondary radiations. As primary cosmic rays also have very high-energy particles that have very small probability of interaction with atmospheric particles, these high-energy primary cosmic ray particles also reach the ground. Thus, at the sea level on the earth there is a mixture of some high-energy primary cosmic ray particles and the secondary radiations generated in the interactions of primary cosmic rays with the atoms/nuclei present in the atmosphere.

12.6.1 The 'Soft' and the 'Hard' Components of Local Cosmic Rays

Bothe and Kohlhorster using the coincidence counting technique studied the absorption of local cosmic rays (cosmic rays at sea level) in different materials. The typical absorption curve for local cosmic rays is shown in figure 12.12. As is indicated by the absorption curve, the local cosmic rays have two distinct components: one which has small penetration power and is readily absorbed by small thicknesses of absorber is called the *soft component* and the other that has very high penetration power and could not be absorbed by very large absorber thicknesses. The component of local cosmic rays that has large penetration power is called the *hard component.*

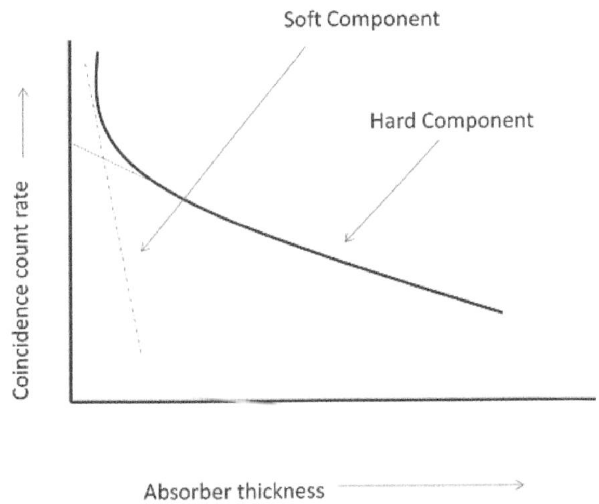

Figure 12.12 *A typical absorption curve for local cosmic rays*

12.6.2 Discovery of Positron

Cosmic rays provide a natural laboratory to study the creation, interactions, and annihilation of elementary particles that are the building block of matter. Many elementary particles were discovered during the course of the study of the cosmic rays. Cloud chamber has been extensively used for the study of the nature of cosmic ray particles. Cloud chamber placed in strong magnetic field and partitioned into two halves by some absorbing material such as a sheet of gold or lead is capable of determining the energy, charge, and the direction of motion of the particle. The energy of the particle and the nature of charge on the particle may be determined from the curvature and the direction of the bending of the path of the particle. The density of drops per unit path length in the cloud chamber picture tells about the energy loss suffered by the particle via ionization. As the energy of the particle is reduced after passing through a layer of absorbing material, the density of drops in the higher energy part before passing through the absorber is less and after passing through the absorber is more. As such, the part of the particle trajectory before absorber has lower density of drops per unit path length, while the lower energy part of the trajectory after passing through the absorbing material has a larger drop density. Thus, by the density of drops per unit path length it becomes possible to identify the direction of motion of the particle.

Anderson carried out experiments on local cosmic rays using such a cloud chamber setup and found that almost half of the tracks registered by the cloud chamber were bent in one direction and the other half on to the other direction. This indicated that there are roughly equal numbers of particles of positive and negative charges in the local cosmic rays. From the analysis of the data of his experiments, Anderson claimed that he has definitely discovered a particle that has a mass equal to the mass of electron but of positive charge. The positive electron was named positron. Almost simultaneously Blackett and Occhialini confirmed Anderson's discovery of positron.

Cloud chamber pictures also showed that sometime a positron and an electron pair suddenly appears in the clod chamber. At about the same time, Dirac has put forward his theory of pair production. According to this theory, a γ ray photon of energy greater than 1.02 MeV (the sum of the rest masses of the electron positron pair) passing through the strong electric field of some nucleus may disappear by creating an electron positron pair. The presence of the strong electric field of the nucleus is required for the conservation of linear momentum in the interaction. The nucleus (in the electric field of which pair production takes place) recoils to conserve the linear momentum. The cloud chamber pictures showing the tracks of positron and electron originating abruptly from same point clearly proved the occurrence of the pair production.

The analysing power of cloud chamber was further enhanced by using pulses from GM counter to trigger the cloud chamber (figure 12.9). It has become possible to study the passage of a specific charged particle that passes through GM counter-1, the pulse from counter-1 being used to trigger the expansion of the cloud chamber and clicking the shutter of photographic camera. In case the particle passes through counter-1, then through the cloud chamber, and on leaving the cloud chamber it goes through counter-2, then there will be coincidence between the pulses of counter-1 and counter-2. However, if the particle after passing through counter-1 is absorbed in the cloud chamber then the pulses from counter-1 and 2 will be in anti-coincidence. As such, using GM counters for triggering the cloud chamber and taking the coincidence and/or anti-coincidence rates, it becomes possible to study particles that are absorbed in the clod chamber (the soft component) and that the particles that penetrate through the cloud chamber (hard component).

12.6.3 Rossi's Experiment on the Softness of Vertical and Slant Cosmic Rays

Using a set of two GM counters in coincidence that may be rotated about the vertical direction, Rossi studied the directional variation of cosmic rays. The expectation was that the intensity of cosmic rays coming from the vertical direction will be largest as they travel through a smallest thickness of the atmosphere and, therefore, have less chance of absorption. The intensity of slant cosmic rays was expected to decrease with the angle the slant direction makes with the vertical. Contrary to the expectations, in the actual experiment Rossi found an increase in the intensity of cosmic radiations with the angle that the slant direction made with the vertical. Next Rossi used a 9-cm thick lead plate between the two GM counters of his cosmic ray telescope. This part of the experiment revealed that the slant cosmic rays contain more particles of lower energy than the vertical cosmic rays. This observation was also contrary to the expectation that the slant cosmic rays will be harder because of the filtering effect of the larger thickness of atmosphere as compared to the vertical cosmic rays.

Rossi, however, said that the results of his experiments may be accounted for if the primary cosmic rays on interaction with atmospheric matter produce *secondary soft* particles and these soft secondary particles are larger in number in slant direction than in the vertical.

12.7 COSMIC RAY SHOWERS

12.7.1 Rossi Transition Curve

With the view to study the production of secondary particles by the interaction of local cosmic rays with atoms and nuclei present in the atmosphere, Rossi carried out some ingenious experiments using a triangular array of GM counters. The detector array (figure 12.13) was shielded from three sides by thick lead shielding to cutoff radiations produced by the radioactive material in the surroundings. However, the thickness of the lead (or of any other absorbing material like iron) on the top of the setup could be changed by increasing or decreasing the number of sheets of the absorbing material of known thicknesses. The absorbing material could also be placed at different distances from the top of the detector setup.

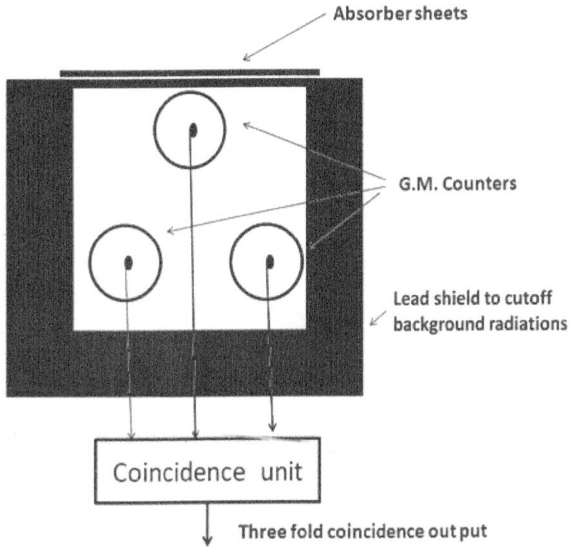

Figure 12.13 *Threefold coincidence setup used by Rossi to study the production of shower particles by local cosmic rays*

The absorption curves obtained by Rossi, called *Rossi transition curves,* for lead (Pb) and iron (Fe) absorbers placed at 1.2 cm and for lead absorber at a distance of 14.6 cm above the top, are shown in figure 12.14. As shown in the figure, Rossi observed a rapid increase of triple (or threefold) coincidences in a triangular arrangement of Geiger counters when some centimetres (1–2 cm) of lead was placed above. As a threefold coincidence may be initiated at least by two or more than two particles emitted simultaneously, increase in the triple coincidence rate means that more and more particles are simultaneously produced as a result of the interaction of local cosmic rays with the absorber material. This simultaneous production of large number of particles from a small volume of space is called *shower production*. Rossi transition curve shows that initially the shower production increases when the thickness of the absorbing material is increased, up to about 2 cm in case of the lead absorber. Only with further increasing absorber thickness did the coincidence rate start to decline. Rossi correctly concluded that soft (low energy) secondary particles were produced by the local cosmic ray particles entering the absorbing material. These secondary particles then suffer increasing absorption with increasing total thickness. The triple coincidence rate decreases very slowly on further increasing the absorber thickness. The tail part of the Rossi transition curve may be attributed to the hard component of the local cosmic rays. Rossi made two important deductions from the nature of the transition curve:

1. Showers are produced essentially by the interaction of the soft component of local cosmic rays with the absorbing material.
2. The nature and properties of the soft and the hard components of the local cosmic rays are quite different.

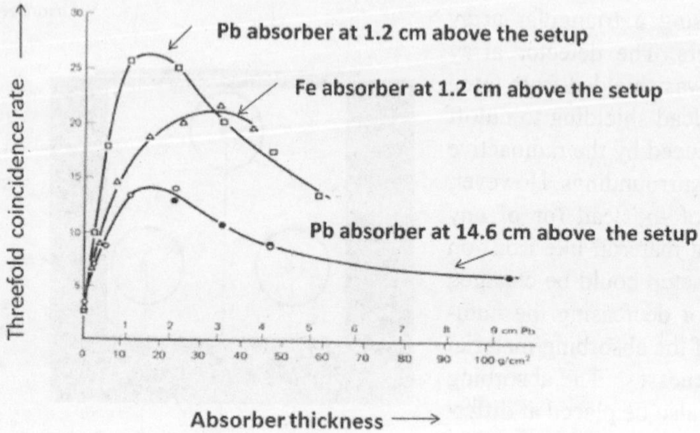

Figure 12.14 *Rossi transition curves for lead (Pb) and iron (Fe) absorbers*

Cloud chamber photographs have also indicated the production of showers when cosmic rays interacted with some matter around the chamber. For example, Blackett and Occhialini observed tracks of several particles that resulted from the interaction of local cosmic rays with the matter near their cloud chamber (figure 12.15), but it was Rossi who for the first time clearly showed that the showers are produced by the soft component of local cosmic rays and that the nature and properties of the soft and the hard components are quite different. Rossi even pointed out that the rise of threefold counting rate in the transition curve is due to the *tertiary soft radiations* produced in the interaction of secondary soft particles (produced in the atmosphere) with the absorbing material.

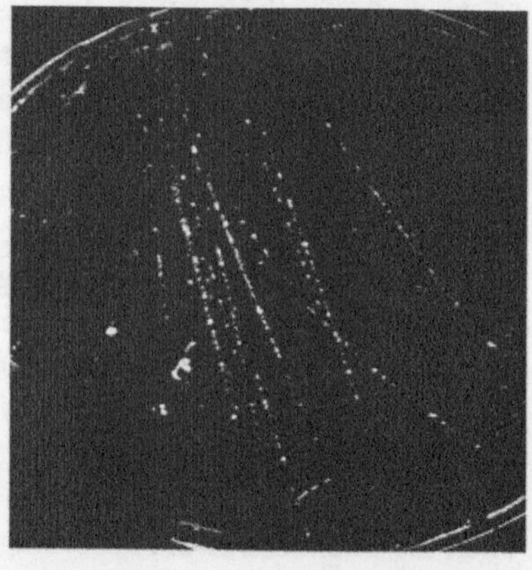

Figure 12.15 *Cloud chamber photo taken by Blackett and Occhialini showing shower particles*

12.7.2 Extensive Air Showers (EAS)

The threefold coincidence counting rate in Rossi's experiment was not zero when the there was no absorbing material at the top. This indicated that showers are produced even in air. In their efforts to further study the nature of the soft component of cosmic rays, Auger and his collaborators made systematic study of showers produced in air. In the course of their research, they were led to extend the study to distances of several tens of meters, progressively arranging counters more and more wide apart in order to reveal more and more dispersed atmospheric showers. They explored the production of large atmospheric showers whose branches could be separated by a distance of several meters and whose origin they thought should be very high in the atmosphere. Auger and collaborators found that such showers could cover a surface of the order of 1000 m^2, implying the presence of several dozens of thousands of particles and involving energies larger than 10^{13} eV, probably beyond 10^{14} eV. In the same year, 1938, Kolhörster and collaborators reported about coincidences recorded between two counters separated by distances going up to 10 m, and remarked that these observations must be related to the secondary rays of the high-altitude radiation, that is to a shower. Auger and his collaborators were since then credited for the discovery of EAS, and all this inaugurated the great season of speculations on the origin of cosmic rays which is still an open question.

12.8 GENERATION OF SHOWERS

Till mid-1930s the understanding about the nature of cosmic rays and its interaction with matter was quite vague; however, experiments carried out till then have clearly established the following:

1. The earth is constantly bombarded from all sides by very energetic positively charged particles, called primary cosmic rays.
2. The primary cosmic rays interact with molecules, atoms, and/or nuclei present in the atmosphere producing air showers. The shower particles are soft and are called secondary cosmic rays.
3. The cosmic rays reaching the sea level, called local cosmic rays, have two distinct components; the soft and the hard components which differ not only in their energies but also in their nature.
4. Soft component of the local cosmic rays on interaction with matter (absorbing material) produce further showers of very soft particles that are absorbed within few cm of lead.
5. Cloud chamber photographs have indicated that the particles that constitute air showers are nearly half of positive and the other half of negative electric charge. Further, the energies of these shower particles were estimated to be of the order of 10^{12} eV.

Although indication was that these shower particles were electrons and positrons, the fact that these (10^{12} eV) electrons and positrons reach the sea level after traversing a distance of about 8 km of air, the thickness of the atmosphere, could not be explained by the recently developed theory of quantum electrodynamics. In 1934, Bethe and Heitler published a landmark paper on the energy loss by charged particle while passing through matter, based on quantum electrodynamics. According to their calculations, a 10^{12} eV electron should have a range of about 2 km of air, 2 m of water, or 4 cm of lead. The fact that the shower electrons travel all the way up to sea level could only mean that either the quantum electrodynamics does not apply to such high energies or the shower particles are not electrons. At a time when quantum electrodynamics was at the verge of losing its validity, particularly at high energies, Bhabha and Heitler put forward their celebrated 'Cascade theory of electron showers' that saved the collapse of the quantum electrodynamics at high energies.

12.8.1 Bhabha–Heitler Cascade Theory of Electron Showers

According to the quantum electrodynamic theory of Bethe and Heitler, there are large cross-sections for the energy loss by an electron or a positron in a single event, by the process of Bremsstrahlung, when they pass through the field of a nucleus. The energy lost by the electron (or positron) is converted into an electromagnetic energy quantum, that is a photon. Thus, in accordance with the Bethe–Heitler theory cosmic ray shower electron (or positron) of high energy

($\approx 10^{12}$ eV) does lose its energy within a short distance in the matter, but the newly created high energy photon, which has a large probability of travelling in the direction of the incident electron, travels a short distance in the matter, and then materializes into a new electron–positron pair. Again, the electron–positron pair has a high probability of travelling in the direction of motion of the parent photon. Both members of the pair after travelling a short distance in the matter lose energy by Bremsstrahlung, creating fresh high-energy γ rays, moving essentially in the direction of motion of the parent pair. This process of conversion into photon and then to electron–positron pair may repeat several times. The net effect thus implies as if the original electron is losing its energy much more slowly than what is predicted by the Bethe–Heitler theory.

These ideas of Bhabha and Heitler provided a natural explanation of cosmic ray showers.

12.8.2 The Hard Component of Local Cosmic Rays

The properties of the soft component of local cosmic rays could be fully accounted for by the Bhabha–Heitler cascade theory of showers as electrons and positrons. The hard component was, however, still in enigma. It was not possible for the quantum electrodynamics to account for the penetrating hard component if they were electrons. Again, it appeared that either the quantum electrodynamics is not applicable to the high-energy penetrating hard component of local cosmic rays or the hard component does not consist of electrons.

Bhabha on the basis of his theoretical calculations showed that breakdown of quantum electrodynamics at higher energy is not likely and the viable alternative is that the constituents of the hard cosmic ray components are not electrons. Bhabha, therefore, proposed in one of his research papers that the constituent of the hard component of local cosmic rays is a new particle that may have one unit of positive or negative charge but is about 100 times heavier than electron. Thus, Bhabha had, in a brilliant piece of cosmic ray phenomenology, predicted the existence of a new particle. These particles are now called muons or mu-mesons and their tracks were clearly observed in 1938 by Anderson and Neddermyer in cosmic rays. The mass of muons is nearly 200 times the mass of an electron. Bhabha also predicted that positive muon should spontaneously decay into a positron and the negative muon into an electron. Not only that, Bhabha also pointed out that decay of muon may be used to test the principle of time-dilation, as predicted by Einstein's theory of relativity. Subsequent experiments determined the half-life of muon as 2.2 μs and also confirmed the principle of time-dilation.

12.8.3 Interaction of Primary Cosmic Rays and Extensive Air Showers

The interaction of primary cosmic rays with the air molecules can be best explained with reference to figure 12.16 that shows the flow chart of processes that may follow the interaction of a primary cosmic ray particle with the nuclei present in the molecules of air. As shown in the figure the end products of these interactions may be divided into three basic categories, namely, the muonic component, the hadronic component and the electromagnetic component. Muonic and hadronic components constitute the hard component of the local cosmic rays at the sea level, while the electromagnetic component constitutes the soft component. As shown in figure, a primary cosmic ray particle (essentially a proton) of high energy may produce π and K-mesons on interaction with the nucleus of atoms present in air at the top or within a short distance from the top of the atmosphere. The charged pions with mean life times of 26 ns and K-mesons with lives of 12 ns decay into muons, which have a much longer half life of 2.2 μs. Neutral pions decay almost immediately (10^{-16} s) into γ rays. The muons can decay into positron or electron depending on the nature of their electric charge but their life time is large enough so that they usually make it all the way to the sea level. These muons at the sea level constitute the hard component of local cosmic rays. Muons may also decay in their flight through atmosphere creating electrons (or positron) that may generate showers in the air. Electron positron pairs produced by the annihilation of γ rays may produce Cerenkov radiations. High-energy hadronic particles such as K and π mesons, neutrons, protons, other light nuclei, created in high-energy interactions of primary cosmic rays, may also penetrate through the atmosphere adding to the hard component of cosmic rays at the sea level. The density of the atmosphere increases with the decrease of height from the seas level, and hence the interaction probability of primary cosmic rays with nuclei in the air increases as they penetrate through the atmosphere. As a result, the generation of secondary cosmic particles increases rapidly on moving from top to the lower parts of the atmosphere. This coupled with the fact that the various interaction processes shown in the figure occur in rapid succession one after the other increasing the number of secondary, tertiary, and higher generation particles and also the radius of their spread. This gives rise to extensive air showers. The number density of particles in the core of the shower is much higher than that in the peripheral region. The schematic development of extensive air showers is also indicated in the figure. At sea level, for every 10,000 muons, there will still be roughly 200 primaries (protons and occasional neutrons), 20 high-energy electrons (E > 1 GeV), and 4 pions. However, there may be up to 1,00,000 low-energy electrons created by the cascade. These particles are absorbed quickly, but if the shower is energetic enough or the shower started low enough, they may still be the most prevalent particles at sea level.

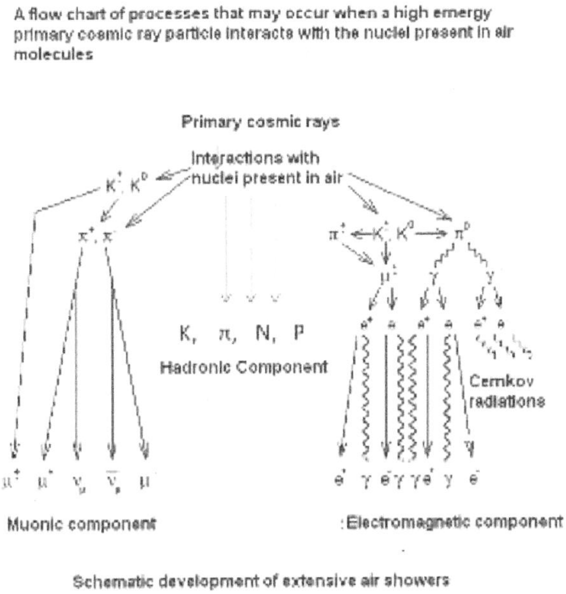

Figure 12.16 *A flow chart of the interaction of primary cosmic rays with the nuclei in the atmosphere*

12.9 SOURCE OF COSMIC RAYS AND THE MECHANISM OF ACCELERATION OF COSMIC RAY PARTICLES

What is the source of cosmic rays and how cosmic ray particles are accelerated to such high energies is still not clear. One of the problems in locating the source of cosmic rays is the fact that primary cosmic rays consist of positively charged particles. Charged particles are readily affected by the magnetic field which changes their direction of motion. Magnetic fields exist throughout the outer space. The galaxy, sun, and the earth each strongly affect the particle's paths. The more energy a particle has, the greater is the radius of curvature. Particles with energies less than

10^{15} eV have paths that are so curved that they are probably trapped within our galaxy and wander throughout. As such, the direction of approach of a cosmic ray particle does not give any information about the source from which the particle is coming.

Black holes, neutron stars, pulsars, supernovae, active galactic nuclei, quasars, and the Big Bang itself, all have been thought to be the source of cosmic rays. The enormous energies that these particles have would require very special mechanism to accelerate them. Shock waves created in any of these intense locales are currently thought to be the most likely mechanism. Even our relatively tranquil sun can and does contribute to the cosmic ray flux, although only a little (about 1%) and only at low energies (up to a few hundred GeV). This is confirmed by the observed 11-year cyclic variation in cosmic ray intensity that synchronizes with the 11-year cycle of solar activity. The sun has a strong magnetic field that is carried away up to and beyond Pluto by the solar wind. Solar magnetic field tends to restrict the lower energy particles of energy below 10^9 eV. However, solar flares create shock waves in solar winds that may accelerate charged particles in solar wind to relativistic energies and may liberate the so-called 'extra solar cosmic rays', which were otherwise trapped in the magnetic field of the sun.

Supernovae explosions also send shock waves, which may accelerate protons already present in the explosion dust to relativistic energies. New results from the Fermi γ ray space telescope, announced in February 2013, provide overwhelming evidence that the shock wave mechanism is responsible for the acceleration of protons up to relativistic energies in supernova remnants. Researchers identified two ancient supernovas whose shock waves accelerated protons nearly to the speed of light, turning them into cosmic rays. As such, it may be concluded that shock wave acceleration of charged particles in the remnants of supernovae is a major source of cosmic rays.

Theories based on acceleration of charged particles, particularly protons, by some mechanism such as the shock waves, are based on the assumption that protons are already present and that they are accelerated to cosmic ray energies. However, theories based on the assumption that very high energy protons are produced directly by the decay of hypothetical super heavy and highly unstable particles are also proposed. These extremely heavy and unstable particles are assumed to have been created just after the Big Bang at the time of the synthesis of matter. Some grand unification theories leave the possibility that these super heavy particles got trapped in the folds (or irregularities) of space–time continuum and escaped instantaneous decay. These particles that are uniformly distributed in the universe, undergo decay when the irregularities in the space–time continuum get quenched, emitting high-energy protons, which constitute cosmic rays.

12.10 HOW COSMIC RAYS AFFECT US?

Human life on earth is influenced by the cosmic rays in many ways. Some important aspects are as follows:

12.10.1 Production of Radio Isotope in the Atmosphere

Several radio isotopes with half-lives ranging from few minutes (^{34}Cl – 32 min) to several million years (^{10}Be – 1.39 million years) are produced by the interaction of primary cosmic rays with the nuclei present in the atmosphere. ^{14}C isotope of half-life 5730 years is also produced by cosmic ray interactions. The rate of production of ^{14}C and the rate of its depletion via natural decay are so balanced that the abundance of 14-carbon in the atmosphere is almost constant for the last hundred thousand years. This has provided us a tool to determine the age of artefacts that lived on this earth few thousand years ago by carbon dating. Living bodies inhale air and, therefore, have a predetermined concentration of ^{14}C. Once dead, they do not inhale and, therefore, the concentration of ^{14}C depletes with time. In some 5730 years from the death, the ^{14}C concentration will reduce to half of its value at the time of death. Thus, by determining the ^{14}C activity it is possible to calculate back the time when the artefact died.

12.10.2 Depletion of Ozone Layer

Cosmic rays also produce ionization of gas molecules of air, which in turn deplete the ozone concentration reducing the protecting shielding cover from harmful ultraviolet rays. Ultraviolet rays are harmful for the skin and may produce skin cancer. There are holes in ozone layer over some countries such as Australia. People living at such places take special care by covering their bodies and using sun-screen creams that protect them from ultraviolet rays.

12.10.3 Malfunctioning of Electronics

Cosmic rays are found to adversely affect the functioning of transistor-based electronics. As the cosmic ray flux on the earth increases with altitude, computerized devices used for the control of airplanes are more likely to be attacked by cosmic rays and malfunction. Several cases of unexplained spikes in control data of planes have been reported. These spikes were attributed to cosmic rays. Artificial satellites sent in space also encounter large flux of galactic cosmic rays. Malfunctioning of Voyager-II was assigned to the adverse affect of cosmic rays on its electronics.

Even on ground, cosmic radiations are found to be capable of changing the state of memory and other electronic elements. With the miniaturization of solid-state elements the problem is becoming more acute. Solid-state electronic manufacturing companies are now thinking of integrating a cosmic ray detection device with the chip so that system may be reset to the initial state when a cosmic ray event is registered.

12.10.4 Radiation Dose

All living organism on the earth is subjected to radiation dose from natural as well as manmade sources. Natural radioactivity present in the environment and cosmic rays contribute to natural sources while radiations from nuclear reactors, nuclear explosions for testing nuclear armaments or for academic studies, reactor meltdown, X-ray photography, radiation therapy and so on for medical purposes comes under the manmade source of radiation dose. It is worth noting that only 11% of the total radiation exposure comes from manmade sources while the remaining 89% is due to natural sources.

Atmosphere provides a good shielding for primary cosmic ray particles of less than 1 GeV energy. Secondary radiations are also absorbed in the air to a large extent and particles such as muons decay out in flight. As such, the radiation exposure due to cosmic rays is minimal on the ground level. On an average, a dose of about 0.4 milli-Sievert (mSv) per year due to cosmic rays is received by human beings on the ground. This is nearly 10% of the total radiation dose of 4 mSv received annually. In the above atmosphere, at altitudes larger than 20 km, where atmospheric absorption is negligible, the dose due to cosmic rays rises to 1.0 mSv at equator and to about 3.0 mSv at poles. Pilots flying at high altitudes are subjected to an additional does of nearly 2.2 mSv.

The radiation environment in deep space is quite different. There is large flux of galactic cosmic rays and also high-energy charged particles, mostly protons, from solar proton events. There are also two radiation belts where the flux of charged particles is high and further increases dramatically during solar storms. These radiations affect human beings, particularly the astronauts passing through radiation belts. Radiations may damage the DNA and may also produce active oxygen-free radicals that may lead to cancer. Thus, cosmic rays projects one of the biggest hurdle for long duration manned space flight, for example up to mars through deep space.

12.10.5 Effect on Global Weather

It is believed that cosmic rays also affect global weather conditions by altering the cloud formation pattern. The cosmic rays are found to generate aerosol particles, condensation around which results in cloud formation. A special experiment named 'CLOUD' has been carried out at CERN-Geneva, to study the effect of cosmic rays on cloud formation. In an extremely clean and closed chamber, air and other gases with water vapours are held at a desired pressure to simulate the atmosphere. The amount and pressure of gases in the chamber is controlled in such a way that the conditions prevailing at different altitudes of atmosphere may be replicated. The cosmic rays in the experiment are simulated by the fast protons provided by a high energy accelerator at CERN. Results of the cloud experiment published in 2012–2013 have indicated that cosmic rays do effect the cloud formation and that the variation in the cosmic ray intensity may influence the cloud formation pattern. Some scientists even believe that cosmic rays also play some role in the global warming, though at present there is no direct evidence for that.

Solved example S-12.1

Assuming earth to be a sphere of radius 6.4×10^6 m and the magnitude of earth's magnetic field at geomagnetic equator to be 1.2×10^{-4} Tesla, calculate the maximum energy of a proton that will just miss reaching the earth.

Solution. The linear momentum P (in GeV/c) of a particle of mass m and velocity V in a magnetic field of strength B (Tesla) is related to its curvature R (meters) by the relation

$$P = mV = 0.3BR \text{ GeV/c}$$

Further, when the momentum is in GeV/c, the proton will be relativistic and its energy will be Pc. Now, we put $R =$ to the radius of earth to get the minimum energy of the proton that will be bent by the amount of the radius of the earth and will not be able to reach the equator.

$$P = 0.3 \times 1.2 \times 10^{-4} \times 6.4 \times 10^6 = 230.4 \text{ GeV/c}$$

Ans: Hence, the maximum energy of the proton should be 230.4 GeV.

Solved example S-12.2

A proton moving with the velocity of 10^6 m/s enters the geomagnetic equator making an angle of 30° with the direction of the magnetic field of 1.2 Tesla. Determine whether the proton will reach the earth's surface or not? The mass of proton is 1.67×10^{-27} kg and charge 1 Coulomb.

Solution. To find whether the proton will reach the surface of earth or not, we calculate the curvature R of the proton in the magnetic field using the expression

$$\frac{m(V_{pep})^2}{R} = qBV_{pep},$$

where V_{pep} is the component of velocity normal to $B = V \sin 30 = \frac{V}{2} = 0.5 \times 10^6$

$$\text{Hence, } R = \frac{V_{pep} \cdot m}{qB} = \frac{0.5 \times 10^6 \times 1.67 \times 10^{-27}}{1 \times 10^{-19} \times 1.2} = 0.69 \text{ cm}$$

We see that the curvature of the particle is only 0.69 cm, which is much small than the radius of the earth. Hence, the proton will reach the earth.

Solved example S-12.3

A proton of linear momentum 20 GeV/c directed towards the centre of the earth when 100 R distance away from the centre of the earth experiences a uniform transverse magnetic field of strength B (Tesla). If the proton just misses the earth, calculate the curvature of the path of the proton and the approximate value of the magnetic field B. Take earth to be a sphere of radius $R = 6.4 \times 10^6$ m.

Solution. Figure S-12.3 shows the path of the proton if it just misses the earth. The first thing to calculate is the curvature X of the path of the proton. The distance Y in the figure is given to be 100 R, where R is the radius of the earth given as 6.4×10^6 m.

As is clear from the figure, $(X + R)^2 = X^2 + Y^2$ or $X^2 + R^2 + 2XR = X^2 + Y^2$, or $2XR = Y^2 - R^2$.

Hence, $X = \frac{(100R)^2 - R^2}{2R}$. As $(100R)^2 >> R^2$, R^2 may be neglected in comparison to $(100R)^2$

Therefore, $X \approx \frac{(100R)^2}{2R} = 500\ R = 500 \times 6.4 \times 10^6 \text{ m} = 32 \times 10^8 \text{ m}$

Hence, the curvature of the proton for just missing earth should be 32×10^8 m.

Next, we calculate the strength B of the transverse magnetic field that will produce this curvature for a proton of linear momentum 20 GeV/c. For this, we use the following relation:

$P\,(\text{GeV/c}) = 0.3\ B\,(\text{Tesla})\ r\,(\text{m})$;

$20 = 0.3 \times B \times 32 \times 10^8$ or

$$B = \frac{20}{0.3 \times 32 \times 10^8} = 2.083 \times 10^{-8} \text{ Tesla}$$

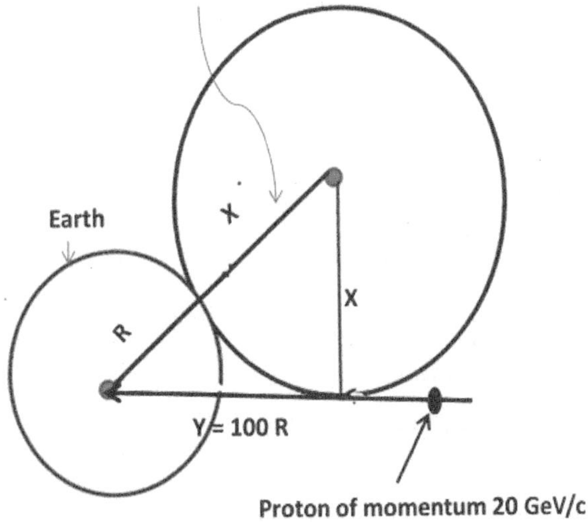

Figure S-12.3 *The trajectory of the proton if it just misses the earth.*

Ans: The curvature of the path of the proton should be 32×10^8 m and the magnitude of the transverse magnetic field be 2.083×10^{-8} Tesla.

Exercise p-12.1: Briefly discus how cosmic rays were discovered, who named them 'cosmic rays' and why they were called the 'birth cry' of elemental synthesis.

The average number of cosmic ray particles hitting the surface of earth is approximately one particle per square centimetre per second and the average energy of these cosmic ray particles is nearly 2.8 GeV. Considering earth as a sphere of radius 6400 km, calculate the energy in watt deposited by cosmic rays on earth per second.

Exercise p-12.2: Describe why there was controversy about the nature of cosmic rays in the early years after their discovery and how it was resolved.

Exercise p-12.3: What is east–west effect? Discuss its cause and the role of bounded and unbounded trajectories of charged particles as introduced by Rossi.

Exercise p-12.4: What experimental evidence finally established that cosmic rays are positively charged particles and not high energy γ rays?

Exercise p-12.5: What is Störmer cone? Discuss how the variation in the intensity of cosmic rays with latitude may be explained.

Show that the linear momentum P of a particle of charge (qe) moving in a circular path of radius r in a transverse magnetic field of strength B is given by P (GeV/c) = 0.3 qB (Tesla) r (meter).

Exercise p-12.6: Do cosmic rays show longitude effect? If yes what is the likely cause?

Exercise p-12.7: What are primary cosmic rays and what is their composition? Briefly discuss the energy distribution of primary cosmic rays.

Exercise p-12.8: Discuss the nature and composition of local cosmic rays. Describe with necessary details an experiment to obtain Rossi transition curve. What is the significance of this curve?

Exercise p-12.9: Give an account of Bhabha–Heitler theory of cosmic ray showers and explain how this theory was able to explain the Rossi transition curve on one hand and the validity of quantum electrodynamics at higher energies, on the other.

Exercise p-12.10: What are extensive air showers and how are they produced?

Exercise p-12.11: Give an outline of the interactions that the primary cosmic rays may have with air molecules/nuclei leading to the various components. Which of these components make the soft and the hard components at sea level?

Exercise p-12.12: Write a note on the source of cosmic rays and the mechanism of their acceleration.

Exercise p-12.13: How cosmic rays may affect human life on earth? Discuss.

Exercise p-12.14: In your opinion, which detector (s) and/or experimental technique(s) made big impact on the study of cosmic rays. Elaborate your answer.

Multiple choice questions

Note: Some questions may have more than one correct alternative. Select all the correct alternatives.

Exercise M-12.1: Cosmic rays were discovered by
(a) Hess (b) Millikan (c) Compton (d) Anderson

Exercise M-12.2: Who called cosmic rays the 'birth cries' of elemental synthesis?
(a) Hess (b) Millikan (c) Compton (d) Anderson

Exercise M-12.3: Initially cosmic rays were confused with high-energy γ rays because
(a) high-energy γ rays may produce low-energy γs and electrons at sea level by Compton scattering with air nuclei
(b) can explain the soft component local cosmic rays
(c) can explain the hard component of local cosmic rays
(d) γ rays were the most penetrating of all the known radiations

Exercise M-12.4: Primary cosmic rays were made up of charged particles and not γ ray was conclusively proved by
(a) the nature of soft component of local cosmic rays
(b) the nature of the hard component of the local cosmic rays
(c) the latitude effect
(d) Rossi transition curve

Exercise M-12.5: What conclusion about the nature of primary cosmic rays would have been drawn if there was latitude effect but no east–west effect?
(a) Cosmic rays are made up of neutral particles
(b) Cosmic rays contain equal number of positively and negatively charged particles in each energy group
(c) Total number of positive particles is equal to the total number of negatively charged particles but the number of positively and negatively charged particles may not be equal in different energy groups
(d) None of these

Exercise M-12.6: Which elementary particle was discovered by Anderson in cosmic ray tracks?
(a) electron (b) positron (c) pi-meson (d) muon

Exercise M-12.7: Cosmic rays at a particular location on the surface of earth coming at some angle from the vertical direction contain more soft particles than those coming from the vertical direction. It is because
(a) primary cosmic rays approaching from slant direction pass through a larger thickness of atmosphere undergoing more interactions.
(b) high-energy component of primary cosmic rays in vertical direction do not undergo any interaction with atmospheric nuclei.
(c) primary cosmic rays contain more soft particles in slant direction than in the vertical direction.
(d) none of these

Exercise M-12.8: 'Isocosms' are

(a) graphs showing geomagnetic parallels.
(b) graphs showing geographical parallels.
(c) graphs showing variation in cosmic ray intensity at sea level.
(d) lines showing points of equal cosmic ray intensity at sea level.

Exercise M-12.9: Apparent latitude effect is due to

(a) the non-coincidence of geomagnetic and geographic poles.
(b) primary cosmic rays are positively charged particles.
(c) the variation of longitude in going from equator to pole.
(d) the variation of temperature in going from equator to pole.

Exercise M-12.10: The most likely mechanism of acceleration of primary cosmic ray protons is

(a) by electric fields in inter galactic space.
(b) by electric and magnetic fields in inter galactic space.
(c) by shock waves
(d) none of these.

Exercise M-12.11: If E_{cut} and λ, respectively, represent the cutoff energy of cosmic ray particles and the geomagnetic latitude of a place on the surface of earth, then E_{cut}

(a) increases with the increase of λ
(b) decreases with the increase of λ
(c) may increase or decrease depending on the value of the half angle of Stormer cone
(d) does not depend on λ

Exercise M-12.12: The minimum energy of cosmic ray particles for which the half angle of Stormer cone is zero

(a) decreases with the increase in the geomagnetic latitude of the place
(b) increases with the geomagnetic latitude of the place
(c) may increase or decrease depending on the cutoff energy
(d) does not depend on the geomagnetic latitude of the place

Solutions to Numerical Problems

CHAPTER 1

Answers to Exercises

p-1.1: (a) 3.3×10^{18} (b) 3.3×10^{17}; other quantities may be calculated accordingly

p-1.2: The behaviour will be very different for $b = 0$. It will nearly be same for $b = 1 \times 10^{0}$ mm

p-1.3: Since force F between the alpha particle and the nucleus is radial, there is no torque in the system, hence angular momentum is conserved. As a matter of fact in all problems of central fields, the angular momentum is conserved

p-1.4: 16×10^{-4} .... [Hint: the probability of scattering by a given angle is proportional to the square of the charge of the incident particle and is inversely proportion to the square of its energy.]

p-1.12: 931.49×10^{-6} MeV; 1.66043×10^{-3} kg

p-1.13: 0.377 MeV

p-1.15: The absolute value of spin = 2.9580 $\hbar$; Possible values of magnetic quantum number: $-\frac{5}{2}; -\frac{3}{2}, -\frac{1}{2}, +\frac{1}{2}, +\frac{3}{2}, +\frac{5}{2}$

Answers to Multiple Choice Questions

M-1.1: (b)

M-1.2: (c)

M-1.3: (c)

M-1.4: (c)

M-1.5: (b)

CHAPTER-2

Answers to Exercises

p-2.1: 1.01×10^{-2} m^2 kg^{-1}

p-2.2: 2.7×10^3 kg/m^3; aluminium

p-2.4: $\left(\dfrac{A_2 \cos\theta}{R_2^2}\right)$

p-2.5: $\dfrac{6m}{X} \times 10^{17}$

p-2.14: 0.00921 ± 0.00024 u; 8.63898 ±0.22512 MeV

p-2.18: [Hint: From the last solved example, we know that $\left(\dfrac{\Delta}{R}\right) = \dfrac{5Q}{4R^2}$, where $\Delta = (a-b)$].

p-2.19: (i) $-\dfrac{q}{12}a^2$ (ii) $+\dfrac{q}{6}a^2$

p-2.30: 538.4 MeV

Answers to Multiple Choice Questions

M-2.1: (a)

M-2.2: (b)

M-2.3: (b)

M-2.4: (a)

M-2.5: (d)

M-2.6: (c)

M-2.7: (b)

M-2.8: (c)

M-2.9: (d)

M-2.10: (d)

CHAPTER-3

Answers to Exercises

p-3.1: 5.81×10^{-37} MeV

p-3.2: 0.72 MeV

p-3.4: 8.14 MeV, 8.20 MeV, and 7.92 MeV

p-3.5: [Hint: Use $\frac{e^2}{4\pi\varepsilon_0}$ =1.44 MeV*F] 3.14 MeV; 1.34 MeV

p–3.6: +1/2, −1/4; −1/2, +1/4

Answers to Multiple Choice Questions

M-3.1: (b)

M-3.2: (b)

M-3.3: (c)

M-3.4: (d)

M-3.5: (a)

M-3.6: (d)

M-3.7: (d)

M-3.8: (d)

M-3.9: (b)

M-3.10: (b)

CHAPTER-4

Answers to Exercises

p-4.4: 3.0 F

p-4.6: 47.59 MeV

p-4.9: 1.6 V

p-4.10: 16.8°

p-4.14: 1 $\hbar$

Answers to Multiple Choice Questions

M-4.1: (b)

M-4.2: (b)

M-4.3: (d)

M-4.4: (c)

M-4.5: (d)

M-4.6: (b)

M-4.7: (b)

M-4.8: (a)

M-4.9: (c)

M-4.10: (c)

CHAPTER 5

Answers to Exercises

p-5.1: $\frac{1}{2}^-$ and $\frac{5}{2}^+$

p-5.4: ${}^{7}_{3}Li$ ($3/2^-$, 2.9 nm); ${}^{14}_{7}N$ (1^+, 0.37 nm); ${}^{41}_{20}Ca$ ($7/2^-$, 2.9 nm)

p-5.11: $a_c = 0.68$

Answers to Multiple Choice Questions

M-5.1: (c)

M-5.2: (a)

M-5.3: (b)

M-5.4: (a), (b), (d)

M-5.5: (c)

M-5.6: (c)

M-5.7: (a)

M-5.8: (a), (b)

M-5.9: (a)

M-5.10: (c)

CHAPTER 6

Answers to Exercises

p-6.1: 197.3 d

p-6.2: 5.26×10^9 year

p-6.3: 2.366×10^{10} s^{-1}

p-6.4: Mode of decay β^-. The ratio of Sr/Y = 3.56×10^5 after 1 hour and 3.82 after 10 years

p-6.5: 5.0×10^8

p-6.6: 28.37 MeV

p-6.9: 3.69 J/s

P-6.11: (a) super allowed (b) allowed Gamow–Teller (c) second forbidden (d) allowed, Fermi
(e) allowed, Gamow–Teller

p-6.17: $3^+, 2^+, 1^{1+}$

p-6.18: 9.8×10^{-7} m/s, downwards

p-6.20: 3.8×10^{-4} eV

Answers to Multiple Choice Questions

M-6.1: (d)

M-6.2: (a)

M-6.3: (d)

M-6.4: (b)

M-6.5: (c)

M-6.6: (a)

M-6.7: (b)

M-6.8: (b)

M-6.9: (a)

M-6.10: (a), (b), (d)

M-6.11: (b)

M-6.12: (b)

M-6.13: (d)

M-6.14: (c)

M-6.15: (c)

M-6.16: (c)

M-6.17: (d)

M-6.18: (c)

M-6.19: (a)

M-6.20: (a)

CHAPTER 7

Answers to Exercises

p-7.1: 0.51 cm^{-1}

p-7.2: 0.33 MeV

p-7.3: 0.55; 0.263 $cm^2\ g^{-1}$

p-7.4: (A) Compton scattering, (B) Pair production, (C) Photoelectric effect

p-7.5: 6.625×10^{-34} J-s

p-7.6: 26.4×10^{-10} m

p-7.7: 4.78×10^{-5}

p-7.8: 2.43×10^{-3} nm

p-7.9: 3.99 MeV

p-7.10: 1

p-7.11: 16, 28

p-7.11: 480 MeV

Answers to Multiple Choice Questions

M-7.1: (c) and (d)

M-7.2: (c)

M-7.3: (b)

M-7.4: (d)

M-7.5: (b)

M-7.6: (b)

M-7.7: (d)

M-7.8: (a)

M-7.9: (a)

M-7.10: (a), (d)

M-7.11: (a), (c)

M-7.12: (c)

M-7.13: (c)

M-7.14: (a)

M-7.15: (b)

CHAPTER-8

Answers to Exercises

p-8.5: 4.37 MeV

Answers to Multiple Choice Questions

M-8.1: (a), (b), (c)

M-8.2: (c)

M-8.3: (a)

M-8.4: (c)

M-8.5: (b)

M-8.6: (a)

M-8.7: (a)

M-8.8: (b)

M-8.9: (c)

M-8.10: (a)

M-8.11: (c)

M-8-12: (a)

M-8.13: (d)

M-8.14: (d)

M-8.15: (b)

M-8.16: (a)

M-8.17: (a)

M-8.18: (d)

M-8.19: (a)

M-8.20: (d)

CHAPTER-9

Answers to Exercises

p-9.1: 16,000 gauss, 24.3 Mc/s

p-9.2: 5.64×10^8 Newton/Coloumb

Answers to Multiple Choice Questions

M-9.1: (c)

M-9.2: (d)

M-9.3: (a)

M-9.4: (a), (b)

M-9.5: (a)

M-9.6: (a)

M-9.7: (d)

M-9.8: (c)

M-9.9: (a)

M-9.10: (a)

CHAPTER-10

Answers to Exercises

p-10.1: 115, 1.5×10^{-4} s

p-10.3: 0.1054 kg

p-10.10: 2.385 MeV

p-10.12: 0.0288 MeV; 22.26×10^8 K

Answers to Multiple Choice Questions

M-10.1: (b)

M-10.2: (d)

M-10.3: (a)

M-10.4: (b)

M-10.5: (c)

M-10.6: (a), (b), and (d)

M-10.7: (a), (b)

M-10.8: (b), (c), and (d)

M-10.9: (c)

M-10.10: (c)

M-10.11: (c)

M-10.12: (b)

M-10.13: (c)

M-10.14: (d)

M-10.15: (d)

CHAPTER-11

Answers to Exercises

p-11.1: (a) allowed by weak interaction (b) not allowed B and Q not conserved (c) allowed by weak interaction (S not conserved) (d) allowed by Strong interaction (e) allowed by strong interaction

p-11.2: $I = 3/2, I_3 = 1/2, S = 0, B = -1, Y = -1, Q = 0$

p-11.3: 0 and 1 (the actual isospin of deuteron is 0)

p-11.4: $Q = 2, S = +2, B = 0, I_3 = +1$

p-11.5: (i) forbidden (ii) $Z^{++} \to K^+ + K^+$ allowed via strong interaction. (iii) $Z^{++} \to p + \bar{p} + \pi^+ + \pi^+$ allowed only via weak interaction.

p-11.6: (c)

p-11.7: (a)

p-11.8: (b)

p-11.9: No, lepton number is not conserved.

p-11.18: (i) $P(r) = -r$; $T(r) = r$ (ii) $P(p) = -p$; $T(p) = -p$; (iii) $P(\sigma) = \sigma$, $T(\sigma) = -\sigma$

p-11.19: To conserve spin, the pions must be emitted in $\ell = 1$ state. Two π^0 in the final state will violate exclusion principle. Hence (a) is allowed.

p-11.20: $p = \dfrac{m_K}{2}\left(\dfrac{m_K^2 - 4m_\pi^2}{2m_K^2 + m_\pi^2}\right)^{1/2}$

p-11.22: (i) $\lambda_c = \dfrac{1}{m_e}$ (ii) $a_0 = \dfrac{1}{\alpha m_e}$

p-11.23: (a) Electron–positron scattering (b) Electron–positron annihilation

Answers to Multiple Choice Questions

M-11.1: (c)

M-11.2: (a)

M-11.3: (a)

M-11.4: (c)

M-11.5: (c) and (d)

M-11.6: (b)

M-11.7: (c) and (d)

M-11.8: (d)

M-11.9: (c)

M-11.10: (a) and (b)

M-11.11: (d)

M-11.12: (a)

M-11.13: (a) and (d)

M-11.14: (d)

M-11.15: (a)

CHAPTER-12

Answers to Exercises

p-12.1: 2.305×10^9 watt

Answers to Multiple Choice Questions

M-12.1: (a)

M-12.2: (b)

M-12.3: (a), (d)

M-12.4: (c)

M-12.5: (b)

M-12.6: (b)

M-12.7: (a)

M-12.8: (d)

M-12.9: (d)

M-12.10: (c)

M-12.11: (b)

M-12.12: (a)

Index